Genetic and Malformation Syndromes in Clinical Medicine

Genetic & Malformation Syndromes in Clinical Medicine

WILLIAM L. NYHAN, M.D., Ph.D.
Professor and Chairman,
Department of Pediatrics
University of California San Diego
La Jolla, California

and

NADIA O. SAKATI, M.D.
Chief of Pediatrics
Ash-Sharq Hospital
Alkhobar, Saudi Arabia

YEAR BOOK MEDICAL PUBLISHERS, INC.
35 East Wacker Drive/Chicago

Library of Congress Catalog Card Number: 75-35058

International Standard Book Number: 0-8151-6486-6

To the Pediatric House Staff
and Medical Students at UCSD
now, in the recent past and in the years to come.

OTHER CONTRIBUTORS

GERSON CARAKUSHANSKY, M.D.
Professor of Genetics and Pediatrics
Universidade Federal do Rio de Janeiro
Rio de Janeiro, Brazil

JOAN SCHINDELER DAVIES
Dermatoglyphic Specialist
University of Miami
Mailman Center for Child Development
Miami, Florida

KENNETH W. DUMARS, M.D.
Associate Professor of Pediatrics
University of California Irvine
College of Medicine
Irvine, California

UTA FRANCKE, M.D.
Assistant Professor of Pediatrics
University of California San Diego
School of Medicine
La Jolla, California

KENNETH LYONS JONES, Jr., M.D.
Assistant Professor of Pediatrics
University of California San Diego
School of Medicine
La Jolla, California

WILLIAM J. LEWIS, M.D.
Department of Pediatrics
U.S. Naval Regional Medical Center
San Diego, California

SHERWOOD A. LIBIT, M.D.
Department of Pediatrics
U.S. Naval Regional Medical Center
San Diego, California

PAMELA J. REED, B.A.
Division of Developmental Disabilities and Clinical Genetics
Department of Pediatrics
University of California Irvine
College of Medicine
Irvine, California

PREFACE

We began this book in self-defense. As biochemists and geneticists, we began increasingly to be called upon to see patients who appeared somehow different from others. Their doctors believed they had some kind of syndrome, but what? If you've ever tried this kind of thing, you know that it begins to seem a lot like general practice. But it is a general practice of rare diseases — all systems and subspecialties are involved. Even if you know the ground pretty well, have seen the occasional case and read the literature, it is difficult to keep all this within your memory. And it grows worse rather than better as the years go by. Therefore, most of us like to have something in the way of notes to look at to refresh our memories.

We therefore began to keep a private compendium. We wrote down some salient features, a short description and some information on genetics and treatment. We collected some pictures and, as scientists, we recorded our own experience in case reports. We collected all of this in loose-leaf notebooks as an aid to diagnosis and teaching. As time went on we added some lists of differential diagnosis. This seemed to be useful. All of this was before the availability of any of the recently published approaches to the same problem. The clinician who sees problems of this sort now has a number of sources to turn to in syndrome identification. The best include Dave Smith's *Recognizable Patterns of Human Malformation*, the *Birth Defects: Original Articles Series* put together by Victor McKusick and Dan Bergsma, the Sidney Gellis and Murray Feingold *Atlas of Mental Retardation Syndromes* and Goodman and Gorlin's *The Face in Genetic Disorders*. Bob Gorlin and Jens Pindborg's *Syndromes of the Head and Neck* is of course a classic. Victor McKusick's *Mendelian Inheritance in Man* is very useful, but of course you have to know what you are doing before you turn to those computerized pages. Similarly very useful to the initiated is the Waardenberg, Franceschetti and Klein *Genetics and Ophthalmology*. Obviously these texts fill a need, and there will be more of them.

We began to see that our own private collection was useful to us, and we thought that it could be useful to others. Its publication followed a series of discussions with Bill Keller and Ken Hoppens of Year Book.

The major difference between our book and those that have preceded it is that ours represents a summary of our own experience. Through this we bring new pictures — as many of them as fiscally possible in color. In this way we have tried to enlarge the visual experience of the reader beyond that which he has encountered before in books or in the literature. We also have provided a somewhat greater coverage of metabolic or enzymatic disease than most other texts on the subject.

Since our collection is obviously organic with us, we would appreciate communications from our readers. We would love to see their patients when we are in their neighborhood, and sooner or later we are close to almost everywhere.

ACKNOWLEDGMENTS

We want to thank many people for their generous help in the preparation of this book. The book is to a large extent a distillate of our experience. We want first to thank our patients. Their understanding of the need for careful documentation and the importance of the dissemination of information in the diagnosis of unusual conditions was an essential element in the genesis of this book. We also want to thank a number of our colleagues who have helped to enlarge our experience. At the University of California San Diego (UCSD), departmental boundaries seem less rigid than in many places we have been. Many of us on the faculty are interested in genetics and development. We like to work together in diagnosis and management. These attitudes are manifest in the help we have received in putting together this book.

Dr. O. W. Jones is Professor of Medicine and Medical Director of the Division of Human Genetics in the Department of Medicine. He is an excellent all-around geneticist. He has established an interesting program in which genetic counseling is brought to the people through a clinic that meets in rotation in all of the regional public health clinics in San Diego County. We have been privileged to see patients with him in these clinics and elsewhere. A number of them are illustrated in this book. Dr. Jones' laboratory does many of the karyotypes at UCSD, and he follows a number of the patients who have cytogenetic abnormalities. He contributed several of the illustrations of karyotypes in this book, particularly those using the Giemsa-banding technic. Dr. Uta Francke, Assistant Professor of Pediatrics, is also expert in cytogenetic problems as well as in biochemical problems. She has written the Introduction to the section on Cytogenetic Syndromes. She also has contributed the beautiful fluorescent chromosomes that illustrate some of our chapters.

Dr. John S. O'Brien, Professor and Chairman of the Department of Neurosciences, is probably the world's authority on the neurolipidoses. He still sees patients in the clinic himself, counsels parents and also runs enzyme assays in the laboratory. He has been very generous in permitting us to study and photograph his patients and in providing us with photographs of patients we have not seen. He also has permitted us to print his latest classification of lipidoses, leukodystrophies and mucopolysaccharidoses. This appears in our Introduction to the section on Mucopolysaccharidoses. Dr. Roger Rosenberg, former Chief of the Division of Pediatric Neurology at UCSD, is now Chairman of the Department of Neurology at the University of Texas Southwestern Medical School in Dallas. He is a superb clinician and scholar and has taught us a great deal about neurogenetics. He has seen many of our patients and helped us to understand their neurologic abnormalities. Dr. William Friedman, Professor of Pediatrics at UCSD, is our Chief of Pediatric Cardiology. We call on him regularly to evaluate the cardiac malformations of our patients. He has been generous in sharing with us his illustrative material, which we have liberally reproduced in this book. Dr. Lawrence Schneiderman, Associate Professor of Community Medicine and Medicine, is an expert clinical geneticist. We have enjoyed considering with him the ethical issues in genetics and, of course, seeing patients with him. Dr. Jerry Schneider, Associate Professor of Pediatrics, and the world's authority of cystinosis, has kindly shared his experience with us. Dr. Douglas Cunningham, Assistant Professor of Pediatrics, is a pediatrician and neonatologist with broad interests in clinical pediatrics. He has provided us with information and illustrations on a number of his patients. Dr. Louis Gluck, Professor of Pediatrics and Reproductive Medicine, is the Director of our Nurseries. Dr. Gluck is a senior neonatologist, expert diagnostician and photographer. He has kindly provided us with a number of illustrations for this book. Among the UCSD faculty we would like to single out for special thanks is Dr. Michael Weller, Assistant Professor of Radiology and Pediatrics and Chief, Division of Pediatric Radiology. Dr. Weller has instructed us about the roentgen features of virtually every condition in this book. We have gone over all of our roentgenograms with him. His sharp eye and articulate nature have made a great impact on the content of this book.

Our close relations with other institutions in the Southern California community have considerably enlarged our experience. We are particularly indebted to Dr. David Allan, former Director of Pediatrics at Mercy Hospital in San Diego. Dr. Allan also is Associate Professor of Pediatrics at UCSD. We run a joint residency in Pediatrics at University Hospital and at Mercy and have enjoyed making rounds with Dr. Allan at Mercy. Many of his patients grace the pages of this book. We are also grateful for a rewarding, continuing relationship with Fairview State Hospital in Costa Mesa, where Dr. Charles H. Fish is the Program Director of the Medical-Surgical Services Program. He is a careful clinician devoted to the care of his patients. We have been privileged to study and photograph a great many of them, some of whom have found their way into this book. We have similarly enjoyed an excellent relationship with Pacific State Hospital in Pomona. This is largely the result of the efforts of Dr. William Morris, who has kindly allowed us to visit his patients with him and to photograph some of them for this book. Our relationships with the U. S. Naval Hospital at Balboa Park in San Diego have been very close. We have a number of joint ventures in education and patient service, and we are indebted to Captain John Schanberger, Chief of Pediatrics at the Naval Hospital, for making this possible. We have been privileged to see and photograph a number of patients at the Naval Hospital whose case reports have gone into this book. We have also seen many patients at the Children's Health Center in San Diego and want to express our appreciation to Dr. David Chadwick, the Medical Director, for this opportunity and for providing us with a number of his photographs. Dr. Raymond M. Peterson, Director of the Children's Health Center at the Regional Center for the Mentally Retarded in San Diego, has graciously introduced us to many of his patients and the many institutions that serve the Regional Center.

We are also indebted to Dr. Vazken M. Der Kaloustian and Dr. Samir S. Najjar of the American University of Beirut for sharing their patients and photographs with us. Dr. Der Kaloustian is Assistant Professor of Pediatrics and runs the Division of Genetics; Dr. Najjar is Associate Professor of Pediatrics and Chairman of the Department. Their experience illustrates the wealth of patients with problems of genetic importance in the Middle East.

We wish particularly to thank our photographers. Donald S. Luczak, Alison Keppie and E. J. Peterson of UCSD have worked closely with us throughout this project. Their sensitivity and their skills have continued to teach us about the optimal use of photographic tools in diagnosis and teaching. It has been a privilege to work with them. We are also indebted to the Department of Medical Photography at Mercy Hospital for all of the photographs of patients there. Our hard-working research librarians have made scholarship easy for us. We were impressed with Jesse Greenstein's statement in the Preface to his *Biochemistry of Cancer* that he never referred to an author's work without having a copy of the publication in front of him. This was hard to do in Greenstein's day. It meant long hours in the library or lugging bound volumes of journals around to the office or home, and libraries do not want us to do that any more. Instead, the Xerox machine and the computer search have really made careful scholarly documentation a part of life. We are greatly appreciative of the efforts of Faith Murphy, Associate

Librarian of the Biomedical Library at UCSD, and of Marjorie Baldwin, Sue Ann Blaise, Laura Renker, Louann Enstad, Gary Gluge and Leon Peterson of the University Hospital Library.

Those who have been at UCSD in recent years know that Dr. Sakati does not drive a car. She bought a new Mustang and managed to get licensed to drive it, but none of us ever felt comfortable with her behind the wheel. A devoted generation of students and house officers have learned genetics and dysmorphology driving Dr. Sakati around Southern California to track down patients.

Without our editorial assistants this book would not have been possible. Eileen O'Farrell Borgwardt worked with Dr. Sakati in what was creative editing at its best. Together they translated ideas and facts into the first draft of many of the chapters. Eileen and Mrs. Dorothy MacElhose typed every page of the book into a finely polished manuscript, which arose from a very actively worked and reworked mixture of type and longhand. Mrs. Ruth Brown helped with some chapters. The partnership this team represents has been invaluable.

WILLIAM L. NYHAN

TABLE OF CONTENTS

Inborn Errors of the Metabolism of Small Molecules

HOMOCYSTINURIA

Cardinal Clinical Features

Ectopia lentis, mental retardation, malar flush, sparse light hair, osteoporosis, genu valgum, arterial and venous thromboembolic phenomena.

Clinical Picture

Homocystinuria is an inborn error of amino acid metabolism in which the primary defect is in the enzyme cystathionine synthase.[1] This enzyme normally catalyzes the formation of cystathionine from serine and homocysteine. Therefore, involved individuals have cystathionine deficiency, and they accumulate homocysteine and methionine. These abnormalities are most readily demonstrated by the examination of fresh urine (Fig. 1). Homocystine is unstable. Therefore, the diagnosis can be missed if attention is not paid to the preservation of urine prior to analysis. In suspected patients the condition can be screened for by using the cyanide nitroprusside test (Fig. 2), which is a test for the excretion of sulfur-containing amino acids.

Patients with homocystinuria generally appear normal at birth. However, typical clinical features have been observed as early as 1 month of age.[2] The most characteristic feature of this disorder is subluxa-

Fig. 1. – M. G. Chromatogram. Urine was collected for 24 hours, and 0.25 ml was placed on a 150 × 0.9 cm column on the Beckman Spinco Automatic Amino Acid Analyzer. There was an enormous peak of homocystine and a moderately large peak of the mixed disulfide of cysteine and homocysteine.

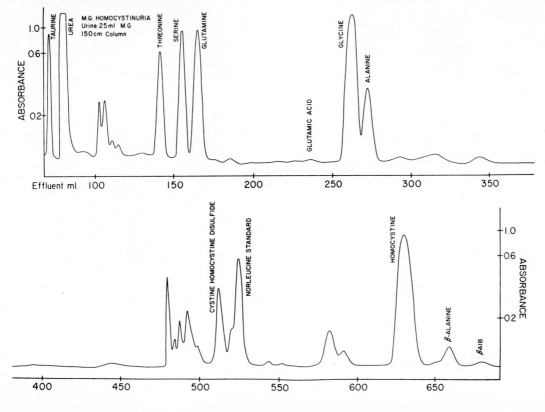

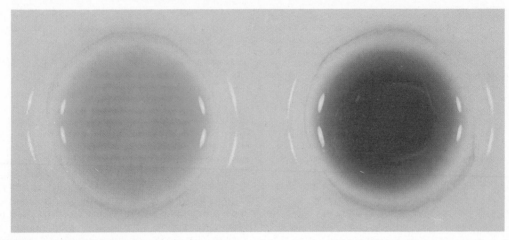

Fig. 2 (top).—Positive cyanide nitroprusside test, indicative of the excretion of larger than normal amounts of a sulfhydryl-containing amino acid.

Fig. 3 (bottom left).—M. G. (Case Report). At 6 years of age, illustrating short stature, poor muscle and subcutaneous tissue mass and genu valgum.
Fig. 4 (bottom right).—M. G. Subluxed lenses were previously removed bilaterally, and glaucoma developed on the left. The patient was fair of skin and hair and had a pronounced malar flush, all of which are unusual for a native of Baja California.

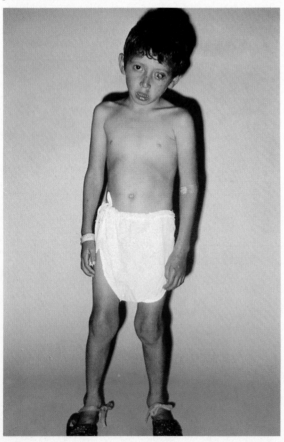

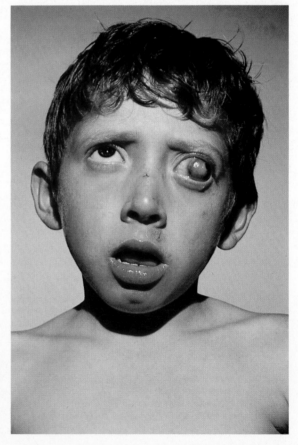

tion of the ocular lens. These patients also may have myopia, glaucoma, cataracts or retinal detachment. Mental retardation is common, although not present in every case. Thrombotic complications, both arterial and venous, probably are the most serious manifestation of the disease and often are the cause of death through pulmonary embolus or renal or cerebral thrombosis. Many patients have seizures and electroencephalographic abnormalities. Fine, light or brittle hair, malar flush and genu valgum are common,[3] and there may be abnormalities of gait. Some patients develop signs of spastic cerebral palsy and most develop osteoporosis. Muscular mass often is small. Patients may fail to thrive.

CASE REPORT

M. G. was a 6-year-old Mexican boy (Fig. 3 and 4) with glaucoma and blindness of the left eye. He was noted to be mentally retarded by 3 years of age. At 4 he had a brief convulsive episode. Shortly thereafter he was noted to have bilaterally subluxed lenses, which were removed surgically. Subsequently the left eye became painful, swollen and red, and enlargement became progressive.

Examination revealed a moderately retarded boy with fair skin, rather fair hair in contrast with his family, and a malar flush. There was iris atrophy, an anterior chamber scar in the right eye and an enormous left eye (see Fig. 4). He had pectus excavatum, genu valgum and an odd gait. There was no arachnodactyly.

X-rays revealed slight thoracolumbar platyspondyly, osteoporosis of the extremities with a prominent trabecular pattern, prominence of the nasal sinuses and genu valgum.

Chemical analysis revealed a positive urinary cyanide nitroprusside test (see Fig. 2), as well as elevated excretion of homocystine, methionine and cysteine homocysteine disulfide. These abnormalities also were found in the plasma. There was no response to treatment with large doses of vitamin B_6.

GENETICS

Homocystinuria is transmitted as an autosomal recessive character. The activity of cystathionine synthase in liver has been found to be lower than that in controls, although, of course, much higher than that in

patients.[4] The frequency of the condition has been estimated at one to 20,000 to one to 40,000 population.

TREATMENT

Experience with treatment has provided evidence of genetic heterogeneity in homocystinuria, since some patients respond to treatment with large doses of pyridoxine and some do not. More recently, patients have been reported to respond to treatment with B_{12}. There is further heterogeneity in that there is another form of homocystinuria in which there are in addition methylmalonic aciduria and a defect in N^5-methyltetrahydrofolate-homocysteine methyltransferase.[6] This condition is due to an inability to convert B_{12} to enzymatically active derivatives and would be expected to respond to large doses of B_{12}.

Dietary treatment has been explored using methionine restriction and cystine supplementation with some success.[7]

REFERENCES

1. Mudd, S. H., Finkelstein, J. D., Irreverre, F., and Laster, L.: Homocystinuria: An enzymatic defect, Science 143:1443, 1964.
2. Gerritsen, T., Vaughn, J. G., and Waisman, H. A.: The identification of homocystine in the urine, Biochem. Biophys. Res. Commun. 9:493, 1963.
3. Carson, N. A. J., Cusworth, D. C., Dent, C. E., Field, C. M. B., Neill, D. W., and Westall, R. D.: Homocystinuria: A new inborn error of metabolism associated with mental deficiency, Arch. Dis. Child. 38:425, 1963.
4. Finkelstein, J. D., Mudd, S. H., Irreverre, F., and Laster, L.: Homocystinuria due to cystathionine synthetase deficiency: The mode of inheritance, Science 146:785, 1964.
5. Barber, G. W., and Spaeth, G. E: The successful treatment of homocystinuria with pyridoxine, J. Pediatr. 75:463, 1969.
6. Mudd, S. H., Levy, H. L., and Abeles, R. H.: A derangement in B_{12} metabolism leading to homocystinemia, cystathioninemia and methylmalonic aciduria, Biochem. Biophys. Res. Commun. 35:121, 1969.
7. Perry, T. L.: Unsolved Problems in Homocystinuria, in W. L. Nyhan (ed.): *Amino Acid Metabolism and Genetic Variation* (New York: McGraw-Hill Book Co., 1967), p. 279.

PHENYLKETONURIA (PKU)

CARDINAL CLINICAL FEATURES

Blue eyes, blond hair, fair skin, mental retardation; eczematoid rash, vomiting in infancy, seizures, hyperactivity, unusual odor, hyperphenylalaninemia, positive urinary ferric chloride ($FeCl_3$) test, phenylalanine hydroxylase deficiency.

CLINICAL PICTURE

Phenylketonuria (PKU) is a disorder of aromatic amino acid metabolism in which phenylalanine cannot be converted to tyrosine (Fig. 1). The full-blown picture of the classic disease should be rare today, as screening and treatment programs become more widespread and more effective. Nevertheless, it is important that we recognize it. Mental retardation is the most important manifestation. Many patients show little else. The intelligence of all but 1% of untreated patients is very low: the IQ usually is under 50.

Vomiting may be a prominent early symptom. It can be of such severity that operations for pyloric stenosis have been performed. Irritability, an eczematoid rash

Fig. 1.—Metabolism of phenylalanine. Site of the phenylalanine hydroxylase defect in PKU and the compounds that accumulate as a consequence.

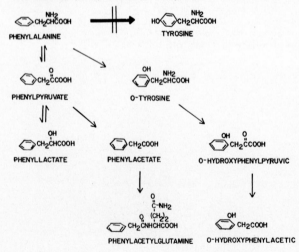

(Fig. 2) and an unusual odor also may be observed very early in life. The odor of the phenylketonuric patient is that of phenylacetic acid (see Fig. 1). It has variously been described as mousy, barny, wolflike or musty. Patients often are good looking. Over 90% are fair haired, fair skinned and blue eyed. However, no amount of pigment in skin, hair or irides (see Fig. 2) excludes the diagnosis. The dermatitis usually is mild and is absent in three fourths of the patients, but it may be a bothersome symptom.

Neurologic manifestations usually are not prominent, but about a third of patients may have all of the signs of cerebral palsy. They are spastic, hypertonic and have increased deep tendon reflexes. Only about 5% have these manifestations to a severe degree, with contractures and limitation of mobility. A second third of the patients have mild neurologic signs such as a unilateral Babinski or hyperactive deep tendon reflexes. A third have no neurologic signs.

Seizures occur in about a fourth of the patients. They usually are neither prominent nor difficult to manage. On the other hand, about 80% have abnormalities of the EEG. Hyperactivity and behavior problems are common. Purposeless movements, rhythmic rocking, stereotypy, tremors and athetosis are seen. Somatic development tends to be normal, but stature often is short. A few patients have minor malformations: these include widely spaced incisor teeth, pes planus, partial syndactyly and epicanthus. Microcephaly is more common.

Phenylalanine hydroxylase, the site of the defect in PKU, is an enzyme found normally only in liver that catalyzes the conversion of phenylalanine to tyrosine. The enzyme complex includes an enzyme that activates a tetrahydropteridine cofactor, which is normal in PKU, and a labile liver enzyme, which has been shown to be lacking in phenylketonuric liver.[3] Heterogeneity in this enzyme in patients with PKU has recently been reported by Barranger et al.[4]

In the presence of a defect in this enzyme, the first compound that accumulates is phenylalanine itself. It is transaminated (see Fig. 1) to form phenylpyruvic acid, the phenylketone for which the disease was named. This is the compound that is responsible for the positive $FeCl_3$ test (Fig. 3), the green color seen on the addition of 10% $FeCl_3$ to the urine of an untreated patient. Phenylpyruvic acid is subsequently converted to phenyllactic acid, phenylacetic acid and phenylace-

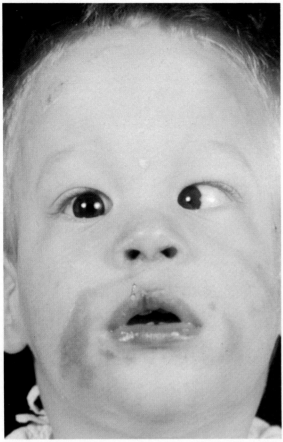

Fig. 2.—Face of a patient with PKU, illustrating the ecze-matoid rash. The brown eyes remind us that all patients with this disease do not have blue eyes.

the most common cause of a positive screening test. In the presence of an elevated phenylalanine and a normal or reduced tyrosine, we have routinely admitted the patient to hospital where protein and phenylalanine intake can be readily monitored and fresh urine specimens readily collected. Most patients with classic PKU have a very rapid rise of serum phenylalanine on a normal diet to levels well over 30 mg/100 ml. Patients with classic PKU also excrete the metabolites phenylpyruvic acid and o-hydroxyphenylacetic acid in the urine. Therefore, we routinely test the urine with $FeCl_3$ and paper chromatography for phenolic acids.

Widespread screening of infant populations has led to the recognition that not all patients with hyperphenylalaninemia have classic PKU.[6] Some variants must certainly represent molecular heterogeneity at the phenylalanine hydroxylase locus, specifying a variant enzyme with partial activity. However, in most instances it is not possible to assess the enzyme in liver, and so the patients tend to be classified in other ways. Most of the variants have phenylalanine concentrations under 15 mg/100 ml, in contrast to most patients with classic PKU in whom concentrations are over 25 mg/100 ml. In terms of dietary criteria, most infants who can tolerate more than 75 mg/kg of phenylalanine/day have a variant.

Transient phenylalaninemia is one of the subgroupings among these patients. This could represent an isolated delay in the maturation of phenylalanine-metabolizing enzymes. Because of this problem all patients with phenylalaninemia should have their dietary tolerance to phenylalanine repeatedly assessed, particularly during the first year.

Increased tolerance to phenylalanine also has been observed in patients described as having a deficiency of phenylalanine transaminase. This was because the performance of a phenylalanine tolerance test led to no phenylketones or other metabolites in the urine.

Most of the variants can only be classified as hyperphenylalaninemic. This includes those patients described from Boston as having Mediterranean hyperphenylalaninemia, since most of the patients with

tylglutamine. Phenylalanine also is hydroxylated in the ortho position, ultimately yielding o-hydroxyphenylacetic acid. These are not abnormal metabolites but normal metabolites that occur in abnormal amounts in PKU. It is current theory that it is this abnormal chemical milieu in which the patient lives that produces the mental retardation and other manifestations of the disease.

The diagnosis of PKU should now be made in the neonatal period. This is initiated through routine screening of all infants after a few days of life and initiation of protein feedings for an elevated concentration of phenylalanine in a drop of blood collected from the heel on filter paper. The most common screening method is that developed by Guthrie,[5] which depends on the inhibition of bacterial growth by β-2-thienylalanine, which is overcome competitively by phenylalanine. Positive tests should be followed up with a quantitative determination of the concentrations of phenylalanine and tyrosine in the blood (Fig. 4). This immediately confirms the phenylalaninemia, and it excludes transient tyrosinemia of the newborn infant,

Fig. 3.—Positive $FeCl_3$ test in a patient with PKU.

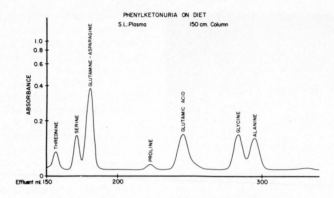

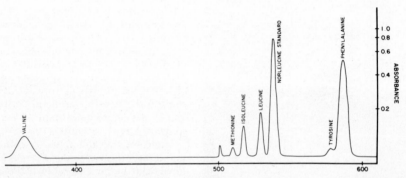

Fig. 4.—Amino acid analysis of the plasma of a patient with PKU. This sample, taken during dietary therapy, reveals a very small tyrosine peak and a very large phenylalanine peak.

classic PKU were of Irish or English ancestry, whereas those with the variant forms were of Italian origin. Similarly, in Israel, phenylalaninemia has been observed in Ashkenazi Jews; however, among 63 patients with PKU, none was among the Ashkenazi, although they make up about half the population.

CASE REPORT

L. S. was a 10½-month-old white male who was admitted to Jackson Memorial Hospital at 9 months of age because of retarded development.

He had developed normally until he was 6 months old. He learned to roll over at 3 months, but he stopped doing so at 4 months and was unable to do so at 10 months. He was followed initially by a careful pediatrician whose records were excellent. There was no blood screening program in his community at the time of his birth. However, his urine was tested for PKU using the Phenistix diaper test and found to be negative at 13 days of age.

He began to have spells in which he drew up his hands and feet, dropped his head forward and was incontinent of urine and feces. These seizures were not prominent during the month prior to admission. By 7 months the patient was unable to sit or hold his rattle. At 9 months, he was still unable to sit or to transfer and appeared retarded.

Physical examination revealed a well-developed and well-nourished blond white male (Fig. 5). He was fair skinned and blue eyed. He was an unhappy, irritable baby who whined constantly and responded poorly to voices. He did not sit without support. He held a bottle only if his hands were placed on it. He appeared quite retarded and was very floppy and hypotonic. The EEG was abnormal in that it showed frequent bursts of irregular paroxysmal activity consistent with a convulsive disorder.

The Guthrie test was carried out on the blood and found to be positive. This was repeated three times. The Phenistix and FeCl$_3$ tests on the urine were positive. The serum concentration of phenylalanine was 37.1 mg/100 ml.

Following documentation of the diagnosis of PKU, treatment was started with a Lofenalac diet. Within 4 days the FeCl$_3$ test became negative. Muscle tone gradually improved and development began to proceed. The child ultimately learned to walk and talk but remained retarded and hyperactive.

GENETICS

Phenylketonuria is transmitted by an autosomal recessive gene. The persistent non-PKU hyperphenylalaninemias also appear to be transmitted as autosomal recessives. Phenylketonuria and hyperphenylalanine-

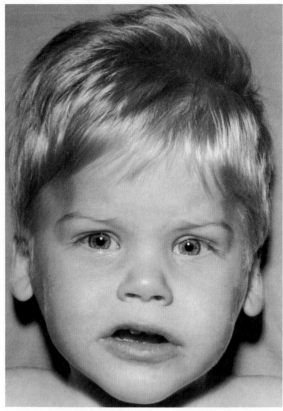

Fig. 5.—L. S. (Case Report). This patient was diagnosed as having PKU at 10 months of age. Pigmentation of skin and hair actually was lighter than apparent in the photograph.

mia assort independently and thus cluster in different families. In each instance heterozygote detection can be demonstrated using phenylalanine tolerance tests. On the other hand, methodology is not sufficiently reliable to guarantee that on any given individual a diagnosis of heterozygosity can be made with certainty.

TREATMENT

Treatment of PKU is the provision of a diet sufficiently low in phenylalanine that the serum concentrations are maintained in a reasonable range and metabolites disappear from body fluids. This requires the provision of enough phenylalanine to meet the normal requirements of this essential amino acid for growth. It also requires frequent quantitative assessment of the concentration of phenylalanine in the blood. Concentrations between 3 mg/100 ml and 15 mg/100 ml appear to be reasonable. Using this approach to therapy, most patients detected by screening programs in the neonatal period should have IQs in the normal range.

REFERENCES

1. Auerbach, V., DiGeorge, A. M., and Carpenter, G. G.: Phenylalaninemia, in W. L. Nyhan (ed.): *Amino Acid Metabolism and Genetic Variation* (New York: McGraw-Hill Book Co., 1967), pp. 11–68.
2. Allen, R. J., Fleming, L., and Spirito, R.: Variations in Hyperphenylalaninemia, in W. L. Nyhan (ed.): *Amino Acid Metabolism and Genetic Variation* (New York: McGraw-Hill Book Co., 1967), pp. 69–96.
3. Jervis, G. A.: Phenylpyruvic oligophrenia deficiency of phenylalanine-oxidizing system, Proc. Soc. Exp. Biol. Med. 82:514, 1953.
4. Barranger, J. A., Geiger, P. J., Huzino, A., and Bessman, S. P.: Isoenzymes of phenylalanine hydroxylase, Science 175:903, 1972.
5. Guthrie, R.: Blood screening for phenylketonuria (Letter to the Editor), J.A.M.A. 178:863, 1961.
6. Hsia, D. Y-Y.: Phenylketonuria, in Steinberg, A. G., and Bearn, A. G. (eds.): *Progress in Medical Genetics* (New York: Grune & Stratton, 1970), Vol. 7, pp. 29–68.

MAPLE SYRUP URINE DISEASE
Branched-Chain Ketoaciduria

CARDINAL CLINICAL FEATURES

Overwhelming illness in the first days of life with lethargy progressive to coma, opisthotonos, convulsions; characteristic maple syrup odor; branched-chain aminoacidemia and aminoaciduria; branched-chain ketoaciduria; deficiency of branched-chain keto acid decarboxylase.

CLINICAL PICTURE

Maple syrup urine disease was first described in 1954 by Menkes et al.,[1] who observed an unusual odor like that of maple syrup in the urine of four infants in a single family who had succumbed to a progressive neurologic disease in the first weeks of life.

Infants with this disease appear normal at birth, but they usually remain well for only a few days. Generally by the end of the first week they become lethargic and feed poorly. A high-pitched cry may be heard early. Neurologic deterioration is progressive and rapid. Patients may have periods of flaccidity in which the deep tendon reflexes and Moro reflex are absent, alternating with hypertonicity. An opisthotonic position is characteristic. Abnormal eye movements may appear. Convulsions occur regularly. These symptoms proceed to abnormalities in respiration, after which coma and death rapidly ensue unless a vigorous therapeutic program is instituted.[2]

Rarely an infant may survive this early phase of the disease. If so, he usually is left with prominent neurologic abnormalities and profound mental retardation. Furthermore, he is always a candidate for an episode of acute overwhelming illness that may be fatal. Such episodes often occur during the catabolic state that accompanies infection.

The characteristic odor may be detected as soon as neurologic symptoms develop. At the same time, it should be recognized that every patient with this disease is not recognizable by smell. Infants with the rest of the clinical picture should be screened for metabolic disease, even if the odor is not detected. The odor may be found in the hair, sweat or cerumen but usually is best found in the urine. Freezing the urine may bring out the smell by concentrating it in an oil that freezes poorly or not at all at the top of the frozen specimen.

The odor is sweet, malty or caramel-like. It does indeed call forth an olfactory image of maple syrup.

The first characterization of maple syrup urine disease as an inborn error of metabolism was made by Westall et al.,[3] who found that the concentrations of the branched-chain amino acids leucine, isoleucine and valine were elevated in the plasma (Fig. 1) and urine of an infant with the disease. These patients also excrete the keto acid products of the transamination of each of these amino acids,[4–6] suggesting the site of the block (Fig. 2). Isovaleric' acid, α-methylbutyric acid and isobutyric acid are not found in the urine.[4] High levels of the branched-chain keto acids also are found in the blood and cerebrospinal fluid. Among the amino acids, the concentrations of leucine are always higher than those of isoleucine and valine in these patients.

Alloisoleucine also is found to accumulate in body fluids in this disease.[7] This amino acid often is mistaken for methionine in the ordinary Moore-Stein analysis[8] on the amino acid analyzer. This created some confusion in the early literature on the disease and in the management of early patients. Alloisoleucine is readily separated from methionine using the Piez-Morris[9] method of analysis. The concentration of alanine in the plasma of these patients usually is decreased.[2] The ferric chloride test on the urine may give a green-gray color. Excretion of the keto acids permits the detection of this disorder using 2,4-dinitrophenylhydrazine, which produces a yellow precipitate of dinitrophenylhydrazones.[10] The individual keto acids can be distinguished by thin layer chromatography[10] or gas liquid chromatography[11] of the dinitrophenylhydrazones formed.

Measurement of the activity of branched-chain decarboxylase in vitro using fibroblasts or leukocytes has been carried out by studying the conversion of leucine-^{14}C, isoleucine-^{14}C, valine-^{14}C or α-ketoisocaproic acid-^{14}C to $^{14}CO_2$. In each measurement the activity has been virtually nil. In contrast, the oxidation of isovaleric acid-^{14}C to $^{14}CO_2$ is normal.

VARIANT FORMS

A number of variants of branched-chain ketoaciduria have now been described.[12] Each is milder in its clinical presentation than classic maple syrup urine

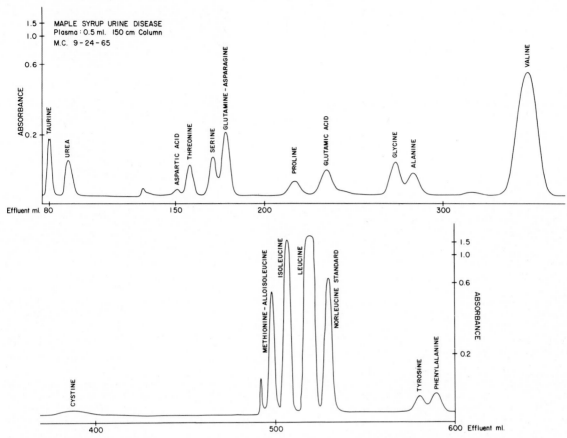

Fig. 1.—Chromatogram of the amino acids of the plasma in a patient with maple syrup urine disease. The large peaks of valine, isoleucine and leucine provide a diagnostic picture.

disease. The first of these forms to be described[13-15] is known as intermittent branched-chain ketoaciduria. Involved individuals may have no problems except in the presence of something like an infection or surgery that produces a special stress.

On the other hand, this disorder, too, can be lethal. Patients with no problems at all over a period of years can suddenly develop coma, convulsions and death following an apparently mild infection.[16] In general, they have intermittent bouts of acute ataxia. Patients

with this disorder have been found to have a partial defect in branched-chain keto acid decarboxylase activity.[12, 17]

A similar incomplete defect in decarboxylase activity, to about 15-25% of the normal level, has been observed in a group of patients considered to have a third form of branched-chain ketoaciduria.[2, 18, 19] All of these patients had mental retardation; hence, some symptomatology is considered to be continuously, as opposed to intermittently, present. The biochemical abnormality in these patients, the accumulation of amino acids and keto acids in body fluids, is always demonstrable, except, of course, when successfully treated.

Dancis[12] has based a classification of variants on protein tolerance. In the classic variant, patients cannot tolerate maintenance requirements of protein; only with an artificial purified amino acid diet can these patients survive. Enzyme levels are 0-2% of normal. In the second variant, protein tolerance is about sufficient to maintain normal growth in infancy, or 1.5-2 gm/kg of protein. Enzyme levels in this group are between 2% and 8% of normal. In the third, an unrestricted diet is tolerated. Attacks of ataxia and

Fig. 2.—Metabolic pathways in the catabolic breakdown of leucine, isoleucine and valine. The symbol designated 1 indicates the site of the defect in maple syrup urine disease.

CATABOLISM OF THE BRANCHED-CHAIN AMINO ACIDS

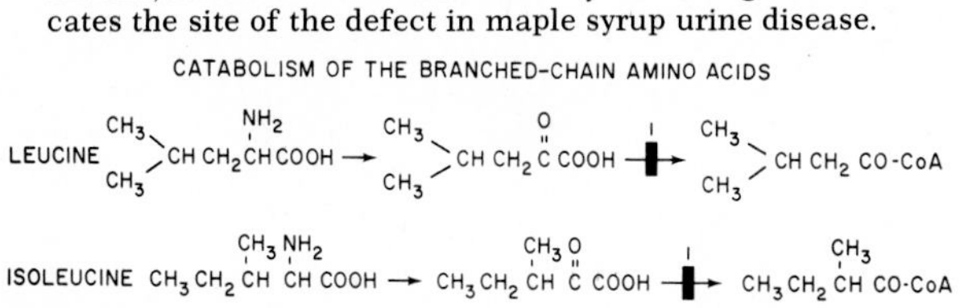

more severe manifestations such as coma or convulsions may occur, especially with infection. Enzyme activity usually is between 8% and 16% of normal.

In another variant, the clinical features are similar to those of the third form, but the biochemical abnormality may be corrected by administration of large doses of thiamine.[20] Therefore, this form is known as thiamine-responsive maple syrup urine disease or thiamine-responsive branched-chain ketoaciduria.

Case Report*

M. B. was the first-born child of Mennonite parents, distantly consanguineous through a common ancestor removed eight generations on the mother's side and nine generations on the father's side of the family. A second male child born to the parents was normal. The patient was the product of an uncomplicated 40-week gestation and vertex delivery. His birth weight was 7 lb, 14 oz. He was considered healthy at birth, but at 3 days of age a blood concentration of leucine of 12 mg/100 ml was discovered by the Oregon State Health Department on routine newborn screening by the use of a Guthrie bacterial inhibition assay. The patient was breast fed and remained clinically well until 4 or 5 days of age, when he became irritable and fed poorly. He then began vomiting and lost weight. At 9 days of age a second Guthrie assay showed the blood leucine to be greater than 20 mg/100 ml. He was referred at 11 days of age to Doernbecher Hospital at the University of Oregon Medical School.

On admission his weight was 6 lb, 9 oz. He was malnourished, moderately dehydrated and opisthotonic (Fig. 3). He was tremulous, but there were no frank convulsive movements and none had been observed by the mother. He had a weak, high-pitched cry. He was hypertonic and had increased deep tendon reflexes, a poor Moro and sucking reflex. Despite the fact that he had not urinated for several hours, he had a strong odor of maple syrup.

His urine gave a strongly positive nitroprusside reaction. The plasma concentrations of amino acids included a leucine

Fig. 3. – M. B. (Case Report). An 11-day-old infant with maple syrup urine disease, shown in the typical opisthotonic position. (Courtesy of Dr. Havelock Thompson of the University of West Virginia.)

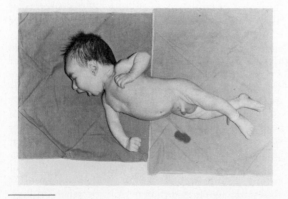

---
*We are indebted to Dr. Havelock Thompson of the University of West Virginia for providing information on this patient.

Fig. 4. – M. B. at 7 months of age. He was an advertisement for very early diagnosis and vigorous therapy. (Courtesy of Dr. Havelock Thompson of the University of West Virginia.)

of 43, isoleucine of 3.3, valine of 7.9 and alloisoleucine of 2.8 mg/100 ml. A diagnosis of maple syrup urine disease was made.

He was fed intravenously at first and then a synthetic formula, as described by Snyderman et al.,[21] was initiated on the 12th day of life. Abnormal amino acid concentrations fell rapidly on treatment, and the infant's clinical condition gradually improved. He became responsive, fed eagerly, and all of the previously abnormal neurologic signs disappeared during this time. The EEG, which showed an abnormal pattern of high voltage spikes initially, became normal and remained so. He gained weight slowly at first, but more rapidly later on.

He continued to thrive, and his growth and development progressed within normal limits. By 6 months his height, weight and head circumference were all above the 50th percentile (Fig. 4). At 2 years and 3 months, his Stanford-Binet performance placed him at 2 years, with a full-scale IQ score of 88. During episodes of infections his leucine concentrations have increased and he has become mildly intoxicated, as evidenced by irritability, restlessness, loss of appetite and at times mild ataxia.

Genetics

The disorder is transmitted as an autosomal recessive trait. This appears to be true of each of the variants studied. Classic maple syrup urine disease has been seen throughout the world: in Chinese, Japanese, Negroes and Indians as well as in white people throughout America, Europe and the Middle East. Heterozygote detection has not reliably been established. The enzyme activity can be measured in cultured amniotic fluid cells.

Treatment

Emergency treatment of an infant in coma should be by exchange transfusion, peritoneal dialysis or both. This may be lifesaving in an infant newly diagnosed or in a later episode accompanying an intercurrent illness.

Chronic management consists of restriction of the

intake of each of the three branched-chain amino acids to those quantities essential for growth and no more. This type of dietary management is much more difficult than that for phenylketonuria. It requires very close regulation of an artificial diet and frequent access to an amino acid analyzer. The best results are, of course, seen in those in whom treatment has been initiated earliest. The largest experience with the management of this disease is Snyderman's, and she has written that there can be little doubt about the beneficial effect of therapy in this disorder.[2, 21]

The milder variants may be manageable by simple restriction of the protein intake to the amounts required for normal growth.

REFERENCES

1. Menkes, J. H., Hurst, P. L., and Craig, J. M.: A new syndrome: Progressive familial cerebral dysfunction with an unusual urinary substance, Pediatrics 14:462, 1954.
2. Snyderman, S. E.: Maple Syrup Urine Disease, in Nyhan, W. L. (ed.): *Heritable Disorders of Amino Acid Metabolism* (New York: John Wiley & Sons, 1974), pp. 17–31.
3. Westall, R. G., Dancis, J., and Miller, S.: Maple syrup urine disease—a new molecular disease, Am. J. Dis. Child. 94: 571, 1957.
4. Dancis, J., Levitz, M., and Westall, R. G.: Maple syrup urine disease: Branched chain keto-aciduria, Pediatrics 25:72, 1960.
5. Mackenzie, D. Y., and Woolf, L. I.: Maple syrup urine disease: Inborn error of metabolism of valine, leucine and isoleucine associated with gross mental deficiency, Br. Med. J. 1:90, 1959.
6. Menkes, J. H.: Maple syrup disease: Isolation and identification of organic acids in the urine, Pediatrics 23:348, 1959.
7. Norton, P. M., Roitman, E., Snyderman, S. E., and Holt, L. E., Jr.: A new finding in maple syrup urine disease, Lancet 1:26, 1964.
8. Spackman, D. H., Stein, W. H., and Moore, S.: Automatic recording apparatus for use in chromatography of amino acids, Anal. Chem. 30:1190, 1958.
9. Piez, K. S., and Morris, L. A.: A modified procedure for the automatic analyses of amino acids, Anal. Biochem. 1:187, 1960.
10. Schmidt, L.: The Biochemical Detection of Metabolic Disease: Screening Tests and a Systematic Approach to Screening, in Nyhan, W. L. (ed.): *Heritable Disorders of Amino Acid Metabolism* (New York: John Wiley & Sons, Inc., 1974), pp. 675–697.
11. Bachmann, C., Nyhan, W. L., and Sweetman, L.: Analysis of Dinitrophenylhydrazones of Aldehydes and Ketones by Gas Chromatpgraphy-Mass Spectrometry, in Mamer, O. A., Mitchell, W. J., and Scriver, C. R. (eds.): *Application of Gas Chromatography-Mass Spectrometry to the Investigation of Human Disease, Proceedings of a Workshop* (Montreal: McGill University-Montreal Children's Hospital Research Institute, 1974), pp. 165–177.
12. Dancis, J.: Variants of Maple Syrup Urine Disease, in Nyhan, W. L. (ed.): *Heritable Disorders of Amino Acid Metabolism* (New York: John Wiley & Sons, Inc., 1974), pp. 32–36.
13. Morris, M. D., Lewis, B. D., Doolan, P. D., and Harper, H. A.: Clinical and biochemical observations on an apparently nonfatal variant of branched-chain ketoaciduria, Pediatrics 28:918, 1961.
14. Kiil, R., and Rokkones, T.: Late manifesting variant of branched-chain ketoaciduria, Acta Paediatr. 53:356, 1964.
15. Van Der Hort, H. L., and Wadman, S. K.: A variant form of branched-chain ketoaciduria, Acta Paediatr. Scand. 60: 594, 1971.
16. Goedde, H. W., Langenbeck, V., Brackertz, D., Keller, W., Rokkones, T., Halvarsen, S., Kiil, R., and Merton, B.: Clinical and biochemical genetic aspects of intermittent branched-chain ketoaciduria, Acta Paediatr. Scand. 59: 83, 1970.
17. Dancis, J., Hutzler, H., Snyderman, S. E., and Cox, R. P.: Enzyme activity in classical and variant forms of maple syrup urine disease, J. Pediatr. 81:312, 1972.
18. Schulman, J. D., Lustberg, T. J., Kennedy, J. L., Museles, M., and Seegmiller, J. E.: A new variant of maple syrup urine disease (branched-chain ketoaciduria), Am. J. Med. 49:118, 1970.
19. Fischer, M. H., and Gerritsen, T.: Biochemical studies on a variant of branched-chain ketoaciduria in a nineteen-year-old female, Pediatrics 48:795, 1971.
20. Scriver, C. R., Clow, C. L., Mackenzie, S., and Delvin, E.: Thiamine responsive maple syrup urine disease, Lancet 1: 310, 1971.
21. Snyderman, S. E., Norton, P. M., Roitman, E., and Holt, L. E., Jr.: Maple syrup urine disease with particular reference to dietotherapy, Pediatrics 34:454, 1964.

METHYLMALONIC ACIDEMIA

Cardinal Clinical Features

Recurrent episodes of ketosis, acidosis, vomiting, dehydration; failure to grow and thrive; hepatomegaly; neutropenia; thrombocytopenia; osteoporosis; hyperglycinemia; methylmalonic acidemia; methylmalonic aciduria.

Clinical Picture

Patients with methylmalonic acidemia usually present with overwhelming illness very early in life. Many of the reported patients have died in the course of such an episode of illness. Consequently, we have thought that this disease is almost certainly more common than diagnosed and that many patients may die with the cause unrecognized. A typical episode is ushered in with ketonuria, followed by vomiting, acidosis, dehydration and lethargy. Without vigorous treatment this leads to coma and death.

These episodes are recurrent. They often follow even minor infections. Furthermore, the untreated patient seems to be unusually prone to infection. Episodes also follow feeding. These patients are intolerant to the usual quantities of dietary protein. More specifically, they are intolerant to the amino acids isoleucine, valine, threonine and methionine.

This multiple amino acid toxicity is consistent with the fundamental defect. All patients with methylmalonic acidemia have a defect in the activity of methylmalonyl CoA mutase, which catalyzes the conversion of methylmalonyl CoA to succinyl CoA (Fig. 1). This enzyme lies on the direct degradative pathway for isoleucine, valine, threonine and methionine. All of these amino acids have been shown to be major sources of methylmalonate in these patients.

Heterogeneity already has been documented in that there are three distinct disorders that present with methylmalonic acidemia. One appears to be an apoenzyme defect in the methylmalonyl CoA mutase reaction itself. A second disorder is characterized by responsiveness to large doses of vitamin B_{12} and has now been shown to be probably due to a defect in the enzymatic conversion of B_{12} to 5'-deoxyadenosylcobalamin. In a third disorder methylmalonic acidemia accompanies elevated concentrations of both homocystine and cystathionine in blood and urine. This suggests a

Fig. 1. — Site of the defect in methylmalonic acidemia. Metabolism of propionic acid.

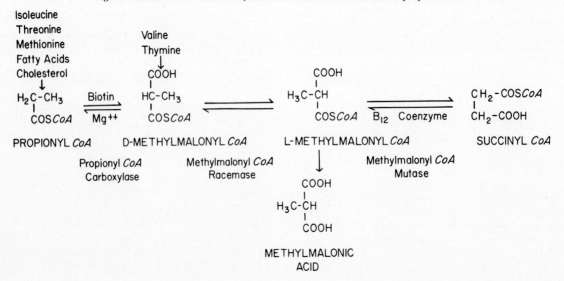

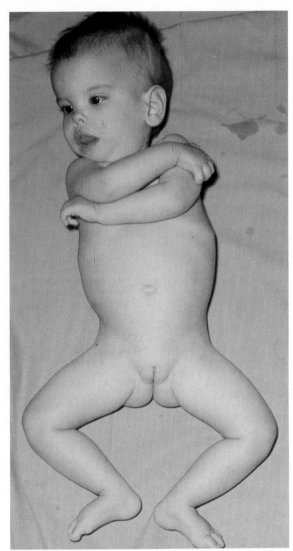

Fig. 2.—L. G. (Case Report). Girl age 14 months with methylmalonic acidemia illustrates the extreme failure to thrive seen in this condition. Her frog position demonstrates the marked hypotonia.

been recorded, and the serum bicarbonate is often 5 mEq/L or less. Ketosis is massive. Hypoglycemia with seizures has been observed during acute acidotic episodes. Elevated concentrations of glycine in the blood and urine may be striking. For instance, concentrations as high as 11.6 mg/100 ml have been observed in the plasma. However, the concentrations of glycine also may be normal, even in the same patient, and concentrations of glycine seem to bear little relationship to clinical condition. Transient thrombocytopenia has been observed. Neutropenia is a regular occurrence except in the case of successful treatment and reduction in the accumulation of methylmalonic acid in body fluids. We have seen one patient with chronic cutaneous moniliasis.

Osteoporosis has been found in all of the patients studied. Growth failure may be striking in this condition. They may present as patients with failure to thrive. Furthermore, they fail to develop during this period of growth failure. They may be markedly hypotonic. Some patients have had hepatomegaly, but not of very great degree.

Fig. 3.—L. G. Close-up of the face. She had a number of minor structural abnormalities including epicanthus, prominent forehead, wide nasal base and high-arched palate. A preauricular skin tag had been removed.

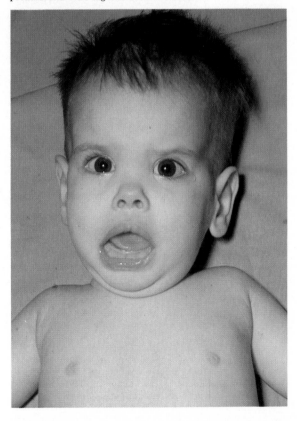

defect in remethylation of homocysteine to methione, and in these patients N^5-methyltetrahydrofolate-homocysteine methyltransferase, the other B_{12}-dependent enzyme in mammalian systems, is defective. This abnormality could result from a failure to transform B_{12} to any coenzymatically active derivatives, either deoxyadenosylcobalamin or methylcobalamin. Methylmalonic aciduria also is seen in acquired B_{12} deficiency, but the order of magnitude is much less than in the inborn methylmalonic acidemias.

During episodes of ketosis these patients become very acidotic. Arterial pH values as low as 6.9 have

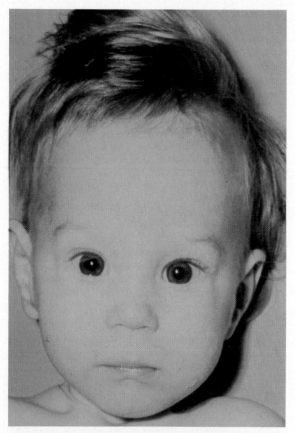

Fig. 4.—S. P. A 9-month-old girl with methylmalonic acidemia. Her similarity in appearance to L. G. (see Fig. 3) was striking.

CASE REPORT

L. G. was the first child of healthy, unrelated parents. Pregnancy was uneventful. By the second day of life she was noted to be a floppy infant who sucked poorly. At 3 months of age, meat was introduced into her diet and she began to vomit. Although she grew in height and weight for the first 3 months, she thereafter failed to gain any weight or to grow an inch, up to the time of admission at 14 months of age. During the interval she had five admissions to another hospital because of episodes of ketosis and acidosis. She had numerous respiratory infections, and some but not all ketotic episodes began with an infection. During these periods she smelled strongly of acetone, and ketones were present in blood and urine. The CO_2 content generally ranged from 10 to 14 mEq/L. Treatment with parenteral fluid therapy containing large amounts of $NaHCO_3$ was effective in reversing acute symptomatology. Neutropenia and relative lymphocytosis were observed repeatedly. At 13 months of age she had seizures, and she was found to have a blood glucose concentration of 18 mg/100 ml. She responded to the administration of glucose.

On physical examination at 14 months she was a small, hypotonic infant who was the size of a 3-month-old baby (Fig. 2). She appeared to perform developmentally at even less than the 3-month level. She held her head in the prone position but

showed little other evidence of psychomotor development. She had a mild epicanthus, a prominent forehead and a wide nasal base (Fig. 3). Another of our patients had a similar facial appearance (Fig. 4). There was a high-arched palate and the remnant of a previously removed preauricular skin tag. The liver was palpable three fingerbreadths below the right costal margin. Monilial lesions were present in the mouth, in the perianal region and on the second and third fingers of the right hand.

Electroencephalography revealed left-sided slow wave asymmetry. The leukocyte count was 7,000/cu mm with 10% polymorphonuclear forms and 76% lymphocytes. The concentrations of electrolytes were normal, but she developed a CO_2 content of 16.5 mEq/L during a mild infection. The urine pH was 5 and routine urinalysis was negative. The plasma concentration of methylmalonic acid was 13 mg/100 ml, and 710 mg of methylmalonic acid was excreted in the urine in 24 hours. The concentration of glycine in the plasma was 4.5 mg/100 ml, and 47 mg of glycine was excreted in the urine in 24 hours. The glycine-to-creatinine ratio in the urine was 857 mg/gm.

As she was found to be unresponsive to B_{12}, dietary therapy was instituted using a diet low in the protein precursors of methylmalonic acid. She reacted slowly but dramatically, the first response being a reduction in the amounts of methylmalonic acid in blood and urine. She began to grow and crossed a series of growth lines for height, weight and head size. Furthermore, she developed rapidly. By 3 years of age her head size, height and weight were normal, as was her IQ.

GENETICS

Definitive genetic information is not available. However, it is clear that each of the forms of methylmalonic acidemia is carried on a rare, autosomal gene.

TREATMENT

Patients with methylmalonic acidemia should first be tested for responsiveness to B_{12}. Those who respond should be treated with the vitamin in doses sufficient to keep concentrations of methylmalonic acid minimal. Patients who do not respond to B_{12} can be treated with a diet designed to keep the precursors of methylmalonic acid at a manageable level. This is complicated because isoleucine, threonine, methionine and valine are all essential for normal growth and development. Therefore, optimal therapy consists of a diet containing the minimal requirements of these amino acids for optimal growth and no more. The rest of the calories can be made up of a diet containing fat and carbohydrate, with or without other amino acids. The management of such a patient is not easy, and it must be controlled with quantitative assay of methylmalonic acid concentrations (Fig. 5). The reward in terms of the development of a normal child is very high.

REFERENCES

1. Oberholzer, V. G., Levin, B., Burgess, E. Ann, and Young, W. F.: Methylmalonic aciduria: An inborn error of metabo-

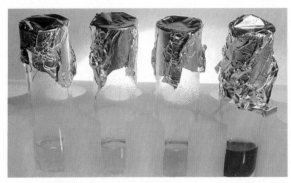

Fig. 5. – Colorimetric assay of the urine for methylmalonic acid. The green color, which develops with p-nitroaniline, provides a useful screening test for the condition. It also can be used for quantitative assay.

lism leading to chronic metabolic acidosis, Arch. Dis. Child. 42:492, 1967.

2. Stokke, O., Eldjarn, L., Norum, K. R., Steen-Johnsen, J., and Halvorsen, S.: Methylmalonic acidemia: A new inborn error of metabolism which may cause fatal acidosis in the neonatal period, Scand. J. Clin. Invest. 20:313, 1967.

3. Lindblad, B., Lindblad, B. S., Olin, P., Svanberg, B., and Zetterström, R.: Methylmalonic acidemia: A disorder associated with acidosis, hyperglycinemia, and hyperlactatemia, Acta Paediatr. Scand. 57:417, 1968.

4. Morrow, G., Barness, L. A., Auerbach, V. H., DiGeorge, A. M., Ando, T., and Nyhan, W. L.: Observations on the coexistence of methylmalonic acidemia and glycinemia, J. Pediatr. 74:680, 1969.

5. Rosenberg, L. E., Lilljeqvist, A. C., Hsia, Y. E.: Methylmalonic aciduria: Metabolic block localization and vitamin B_{12} dependency, Science 162:805, 1968.

6. Rosenberg, L. E., Lilljeqvist, A. C., and Hsia, Y. E.: Methylmalonic aciduria: An inborn error leading to metabolic acidosis, long-chain ketonuria and intermittent hyperglycinemia, N. Engl. J. Med. 278:1319, 1968.

7. Mudd, S. H., Levy, H. L., and Abeles, R. H.: A derangement in B_{12} metabolism leading to homocystinemia, cystathioninemia and methylmalonic aciduria, Biochem. Biophys. Res. Commun. 35:121, 1969.

8. Morrow, G., Mellman, W. J., Barness, L. A., and Dimitrov, N. V.: Propionate metabolism in cells cultured from a patient with methylmalonic acidemia, Pediatr. Res. 3:217, 1969.

ALBINISM

Cardinal Clinical Features

Oculocutaneous: Absence of normal pigmentation in skin and hair, translucent irides, hypopigmented ocular fundus, nystagmus, photophobia. *Ocular:* Eyes as in oculocutaneous albinism, but skin normally pigmented. *Partial (localized cutaneous albinism):* Congenital patterned absence of melanin over the body, usually ventral and often involving parts of eyebrows or lashes but sparing the eyes; white forelock.

Clinical Picture

Oculocutaneous Albinism

This is a common genetic disease. It was proposed by Garrod[1] in 1908 as an inborn error of metabolism in which there was a defect in an intracellular enzyme of melanin synthesis. Albinism is found in fish, birds and mammals, including all races of man. It is a heritable defect in the pigment cell (melanocyte) system of the skin and eyes. The melanocytes are specialized cells in which biosynthesis of melanin occurs in the melanosomes. The cells arise from melanoblasts, which develop early in embryogenesis in the neural crest.

Melanin is a product of the metabolism of tyrosine (Fig. 1). The first two steps, conversions to dopa and dopaquinone, are catalyzed by the same enzyme, tyrosinase, which is a copper-containing oxidase. The remaining steps of melanogenesis can proceed nonenzymatically. Melanin exists in nature as a polymer of high molecular weight. Only part of the repeating structure is illustrated in Figure 1. The entire transformation takes place in the melanosome. Melanosomes are present in the melanocytes of human albinos. Tyrosinase can be studied in tissues radioautographically following incubation with ^{14}C-labeled tyrosine. It also is possible to visualize pigment formation directly following incubation with substrate quantities of tyrosine or dopa. Either of these tests is conveniently performed using the hair follicle, which can be incubated directly following careful plucking of the hair. Studies of tyrosinase in vitro have provided evidence for genetic heterogeneity in this condition that had been suspected on the basis of phenotypic and genetic evidence. Some albinos lack tyrosinase, the disease thus being readily understandable as an inborn error of metabolism. Other albinos are tyrosinase positive.

Tyrosinase-negative oculocutaneous albinism. — In these patients the hair is snow-white or milk-white (Figs. 2 and 3). The skin has a pink or reddish

Fig. 1. — Formation of *melanin* from *tyrosine.*

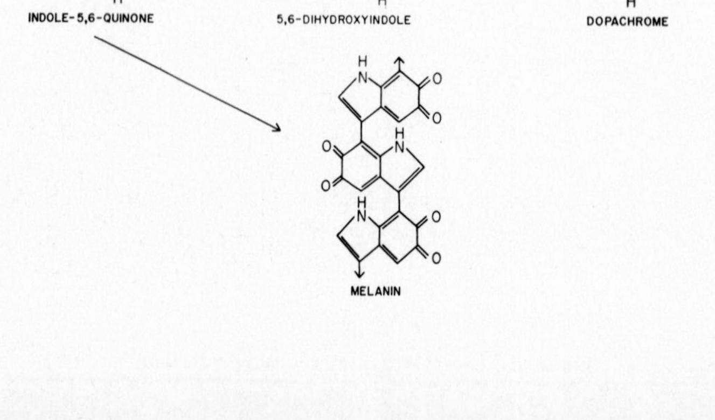

18

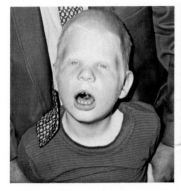
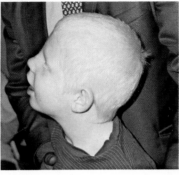

Fig. 2 (left).—T. G. (Case 1). Oculocutaneous albinism. This boy was one of two brothers, both albino and both severely retarded. He had snow-white hair and pink-white skin. The irides were blue, and there was a pupillary red reflex.

Fig. 3 (center).—T. G. The lateral projection illustrates the beautiful white eyelashes.

Fig. 4 (right).—Negro woman with oculocutaneous albinism. There was a certain amount of pigment in the face, and the hair was yellow.

hue. The iris is blue or gray, and the pupils display a red reflex at all ages. No visible pigment can be seen in the fundus. Patients with this form of albinism have severe nystagmus and photophobia; they squint even in moderate light. Visual acuity is reduced and grows worse with age. Many patients are legally blind.

TYROSINASE-POSITIVE OCULOCUTANEOUS ALBINISM.—In this form of albinism the hair is white or yellow-white in infancy and changes to yellow or even red with age. The skin is white or cream colored. In the Negro it usually is easy to distinguish this yellow-haired, yellow-skinned phenotype from the tyrosinase-negative type (Fig. 4). In the Caucasian there is considerable overlap, and the skin may be pink in this form. Pigmented nevi and freckles are common in tyrosinase-positive albinos. The iris color is blue early in life, but brown pigment increases with age, especially at the pupillary border. Nystagmus and photophobia usually are less severe than in tyrosinase-negative patients.

A number of additional ocular abnormalities are common to both types of albinism. These include strabismus, central scotomas and partial aniridia. The intelligence of persons with oculocutaneous albinism usually is normal.

YELLOW MUTANT FORM OF OCULOCUTANEOUS ALBINISM.—A third type of oculocutaneous albinism has recently been identified.[3, 4] In these patients the tyrosinase response may be confusing but usually is negative. Nevertheless patients have considerable pigment and more often show resemblance to the tyrosinase-positive albino. Affected individuals have generalized albinism at birth, but over the first few years they de-

velop relatively normal skin pigmentation and yellow hair. Persistent ocular manifestations such as retinal hypopigmentation and red reflex are, however, present in adult life. The irides are blue in infancy but may develop pigment with age.

The differential diagnosis includes the Chediak-Higashi syndrome, in which tyrosinase-positive oculocutaneous pigment dilution is associated with giant peroxidase-positive lysosomal granules in the leukocytes; neutropenia; thrombocytopenia; folic acid-deficient anemia and susceptibility to infections and to lymphoma or leukemia. The melanocytes contain giant melanosomes. In the Cross syndrome hypopig-

Fig. 5.—B. S. (Case 2). Partial albinism. The patient had a white forelock and a larger patch of white hair on the back of the head.

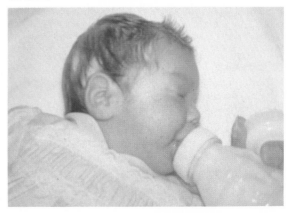

mentation is associated with a relative deficiency of melanosomes and with microphthalmia.

Ocular Albinism

Albinism in this condition is confined to the eyes and does not affect the hair or skin color. The iris is light blue but may develop pigment with age, whereas the retina does not. Head nodding usually is confined to infancy. Nystagmus also may improve with age.

Partial Albinism

Cutaneous albinism occurs without deafness and is localized to areas in the skin and hair without involvement of the eye. There usually is a white forelock, often triangular with the apex pointing backward. The white area always includes the skin beneath the forelock. There may be other patches of depigmented hair. Lack of pigment in the skin tends to be ventral in distribution, occurring in patches over the anterior trunk or extremities. This condition should be distinguished from the Waardenberg syndrome (see section on Dysmorphic Syndromes with Prominent Facial and Ocular Abnormalities), in which partial albinism is associated with deafness, lateral displacement of medial canthi and heterochromia iridis.

CASE REPORTS

CASE 1.—(Oculocutaneous albinism). T. G., a 10-year-old white male, was mentally retarded and had been hospitalized at the Fairview State Hospital, Costa Mesa, California, since the age of 4 years. He was the product of a normal pregnancy and of a full-term spontaneous delivery. The father was not known. The patient was noted to be an albino at birth. A brother also was albino and retarded.

The patient sat at 6 months, walked at 20 months and spoke a few sentences at 18 months, but at 20 months he ceased verbal communication. Toilet training was not successful. He was autistic and made no contact with his peers. He had a febrile convulsion at 20 months of age. Visual impairment was severe. The IQ score was 52 on the Cattell Infant Intelligence Scale.

The hair was snow-white and the skin pink-red (see Figs. 2 and 3). The irides were a pale gray. There was a red reflex as well as severe nystagmus and photophobia. Funduscopy revealed a lack of retinal pigmentation. Visual acuity was less than 20/200, which is in the range of the legally blind.

CASE 2.—(Partial albinism). B. S. was a 1-month-old baby girl referred for evaluation of a white forelock and a white occipital lock (Fig. 5). There were no other complaints. The child was the product of a normal pregnancy and delivery. Birth weight was 6 lb. This baby was the tenth child, and there was no similar condition in the family. The father was 45 years old at the time of the patient's birth, the mother 35.

Examination revealed a white occipital lock as well as a white forelock. The skin beneath the white forelock was not pigmented. The medial part of the eyebrow was white. There was no eye involvement and no heterochromia of the iris. Hearing was normal.

GENETICS

Oculocutaneous albinism is transmitted as an autosomal recessive trait. In groups of albinos there is a history of increased parental consanguinity. It is clear that more than one gene is involved. If generalized oculocutaneous albinism were due to a single recessive gene, all offspring of the mating of two albinos should be albinos; however, normal offspring have been reported from such matings. These observations indicate that not only is there more than one gene, which also was indicated by data on tyrosinase-positive and -negative albinism, but the genes are not allelic.

Ocular albinism is transmitted as an X-linked recessive trait. Females heterozygous for the gene have translucent irides and a mosaic pattern of patches or clumps of pigmentation in the retina.

Partial albinism is transmitted as an autosomal dominant. Sporadic cases such as Case 2 appear to represent new mutations. The advanced age of the father of Case 2 is consistent with this hypothesis.

TREATMENT

Avoidance of solar radiation; the use of tinted glasses.

REFERENCES

1. Garrod, A. E.: Inborn errors of metabolism, the Croonian lectures, lecture II, Lancet 2:73, 1908.
2. Fitzpatrick, T. B., and Quevedo, W. C., Jr.: Albinism, in Stanbury, J. B., Wyngaarden, J. B., and Fredrickson, D. S. (eds.): *The Metabolic Basis of Inherited Disease* (2d ed.; New York: McGraw-Hill Book Co., 1966), pp. 324–340.
3. Witkop, C. J., Jr.: Albinism, in Harris, H., and Hirschhorn, K. (eds.): *Advances in Human Genetics* (New York: Plenum Press, 1971), pp. 61–144.
4. Nance, W. E., Tuckson, C. E., and Witkop, C. J., Jr.: Amish albinism, a distinctive autosomal recessive phenotype, Am. J. Hum. Genet. 22:579, 1970.
5. Witkop, C. J., Jr., Nance, W. E., and Rawls, R. F.: Autosomal recessive oculocutaneous albinism in man, evidence for genetic heterogeneity, Am. J. Hum. Genet. 22:55, 1970.
6. Fitzpatrick, T. B., Seiji, M., and McGugan, A. D.: A melanin pigmentation, N. Engl. J. Med. 265:3747, 1961.
8. Knox, W. E.: Sir Archibald Garrod's inborn errors of metabolism, III. Albinism, Am. J. Hum. Genet. 10:249, 1958.
9. Falls, H. F.: Sex-linked ocular albinism displaying typical fundus changes in female heterozygote, Am. J. Ophthalmol. 34:41, 1951.
10. Cross, H. E., McKusick, V. A., and Breen, W.: A new oculocerebral syndrome with hypopigmentation, J. Pediatr. 70: 3, 1967.
11. Windhorst, D. B., Clickson, A. S., and Good, R. A.: Chediak-Higashi syndrome, hereditary gigantism of cytoplasmic organelles, Science 151:81, 1966.

ALKAPTONURIA

Cardinal Clinical Features

Dark pigment deposits in the sclerae and elsewhere, early osteoarthritis, dark urine, homogentisic aciduria, defective homogentisic acid oxidase.

Clinical Picture

Alkaptonuria was recognized by Garrod[1,2] around the turn of the century as an inborn error of metabolism. In fact, it was from his studies of patients with alkaptonuria and their families that he conceived the idea that an inborn error of metabolism results from an alteration in an enzyme that is itself the consequence of a single genetic event. This was the first enunciation of what has come to be known as the "one gene–one enzyme" hypothesis.

Alkaptonuria, or the excretion of dark urine, is a consequence of the excretion of large amounts of homogentisic acid in the urine. This compound is a normal intermediate in the catabolism of the aromatic amino acids phenylalanine and tyrosine (Fig. 1). It accumulates because of a defect in the activity of homogentisic acid oxidase.[3] This enzyme, found in mammalian systems only in liver and kidney, has been shown to be defective in both tissues in alkaptonuria. It catalyzes the conversion of homogentisic acid to maleylacetoacetic acid, which is ultimately converted to fumaric and acetoacetic acids.

The urine of an alkaptonuric individual usually appears normal when passed but turns dark on standing. Thus, it is not surprising that most individuals live many years, usually well into adulthood, without recognizing that they are alkaptonuric. The pigment may be brought out by the addition of alkali to the urine (Fig. 2). This is sometimes seen in infants when an alkaline soap is placed in contact with a wet diaper. In some patients the diagnosis is suggested by the presence of reducing substance in the urine turned up on routine urinalysis or screening for sugar. Homogentisic acid also will reduce silver in photographic emulsion, and this property has been used as a qualitative test for the disease. In fact, alkaptonuric urine has been used, as a demonstration, to develop photographs. Homogentisic acid can be identified by paper chromatography.[4] A specific enzymatic analysis permits its quantitation in urine and other body fluids.[5]

Patients with alkaptonuria have no symptoms as children or young adults. With age they develop pigmentation of the sclerae or of the cartilage of the ear. These pigment deposits should be visible by 30 years of age. Actually, deposition may be widespread throughout the cartilage and fibrous tissue of the body. This may be seen at surgery and, of course, with exposure to the air, additional pigment is formed during the operation. Pigment also may be seen in the buccal mucosa and nails. Some patients develop extensive dusky coloration of the skin, especially over the cheeks, forehead, axillae and genital regions. The sweat may be dark and the cerumen often is brown or black. This widespread deposition of pigment is known as ochronosis. In contrast to its rather subtle clinical presentation, it may be very impressive in an old person at operation or autopsy. The pigment is thought to be a polymer derived from homogentisic acid.

Ultimately, these patients develop ochronotic arthritis. This disorder resembles osteoarthritis roentgenographically, except for its occurrence at an earlier age. Some of its clinical features are reminiscent of rheumatoid arthritis. The earliest symptoms usually are in the hip or knee. Acute periods of inflammation are part of the disease. Limitation of motion is seen early, par-

Fig. 1.—Aromatic amino acid metabolism, site of the defect in alkaptonuria.

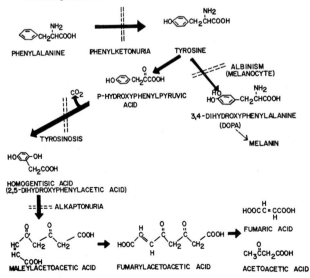

21

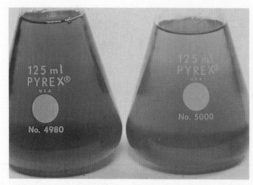 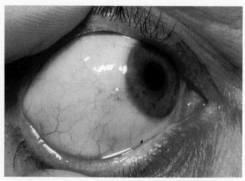

Fig. 2 (left). – Alkaptonuric urine. The flask on the right contains fresh urine but, even so, has darkened. The flask on the left was alkalinized with NaOH. The resultant black precipitate has the typical appearance and is easy to see on careful inspection; on the other hand, it is easy to appreciate that this diagnosis is often missed.

Fig. 3 (right). – A. B. (Case Report). Ochronotic pigment in the sclera of woman with alkaptonuria.

ticularly in the weight-bearing joints. Ultimately, marked limitation of motion is the rule, and ankyloses are common.

The roentgenographic appearance may be pathognomonic. The intervertebral disks undergo marked degeneration. The disk spaces become narrow, and calcium is deposited. There is a variable degree of fusion of the vertebral bodies. In contrast to rheumatoid disease, there is little osteophyte formation or calcification of the intervertebral ligaments. In contrast to osteoarthritis, the large joints at the hip and shoulder are most commonly involved in ochronosis, whereas the sacroiliac joint may be uninvolved. In the involved joints there are degenerative osteoarthritic changes, occasional free intra-articular bodies and calcification of the surrounding tendons. The arthritis of this disease is severe, and patients may even be bedridden as a consequence. Calcification of ear cartilage is another roentgenographic characteristic of the disease.

Patients with alkaptonuria also have a high inci-

Fig. 4. – A. B. Knee in ochronotic arthritis.

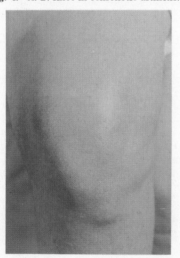

dence of heart disease. This includes the presence of mitral and aortic valvulitis at autopsy, calcifications of cardiac valves and the occurrence of myocardial infarction as a common cause of death. Rupture of the intervertebral disk has been reported.

CASE REPORT

A. B., a 53-year-old white female, was admitted to the Jackson Memorial Hospital, Miami, Florida, with a diagnosis of degenerative disease of the hip. She had had pain over the right hip and knee and other arthritic symptoms for a number of years. She could no longer voluntarily move her right lower extremity. At 33 years of age she had had a fusion of L4 for degenerative disease of the disk. At operation it was noted that many of her tissues were black, and that they became more black with exposure to the air. She had known that she was alkaptonuric since age 11. Her brother was 12 at the time and was diagnosed first on the basis of a brown-black staining of the underwear. The diagnosis of alkaptonuria was documented at the Massachusetts General Hospital.

Examination revealed ochronotic deposits in the sclerae (Fig. 3), ears and oral mucosa. There was severe degenerative arthritis of both legs (Fig. 4).

Laboratory examination revealed a blood glucose of 132 mg/100 ml, hemoglobin of 13.8 gm/100 ml, hematocrit of 41 and a WBC count of 10,000/cu mm. The urine had a 3+ test for reducing substance. Alkaptonuria was evident (see Fig. 2). Paper chromatography revealed a pronounced homogentisic acid spot. X-rays of the hips revealed some narrowing of the right acetabular space (Fig. 5). There was generalized osteoporosis. The intervertebral disks were calcified (Fig. 6). Pathologic examination of bone around the joint revealed granular deposits, which stained black with crystal violet.

GENETICS

Alkaptonuria is inherited as an autosomal recessive trait. A high incidence of consanguinity was originally noted by Garrod.[1] Heterozygote detection should be

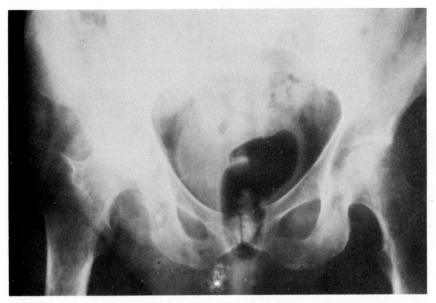

Fig. 5.—A. B. Roentgenogram of the hip illustrating the advanced, early-onset osteoarthritis characteristic of this disease.

possible only through liver biopsy, which has not seemed justified.

TREATMENT

No treatment directed at the accumulation of homogentistic acid has been reported. The arthritis may require orthopedic treatment.

REFERENCES

1. Garrod, A. E.: The incidence of alkaptonuria: A study in chemical individuality, Lancet 2:1616, 1902.
2. Garrod, A. E.: *Inborn Errors of Metabolism* (London: Oxford University Press, 1923).
3. La Du, B. N., Zannoni, V. G., Laster, L., and Seegmiller, J. E.: The nature of the defect in tyrosine metabolism in alcaptonuria, J. Biol. Chem. 230:251, 1958.
4. Knox, W. E., and LeMay-Knox, M.: The oxidation in liver of L-tyrosine to acetoacetate through p-hydroxyphenylpyruvate and homogentisic acid, Biochem. J. 49:686, 1951.
5. Seegmiller, J. E., Zannoni, V. G., Laster, L., and La Du, B. N.: An enzymatic spectrophotometric method for the determination of homogentisic acid in plasma and urine, J. Biol. Chem. 236:774, 1961.
6. La Du, B. N.: Alcaptonuria, in Stanbury, J. B., Wyngaarden, J. B., and Fredrickson, D. S. (eds.): *The Metabolic Basis of Inherited Disease* (2d ed.; New York: McGraw-Hill Book Co., 1966), pp. 303–323.
7. Osler, W.: Ochronosis: The pigmentation of cartilages, sclerotics and skin in alkaptonuria, Lancet 1:10, 1904.
8. Bunim, J. J., McGuire, J. S., Jr., Hilbish, T. F., Laster, L., La Du, B. N., Jr., and Seegmiller, J. E.: Alcaptonuria: Clinical staff conference, National Institutes of Health, Ann. Intern. Med. 47:1210, 1957.

Fig. 6.—A. B. Roentgenogram of the osteoarthritic spine in alkaptonuria, showing resemblance to bamboo.

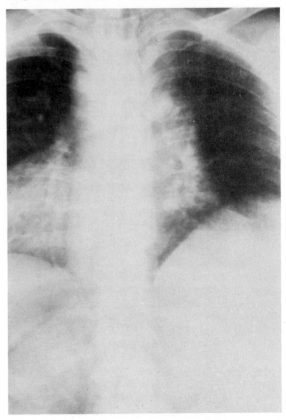

LESCH-NYHAN SYNDROME

CARDINAL CLINICAL FEATURES

Hyperuricemia, mental retardation, spastic cerebral palsy, choreoathetosis, self-mutilative biting behavior. Patients may have any of the manifestations of gout, including urinary tract stone disease, nephropathy, arthritis, tophi. Activity of the enzyme hypoxanthine-guanine phosphoribosyl transferase is virtually completely absent.

CLINICAL PICTURE

The Lesch-Nyhan syndrome occurs exclusively in males. Involved infants appear normal at birth and usually develop normally for the first 6–8 months. Crystalluria, hematuria or renal tract stone disease may develop during these early months of life, but in most patients the neurologic examination is negative.

The onset of cerebral manifestations is with athetosis (Fig. 1). Infants who have been sitting and holding up their heads will begin to lose these abilities. In the established patient the motor defect is of such severity that he can neither stand nor sit unassisted. Patients may learn to sit in a chair only if they are securely fastened around the chest. Movements are of both choreic and athetoid form. Spasticity usually is severe. Deep tendon reflexes are increased. Babinski responses are often but not always found. Mental retardation is prominent. In most patients the IQ is less than 50. A few patients have been observed in whom there has been near normal intelligence. However, even in these instances the severity of the motor defect prevents much in the way of function.

Behavioral manifestations are bizarre. These patients display a generalized, compulsive, aggressive behavior. Most of this aggressive activity is directed against the patient himself. The motor abnormality prevents much aggression against others. The most characteristic feature is self-destructive biting of the lips (Fig. 2) and fingers. Unlike many other retarded patients who engage in self-mutilative behavior, these patients bite with a ferocity that leads to significant loss of tissue. Partial amputation of fingers has been observed. These patients do not have sensory abnormalities; they scream in pain when they bite. Because they also will bite others, parents and medical personnel caring for them become expert in avoiding this as well as the hitting, kicking and butting with the head that are also characteristic. As the patient becomes older, he learns to become aggressive with speech.

Hyperuricemia is present in virtually all patients. The concentration of uric acid in the plasma usually is between 9 and 12 mg/100 ml, which reflects the limits of solubility of urate in plasma. Patients with some degree of glomerular insufficiency have higher concentrations of uric acid, and some who are very efficient at excreting urate may have lower values, occasionally in the normal range. All patients have large amounts of uric acid in the urine. Twenty-four-hour excretions of 600–1,000 mg are the rule in patients weighing in the vicinity of 30 lb. Excretion of uric acid exceeds 1 mg/mg of creatinine.

Patients with such large amounts of uric acid in body fluids may develop any of the clinical manifestations of gout. Among the first signs is crystalluria; a history of orange crystals or sand in the diapers is common. There may be frequent episodes of hematuria. Urinary tract stone disease is prominent and may be seen in the first months of life. In the absence of treatment, most patients develop urate nephropathy, and death from renal failure at less than 10 years of age is the usual outcome. Patients surviving longer than 10 years usually develop tophi (Fig. 3). We have seen acute gouty arthritis in three patients over 12 years of age. The expected response to colchicine was obtained.

Biochemical investigation has yielded evidence of a markedly increased rate of synthesis of purines de novo. This overproduction of the purine ring approximates 20 times the normal value. The molecular defect is in the activity of the enzyme, hypoxanthine-guanine phosphoribosyl transferase (HGPRT). This enzyme catalyzes the reaction of hypoxanthine or guanine with phosphoribosyl pyrophosphate (PRPP) to form their respective nucleotides as follows:

$$\begin{array}{ccc}
\text{hypoxanthine} & & \text{inosine monophosphate} \\
 & \xrightarrow{\text{PRPP}} & \\
\text{guanine} & & \text{guanine monophosphate}
\end{array}$$

It is present in all cells of the body. The defect is readily detectable in erythrocytes and in fibroblasts in cell culture. In the erythrocytes, quantitative assays reveal no activity. The enzyme is present in amniotic cells, and the disease has been diagnosed in utero.

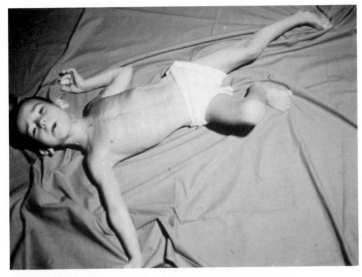

Fig. 1.—A 5-year-old boy with the Lesch-Nyhan syndrome illustrates the characteristic spasticity and athetoid posturing. Loss of tissue around the lip is evident.

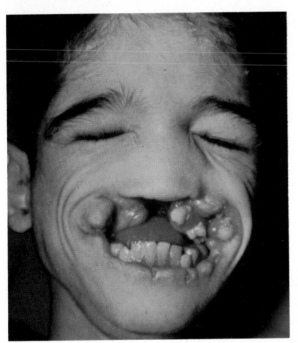

Fig. 2.—A 14-year-old boy, illustrating an extreme degree of mutilation around the face.

CASE REPORT

M. J. was born to gravida III, para III mother after a normal full-term pregnancy and delivery. Birth weight was 5 lb, 12 oz. The neonatal period was uneventful. At age 5 months the patient was noted to be developing poorly and losing weight. Previously he had held his head up well and had started to sit by himself. At about that time he began to have repeated vomit-

ing. At 9 months he had developed spasticity of the extremities, and at 2½ years of age he began to bite his lips and fingers.

He continued to gain weight poorly and to have problems with feeding and vomiting. He required almost constant restraint and vigilance in order to prevent self-mutilating activity. He had been on treatment with allopurinol, 50 mg, b.i.d., since the time of diagnosis. At 7 years of age he was unable to stand, sit or walk. He said only simple words such as "hi" and "all right" and seemed to understand relatively simple speech. Family history revealed no other members with this disorder. A sister and brother, 1 and 2 years older respectively, than the patient were well.

Physical examination revealed a 7-year-old white male with severe retardation of growth and development. Height was 33 inches (50th percentile for 1½ years), and weight 19 lb, 12 oz (50th percentile for 9 months). The upper lip was excoriated,

Fig. 3.—A 17-year-old boy with prominent tophaceous deposits in the ears. The violaceous inflammatory reaction is unusual around tophi. It cleared dramatically with use of colchicine.

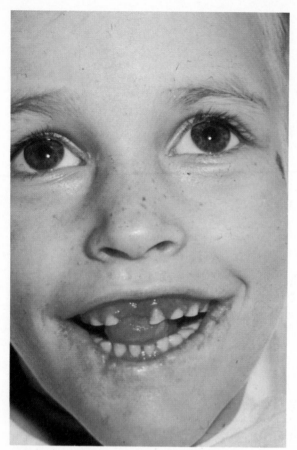

Fig. 4.—M. J. (Case Report). The degree of mutilation of the lip is relatively mild.

the lower lip mutilated (Fig. 4). Dentition was delayed. There were no tophi. The right thumb and left fifth finger were macerated and excoriated and the distal phalanx of the left finger was missing. Deep reflexes were symmetrically hyperactive. There were no pathologic reflexes. However, Babinski responses had been obtained 1 year previously. Athetoid movements of the head, trunk and extremities were noted, particularly on intention. When disturbed, he intended to assume an opisthotonic position. His IQ was estimated at 40–50.

Laboratory data included a serum uric acid concentration of 7.8 mg/100 ml. With attention to allopurinol administration in hospital, this was reduced to 3.5 mg/100 ml. BUN was 24 mg/100 ml on admission and 16 mg/100 ml thereafter. Serum electrolytes were normal. Urinalyses were normal, except for crystals seen on microscopy. Urinary excretion of urate at a time when the patient was receiving no allopurinol and no exogenous purines was 467 mg/24 hours or 3 mg/mg creatinine. The activity of HGPRT in erythrocytes was zero. Bone age was markedly retarded, and there was lateral subluxation of both hips. An intravenous pyelogram was normal. The bone marrow appeared normal. Peripheral blood revealed hemoglobin 12.7, hematocrit 41 and WBC 5,800, with a normal differential. The EEG was borderline, with considerably slower than normal activity.

GENETICS

HGPRT is determined by a gene on the X chromosome. The disease is transmitted as an X-linked recessive trait.

The heterozygous carrier female can be detected by study of the enzyme in individual cells in cultures of fibroblasts from skin. The two populations of cells specified in the Lyon hypothesis have been demonstrated in heterozygotes for this syndrome. Diagnosis may be made by cloning the total cell population or by pharmacologic selection using thioguanine or azaguanine.

A simpler method of heterozygote diagnosis involves the study of individual hair follicles, which can be tested directly following careful plucking of hair from the head.[6]

TREATMENT

The most effective method of control is that afforded in families at risk by heterozygote detection and intrauterine diagnosis. In this way, in families already known to carry the disease, therapeutic abortion may be performed following specific prenatal diagnosis.

Aspects of the syndrome that can be related directly to uric acid itself can be effectively managed with allopurinol. In this way nephropathy can be avoided and tophi can be prevented, or resorbed once developed, as may urinary tract stones if they have not incorporated too many calcium salts. Arthritis should not develop in treated patients. On the other hand, no treatment has been discovered that alters the neurologic or behavioral manifestations of the disease. Self-mutilation can be ameliorated by the removal of primary teeth and by a vigorous program of physical restraint.

REFERENCES

1. Lesch, M., and Nyhan, W. L.: A familial disorder of uric acid metabolism and central nervous system function, Am. J. Med. 36:561, 1964.
2. Seegmiller, J. E., Rosenbloom, F. M., and Kelley, W. N.: Enzyme defect associated with a sex-linked human neurological disorder and excessive purine synthesis, Science 155:1682, 1967.
3. Bland, J. H. (ed.): Seminars on the Lesch-Nyhan Syndrome, Fed. Proc. 27:1019, 1968.
4. Migeon, B. R., Der Kaloustian, V. M., Nyhan, W. L., Young, W. J., and Childs, B.: X-linked hypoxanthine-guanine phosphoribosyl transferase deficiency: Heterozygote has two clonal populations, Science 160:425, 1968.
5. Bakay, B., Telfer, M. A., and Nyhan, W. L.: Assay of hypoxanthine-guanine and adenine phosphoribosyl transferases. A simple screening test for the Lesch-Nyhan syndrome and related disorders of purine metabolism, Biochem. Med. 3:230, 1969.
6. Gartler, S. M., Scott, R. C., Goldstein, J. L., and Campbell, B.: Lesch-Nyhan syndrome: Rapid detection of heterozygotes by use of hair follicles, Science 172:572, 1971.

PRIMARY HYPEROXALURIA
Glycolic Aciduria, Glyceric Aciduria

Cardinal Clinical Features

Hyperoxaluria, urinary tract stones, nephrocalcinosis, nephropathy, renal failure.

Clinical Picture

The clinical manifestations of oxalosis are entirely due to the deposition of calcium oxalate in body tissues. Oxalate may be widely distributed in tissues, but in most patients its pathologic effects are exclusively renal.[1] The onset of symptoms usually is with urinary calculi. Patients may pass calculi, gravel or crystals. Small concretions may block the male urethra in an infant, and some meatotomies have been performed for this reason. Stones also may block a ureter, leading to obstruction or renal colic.

These patients regularly develop nephrocalcinosis. Urinary tract infection is common, and ultimately renal failure occurs. Most reported patients have died early in life, usually before age 15, and few have survived over 30.

Linear growth may be normal or, in those in whom uremia and acidosis develop in early childhood, growth may be impaired. No other clinical characteristics are common. A few patients have had acute attacks of arthritis thought to be gout. These patients had hyperuricemia, as do many with hyperoxaluria. Presumably this represents a competition with uric acid for excretion by some metabolite excreted in hyperoxaluria. It is, of course, possible that deposition of calcium oxalate crystals in the joint space could cause pseudogout. Certainly, oxalate crystals have been found in synovial membrane. It has been suggested that oxalate deposits in the myocardium may cause disordered rhythm. Two patients with complete atrioventricular block have been reported.

Crystals may be found in any tissue. The bone marrow often has been aspirated in order to confirm a suspected diagnosis, but this approach usually has been negative. Deposition of calcium oxalate in the kidney may be massive; concentrations of 5% of the dry weight of the kidney have been recorded. The responsibility of oxalate for the renal damage is supported by its occurrence following administration of large amounts of oxalate to experimental animals and following accidental ingestions in man. The crystals found in the kidney are doubly refractile and often are arranged in radiating sheaves, forming rosettes. The roentgen appearance of the nephrocalcinotic kidneys may be striking.

A requirement for diagnosis is the determination that large amounts of oxalate are being excreted in the urine.[1,2] Normally, 20–55 mg/1.73 m^2 of oxalate is excreted in 24 hours. Patients with hyperoxaluria excrete more than this—usually over 100 mg/1.73 m^2.

Genetic Heterogeneity

It is now recognized that there are two forms of primary hyperoxaluria. The classic form (described above) is sometimes known as hyperoxaluria type I. It is characterized by increased excretion of glyoxylic and glycolic acids, as well as oxalic acid, and therefore also has been called glycolic aciduria. In type II L-glyceric acid is excreted in large amounts in the urine and therefore it is known as L-glyceric aciduria. The initial description of this second disorder was made in 1968,[3] and only four patients have been described. They have a clinical picture of recurrent nephrolithiasis similar to that of patients with type I, but the disease appears to be milder. None of the patients have died of the disease and the original patient was 38 years old at the most recent report. The degree of hyperoxaluria is similar in the two types of disease—usually in the range of 100–300 mg/24 hours in the adult.

Differentiation between the two types requires analysis of the organic acids of the urine. This is conveniently done using gas liquid chromatography or liquid partition chromatography on silicic acid columns.[4] Glycolic acid and glyoxylic acid are found in the type I disorder but not in type II. In this latter form, glyceric acid is excreted in amounts up to 1,000 mg/24 hours.

In the type I disease, abnormal glyoxylate metabolism (Fig. 1) has been demonstrated by administration of ^{14}C-glyoxylate to patients who convert this compound inefficiently to CO_2 and in increased amounts to glycolate and oxalate. Similarly, the metabolism of ^{14}C-glycolate to CO_2 is lower than normal, whereas its incorporation into oxalate is considerably greater. It is

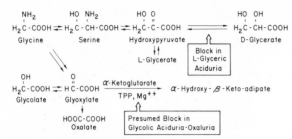

Fig. 1. — Metabolic pathways involved in production of oxalate.

thought that the defect is in the carboligase reaction, in which there is synergistic decarboxylation of glyoxylate to CO_2 and α-hydroxy-β-ketoadipate in the presence of α-keto-glutarate (keto-glutarate: glyoxylate carboligase). This enzyme activity has been reported to be reduced in the soluble fraction but not in the mitochondrial fraction of tissues from five patients with the disease.[5] An accumulation of glyoxylate in the presence of such a defect would lead secondarily to conversion to glycolate and oxalate.

In L-glyceric aciduria (see Fig. 1) a defect has been found in D-glyceric acid dehydrogenase in leukocytes of four patients. In the absence of this activity, hydroxypyruvate, an intermediate in gluconeogenesis from serine, accumulates. Hydroxypyruvate is then converted to L-glyceric acid. The mechanism of the hyperoxaluria in this type of disease is not clear. There is some evidence that reduction of hydroxypyruvate to L-glycerate in the presence of lactic dehydrogenase and nicotinamide adenine nucleotide dehydrogenase may be coupled with oxidation of glyoxylate to oxalate. On the other hand, hydroxypyruvate-2-^{14}C recently has been reported to be converted to oxalate in rat liver.[6]

CASE REPORT

M. E., an 8-year-old white female who was studied in the Clinical Research Unit of the University of Miami (Florida), had been found to have bilateral renal stones some 4 years previously. She came from rural North Carolina, but consanguinity could not be established.

Three years prior to admission she had had a retrograde pyelogram that demonstrated bilateral renal calculi (Figs. 2–4). Cystoscopy showed inflammatory polyps in the area of the bladder neck and trigone. A cystourethrogram was normal. A renogram was indicative of damage to the left kidney. BUN was 30 and creatinine 1.3 mg/100 ml. Serum electrolytes were normal and alkaline phosphatase was 4.1 Bodansky units. A urinary tract infection was found, and antibiotic therapy initiated. One month later she underwent a left pyelolithotomy. At this time bone age was normal, and BUN 16 mg/100 ml. The calculi removed consisted of calcium oxalate and contained calcium phosphate.

Follow-up and therapy were apparently difficult to accomplish with this family. When the patient was next seen 16 months later, the left kidney was virtually nonfunctional on renogram. The right renogram appeared normal. An open biopsy of the kidney revealed oxalate crystals, chronic inflammation, fibrosis and atrophy. A 24-hour urinary excretion of oxalate was 126 mg.

A year later at 8 years of age, the patient was in renal failure and was admitted to hospital following a seizure. Examination of the chest revealed cardiomegaly. BUN was 119 mg/100 ml and CO_2, 9 mEq/L. Alkaline phosphatase was 30 Bodansky units and the serum calcium, 7 mg/100 ml. Urinalysis revealed many leukocytes in clumps and 10–12 RBC/hpf. A creatinine clearance was markedly reduced. Hemoglobin was 4.4 gm/100 ml, hematocrit 13. BUN rose within a week to 167 mg/100 ml, and the anemia could only very transiently be improved by erythrocyte transfusion. The patient became increasingly edematous. BUN rose to 188 mg/100 ml and serum potassium to 7.3 mEq/L. Peritoneal dialysis reduced the BUN to 7.2 mg/100 ml and potassium to 5.2 mEq/L, but these values rapidly returned to previous levels. After about a month of repeated dialysis, the parents elected to take her home.

GENETICS

Hyperoxaluria is inherited as an autosomal recessive character.[1, 7] Consanguinity in reported families was ten- to twentyfold the rate in the general population.

TREATMENT

General therapy, used in any form of renal stone disease, includes a sufficiently high fluid intake to keep

Fig. 2. — M. E. (Case Report). Roentgenogram of the abdomen illustrates diffuse nephrocalcinosis and small size of left kidney and a large opaque stone on the right.

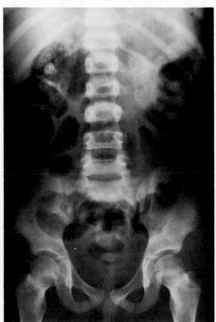

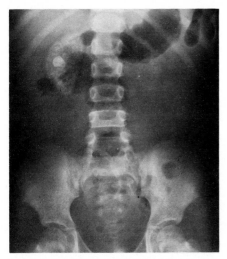

Fig. 3.—M. E. In this view nephrocalcinosis of the right kidney is better visualized.

the urine dilute, reduced calcium intake and appropriate antibiotic treatment for infection.

Therapy directed to the reduction of oxalate excretion, including calcium carbimide, has not been effective. Renal transplantation also has failed because of rapid accumulation of calcium oxalate in the transplanted kidney. Chronic hemodialysis is effective.

REFERENCES

1. Hockaday, T. D. R., Clayton, J. E., Frederick, E. W., and Smith, L. H., Jr.: Primary hyperoxaluria, Medicine 43: 315, 1964.
2. Smith, L. H., Jr., and Williams, H. E.: Heritable Disorders of Oxalate Metabolism, in Nyhan, W. L. (ed.): *Heritable Disorders of Amino Acid Metabolism* (New York: John Wiley & Sons, Inc., 1974), p. 187.
3. Williams, H. E., and Smith, L. H., Jr.: L-glyceric aciduria: A new genetic variant of primary oxaluria. N. Engl. J. Med. 278:233, 1968.

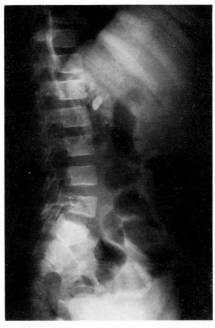

Fig. 4.—M. E. Lateral roentgenogram illustrates the stone and the nephrocalcinosis.

4. Sweetman, L.: Liquid partition chromatography and gas chromatography-mass spectrometry in identification of acid metabolites of amino acids, in Nyhan, W. L. (ed.): *Heritable Disorders of Amino Acid Metabolism* (New York: John Wiley & Sons, Inc., 1974), p. 420.
5. Koch, J., Stokstad, E. L. R., Williams, H. E., and Smith, L. H., Jr.: Deficiency of 2-oxoglutarate: Glyoxylate carboligase activity in primary hyperoxaluria, Proc. Nat. Acad. Sci. USA 57:1123, 1967.
6. Richardson, K. E., and Liao, L. L.: Formation of oxalate from hydroxypyruvate by isolated perfused rat liver, Fed. Proc. 32:565, 1973.
7. Witzleben, C. L., and Elliott, J. S.: Hereditary hyperoxaluria, Am. J. Dis. Child. 111:56, 1966.

GALACTOSEMIA

Cardinal Clinical Features

Failure to thrive, hepatomegaly, icterus, renal Fanconi syndrome, cataracts, mental retardation, deficiency of galactose-1-phosphate uridyl transferase.

Clinical Picture

Manifestations of the disease[1-3] generally appear within days of birth or of the initiation of milk feedings, and they increase in severity in the first months of life. Hepatomegaly (Fig. 1) is the most constant finding. Vomiting and jaundice are the most common initial symptoms. Jaundice may develop as early as a few days after milk feedings are begun. Anorexia, failure to gain weight or to grow or even weight loss ensues. In the absence of treatment parenchymal damage to the liver is progressive to typical Laennec's cirrhosis. Patients may have edema, ascites, hypoprothrombinemia and bleeding. Splenomegaly is common.

The development of lenticular cataracts is a characteristic feature of the disease and occurs in infants who have received milk for 3–4 weeks. Cataracts may occur as early as after a few days.

Mental retardation probably is the most important manifestation of the disease. It is most impressive in patients not diagnosed or treated until a number of months have elapsed. Untreated patients also are often hyperactive. Bulging of the anterior fontanel may be an early sign of the disorder.

Patients with galactosemia are unusually susceptible to infection. Many patients have had a fulminant course, with septicemia followed by an early demise. Concomitants of sepsis such as osteomyelitis, meningitis and gangrene of the toes have been observed.

Renal abnormalities usually are first detected in the laboratory, although some patients may have frequency. The consequence of the disease is renal tubular dysfunction, manifest by glycosuria, a generalized aminoaciduria, proteinuria and acidosis.

Recognition of the disease usually is possible through the presence of reducing substance in the urine. It is important to emphasize that testing of the urine with glucose oxidase (Clinistix, Tes-tape) will not detect galactose; this is a strong argument for the continued use of the older methods for the screening of urine (Benedict's or Fehling's test; Clinitest). Furthermore, galactosuria is dependent on dietary intake of lactose. Thus, in a patient who is admitted acutely

Fig. 1.—Young infant with galactosemia. The hepatomegaly is *outlined*. Failure to thrive is evident.

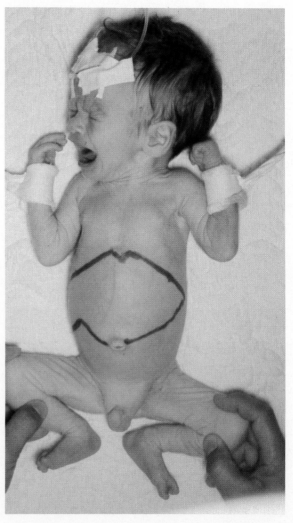

30

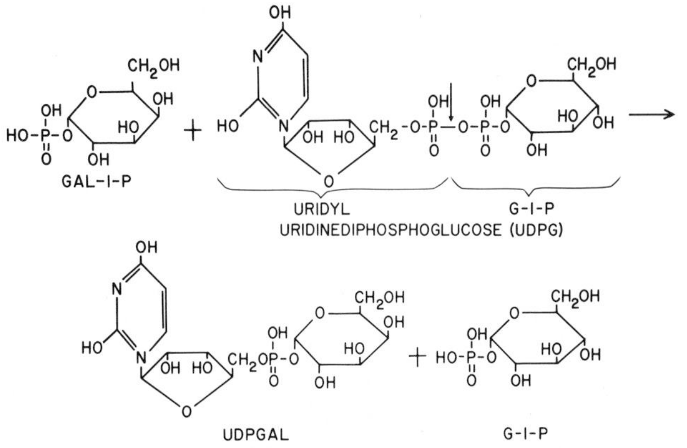

Fig. 2.—Galactose-1-phosphate uridyl transferase. The *vertical arrow* indicates the bond split in the uridyl transferase reaction. This enzyme is deficient in patients with galactosemia.

ill and treated with parenteral fluid therapy, the disease may not be recognizable because he has not received galactose for 24–48 hours.

Characterization of the reducing substance found in a urine sample proceeds in a number of ways, but today it usually is done by paper chromatography. Of course, an infant's urine that contains reducing substance that is glucose oxidase negative is galactosuric until proved otherwise. Following suspicion, it may be easiest to go directly to an assay of the relevant enzyme, galactose-1-phosphate (Gal-1-P) uridyl transferase (Fig. 2).

The structure of galactose is identical to that of glucose except for the position of the hydroxyl on carbon 4. Lactose, the principal sugar of mammalian milks, is the predominant dietary source of galactose. It is a disaccharide in which glucose and galactose are linked in an α-1,4-glucosidic bond in which an oxygen bridge connects carbon 1 of galactose and carbon 4 of glucose. The utilization of dietary galactose depends first on galactokinase, which catalyzes the following reaction:

$$\text{Galactose} + \text{ATP}^* \xrightarrow{\text{galactokinase}} \text{Gal-1-P} + \text{ADP}^*$$

Gal-1-P is then converted to glucose-1-phosphate in a series of two reactions in which uridine diphosphoglucose (UDPG) functions catalytically. The first of these is the uridyl transferase reaction (see Fig. 2). This is followed by an epimerase reaction:

$$\text{UDP Gal} \rightleftharpoons \text{UDPG}$$

This reaction is catalyzed by UDP Gal-4-epimerase.

In patients with galactosemia the enzyme defect is in the uridyl transferase[4] (see Fig. 2). The abnormality

---

*ATP = adenosine triphosphate; ADP = adenosine diphosphate.

can be detected in the erythrocyte, even in cord blood. It also can be detected in fibroblasts and amniotic cells in culture.

The other enzymes involved in galactose metabolism are normal in these patients. Gal-1-P accumulates in tissues and may be responsible for many of the clinical manifestations of the disease. At any rate, therapeutic measures that result in reduction of intracellular concentrations of Gal-1-P lead to the prevention or disappearance of symptoms (Fig. 3). The abnormalities in the lens are probably produced by galactitol. This by-product of galactose accumulation is present in urine and in tissues. It is found in the absence of Gal-1-P in patients with galactokinase deficiency.[5] Patients with this disorder have cataracts and galactosuria but none of the other manifestations found in patients with galactosemia.

CASE REPORT

A. J., a 21-month-old white female, was referred for evaluation of galactosemia. She had received a milk-free diet since the fifth day of life and was believed by her parents to be well.

The patient was the product of a normal pregnancy and delivery. Birth weight was 9 lb, 14 oz. At birth she was noted to have generalized edema. At 2 days of age a heart murmur was heard and chest film revealed cardiomegaly. She was referred to Emory University on the fifth day of life, where reducing substance was found in the urine. There was no jaundice or hepatosplenomegaly. There never was any hypoglycemia. The infant was treated with a diet based on soy bean formula. She has received no milk or milk products since then. Cataracts have never been observed.

On the ninth day of life, the child developed a seizure and was diagnosed as having *Escherichia coli* sepsis and meningitis. She was treated successfully and discharged in good

Fig. 3.—Teen-ager with treated galactosemia. There were no clinical manifestations of the disease.

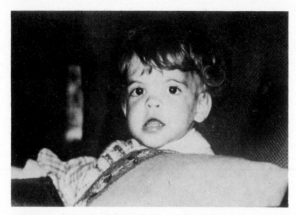

Fig. 4.—A. J. (Case Report). The child had classic galactosemia but no clinical manifestations of the disease.

health 1 month later. Gal-l-P uridyl transferase activity was found to be absent in the patient's blood.

The infant smiled at 2 months, sat at 6 months and walked at 15 months. She had always been active and alert. She had a vocabulary of 10 words at 21 months. There were no further convulsions.

The mother was 23 years old and well. A glucose tolerance test had been carried out because she had had two infants with birth weights over 9 lb. The test was normal. The father was 24 years old and well. One sibling, 4 years old, was normal. There was no history of consanguinity.

Physical examination revealed a well-developed, well-nourished female infant (Fig. 4). Weight at 21 months was 24 lb, height 32 inches. She had a moderate hypertelorism and slight epicanthal folds. The eyes were normal. There were no cataracts. The heart was normal. There were no murmurs. There was no hepatomegaly or splenomegaly. Neurologic examination was negative. Blood was assayed for Gal-l-P uridyl transferase; no activity was found. It was concluded that she had galactosemia under good control.

GENETICS

Galactosemia is inherited as an autosomal recessive trait. Heterozygotes can be detected using the uridyl transferase assay of the erythrocyte.[1, 6] The mean for a group of heterozygotes is approximately half that of normal individuals, whereas values in homozygotes approximate zero.

Polymorphism has been established in the case of the Gal-l-P uridyl transferase, which complicates the screening of large populations for heterozygotes. The Duarte variant,[7] which produces no clinical disease, has an activity of about 50% of normal in the homozy-gote. Family study can elucidate the problem in any individual instance, since heterozygotes for the Duarte variant have about 75% of normal activity. The Duarte variant also can be distinguished electrophoretically.

TREATMENT

The treatment for galactosemia is exclusion of galactose from the diet. This involves the elimination of milk and its products. The mainstay of the diet for an infant is the substitution of casein hydrolysate (Nutramigen) for milk formulas. Soy bean preparations also may be used. Education of the parents, and of the child as he grows older, on the galactose content of foods is important. The determination of the Gal-l-P content of erythrocytes is useful in monitoring adherence to the diet. When this is not available, the serum bilirubin and transaminase can be employed.

A growing experience with early treatment supports the concept that effective treatment instituted in the first weeks of life can prevent all manifestations of the disease. At the other end of the scale, mental retardation, once established, is irreversible. Cataracts are reversible if treatment is started within the first 3 months of life. Hepatic and renal manifestations of the disease are reversible.

REFERENCES

1. Donnell, G. N., Bergren, W. R., and Koch, R.: Abnormal galactose metabolism in man, in Farrell, G. (ed.): *Congenital Mental Retardation* (Austin and London: University of Texas Press, 1969), p. 87.
2. Donnell, G. N., Bergren, W. R., and Cleland, R. S.: Galactosemia, Pediatr. Clin. North Am. 7:315, 1960.
3. Holzel, A., Komrower, G. M., and Schwarz, V.: Galactosemia, Am. J. Med. 22:703, 1957.
4. Isselbacher, K. J., Anderson, E. P., Kurahashi, K., and Kalckar, H. M.: Congenital galactosemia, a single enzymatic block in galactose metabolism, Science 123:635, 1956.
5. Gitzelmann, R.: Hereditary galactokinase deficiency, a newly recognized cause of juvenile cataracts, Pediatr. Res. 1:14, 1967.
6. Donnell, G. N., Bergren, W. R., Bretthauer, R. K., and Hansen, R. G.: The enzymatic expression of heterozygosity in families of children with galactosemia, Pediatrics 25:572, 1960.
7. Beutler, E., Baluda, M. C., Sturgeon, P., and Day, R.: A new genetic abnormality resulting in galactose-l-phosphate uridyltransferase deficiency, Lancet 1:353, 1965.

COMBINED IMMUNODEFICIENCY DISEASE AND ADENOSINE DEAMINASE DEFICIENCY

CARDINAL CLINICAL FEATURES

Deficiency of immunoglobulins and of cell-mediated immunity; clinical immunodeficiency triad of persistent diarrhea, progressive pneumonopathy, extensive moniliasis; skeletal abnormalities; deficiency of adenosine deaminase.

CLINICAL PICTURE

An association of adenosine deaminase deficiency with combined immunodeficiency disease, recently reported,[1-4] has raised the exciting possibility of an etiologic relationship of altered purine metabolism to developmental immunobiology. Patients with adenosine deaminase deficiency appear to have a distinct syndrome.

These patients, in common with other patients with combined immunodeficiency disease, have both defective immunoglobulins or B cell function and defective cell-mediated immunity or T cell function. Twelve patients from 11 families have been described[4] in whom there was deficiency of adenosine deaminase. Some heterogeneity has been encountered in the clinical manifestations of the disease, four patients being under 1 year of age, and two under 4. Of those alive at the time of report, most were very young infants. The sexes were equally involved. Failure to thrive was uniform. Recurrent or persistent diarrhea was prominent. More than 75% of patients have had extensive candidiasis, and more than 50% pneumonia. All who died did so of infections with opportunistic agents, some viral and some bacterial, not usually productive of severe infections in ordinary individuals.

All of the immunoglobulins in these patients are decreased in concentration, IgG, IgM, IgA and others, but often to a variable degree. Antibody response to the injection of an immunizing antigen is faulty. Delayed hypersensitivity, too, is deficient, and skin tests for Candida are negative in patients known to have had candidal infection. The response of lymphocytes to phytohemagglutinin or the formation of T cell rosettes is faulty.

Roentgenographic examination has revealed some impressive features of the syndrome. Osteoporosis may be profound. One patient had compression fractures of two vertebral bodies. The sacroiliac notch may be large, as in achondroplasia, and the ilium flares outward, resembling Mickey Mouse ears. The acetabular angle is reduced, as in achondroplasia. The pubis is short and the ischium squared off. The ribs flare and enlarge anteriorly. Cupping of the costochondral ends can occur in any process such as rickets that results in disordered cartilaginous growth. The spine reveals some platyspondly and an appearance, as in the ribs, reminiscent of mucopolysaccharidosis. In contrast to the spine of the achondroplast, the interpedicular distance in these patients does not decrease from L1 to L5. Most of these patients have pulmonary changes of a chronic sort such as that seen in infections with *Pneumocystis carinii*. The upper mediastinum is narrow, and no thymus shadow can be seen. In lateral views there may be retrosternal radiolucency.

Examination of the thymus in patients who have died of combined immunodeficiency disease reveals a very small organ with little differentiation into lobules. No Hassell's corpuscles are seen. There is no central medullary area and no differentiation into cortex and medulla. Huber,[4] who analyzed tissues from four patients with adenosine deaminase deficiency and five without, all of whom had died of combined immunodeficiency disease, believed that he could distinguish between the two groups. The enzyme-positive patients appeared to have failed to develop in early embryonic life, whereas the adenosine deaminase-deficient patients had what he called extreme involution. The thymus in these patients appeared to have known better days.

Adenosine deaminase is an enzyme involved in purine interrelations, which converts adenosine to inosine (Fig. 1). This is an important reaction because adenosine is not a substrate for nucleoside phosphorylase, which converts inosine and guanosine to their bases hypoxanthine and guanine. Adenosine deaminase is widely distributed in animal tissues. This enzyme is determined by allelic autosomal genes, and electrophoretic heterogeneity among populations of normal individuals has been extensively studied. This is the enzyme that limits the action of adenine arabinoside by converting it to hypoxanthine arabinoside; thus, this should be an unusually effective antiviral

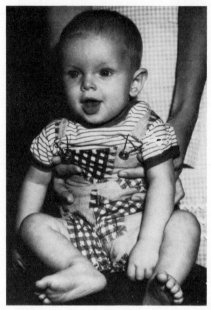

Fig. 1.—The adenosine deaminase reaction.

agent in patients with adenosine deaminase deficiency. Adenosine deaminase may be assayed in the erythrocyte.

CASE REPORT*

M. R. (Fig. 2) was the second child in the family. The first child, a girl, had died at 2 months of age with meningitis, pneumonia, oral candidiasis and otitis media. At autopsy she had had severe lymphoid hypoplasia and there were no Peyer's patches or gut-associated lymphoid tissue; there were few lymphocytes in the spleen and the thymus was rudimentary.

Examination of the patient's cord blood revealed a profound deficiency of lymphoid cells, presence of a very rare cell (three of 100,000 scanned) positive for IgM and absence of IgA-bearing cells. The majority of cells stained for surface IgG, but all had the morphologic characteristics of monocytes. Repeated examination of peripheral blood over the next 3 months showed a complete absence of small lymphocytes. Examination of a red-cell lysate showed an absence of adenosine deaminase activity.

The patient gained weight poorly and developed a slightly grotesque appearance, with a large protruding tongue, absent eyebrows, sparse scalp hair and an absence of full extension at the knees and elbows. Spleen and lymph nodes never were palpable. At 1 month of age he developed otitis media and subsequently a persistent dry cough and a minimal interstitial

Fig. 2.—M. R. (Case Report). A 3-month-old boy with combined immunodeficiency disease and adenosine deaminase deficiency. He had failed to thrive, had sparse facial and scalp hair and had begun to have repeated infections. (Courtesy of Dr. R. Keightley of the University of Alabama.)

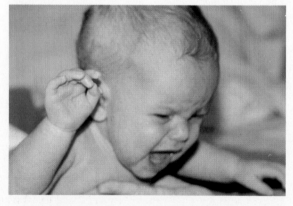

*We are indebted to Dr. R. Keightley of the University of Alabama for providing us with details of this patient.

Fig. 3.—M. R. At 10 months of age, 7 months following transplantation with fetal liver, he was thriving and looked well. (Courtesy of Dr. R. Keightley of the University of Alabama.)

infiltrate. He was found to have marked flaring of the rib ends, widened metaphyses of the long bones and a dysplastic pelvis.

When the patient was 3 months old, he was transplanted with fresh fetal liver. Following transplantation, T and B cells appeared and gradually increased in number. IgM- and IgA-bearing B lymphocytes were seen, followed by the appearance of circulating IgM and IgA. Ten months posttransplantation the patient's absolute lymphocyte count ranged between 800/cu mm and 1000/cu mm. Superficial lymph nodes became readily palpable. His skin was clear and he was free of active infection. Adenosine deaminase activity was found in lymphocytes.

During the first month following transplantation, a striking improvement was noted in the patient's physical well-being and mental development. There was considerable weight gain and hair growth. At 10 months of age he was a happy, alert little boy (Fig. 3). His weight at 1 year of age was 19 lb (3d percentile); length and head circumference were normal for age. Since the day of transplantation he had been hospitalized for a total of only 15 days.

GENETICS

Combined immunodeficiency disease and adenosine deaminase deficiency appear to be transmitted as an autosomal recessive disease. Other forms of combined immunodeficiency disease may be X-linked.

TREATMENT

Patients with this disease are incredibly susceptible to infection. The survivors are those that have been successfully treated with marrow transplantation[3] or with transplantation with fetal liver.[4]

REFERENCES

1. Giblett, E. R., Anderson, J. E., Cohen, F., Pollara, B., and Meuwissen, H. J.: Adenosine-deaminase deficiency in two patients with severely impaired cellular immunity, Lancet 2:1067, 1972.
2. Pollara, B., Pickering, R. J., and Meuwissen, H. J.: Combined immunodeficiency disease and adenosine deaminase deficiency, an inborn error of metabolism, Pediatr. Res. 7: 362, 1973.
3. Meuwissen, H. J., Moore, E., and Pollara, B.: Maternal marrow transplant in a patient with combined immune deficiency disease (CID) and adenosine deaminase (ADA) deficiency, Pediatr. Res. 7:362, 1973.
4. Meuwissen, H. J., Pickering, R. J., Pollara, B., and Porter, I. H., (eds.): *Combined Immunodeficiency Disease and Adenosine Deaminase Deficiency* (New York: Academic Press, 1975).

Mucopolysaccharidoses

INTRODUCTION
The Mucopolysaccharidoses

The mucopolysaccharidoses are a group of genetically determined diseases in which acid mucopolysaccharides are stored in the tissues[1] and excreted in large quantities in the urine.[2] The storage of this material leads to effects on a wide variety of tissues and to remarkable changes in morphology. Among the most striking effects is the alteration in the appearance of the patient, most dramatically represented in the Hurler syndrome. Mental retardation and early demise are the most devastating consequences of the abnormal storage. However, all patients with mucopolysaccharidosis do not experience these consequences. The disorders that make up this group represent a spectrum of clinical manifestation. A summary of the clinical features of the major mucopolysaccharidoses is given in Table 1.

Research in this field has been proceeding with rapidity, and it is now possible in a majority of the classic mucopolysaccharidoses to delineate the molecular defect in enzyme activity that is the primary product of the abnormal gene and the fundamental cause of all of the other manifestations of the disease. This information also is summarized in Table 1. These disorders provide an important model of the interaction of structure and function in man. They illustrate clearly that a molecular defect can lead to structural malformation syndromes.

A clinical hallmark of many of the mucopolysaccharidoses is the set of skeletal malformations known as dysostosis multiplex. This is most dramatically seen in the Hurler syndrome. It also is seen in generalized GM_1 gangliosidosis and in the mucolipidosis, I cell disease. The roentgen appearance of dysostosis multiplex is an effective way to screen clinically for mucopolysaccharidosis.

Laboratory screening for mucopolysaccharidosis usually begins with a test for the presence of large amounts of mucopolysaccharide in the urine. A variety of spot tests has been developed such as that using toluidine blue or azure A (Fig. 1). Total excretion of mucopolysaccharide is increased, as quantitated by turbidity assay using acid albumin or cetylpyridinium chloride.[2] The type of mucopolysaccharide excreted in excess can be determined by chromatographic separation on paper or thin layer plates[3] (Fig. 2). These procedures use an azure A or similar stain to locate the

mucopolysaccharide bands. They can be highly reproducible in some hands, but many technicians have trouble with these procedures. More reliable separations can be carried out using chromatography on anion exchange columns[3] (Figs. 3 and 4). Patients with the Hunter and Hurler syndromes excrete dermatan sulfate, in which the repeating sugar contains iduronic acid. The mucopolysaccharides are now referred to by chemists as glycosaminoglycans, but this term is unlikely to replace the older one in the clinical literature.

Recent advances in the understanding of the mucopolysaccharidoses have followed the growth of fibroblasts from these patients in cell culture and the recognition that there is phenotypic expression of the disease in the fibroblast. The first evidence of this was the demonstration by Danes and Bearn[4] that fibroblasts derived from children with mucopolysaccharidosis stained histochemically with metachromatic stains. Quantitative expression has been documented by Matalon and Dorfman[5] by quantitation of dermatan sulfate, which they have found to accumulate in cultured fibroblasts derived from patients with the Hunter-Hurler phenotype. Characterization of the mucopolysaccharidoses as disorders in the degradation of intracellular acid mucopolysaccharide began with the studies of Fratantoni, Hall and Neufeld[6] using ^{35}S-labeled sulfate. $^{35}SO_4$ is taken up by the cells of patients, just as it is by normal cells. However, in the patients, in contradistinction to controls, there was no turnover; these cells simply accumulated the label and kept it.

In what is now a landmark series of experiments, Fratantoni, Hall and Neufeld[7] mixed normal fibroblasts with those of patients with the Hurler or Hunter syndrome and found that the kinetics of $^{35}SO_4$ incorporation were now normal. Furthermore, it was possible to restore normal kinetics in Hurler cells by mixing them with Hunter cells and vice versa. It also was found that medium in which normal cells or Hunter cells had grown could cure or correct the defect in Hurler cells. This research has now been extensively pursued,[8] and corrective factors have been identified for each of the classic mucopolysaccharidoses. The Morquio syndrome is the exception. In fact, demonstration of two different corrective factors first permitted the distinction of Sanfilippo types A and B. These

TABLE 1.—CLINICAL AND LABORATORY FEATURES OF THE MUCOPOLYSACCHARIDOSES

DISORDER	INHERITANCE	MENTAL RETARDATION	EYE	FIRST SYMPTOM	HEPATO-SPLENO-MEGALY	BONY DEFECT	COMPOUND STORED, EXCRETED	ENZYME DEFECT
I$_H$ Hurler syndrome	Autosomal recessive	+	Corneal clouding	Coarse facial features, motor weakness, mental retardation	+	+	Dermatan sulfate, heparan sulfate	α-L-iduronidase
I$_S$ Scheie syndrome	Autosomal recessive	−	Corneal clouding	Coarse facies	−	+	Dermatan, and heparan sulfate	α-L-iduronidase
II Hunter syndrome	X-linked recessive	+	Cornea clear	Weakness, coarse features, mental retardation, aggressive behavior	+	+	Dermatan, and heparan sulfate	L-iduronosulfate sulfatase
III Sanfilippo type A	Autosomal recessive	+	Cornea clear	Hyperkinetic behavior, mental retardation	+	+	Heparan sulfate	Heparan sulfate sulfatase
III Sanfilippo type B	Autosomal recessive	+	Cornea clear	Hyperkinetic behavior, mental retardation	+	+	Heparan sulfate	α-D-acetyl-glucosaminidase
IV Morquio syndrome	Autosomal recessive	+	Corneal clouding	Bony deformities	−	+	Keratan sulfate	
VI Maroteaux-Lamy syndrome	Autosomal recessive	−	Corneal opacities	Growth retardation	+	+	Dermatan sulfate	Maroteaux-Lamy corrective factor

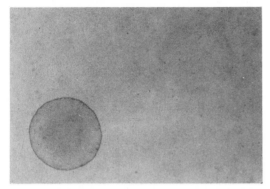

Fig. 1.—Azure A spot test for mucopolysacchariduria. The lower spot on the left is positive. The other spot is negative.

studies have permitted the ready identification of patients by the response of their cells to corrective factor. A patient, for instance, with the Hunter syndrome is one with cells that do not respond to corrective factor from Hunter cells. An exception to the general rule is that fibroblasts from patients with the Scheie syndrome cannot be corrected by the factor from Hurler cells.[9] Thus, it appeared that the genes for these two conditions might be allelic and might represent different defects in the same enzyme protein.

It is now known that the defect in the Hurler cell is in α-L-iduronidase.[10, 11] Hurler corrective factor has been shown to have iduronidase activity.[12] Defective activity of α-L-iduronidase also has been demonstrated in fibroblasts of a patient with the Scheie syndrome.[12]

The Hunter corrective factor[13] appears to be an iduronate sulfate sulfatase,[14, 15] which releases sulfate from the iduronic acid found in the dermatan sulfate that accumulates in Hunter fibroblasts.[16] Similarly, in

Sanfilippo type A disease, the defect is in a sulfatase, heparan sulfate sulfatase.[17] Sanfilippo type B disease is due to a defect in α-N-acetylglucosaminidase activity.[18] Acetyl glucosaminidase activity also has been found in purified type B corrective factor.[19] Heparan sulfate, which is excreted in the Sanfilippo syndrome, contains the α-N-acetylglucosaminide linkage.

REFERENCES

1. Brante, G.: Gargoylism: A mucopolysaccharidosis, Scand. J. Clin. Lab. Invest. 4:43, 1952.
2. Dorfman, A., and Lorincz, A. E.: Occurrence of urinary mucopolysaccharides in the Hurler syndrome, Proc. Nat. Acad. Sci. USA 43:443, 1957.
3. Schmidt, L.: The Biochemical Detection of Metabolic Disease: Screening Tests and a Systematic Approach to Screening, in Nyhan, W. L. (ed.): *Heritable Disorders of Amino Acid Metabolism*, (New York: John Wiley & Sons, Inc., 1974).
4. Danes, B. S., and Bearn, A. G.: Hurler's syndrome: Demonstration of an inherited disorder of connective tissue in cell culture, Science 149:989, 1965.
5. Matalon, R., and Dorfman, A.: Hurler's syndrome: Biosynthesis of acid mucopolysaccharides in tissue culture, Proc. Nat. Acad. Sci. USA 56:1310, 1966.
6. Fratantoni, J. C., Hall, C. W., and Neufeld, E. F.: The defect in Hurler's and Hunter's syndromes: Faulty degradation of mucopolysaccharides, Proc. Nat. Acad. Sci. USA 60:699, 1968.
7. Fratantoni, J. C., Hall, C. W., and Neufeld, E. F.: Hurler and Hunter syndromes. I. Mutual correction of the defect in cultured fibroblasts, Science 162:570, 1968.
8. Neufeld, E. F., and Cantz, M. J.: Corrective factors for inborn errors of mucopolysaccharide metabolism, Ann. N.Y. Acad. Sci. 179:580, 1971.
9. Wiesmann, U., and Neufeld, E. F.: Scheie and Hurler syndromes: Apparent identity of the biochemical defect, Science 169:72, 1970.

Fig. 2.—Thin layer chromatography of the mucopolysaccharides of the urine of patients with the Hurler and the Sanfilippo syndromes and of a normal child. (Reprinted with permission from Nyhan.[3])

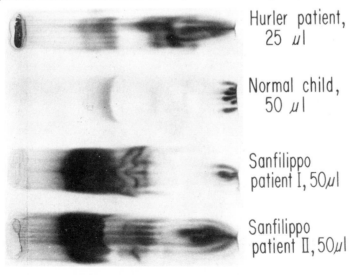

Hurler patient, 25 μl

Normal child, 50 μl

Sanfilippo patient I, 50μl

Sanfilippo patient II, 50μl

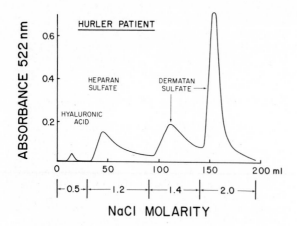

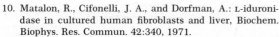

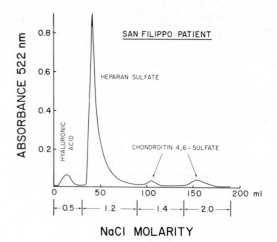

Fig. 3. — Elution pattern of a patient with the Hurler syndrome obtained by column chromatography with increasing sodium chloride concentration. (Reprinted with permission from Nyhan.[3])

Fig. 4. — Elution pattern obtained by column chromatography of urine from a patient with the Sanfilippo syndrome with increasing sodium chloride concentration. (Reprinted with permission from Nyhan.[3])

10. Matalon, R., Cifonelli, J. A., and Dorfman, A.: L-iduronidase in cultured human fibroblasts and liver, Biochem. Biophys. Res. Commun. 42:340, 1971.
11. Matalon, R., and Dorfman, A.: Hurler's syndrome, an α-L-iduronidase deficiency, Biochem. Biophys. Res. Commun. 47:959, 1972.
12. Bach, G., Friedman, R., Weissmann, B., and Neufeld, E. F.: The defect in the Hurler and Scheie syndromes: Deficiency of α-L-iduronidase, Proc. Nat. Acad. Sci. USA 69:2048, 1972.
13. Cantz, M., Chrambach, A., Bach, G., and Neufeld, E. F.: The Hunter corrective factor, J. Biol. Chem. 247:5456, 1972.
14. Bach, G., Cantz, M., Okada, S., and Neufeld, E. F.: Enzymatic defect in the Hunter syndrome, (Abstr.) Fed. Proc. 1471, 1973.

15. Bach, G., Eisenberg, F., Cantz, M., and Neufeld, E. F.: The defect in the Hunter syndrome: Deficiency of sulfoiduronate sulfatase, Proc. Nat. Acad. Sci. USA 70:2134, 1973.
16. Sjoberg, I., Fransson, L. A., Matalon, R., and Dorfman, A.: Hunter's syndrome: A deficiency of L-idurono-sulfate sulfatase, Biochem. Biophys. Res. Commun. 54:1125, 1973.
17. Kresse, H., and Neufeld, E. F.: The Sanfilippo A corrective factor, J. Biol. Chem. 247:2164, 1972.
18. O'Brien, J. S.: Sanfilippo syndrome: Profound deficiency of alpha-acetylglucosaminidase activity in organs and skin fibroblasts from type-B patients, Proc. Nat. Acad. Sci. USA 69:1720, 1972.
19. Von Figura, K., and Kresse, H.: The Sanfilippo B corrective factor: A N-acetyl-α-D-glucosaminidase, Biochem. Biophys. Res. Commun. 48:262, 1972.

HURLER SYNDROME
Mucopolysaccharidosis Type I

CARDINAL CLINICAL FEATURES

Coarse features, dwarfism, severe mental retardation, corneal clouding, hepatomegaly, dysostosis multiplex, widespread accumulation of mucopolysaccharide, deficiency of iduronidase.

CLINICAL PICTURE

The classic syndrome was first described in 1919 by Hurler[1] in Munich. This is the prototype of the mucopolysaccharidoses. It has been classified as mucopolysaccharidosis I by McKusick.[2] Patients appear normal at birth and usually develop normally for some months, after which they deteriorate progressively. They may present first for repair of inguinal hernias or chronic rhinitis.[3] In these patients the diagnosis is seldom suspected at that time. However, as the first year of life proceeds, the characteristic, increasingly grotesque appearance develops. In the established syndrome the facial features are coarse (Fig. 1). The head is large, bulging and scaphocephalic, and there is hyperostosis of the sagittal sutures. The face is flat, the nose and nostrils wide, the bridge flattened. The lips are large and thickened. The mouth often is open and the enlarged tongue protrudes. There is hypertrophy of the gums, hypertrophy of the bony alveolar ridge and small, widely spaced teeth. Patients are generally hirsute, and the hair is thick and coarse. They have thick, bushy eyebrows and low hairlines with a large amount of forehead hair.

A cloudy cornea is a hallmark of this syndrome. Clouding of the cornea, which has a ground-glass appearance, may be progressive and may lead to blindness. Patients usually have hypertelorism. Nystagmus and strabismus are occasionally seen. Deafness occurs rather frequently.

These patients develop a severe degree of mental retardation. They may reach a peak of intellectual development at about 2 years of age or earlier, after which there is steady regression. The skin is thick and lanugo is plentiful. Behavior usually is quite pleasant, and these often are lovable children despite their appearance.

The patients are generally short. Maximum height in one large series was 42 inches.[3] The neck is short, and the head appears to rest directly on the thorax. The lower rib cage flares. Kyphosis with gibbus in the lower thoracic or upper lumbar area is characteristic.

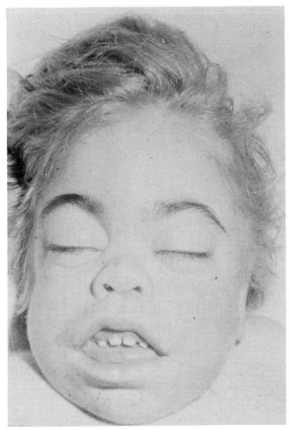

Fig. 1.—Hurler syndrome. The coarse features and prominent lips and tongue are striking.

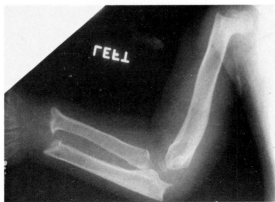

Fig. 2.—D. D. The upper extremities illustrate the lack of normal modeling and tubulation of the diaphyses, making these bones short and stubby. There was a varus deformity of the humerus. The ulnar semilunar notch was shallow and the radioulnar inclination abnormal.

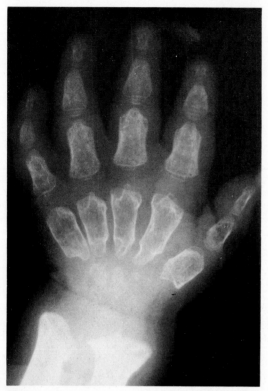

Fig. 3.—D. D. Dysostosis multiplex. The radial and ulnar articular surfaces were angulated toward each other. There were marked irregularity and retarded ossification of the carpal bones, as well as coarsening of the trabeculae of phalanges and metacarpals. The metacarpals were broadened at their distal ends and tapered at the proximal ends with a hook-like deformity. The phalanges, especially the distal ones, were short, and the proximal and middle phalanges were characteristically thick and bullet shaped.

The abdomen is protuberant. The liver and spleen become very large and very hard. Umbilical hernias and hydroceles, as well as inguinal hernias, are common. Recurrence following surgical hernia repair is fairly frequent. The joints become stiff, and mobility may be severely limited, especially at the elbows. The hands become broad, the fingers stubby. This and the limitation of extension and positioning in flexion produces the characteristic claw hand.

Cardiac complications are prominent late findings. Cardiac murmurs may be due to valvular disease of the mitral, aortic, tricuspid or pulmonary valves. There also may be thickening of the valves of the coronary arteries, leading to angina pectoris, or myocardial infarct. Patients may develop congestive heart failure, and death may be due to cardiac disease.

Survival is not expected beyond the first decade. Patients may die of pneumonia as well as cardiac complications. They tolerate anesthesia very poorly.[3] The roentgenographic appearance of these patients is striking. The Hurler patient provides the prototype for

the roentgenographic syndrome of dysostosis multiplex. The shafts of the bones widen. The cortical walls are thickened externally during the first year of life, but later they become thin as the medullary cavity dilates. The bones of the upper extremities become short and stubby (Fig. 2) and taper toward the ends, often with enlargement of the midportions. Lack of normal modeling and tubulation characterizes all of the bones, especially those of the upper extremities. The ends of the radius and ulna angulate toward each other (Fig. 3), so that the hands are clawed. The shape of the bones is pathognomonic of dysostosis multiplex. The metacarpals are broad at their distal ends and taper at their proximal ends. The phalanges are thickened and bullet-shaped. The lower extremities show moderate or minimal enlargement of the shafts. The skull is large and the sella turcica shoe shaped. The lower ribs are broad and spatulate (Fig. 4), the clavicles hypoplastic and thick. The vertebrae are hypoplastic, scalloped posteriorly and beaked anteriorly, especially at the thoracolumbar junction (Fig. 5). Hypoplasia of the odontoid is characteristic. In this area there is anterior vertebral wedging, and this leads to the thoracolumbar gibbus, with typically a hooked-shaped vertebra at the gibbus.

At autopsy the brain weight is increased, indicating

Fig. 4.—D. D. The ribs were classic. The spatula shape was caused by a generalized widening of the ribs, which spared the relatively narrow proximal portions.

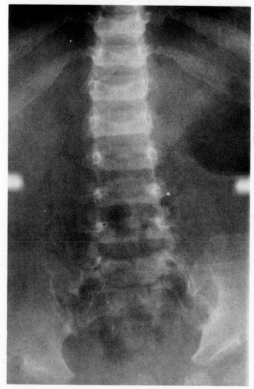

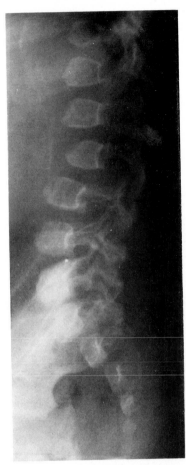

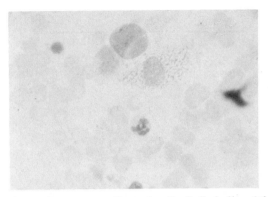

Fig. 6.—Bone marrow illustrating the Reilly bodies of the Hurler syndrome. The histiocyte in the center of the field was full of these inclusions. (Courtesy of Dr. Faith Kung of the University of California San Diego.)

Fig. 5.—D. D. The anteroposterior distance was diminished in the vertebral bodies and there was marked posterior scalloping. The pedicles of the lumbar spine were elongated. There was a marked thoracolumbar gibbus with inferior beaking of T12, L1 and L2.

that the increase in head size in these patients is a consequence of an abnormally large brain. Thickening of the meninges also is seen, and some patients develop hydrocephalus. Electron microscopic examination of the brain has revealed the presence of Zebra bodies resembling those of Tay-Sachs disease and interpreted as indicating the accumulation of ganglioside.[5] The material that accumulates in the brain in the mucopolysaccharidoses is predominantly ganglioside.[6] Large vacuolated cells are found in many tissues. Characteristic granules (Reilly bodies) are found in the polymorphonuclear leukocytes (Fig. 6). The mucopolysaccharide found in the tissues is dermatan sulfate.[6, 7] Large quantities of dermatan sulfate and heparan sulfate are excreted in the urine. In the Hurler syndrome these two compounds are excreted in an approximate ratio of 2:1. Mucopolysaccharide also accumulates in the brain. Metachromasia may be demonstrated in cultured fibroblasts by a pink stain

with toluidine blue. Quantitative analyses have revealed increased amounts of dermatan sulfate in fibroblasts of patients with the Hurler syndrome.[8] These cells accumulate radioactive sulfate in stored mucopolysaccharide because of defective degradation.[9] The defective enzyme in the Hurler syndrome is α-L-iduronidase.[10]

This enzyme also is defective in the Scheie syndrome, which is characterized by coarse features, diffuse corneal clouding and short stubby, often contracted fingers in patients with normal intelligence. The gene appears to be allelic with that which causes the Hurler syndrome.

A deficiency of β-galactosidase has been reported in the Hurler syndrome.[11] This could relate to the excessive cerebral storage of gangliosides.

CASE REPORTS

CASE 1.—D. D. was admitted at 7 years of age to University Hospital. She previously had been diagnosed as having the Hurler syndrome. She was the product of a normal pregnancy and delivery, the first child of a healthy, young couple. The mother subsequently divorced the patient's father, remarried and had three normal children. The patient's birth weight was 9 lb. Her development during the first year of life appeared normal, but she had repeated upper respiratory tract infections and a chronic nasal discharge. She sat at 5 months and walked at 11 months. Mental and motor development began thereafter to regress, and she was diagnosed at the age of 18 months. At the time of admission the child was legally blind and could speak only a few words and phrases. She could feed herself and could turn on television.

Physical examination revealed the typical features of the Hurler syndrome (Figs. 7–9). The child was short and had a large head and protuberant abdomen. She weighed 42 lb and was 35½ inches tall. The head circumference was 55 cm. She had very coarse features, and the hair was coarse and dry. She had frontal bossing and hypertelorism. The eyes were prominent, and there was bilateral diffuse steamy clouding of the cornea (Fig. 10). The nasal bridge was depressed and the nares large. She had a prominent mucoid rhinorrhea. The lips were

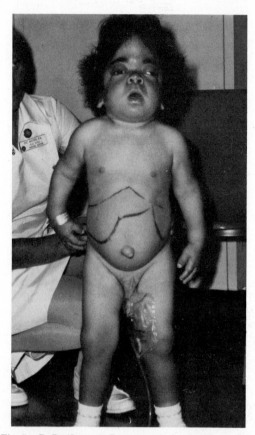

Fig. 7.—D. D. (Case 1). Patient with the Hurler syndrome. The features were coarse. She had a large head, prominent eyes, hypertelorism and a depressed nasal bridge. The abdomen was protuberant and there was an umbilical hernia. (Courtesy of Dr. John S. O'Brien of the University of California San Diego.)

large and full, and the teeth peg-like and widely spaced. The gums were hypertrophic and the tongue appeared large. The ears were thickened and stiff and the neck short. The chest was barrel shaped. There was a grade II/VI cardiac systolic murmur. The abdomen was protuberant and there was an umbilical hernia. A very hard liver was palpable 9 cm below the costal margin, and the spleen was palpable 7 cm below the costal margin. A lumbar gibbus was present (see Fig. 8). The hands were short and broad. The fingers were stubby and had limited joint mobility (see Fig. 5). There was some limitation of motion of all the large joints, especially the elbows. The skin was thick and inelastic.

The patient's IQ was 30. Skeletal survey revealed the characteristic features of dysostosis multiplex (see Figs. 2–5). The skull was enlarged (Fig. 11). There was calcification in the choroid plexus of the left lateral ventricle, and there was erosion and massive enlargement of the sella turcica. The condyles of the mandible were flattened and misshapen. A thoracolumbar gibbus was present. The pelvis showed general hypoplasia and flaring of the iliac wings. There was bilateral coxa valga (see Fig. 5). Bone marrow showed typical Reilly bodies (see Fig. 6). Smear of the peripheral blood revealed in-

clusion bodies in the white cells (Fig. 12). Analysis of the mucopolysaccharides of the urine revealed an increase in total excretion and the Hurler pattern of increase in excretion of dermatan sulfate and heparan sulfate.

CASE 2.—K. B. was a 1-year-old white female referred to University Hospital for confirmation of a diagnosis of the Hurler syndrome. At 2 weeks of age she was noted by a physician to have mottled skin. At 10 months of age he found her to have cardiomegaly and judged that she had the Hurler syndrome. She had sat alone at 5 months, stood at 7 months and was walking with assistance at 8 months. She had a history of chronic rhinitis. A 5-year-old sister was normal, and there was no consanguinity.

Physical examination revealed an irritable, white female with gray, mottled skin. Her height was 30 inches, normal for her age. Her weight was 29 lb, 9 oz. Head circumference was 46 cm. She had a depressed nasal bridge, hypertelorism, thick lips and an enlarged tongue. There was noisy breathing and a nasal discharge. The corneas were distinctly cloudy. The abdomen was soft, and hepatosplenomegaly was prominent. She had short, broad hands and walked on her tiptoes. She had

Fig. 8.—D. D. The gibbus.

Fig. 9.—D. D. The hands were typical of the Hurler syndrome. They were short, broad and claw shaped.

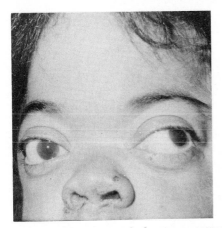

Fig. 10.—D. D. Close-up reveals the steamy corneas.

Fig. 11.—D. D. The cranium was very large. Both occiput and frontal area were prominent. Calcification of the choroid plexus of the left lateral ventricle is seen. There was an enormous enlargement of the sella, with erosion of the clinoid processes. The mandibular rami were short and there was increased angulation at the junction of the body and the varus, and flattening of the condyles.

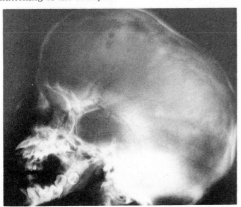

evidence of severe cardiomyopathy but no congestive heart failure. Roentgenograms obtained at 3½ weeks of age showed broad ribs, which flared more widely at their anterior ends but were considerably thickened posteriorly. The clavicles were slightly short and broad. At 10 months of age the spine showed inferior beaking of the body of L2. The bones of the hands were broad and thick. The proximal portions of the metacarpals were pointed.

The ends of the radius and ulna showed a mild modeling deformity. The heart was markedly enlarged. Mucopolysaccharide concentration in the urine was 150 mg/ml. The control range was 2–10. The activity of β-galactosidase in the skin was 3.3 mμM/mg of protein/hour, compared with a control mean value of 25.

GENETICS

The Hurler syndrome is determined by an autosomal recessive gene. Parental consanguinity has commonly been reported. The carrier state can be detected.

TREATMENT

There is no effective therapy for these patients, and prognosis is for demise, usually before 10 years of age. Enzyme therapy is under study in this disorder, and infusion of plasma[12] or leukocytes[13] has been reported to be followed by some clinical changes as well as changes in the pattern of excretion of urinary mucopolysaccharides. Prenatal diagnosis via amniocentesis has been carried out.[14]

REFERENCES

1. Hurler, G.: Ueber einen Typ multipler Abartungen, Vorwiegend, am Skelettsystem, Z. Kinderheilk. 24:220, 1919.
2. McKusick, V. A.: *Heritable Disorders of Connective Tissue* (4th ed.; St. Louis: C. V. Mosby Co., 1972), p. 521.
3. Leroy, J. G., and Crocker, C.: Clinical definition of the Hurler-Hunter phenotypes. A review of 50 patients, Am. J. Dis. Child. 112:518, 1966.

Fig. 12.—D. D. Peripheral blood, illustrating inclusions in a monocyte.

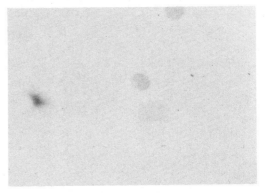

4. Caffey, J.: Gargoylism (Hunter-Hurler disease, dysostosis multiplex, lipochondrodystrophy), Am. J. Roentgenol. Radium Ther. Nucl. Med. 67:715, 1952.

5. McKusick, V. A.: The nosology of the mucopolysaccharidoses, Am. J. Med. 47:730, 1969.

6. Dorfman, A., and Matalon, R.: The Hurler and Hunter syndromes, Am. J. Med. 47:691, 1969.

7. Muir, H.: The structure and metabolism of mucopolysaccharides (glycosaminoglycans) and the problem of the mucopolysaccharidoses, Am. J. Med. 47:673, 1969.

8. Matalon, R., and Dorfman, A.: Acid mucopolysaccharides in cultured human fibroblasts, Lancet 2:838, 1969.

9. Fratantoni, J. C., Hall, C. W., and Neufeld, E. F.: The defect in Hurler's and Hunter's syndromes: Faulty degradation of mucopolysaccharides, Proc. Nat. Acad. Sci. USA 60:699, 1968.

10. Bach, G., Friedman, R., Weissmann, B., and Neufeld, E. F.: The defect in the Hurler and Scheie syndromes: Deficiency of α-L-iduronidase, Proc. Nat. Acad. Sci. USA 69: 2048, 1972.

11. MacBrinn, M., Okada, S., Woollacott, M., Patel, V., Ho, M. W., Tappel, A. L., and O'Brien, J. S.: Beta-galactosidase deficiency in the Hurler syndrome, New Engl. J. Med. 281: 338, 1969.

12. Di Ferrante, N., Nichols, B. L., Donnelly, P. V., Neri, G., Hrgovcic, R., and Berglund, R. K.: Induced degradation of glycosaminoglycans in Hurler's and Hunter's syndromes by plasma infusion, Proc. Nat. Acad. Sci. USA 68:303, 1971.

13. Knudson, A. G., Jr., Di Ferrante, N., and Curtis, J. E.: Effect of leukocyte transfusion in a child with type II mucopolysaccharidosis, Proc. Nat. Acad. Sci. USA 68:1738, 1971.

14. Fratantoni, J. C., Neufeld, E. F., Uhlendorf, B. W., and Jacobson, C. B.: Intrauterine diagnosis of the Hurler and Hunter syndrome, N. Engl. J. Med. 280:686, 1969.

HUNTER SYNDROME
Mucopolysaccharidosis Type II

CARDINAL CLINICAL FEATURES

The coarse features of a typical mucopolysaccharidosis, the skeletal deformity of dysostosis multiplex, stiff joints, mental retardation, hepatosplenomegaly, cardiomegaly and thickened skin, often with nodular skin lesions over scapular area or arms.

CLINICAL PICTURE

Mucopolysaccharidosis type II was first described by Hunter in 1917 in two brothers.[1] Patients with the Hunter syndrome have clinical features similar to those of the Hurler syndrome, but they usually are less severely affected. The onset of the disease also is a little later, most often around 2–4 years of age. However, patients have chronic respiratory symptoms prior to that and often present for hernia repair early. Mental development usually continues until at least 2 years of age.

Characteristic coarse features (Figs. 1 and 2) include a flat nose with wide nostrils and a depressed bridge, thick lips, hypertrophic gums and a large tongue. Patients are generally hirsute and have low hairlines. The superciliary ridges become very prominent. The patients have noisy breathing and appear to have a constant rhinitis. Hearing loss is common but often is not severe.[2] Shortness of stature is not so pronounced as in the Hurler syndrome. These patients have stiff joints. The hands are broad with stubby fingers and a clawhand appearance (see Fig. 1). The liver and spleen are large and hard. Inguinal and umbilical hernias are common. Nodular skin lesions sometimes are seen in the scapular area or on the arms; these lesions are characteristic of mucopolysaccharidosis type II. Patients with the Hunter syndrome tend to develop high coloration "like that of a middle-aged farmer who is fond of malt liquor."[3] Two important negative findings in the Hunter syndrome, the absence both of a cloudy cornea (see Fig. 2) and gibbus,[4] distinguish it from the Hurler syndrome. However, these patients may develop a rounded kyphosis. Retinitis pigmentosa may occur in this syndrome, and it may cause blindness. Papilledema may be seen, usually in association with hydrocephalus.

Mental deterioration is progressive. The behavior of the patient with the Hunter syndrome may be characteristic.[2] From 2–6 years of age he may develop primitive, uncontrolled activity in which he throws toys and

Fig. 1.—C. T., a patient with the Hunter syndrome, illustrates the most severe form of the disease. The coarse features, thick lips and clawhands were as marked as in the Hurler syndrome.

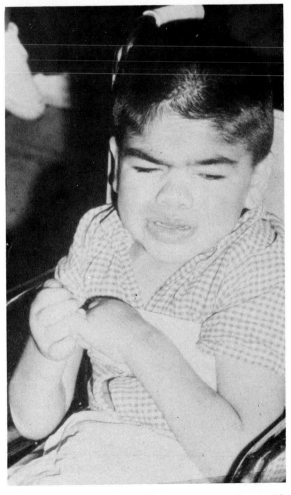

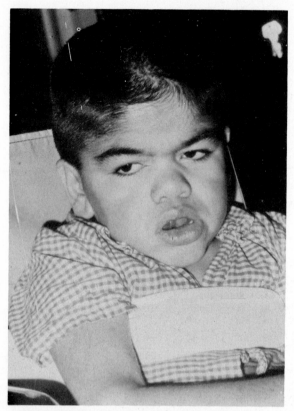

Fig. 2.—C. T. The corneas were clear. The patient had hypertelorism and a large nose with a depressed nasal bridge.

seems to enjoy self-created noise. He may be hyperkinetic. Rough, aggressive play may be dangerous to pets or younger siblings. These patients often are stubborn, fearless and unresponsive to discipline. Eating habits may be awkward, and pica is common. One patient was reported to have lead poisoning. The management of such a child is difficult and institutionalization is common. Every child does not manifest abnormal behavior, and some patients with this syndrome maintain fair cerebral function into late childhood. It has been thought that there might be two forms of this syndrome, one of which is milder and may be compatible with long survival. Most patients deteriorate progressively after 5 or 6 years of age. Physical activity decreases. Speech is reduced. Difficulty in ingesting solid food is progressive, and there is loss of weight. Respiratory infections become more frequent and more severe and may be the cause of death. Cardiac complications such as congestive failure or coronary insufficiency also may lead to death. Generalized seizures may occur in the final months of life.

Roentgenographic study early shows external thickening of the cortex of the bones. With increasing age the cortical walls become thinner as the marrow cavi-

ties expand.[5] The findings are similar to those of the Hurler syndrome, but they tend to be less dramatic. The skull is large and the sella shoe shaped. The lower ribs are broad and spatulate. There is hypoplasia of the vertebrae and slight breaking of L2.

Fundamental to the clinical phenotype is the excessive intracellular accumulation of acid mucopolysaccharides. Large vacuolated cells are present in many tissues. These patients, like those with the Hurler syndrome, excrete large amounts of dermatan sulfate in the urine and heparan sulfate[6, 7] in approximately equal amounts. Fibroblasts in tissue culture from patients with the Hunter syndrome show metachromatic staining and contain large amounts of mucopolysaccharide.[7, 8] Hunter cells accumulate labeled sulfate like typical mucopolysaccharidosis cells and are altered by the corrective factor derived from normal

Fig. 3.—G. B. (Case Report). A mild case of the Hunter syndrome in a 6-year-old boy. He had coarse features and hypertelorism. The lips were somewhat thickened.

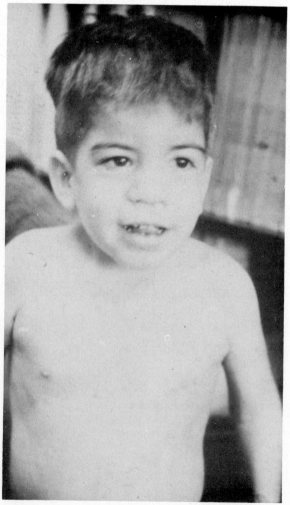

cells or those of patients with other mucopolysaccharidoses.[9] The enzyme defect appears to be in sulfoiduronate sulfatase.[10]

β-Galactosidase activity is diminished in the skin and other tissues of patients with the Hunter syndrome.[11] This defect, also seen in the Hurler syndrome, appears to be secondary to the primary defect. It could relate to the accumulation of ganglioside and other lipids found in the brain of these patients.

CASE REPORT

G. B., a 5-year-old boy (Fig. 3) was referred to the Children's Health Center, San Diego, with a chief complaint of slow growth. His mother related that he had been in good health and had no problems in development until he was 2 years old. At that time she noticed that his speech was delayed. He looked to her at that time like a chubby baby with short stature and abdominal distention. At 3½ years of age the bone age was found to be 18 months. At 5 years he attended kindergarten and seemed to have a normal intellect. However, his physical development lagged consistently, and he could not keep up with the physical activities of his peers.

His past history was unrevealing for significant illness. He had been born after 8 months' gestation and weighed 6½ lb. He had had a hernia repair at 7 weeks of age and a tonsillectomy and adenoidectomy at age 4 years.

His immediate family history was not informative. He was the product of a twin pregnancy, but the other fetus died early in pregnancy. The family tree is illustrated in Figure 4. A 3-month-old male sibling died of congestive heart failure. At autopsy only diffuse cardiac hypertrophy was noted. The mother had a history of a ventricular septal defect, which was surgically repaired at 17 years of age. The mother related that she had had a great uncle who was said to look just like the patient (Fig. 5).

Physical examination revealed an unusual-appearing child

with coarse features. He was 41¾ inches tall, which was in the third percentile, and weighed 45 lb, which was in the 50th percentile for age 5 years. There were no corneal lesions. The nose was broad and the tongue large. The neck was short and thick. There was an increase in the anteroposterior diameter of the chest. There were no cardiac murmurs. The abdomen was distended, and a large liver was palpable 6 cm below the costal margin in the midclavicular line. A splenic tip was palpable. All of the joints exhibited a moderate to marked degree of limitation of motion. The patient had spade-like hands that were broad and short. All of the fingers were short. There was a slight lumbar lordosis and a dorsal kyphosis. The neurologic examination was normal.

Fig. 5.–G. B. The great-uncle in generation I of the pedigree shown in Figure 4. Shortness of stature and coarse features as illustrated, as well as mental retardation and death early in life, suggest that he too had the Hunter syndrome. (Courtesy of Dr. David Chadwick of the Children's Health Center, San Diego.)

Fig. 4.–G. B. Pedigree of the family, illustrating X-linked inheritance. Heterozygotes in generations II and III were identified by Dr. John S. O'Brien of the University of California San Diego, by β-galactosidase assay.

Pedigree: Hunter's Syndrome   (MPS Type 2)

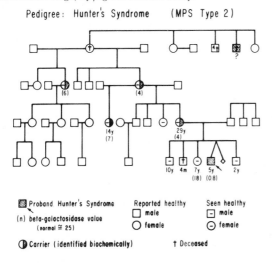

Roentgenographic examination revealed minimal cardiac enlargement, normal ribs and an enlarged sella turcica. There were hypoplastic vertebral bodies at T11 and T12, which showed beaking. EEG and ECG were normal. The activity of β-galactosidase in the skin was determined in Dr. John O'Brien's laboratory, and showed a level of 0.8 units, which is markedly reduced. The mother was found to have an intermediate activity, and the father was normal.

GENETICS

The Hunter syndrome is inherited as an X-linked recessive trait. Heterozygous female carriers of the Hunter gene have been recognized by cloning of fibroblasts followed by staining for metachromasia.[8] Two clonal populations, one normal and the other abnormal, have been demonstrated, as specified by the Lyon hypothesis.

TREATMENT

There is no effective treatment for this disease. Some transient clinical improvement, as well as changes in the excretion of mucopolysaccharide in the urine following the infusion of plasma or leukocytes, provides a stimulus for further research on enzyme replacement therapy.[12] Prenatal diagnosis following amniocentesis has been reported.[13]

REFERENCES

1. Hunter, C.: A rare disease in two brothers, Proc. R. Soc. Med. 10:104, 1917.
2. Leroy, J. G., and Crocker, A. C.: Clinical definition of the Hurler-Hunter phenotypes, Am. J. Dis. Child. 112:518, 1966.
3. Weber, F. P.: Discussion (ref. 1).
4. McKusick, V. A.: *Heritable Disorders of Connective Tissue* (4th ed.; St. Louis: C. V. Mosby Co., 1972), pp. 556–574.
5. Caffey, J.: Gargoylism (Hunter-Hurler disease, dysostosis multiplex, lipochondrodystrophy), Am. J. Roentgenol. Radium Ther. Nucl. Med. 67:715, 1952.
6. Muir, H.: The structure and metabolism of mucopolysaccharides (glycosaminoglycans) and the problem of the mucopolysaccharidoses, Am. J. Med. 47:673, 1969.
7. Dorfman, A., and Matalon, R.: The Hurler and Hunter syndromes, Am. J. Med. 47:691, 1969.
8. Danes, B. S., and Bearn, A. G.: Hurler's syndrome: A genetic study of clones in cell culture with particular reference to the Lyon hypothesis, J. Exp. Med. 126:509, 1967.
9. Fratantoni, J. C., Hall, C. W., and Neufeld, E. F.: Hurler and Hunter syndromes. I. Mutual correction of the defect in cultured fibroblasts, Science 162:570, 1968.
10. Bach, G., Eisenberg, F., Cantz, M., and Neufeld, E. F.: The defect in the Hunter syndrome: Deficiency of sulfoiduronate sulfatase, Proc. Nat. Acad. Sci. USA 70:2134, 1973.
11. Gerich, J. E.: Hunter's syndrome. Beta-galactosidase deficiency in skin, New Engl. J. Med. 280:799, 1969.
12. Knudson, A. G., Di Ferrante, N., and Curtis, T. E.: Effect of leukocyte transfusion in a child with type II mucopolysaccharidosis, Proc. Nat. Acad. Sci. USA 68:1738, 1971.
13. Fratantoni, J. C., Neufeld, E. F., Uhlendorf, W., and Jacobson, C. B.: Intrauterine diagnosis of the Hurler and Hunter syndromes, New Engl. J. Med. 280:686, 1969.

SANFILIPPO SYNDROME
Mucopolysaccharidosis Type III

CARDINAL CLINICAL FEATURES

Mental retardation, mild skeletal deformity, excessive urinary excretion of heparan sulfate. The Sanfilippo syndrome type A is due to a deficiency of heparan sulfate sulfatase and type B to a deficiency of α-acetylglucosaminidase.

CLINICAL PICTURE

This syndrome first surfaced with the report of Harris[1] in 1961 of a mildly retarded 6-year-old girl with hepatosplenomegaly and a normal skeletal survey who excreted large amounts of heparan sulfate in the urine. Sanfilippo and colleagues[2,3] in 1962 and 1963 described eight children with a wide range in degree of mental retardation, all of whom had heparan sulfate mucopolysacchariduria. Some of these patients had similarities in appearance and in roentgenographic findings to patients with the Hurler and Hunter syndromes.

In the Sanfilippo syndrome it is the central nervous system that is most prominently involved; the visceral and skeletal features are much less striking.[4] These patients are characteristically normal in appearance at birth and appear to develop normally during the first year. They usually are referred after 1 or 2 years of age for slowness in development. They may have difficulty in feeding, especially with solids, as well as repeated respiratory infections from the beginning. Mental retardation becomes progressively more obvious with time. These patients do not have a problem with linear growth and muscle strength is good. Management may be a problem.[5] Behavior tends to become worse as, with age, patients become increasingly stubborn and withdrawn. They may be hyperactive and aggressive children. They may be crude in their personal habits and eat unusual objects such as cigarette butts. They can be destructive and may harm siblings. They often are so difficult to handle that institutionalization is necessary. Skills learned during the first years are lost, including speech and toilet training. Neurologic problems are progressive. The gait becomes clumsy and coordination poor. Deep tendon reflexes become accentuated. Purposeless athetoid-like movements may develop. The patients drool constantly and are difficult to feed. They may have seizures. Finally, they become bedridden. They usually die before their 20th birthday and often before the 10th[4] but survival into the third or fourth decade is possible.[5]

In appearance the Sanfilippo patient usually has some coarseness of features, but he is not readily recognizable as a patient with mucopolysaccharidosis. Screening of the urine for metabolic disease in institutions for the retarded is likely to bring to light previously undiagnosed patients with this syndrome.

The bridge of the nose may be slightly flattened and the lips somewhat thick (Fig. 1). The eyebrows often are bushy. The dull, rigid facies is a consequence of cerebral deterioration, as contrasted with the local tissue changes in the Hurler syndrome. Patients may have mild limitation of joint mobility. Hepatomegaly may be mild or undetectable. There is no gibbus, and the corneas are clear. Cardiac abnormalities usually have not been observed in these patients.[5] However, a patient with severe incapacitating involvement of the

Fig. 1.—A markedly retarded girl with Sanfilippo disease.

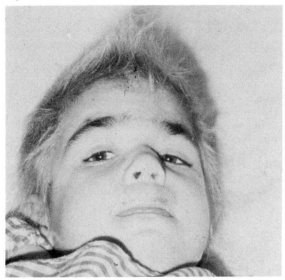

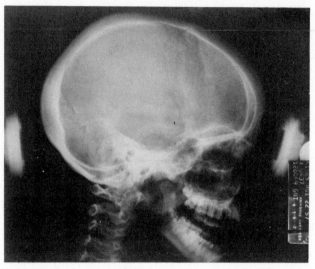

Fig. 2.—Roentgenogram of the skull of a 16-year-old boy with Sanfilippo disease type B. The marked increase in thickening of the diploic space, seen best posteriorly, is characteristic of this disease and not seen in other mucopolysaccharidoses. The sella turcica was normal.

mitral valve has been reported.[6] Hernias may develop and may recur after correction.

Roentgenographic findings are of a mild dysostosis multiplex.[5] (Figs. 2–4). Most patients with this syndrome have a thickening and increased density of the calvaria in the posterior parietal and occipital areas.[7] The mastoids are sclerotic. The sella turcica appears normal. Among patients with dysostosis multiplex, those with this syndrome have the mildest bony changes. Those with I cell disease, or mucolipidosis II and GM_1 gangliosidosis, and, among patients with mucopolysaccharidosis, those with the Hurler syndrome have the most severe bony changes.

Blood smears reveal the presence of metachromatic inclusion bodies in the lymphocytes that are characteristically coarser and sparser than those seen in the Hurler syndrome. Inclusions are also seen in cells of the bone marrow. Chondrocytes in cartilage biopsied from the iliac crest and the ribs have been reported to be vacuolated.[8] The diagnosis usually is confirmed by finding increased quantities of mucopolysaccharide in the urine. In this syndrome it is specifically heparan sulfate that is excreted in excess.[9]

Patients with this disorder accumulate gangliosides in the brain.[10, 11] These may include GM_2 and GM_3[10] or there may be a great increase in GM_1.[11] The electron microscopic appearance of the neurons may be like that in Tay-Sachs disease.[10] Patients also may have zebra bodies, and they accumulate mucopolysaccharide in the brain as well as in the peripheral tissues.

Fibroblasts derived from patients with the Sanfilippo syndrome accumulate $^{35}SO_4$. The existence of two types of disease was first recognized through cross-correction studies.[12] Patients studied fell into two groups, and those of each group could correct the

Fig. 3.—Roentgenogram of the thoracolumbar spine of same patient (Fig. 2) reveals mild platyspondyly. The ribs were spatulate, with posterior narrowing characteristic of dysostosis multiplex, but the process in this 16 year old was significantly milder than that seen in other mucopolysaccharidoses.

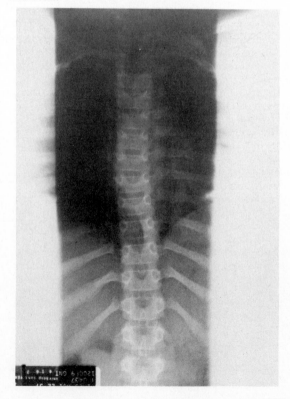

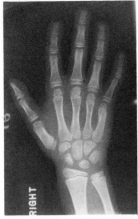 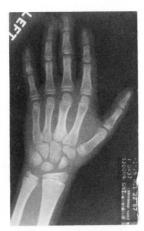

Fig. 4.—Roentgenogram of the hand of patient discussed in Figures 2 and 3 looks normal. This highlights the difference in the bones of the patient with Sanfilippo disease from the more severe examples of dysostosis multiplex.

other. The correction factors are the enzymes whose activity is lacking in each of the types. In type A Sanfilippo cells, Neufeld and colleagues[13] found that the defective enzyme is heparan sulfate sulfatase. In type B the defect was found to be in α-acetylglucosaminidase.[14, 15] The heparan sulfate molecule may be represented as follows:

$$\text{glucosamine}\overset{b}{-}\text{hexuronic acid}\overset{b}{-}\text{glucosamine}\overset{b}{-}\text{hexuronic acid}$$
$$\underset{SO_4}{\overset{|\,a}{}} \qquad\qquad \underset{SO_4}{\overset{|\,a}{}}$$

Heparan sulfate sulfatase catalyzes the breakdown of this molecule by splitting off the sulfate groups designated "a". α-Acetylglucosaminidase catalyzes its breakdown at the glucosamine to hexuronic acid linkage designated "b". Thus, the defects in Sanfilippo disease types A and B both lead to the accumulation of the same mucopolysaccharide. Phenotypically the patients of types A and B cannot be distinguished.

CASE REPORT*

D. M. was a 21-year-old mentally retarded male whom we first saw at Fairview State Hospital, Costa Mesa, where he had been diagnosed as having the Sanfilippo syndrome (Figs. 5 and 6). He was one of four children. Two siblings also had the disease (Fig. 7); the third was normal. The patient had appeared normal at birth and weighed 7 lb, 10 oz. He seemed to develop normally during the first 2 years of life. When he was 3 years old his voice became guttural, and he no longer used the words he had learned earlier. He never was toilet trained. Seizures developed and mental deterioration was progressive.

At the time of this examination he was severely retarded

*We are indebted to Dr. Charles H. Fish of Fairview State Hospital, Costa Mesa, California, for the opportunity to study and photograph this patient and others illustrated in the figures.

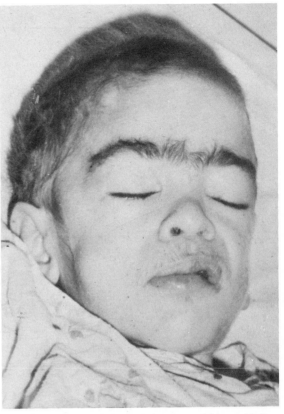

Fig. 5.—D. M. (Case Report). A 16-year-old boy with Sanfilippo disease type B, who was markedly retarded. The features were coarse, but not at all like the features of the Hurler or Hunter syndrome. The eyebrows were bushy and the hairline low.

Fig. 6.—D. M. At age 20. The thick lips, large nose and depressed nasal bridge are evident.

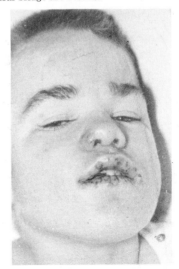

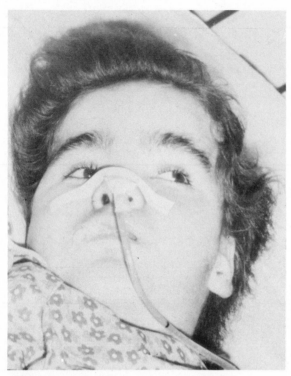

Fig. 7.—L. M., the sister (age 13) of D. M., also had Sanfilippo disease type B.

and had no speech. He was uninterested in his surroundings, bedridden and paraplegic. He had mildly coarse features, thick lips and a large tongue. The hair was coarse, the eyebrows thick and bushy. The corneas were not cloudy. He had a stuffy nose. There was mild hepatosplenomegaly. He was severely spastic and had hyperactive deep tendon reflexes.

He excreted large amounts of mucopolysaccharide in the urine, which column chromatography and thin layer chromatography revealed to be heparan sulfate. Cultured skin fibroblasts were initially sent to Dr. E. F. Neufeld, who demonstrated that he had a defect in the degradation of acid mucopolysaccharide. This defect was corrected by media in which cells of a patient with Sanfilippo A were grown, but it was not corrected by media that had contained cells of Sanfilippo B. Enzyme assay of skin fibroblasts and leukocytes in Dr. John O'Brien's laboratory showed a profound deficiency of α-acetylglucosaminidase. His mother had 50% of normal activity.

GENETICS

Both the A and B types of Sanfilippo syndrome are transmitted in an autosomal recessive fashion. Multiple affected siblings have been observed in several families. Detection of heterozygotes can be accomplished through enzyme assay. Intrauterine diagnosis of the affected fetus also is feasible through assay of the relevant enzyme in amniotic cells in culture.

TREATMENT

There is no effective treatment. Enzyme replacement therapy has been attempted using partially purified enzyme, after which there were some distinct changes in urinary mucopolysaccharides, but the systemic pyrogenic reaction to a single injection was formidable.

REFERENCES

1. Harris, R. C.: Mucopolysaccharide disorder: A possible new genotype of Hurler's syndrome, (Abstr.) Am. J. Dis. Child. 102:741, 1961.
2. Sanfilippo, S. J., and Good, R. A.: Urinary acid mucopolysaccharides in the Hurler syndrome and Morquio's disease, J. Pediatr. 61:296, 1962.
3. Sanfilippo, S. J., Podosin, R., Langer, L., and Good, R. A.: Mental retardation associated with acid mucopolysacchariduria (heparitin sulfate type), (Abstr.) J. Pediatr. 63:837, 1963.
4. Leroy, J. G., and Crocker, A. C.: Clinical definition of the Hurler-Hunter phenotypes. A review of 50 patients, Am. J. Dis. Child. 112:518, 1966.
5. McKusick, V. A.: *Heritable Disorders of Connective Tissue*, (4th ed.; St. Louis: C. V. Mosby Co., 1972), p. 521.
6. Herd, J. K., Subramanian, S., and Robinson, H.: Type III mucopolysaccharidosis: Report of a case with severe mitral valve involvement, J. Pediatr. 82:101, 1973.
7. Spranger, J., Teller, W., Kosenow, W., Murken, J., and Eckert-Husemann, E.: Die HS-mucopolysaccharidose von Sanfilippo (Polydystrophe oligophrenie) bericht über 10 Patienten, Z. Kinderheilkd. 101:71, 1967.
8. Silberberg, R., Rimoin, D. L., Rosenthal, R. E., and Hasler, M. B.: Ultrastructure of cartilage in the Hurler and Sanfilippo syndromes, Arch. Pathol. 94:500, 1972.
9. Gordon, B. A., and Haust, M. D.: The mucopolysaccharidoses types I, II and III: Urinary findings in 23 cases, Clin. Biochem. 3:203, 1970.
10. Wallace, B. J., Kaplan, D., Adachi, M., Schneck, L., and Volk, B. W.: Mucopolysaccharidosis type III. Morphologic and biochemical studies of two siblings with Sanfilippo syndrome, Arch. Pathol. 82:462, 1966.
11. Dekaban, A. S., and Patton, V. M.: Hurler's and Sanfilippo's variants of mucopolysaccharidosis, Arch. Path. 91:434, 1971.
12. Kresse, H., Wiesmann, U., Gantz, M., Hall, C. W., and Neufeld, E. F.: Biochemical heterogeneity of the Sanfilippo syndrome: Preliminary characterization of two deficient factors, Biochem. Biophys. Res. Commun. 42:892, 1971.
13. Kresse, H., and Neufeld, E. F.: The Sanfilippo A corrective factors, J. Biol. Chem. 247:2164, 1972.
14. O'Brien, J. S.: Sanfilippo syndrome: Profound deficiency of alpha-acetylglucosaminidase activity in organs and skin fibroblasts from type-B patients, Proc. Nat. Acad. Sci. 69:1720, 1972.
15. von Figura, K., and Kresse, H.: The Sanfilippo B corrective factor: A N-acetyl-a-D-glucosaminidase, Biochem. Biophys. Res. Commun. 48:262, 1972.

MORQUIO SYNDROME
Mucopolysaccharidosis Type IV, Morquio-Brailsford Disease

CARDINAL CLINICAL FEATURES

Short-trunked dwarfism, pectus carinatum, dorsolumbar kyphosis, genu valgum, corneal clouding, dental anomalies, aortic valve damage, keratosulfaturia.

CLINICAL PICTURE

The syndrome was described in Uruguay by Morquio in 1929[1] in four affected siblings who were the products of a marriage of first cousins of Swedish origin. A fifth sibling was normal. In the same year in England, Brailsford[2] described a similar patient. In 1961 Maroteaux and Lamy[3] found the excretion of keratosulfate in the urine of patients with this disease.

The most characteristic feature of this syndrome is a severe dwarfism, which is particularly short trunked, although the long bones also are involved. The neck is short and the head appears to sit directly on a barrel-chest, which has pectus carinatum so pronounced that the sternum may seem to come to a point (Fig. 1). The upper part of the sternum may be almost horizontal. There also is dorsolumbar kyphosis. The arms are long in comparison with the trunk, and the fingertips may reach below midthigh. These patients also have a most pronounced genu valgum (Fig. 2). They often have a semi-crouching stance. The joints become enlarged and prominent, but there may be extreme mobility and hyperextension of the joints, particularly at the wrists where there may be marked ulnar deviation. Pes planus also is seen. These skeletal changes are not obvious during the first year of life because intrauterine growth and early extrauterine development are normal. The deformities are progressive and become exaggerated with age.[4] Growth is markedly slow after about 5 years of age.

These patients have fine corneal opacities, which usually are visible only on slit lamp examination but may cause a hazy cloudiness of the cornea.[3, 5] The enamel is hypoplastic both in deciduous and permanent teeth. The teeth develop a gray or yellowish color and the enamel becomes flaky or fractured. Molars often have sharp cusps. The teeth readily develop car-

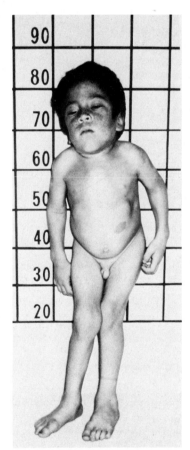

Fig. 1.—A 5-year-old boy with the Morquio syndrome. He had short-trunked dwarfism as well as deformity of the extremities. The neck was very short. He had a pectus carinatum and marked genu valgum. There were angulations at the elbow and wrist. (Courtesy of Dr. David Rimoin.)

ies. Progressive sensorineural deafness begins usually in the second decade and is present uniformly after 20 years. A regular later manifestation of the disease is aortic regurgitation. Inguinal hernia probably is more common than in normal individuals. The head is nor-

57

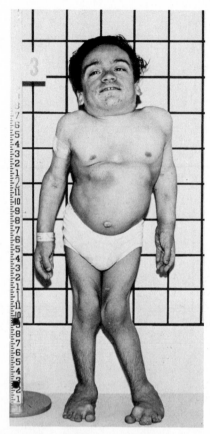

Fig. 2.—A 16-year-old with the Morquio syndrome. The trunk appeared more disproportionately short than in the patient in Figure 1. There seemed to be essentially no neck, with the head resting on the deformed thorax. The flaring of the rib cage was pronounced and irregular. Genu valgum was very marked. (Courtesy of Dr. David Rimoin.)

mal, and the patients have a normal range of intelligence.

Roentgenographic findings[6] in the Morquio disease vary with the age of the patient. The most characteristic and consistent finding is the universal platyspondyly or vertebra plana, which produces the short spine (Fig. 3). The vertebral bodies usually are oval shaped in the younger affected child, becoming flatter and more rectangular in later childhood and flat in the adult. The intervertebral spaces are deep in all age groups. The cervical spine is striking in that the odontoid process of C_2 is either absent or hypoplastic. The remainder of the cervical vertebrae are flat. The thoracic and lumbar vertebrae show flattening and anterior beaking or tonguing. L1 is often short and displaced posteriorly, accounting for the gibbus. Patients with this syndrome always have a marked coxa valga deformity (Fig. 4). The pelvis is narrow. With age the anterior portions of the ribs become wide and spatula shaped. The clavicle is normal. The sternum protrudes. The acetabular cavity is large and becomes

larger with age. The femoral head becomes progressively flattened and fragmented; it may be completely resorbed. The femoral neck initially loses its angle and later becomes thickened. The distal femur is wide, as is the proximal tibia. The lateral aspect of the proximal tibia shows a deficiency of ossification. These changes contribute to the production of the genu valgum. The distal end of the humerus is wide and irregular, as are the proximal ulna and radius, in the same manner as the corresponding bones of the lower extremities. The growth plates of the distal ulnae and radii are slanted toward each other, the ulna usually being somewhat shorter than the radius. The ossified carpal bones are small and may be reduced in number. The metacarpals are short and their distal metaphyses are widened. Osteoporosis is common in the adult patient.

A very serious and common complication of the bony deformity in this syndrome is that the spinal cord may be compressed following atlantoaxial subluxation or dislocation.[4, 7] This is a major cause of death. Manipulation of the head for intubation may be particularly dangerous in these patients for this reason, but subluxation may occur during sleep. These patients may die in their sleep. Propensity for subluxation is due to the hypoplasia of the odontoid process and to the general laxity of the ligaments. These features are present in all patients with the disease, who therefore may all

Fig. 3.—Roentgenogram of the spine of patient shown in Figure 1. The vertebral bodies were very flat—the vertebra plana that characterizes this disease—and were beaked anteriorly. The second vertebra from the top in this figure was hypoplastic and displaced posteriorly. This is the genesis of the gibbus in this syndrome. (Courtesy of Dr. David Rimoin.)

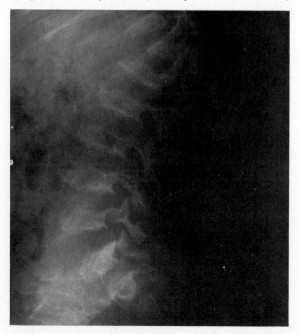

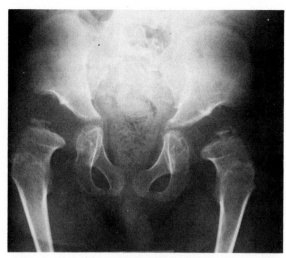

Fig. 4.—Roentgenogram of the pelvis in the same patient (Fig. 1). The capital femoral epiphyses were flattened and irregular. There was a coxa valga deformity. The lateral margin of the acetabulum was hypoplastic, creating in essence a large acetabulum extending to the anterior superior iliac spine. (Courtesy of Dr. David Rimoin.)

expect sooner or later to experience a complication of compression of the cord. Neurologic manifestations may include weakness and difficulty in walking, uselessness of the legs on awakening or spastic paraplegia. Loss of vibratory sensation in the lower extremities may be an early sign, and many patients have hyperactive deep tendon reflexes. Spinal cord compression also may occur at the level of the thoracolumbar gibbus.

Patients with the Morquio disease have increased concentrations of keratosulfate in the urine.[3, 8, 9] Levels of this acid mucopolysaccharide, which does not contain uronic acid, are often two to three times the normal amount. With age the concentrations of this compound in urine decrease in patients with the Morquio disease, as they do in normal children. Metachromatic granules may be seen in the polymorphonuclear leukocytes. Danes and Grossman[10] also have reported metachromasia in the fibroblasts of these patients and in the cells of heterozygous carriers.

DIFFERENTIAL DIAGNOSIS

The Morquio syndrome may be considered a spondyloepiphyseal dysplasia as well as a mucopolysaccharidosis and disorder of connective tissue. Among the other forms of spondyloepiphyseal dysplasia, from which it must be differentiated, is the nonkeratosulfate-excreting Morquio syndrome, which was described by McKusick[4] and Norum[11] in four first cousins in a highly inbred kindred. The skeletal deformities and other symptoms in these patients are similar to

those in the Morquio syndrome but less severe. These patients have platyspondyly, genu valgum, flat feet, pectus carinatum and flat, fragmented femoral heads, but they do not have keratosulfaturia.

Spondyloepiphyseal dysplasia tarda[12] is an X-linked disorder that has been confused with the Morquio syndrome. It was distinguished from it by Maroteaux and colleagues in 1957. It is characterized by a short-trunked dwarfism that appears between 5 and 10 years of age; ultimate height usually is between 52 and 62 inches. Roentgenographic findings include platyspondyly with central humps of dense bone,[12] mild epiphyseal irregularities and flat, fragmented femoral heads. The initial appearance of the spine suggests ochronosis, but then it is apparent that the dense calcification is in the vertebral bodies, not in the intervertebral disks. Patients often complain of back and hip pain. They do not have extraosseous abnormalities.

The Morquio disease also must be distinguished from spondyloepiphyseal dysplasia congenita, a skeletal disorder transmitted as an autosomal dominant (see section on Dwarfism with Short Trunk). Patients with this syndrome are recognizable at birth, whereas those with the Morquio disease are not, and they lack the enamel hypoplasia and corneal clouding. Instead they have severe myopia and retinal detachment. Patients with costovertebral dysplasia (see section on Dwarfism with Short Trunk) may appear at first to be examples of the Morquio syndrome. They are readily distinguished by the roentgenographic appearance of the vertebrae.

GENETICS

The Morquio syndrome is transmitted by an autosomal recessive gene. Multiple involved siblings have been reported to have been born to normal couples on several occasions. Consanguinity between the parents of patients is high.

TREATMENT

Surgical fusion of the cervical spine may be lifesaving in the prevention of spinal cord compression. There is a tendency for the prognosis to be better in females, possibly due to a lesser likelihood of vigorous activity leading to cord compression. Osteotomies may be useful in correction of the genu valgum. The instability of the wrists, which makes working with the hands very difficult, may be aided by the use of wrist splints.

REFERENCES

1. Morquio, L.: Sur une forme de dystrophie osseuse familiale, Arch. Med. Enfants. 32:129, 1929.

2. Brailsford, J. F.: Chondro-osteo-dystrophy: Roentgeno-graphic and clinical features of a child with dislocation of vertebrae, Am. J. Surg. 7:404, 1929.

3. Maroteaux, P., and Lamy, M.: Opacités cornéennes et trouble metabolique dans la maladie de Morquio, Rev. Fr. Etud. Clin. Biol. 6:48, 1961.

4. McKusick, V. A.: The Mucopolysaccharidoses, in *Heritable Disorders of Connective Tissue* (St. Louis: C. V. Mosby Co., 1972), p. 583.

5. Von Noorden, G. K., Zellweger, H., and Ponseti, I. V.: Ocular findings in Morquio-Ullrich's disease, Arch. Ophthalmol. 64:585, 1960.

6. Langer, L. O., Jr., and Carey, L. S.: The roentgenographic features of the KS mucopolysaccharidosis of Morquio (Morquio-Brailsford's disease), Am. J. Roentgenol. Radium Ther. Nucl. Med. 97:1, 1966.

7. Blaw, M. E., and Langer, L. O.: Spinal cord compression in Morquio-Brailsford disease, J. Pediatr. 74:593, 1969.

8. Humbel, R., Marchal, C., and Fall, M.: Diagnosis of Morquio's disease: A simple chromatographic method for the identification of keratosulfate in urine, J. Pediatr. 81:107, 1972.

9. Linker, A., Evans, L. R., and Langer, L. O.: Morquio's disease and mucopolysaccharide excretion, J. Pediatr. 77: 1039, 1970.

10. Danes, B. S., and Grossman, H.: Bone dysplasias, including Morquio's syndrome, studied in skin fibroblast cultures, Am. J. Med. 47:708, 1969.

11. Norum, R. A.: Nonkeratosulfate-excreting Morquio's Syndrome in Four Members of an Inbred Group, in *Skeletal Dysplasias, Clinical Delineation of Birth Defects*, Part IV (New York: The National Foundation-March of Dimes, 1969), p. 334.

12. Langer, L. O.: Spondyloepiphyseal dysplasia tarda. Hereditary chondrodysplasia with characteristic vertebral configuration in the adult, Radiology 82:833, 1964.

Abnormalities of Lipid Metabolism

SANDHOFF DISEASE
GM$_2$ Gangliosidosis Type II

Cardinal Clinical Features

Progressive cerebral deterioration starting after age 6 months, blindness, cherry red macular spots, hyperacusis, accumulation of GM$_2$ ganglioside, deficiency of hexosaminidase A and B.

Clinical Picture

The Sandhoff disease is very similar to Tay-Sachs disease (see chapter, this section). The two conditions were distinguished by Sandhoff et al.[1] in 1968 in the study of a patient who was unusual in that he stored ganglioside not only in the brain but also in other viscera. They furthermore found that, in contrast to patients with Tay-Sachs disease, the activity of total hexosaminidase was deficient.[1]

It has not been possible to distinguish the Sandhoff disease from Tay-Sachs disease clinically.[2] It may be suspected in non-Jewish patients with the Tay-Sachs phenotype. Individual patients appear normal at birth and appear to develop normally (Fig. 1) until 6–9 months of age, when signs of motor weakness begin to become evident. Abilities that have been learned are progressively lost. These might include the ability to grasp objects or to sit, crawl or hold up the head. Patients never learn to walk. Many of these infants have doll-like faces with long eyelashes and fine hair, pale translucent skin and pink coloring. Cherry red macular spots are seen bilaterally. Blindness is progressive, and optic atrophy develops. The patients develop hyperacusis, or an exaggerated startle response to noise, which may be seen even quite early. The size of the head increases abnormally, producing macrocephaly. Seizures are commonly encountered and may be generalized or myoclonic. Patients are spastic and weak. Mental deterioration continues (Fig. 2) until the patient is rigid, decerebrate and completely blind. Such a patient must be tube fed. Death occurs usually between the ages of 1 and 3 years, usually from bronchopneumonia or aspiration.

At autopsy the visceral organs may be somewhat heavier than those of patients with classic Tay-Sachs disease.[2] Some patients may have clinical hepatosplenomegaly, but most do not. Renal tubular cells show lipid deposits, unlike those of Tay-Sachs disease,[3] and there may be occasional lipid-laden histiocytes in viscera. In the brain there is a typical neuronal lipidosis and the membranous cytoplasmic bodies characteristic of Tay-Sachs disease are seen in the electron microscope.

Patients with the Sandhoff disease accumulate the GM$_2$ ganglioside in the brain.[1] The amounts found are 100–300 times the normal concentrations[3] and quite similar to those of Tay-Sachs disease. In contrast to patients with Tay-Sachs disease, these patients also accumulate the asialo derivative of GM$_2$ and globoside in the brain and in other tissues, especially the liver, kidney and spleen.[4,5] Globoside also may be demonstrated in urinary sediments and plasma.[6] These compounds are all structurally related (Fig. 3). The asialo derivative differs from GM$_2$ in the absence of the N-acetylneuraminic side chain, whereas the globoside contains an extra galactose moiety. Asialo GM$_2$ may be present in the brain in amounts 100 times normal.[4,5]

The Sandhoff disease is characterized by the lack not only of hexosaminidase A, as in Tay-Sachs,

Fig. 1.—A patient with the Sandhoff disease at 1 year of age. At this time he looked normal. (Courtesy of Dr. John S. O'Brien of the University of California San Diego.)

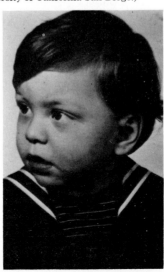

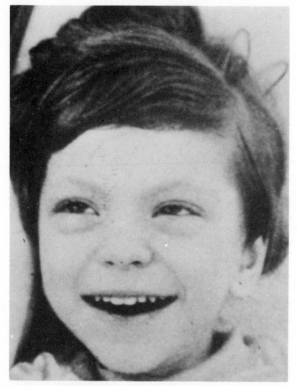

1–3% of normal.[5] Hexosaminidase A is a glycoprotein, which can be converted in the presence of neuraminidase to hexosaminidase B. Thus, it is thought that the two isozymes have the same protein molecule, and that carbohydrate side chains are added to produce the A enzyme. In the Sandhoff disease the residual hexosaminidase in liver has been shown to have an increased K_m and pH optimum, indicating that there is a structural gene alteration.[7] It has been postulated that the Sandhoff disease might result from a mutation in the structural gene that produces hexosaminidase B, and Tay-Sachs from a defect in an asialotransferase responsible for converting the B to the A form.[9] Hybridization of fibroblasts from a patient with Tay-Sachs disease with those of a patient with the Sandhoff disease revealed a complementation in which hexosaminidase, present in neither parental strain, appeared.[10]

Five disorders of ganglioside storage are now delineated.[3, 4] A classification and some of their features is shown in Table 1. All are progressive cerebral degenerative diseases. The cherry red macular spot is a prominent feature in many of these disorders. All are fatal. All are autosomal recessive. Neuronal lipidosis is a common histologic feature and results from the storage of ganglioside. In three conditions this is GM_2, and in two, GM_1. Deficiency of specific lysosomal hydrolase enzymes provides in each a molecular explanation for the disease.

Fig. 2.—At 3 years of age appearance of patient shown in Figure 1 indicated cerebral degeneration. He still focused and smiled but appeared weak, and the gaze was vacuous. (Courtesy of Dr. John S. O'Brien of the University of California San Diego.)

but of hexosaminidase B as well.[2, 5, 7, 8] The enzyme defect has been demonstrated with natural substrate as well as with the artificial nitrophenylacetylglucosaminide.[1, 2] The activity of each enzyme is about

CASE REPORT*

The patient was seen at 1 year of age at Los Angeles Children's Hospital because of slow motor development. She was the first born of healthy, young Mexican-American parents.

*We are indebted to Dr. John O'Brien of the University of California San Diego for details of this patient. (Reprinted with permission from Pediatr. Res.[5])

Fig. 3.—Structures of the gangliosides that accumulate in the Sandhoff disease.

In Brain	Ceramide	NANA | -Glu-Gal-GalNAc	GM_2
	Ceramide	-Glu-Gal-GalNAc	Asialo GM_2 (GA_2)
In Viscera	Ceramide	-Glu-Gal-Gal-GalNAc	Globoside GL_4

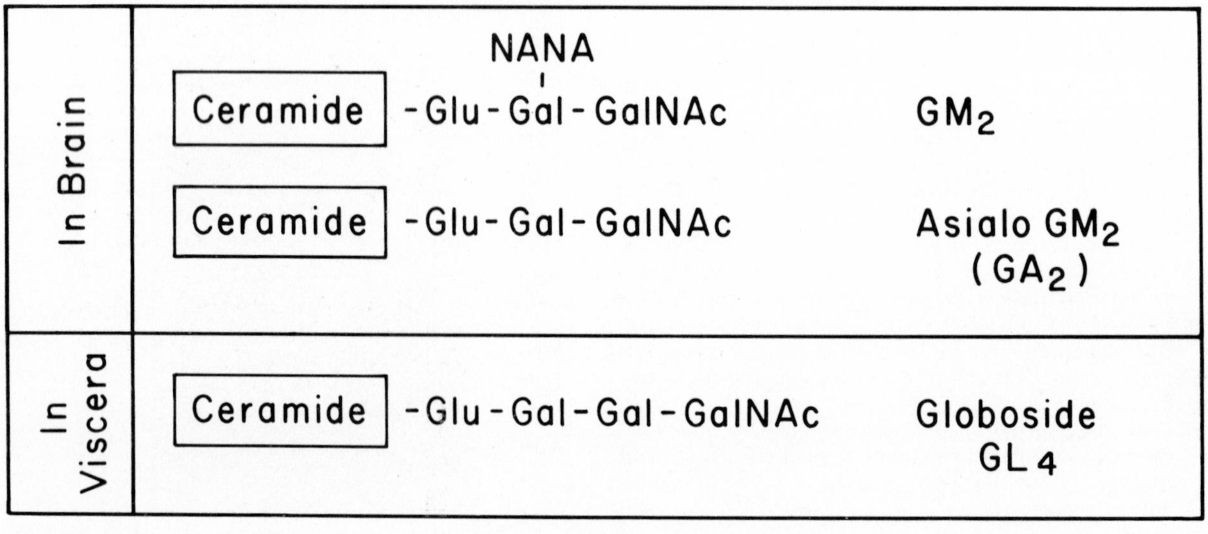

TABLE 1.—GANGLIOSIDE-STORAGE DISEASES

DISORDER	CHEMICAL CLASSIFICATION	GANGLIOSIDE STORED IN BRAIN	AGE AT ONSET	AGE AT DEATH (YR)	ENZYME DEFECT	CARRIER DETECTION	PRENATAL DIAGNOSIS
Tay-Sachs disease	GM_2-gangliosidosis, type I	GM_2	3–6 mo	2–4	Hexosaminidase A	+	Established
Sandhoff disease	GM_2-gangliosidosis, type II	GM_2 asialo-GM_2	3–6 mo	2–4	Hexosaminidase A & B	+	Possible
Juvenile GM_2 gangliosidosis	GM_2-gangliosidosis, type III	GM_2	2–6 yr	5–15	Partial deficiency hexosaminidase A	+	Possible
Generalized gangliosidosis	GM_1-gangliosidosis, type I	GM_1	Birth	$1/2$–2	β-galactosidase	+	Possible
Juvenile GM_1 gangliosidosis	GM_1-gangliosidosis, type II	GM_1	$1/2$–2 yr	3–10	β-galactosidase	+	Possible

She appeared well and developed normally for 6 months, after which she began to have difficulty holding objects and to have a weak grasp. A marked extensor response to sound was first noted at about 6 months. By 12 months she could not sit or crawl. There was a head lag on elevating the trunk. The optic discs were pale, and in each macula there was a dark red spot surrounded by a circular, starkly pale halo. There was no hepatosplenomegaly. She was very hypotonic, and deep tendon reflexes were hypoactive. After 1 year she deteriorated rapidly. Sucking and swallowing became so inefficient that she was fed by gavage. She began having four to five generalized seizures a day. The head enlarged progressively, and on x-ray there was widening and blurring of the sutures. By 18 months she was blind and decerebrate, completely unresponsive. At 25 months she developed lobar pneumonia and died. At autopsy, vacuolated histiocytes were found in many viscera. The neurons were swollen and contained PAS-positive material. GM_2 and its asialo derivative were found in large amounts in the brain and globoside in the liver, spleen and kidney. Hexosaminidase A and B activities were reduced to 1–3% of normal.

GENETICS

The Sandhoff disease is a rare genetic disorder that is thought to be transmitted as an autosomal recessive trait. All patients so far reported have been non-Jewish. Detection of the carrier is possible by enzyme assay, as the heterozygote has amounts of hexosaminidase A and B in leukocytes, skin,[3] cultured fibroblasts and serum[4] that are intermediate between normal and patient concentrations. Serum assay makes screening for heterozygosity convenient. Intrauterine diagnosis following amniocentesis is feasible. In one pregnancy at risk monitored in this way, the fetus was found to have normal levels of hexosaminidase A and B and so did the normal baby at birth.[3]

The fact that Tay-Sachs and Sandhoff diseases cannot be distinguished clinically raises the question that some of the non-Jewish patients recorded as having Tay-Sachs disease might indeed have had the Sandhoff disease. This might alter available estimations of gene frequencies of Tay-Sachs disease among non-Jews. The differentiation between the two diseases is now readily accomplished by enzyme assay. About half of the non-Jewish patients so far studied have turned out to have the Sandhoff disease.[5]

TREATMENT

None.

REFERENCES

1. Sandhoff, K., Andreae, U., and Jatzkewitz, H.: Deficient hexosaminidase activity in an exceptional case of Tay-Sachs disease with additional storage of kidney globoside in visceral organs, Life Sci. 7:283, 1968.
2. Sandhoff, K., Harzer, K., Wässle, W., and Jatzkewitz, H.: Enzyme alterations and lipid storage in three variants of Tay-Sachs disease, J. Neurochem. 18:2469, 1971.
3. O'Brien, J. S.: Ganglioside-storage diseases, N. Engl. J. Med. 284:893, 1971.
4. O'Brien, J. S., Okada, S., Ho, M. W., Fillerup, D. L., Veath, L., and Adams, K.: Ganglioside storage diseases, Fed. Proc. 30: 956, 1971.
5. Okada, S., McCrea, M., and O'Brien, J. S.: Sandhoff's disease (GM_2 gangliosidosis type 2): Clinical, chemical, and enzyme studies in five patients, Pediatr. Res. 6:606, 1972.
6. Krivit, W., Desnick, R. J., Lee, J., Moller, J., Wright, F., Sweeley, C. C., Snyder, P. D., and Sharp, H. L.: Generalized accumulation of neutral glycosphingolipids with GM_2 ganglioside accumulation in the brain, Am. J. Med. 52:763, 1972.
7. Tateson, R., and Bain, A. D.: GM_2 gangliosidoses: Consideration of the genetic defects, Lancet 2:612, 1971.
8. Suzuki, Y., Jacob, J. C., Suzuki, K., and Kutty, K. M.: GM_2 gangliosidosis with total hexosaminidase deficiency, Neurology 21:313, 1971.
9. Snyder, P. D., Krivit, W., and Sweeley, C. C.: Generalized accumulation of neutral glycosphingolipids with GM_2 ganglioside accumulation in the brain, J. Lipid Res. 13:128, 1972.

TAY-SACHS DISEASE
Amaurotic Familial Idiocy, GM$_2$ Gangliosidosis Type I, Hexosaminidase A Deficiency

CARDINAL CLINICAL FEATURES

Infantile degeneration of the brain and retina, cherry red macular spot, hyperacusis, macrocrania, storage of the GM$_2$ ganglioside in the brain, deficiency of hexosaminidase A.

CLINICAL PICTURE

Patients with Tay-Sachs disease appear normal at birth. They usually continue to look alert and healthy until about 6 months of age. The onset may be between birth and 10 months of age. The earliest clinical manifestation is an exaggerated extensor startle response to sound. This usually is present by 1 month of age, but it may not be appreciated early, as it can be seen in some normal babies, usually disappearing by about 4 months. In contrast, in the baby with Tay-Sachs disease this hyperacusis becomes more prominent. It is brought on even by very gentle sound stimuli. It may be followed by clonus.

Parents often notice weakness as the first clinical sign. The infant may begin to sit less well or to develop poor head control. Physical examination reveals hypotonia. This is progressive, and by 1 year of age few of these patients can sit without support. By 8 months of age the baby may look sleepy, or less alert. There may be nystagmus and a fixed or roving gaze. Examination of the fundus reveals the typical cherry red spot in the macula (Fig. 1). This usually is present as early as 2 months of age. In looking for this, it is important to remember that the white degeneration of the macula is bigger and more impressive than the red spot in the middle. It looks very much like a fried egg.

Cerebral and macular degenerations are progressive. The infant becomes blind and decerebrate and usually must be fed by tube. Muscle tone is increased and there is hyperreflexia. Convulsions and myoclonic jerks are common. These patients generally die at about 2 years of age, usually of pneumonia. They often have a clear, translucent, doll-like skin. After about 15 months the head size usually enlarges. The brain weight at the time of death may be 50% greater than normal.

Pathologic changes are restricted to the nervous system, where the neurons are swollen or "ballooned." Electron microscopy of these cells reveals lamellar membranous bodies. These cytoplasmic inclusions are round and consist of concentric layers of accumulated ganglioside.

The ganglioside stored in Tay-Sachs disease is a glycosphingolipid with a terminal hexosamine, N-acetylgalactosamine (Fig. 2). The sphingolipids all contain the long-chain base sphingosine, which has the following structure:

$$CH_3 \, (CH_2)_{12} \, CH = CH \; CH \; CH \; CH_2OH$$
$$\qquad\qquad\qquad\quad\; | \quad\; |$$
$$\qquad\qquad\qquad\quad OH \; NH_2$$

This compound is acetylated with long-chain fatty acids on the amino group on carbon 2 to form ceramide, which makes up the base unit of all of the sphingolipids. In the gangliosides, as well as in the cerebrosides and other glycolipids, a sugar is linked glycosidically to carbon 1. In the parent ganglioside, GM$_1$, the glycolipid that accumulates in generalized gangliosidosis, ceramide is linked successively to glucose; galactose, to which N-acetylneuraminic acid is attached; N-acetylgalactosamine; and galactose. This terminal galactose is cleaved by a β-galactosidase, which is defective in generalized gangliosidosis, to yield the Tay-Sachs lipid, GM$_2$.[2] This is normally converted to GM$_3$ by cleavage of the terminal N-acetylgalactosamine.

It was logical to consider that the problem was a defect in hexosaminidase. However, hexosaminidase activities were elevated rather than depressed in the tissues of patients with Tay-Sachs disease. This problem was resolved by the demonstrations by Okada and O'Brien[3] (1) that hexosaminidase could be separated by starch gel electrophoresis into two components, designated A and B, and (2) that hexosaminidase A was absent in patients with Tay-Sachs disease (Fig. 3). The enzyme can be demonstrated in all tissues of the body including the plasma and in fibroblasts and amniotic fluid cells in culture. Its activity is absent in all of these in Tay-Sachs disease.

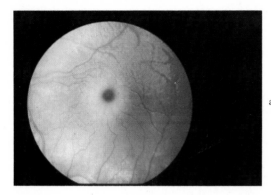

Fig. 1. — The cherry red spot. Photograph of the fundus of an infant with Tay-Sachs disease.

Fig. 2. — Metabolic pathways of glycosphingolipid metabolism. The site of the defects in a number of conditions is illustrated, as well as that of Tay-Sachs disease. Abbreviations used are *Cer*, ceramide; *Glu*, glucose; *Gal*, galactose; *NANA*, N-acetyl-neuraminic acid; and *GalNAc*, N-acetylgalactosamine.

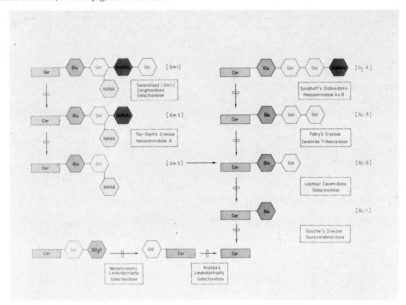

Fig. 3. (Left) — Hexosaminidase *A* and *B*, as separated by starch gel electrophoresis. The violet areas of activity resulted from incubation with the naphthol AS-BI derivatives of β-D-N-acetylglucosamide. The enzymes shown were from liver. The picture from blood and other organs is identical. The designations used are *type I*, Tay-Sachs; *II*, Sandhoff; and *III*, juvenile GM₂ gangliosidosis. or GM₂ gangliosidosis type III. (Courtesy of Dr. John S. O'Brien of the University of California San Diego.)

Fig. 4. (Right) — A. S. (Case Report). Patient at 2 years of age. He was blind and decerebrate at this time. Hexosaminidase A activity was absent.

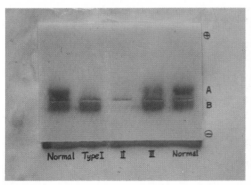

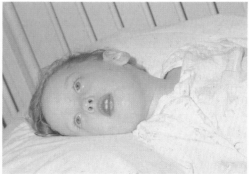

CASE REPORT

A. S., a male, was the first born child of healthy, unrelated Jewish parents. All four grandparents had come from Eastern Europe. There was no history of Tay-Sachs disease in the family.

The patient weighed 4 lb, 10 oz and appeared normal at birth. He did well during the first few months, but he never sat without support. He turned over, reached for bright objects, grasped them and transferred. The parents noted the onset of visible motor weakness at about 6 months. By 9 months he was generally weak and appeared to be growing weaker. Examination revealed bilateral cherry red macular spots.

Vision diminished progressively after 1 year of age. During this period he went progressively downhill. Seizures began at 16 months. Spasticity and generalized paralysis were noted at that time. Difficulty in swallowing was progressive. The child had to be fed by gavage. He had had numerous episodes of pneumonia.

Examination at 2 years of age revealed a picture of decerebate rigidity (Fig. 4). He had long eyelashes, red hair and clear, translucent skin. He had a wandering, sightless gaze. He lay in a tonic neck reflex position, unresponsive and out of contact. He had marked hyperacusis, having an exaggerated startle response to quiet speech anywhere in the room. The head measured 53 cm, evidence of definite macrocephaly. The deep tendon reflexes were hyperactive. There was ankle clonus and a positive Babinski response bilaterally. Cherry red spots were readily visualized in both macular areas (see Fig. 1). Electroencephalography revealed high voltage seizure activity and an absence of driving on photic stimulation. Assay of the plasma revealed no activity of hexosaminidase A.

GENETICS

Tay-Sachs disease is carried on an autosomal recessive gene. Heterozygous carriers of the gene have intermediate activities of hexosaminidase A in their plasma.[4] These values average 65% of normal. The gene is common in Ashkenazi Jews. Its frequency in that population has been calculated to be one in 30. This is sufficiently high that it is practical to undertake programs of heterozyote detection in entire popula-

tions of Ashkenazi Jews. Such a program could virtually eliminate this disease.

Tay-Sachs disease can be detected prenatally.[5] The enzyme is measurable directly in amniotic fluid and in uncultured amniotic cells. The most reliable assay is of amniotic fluid cells in culture. Thirty pregnancies at risk have now been monitored in Dr. O'Brien's laboratory. Eight have been diagnosed prenatally as having Tay-Sachs disease. In seven this was early enough, therapeutic abortion was carried out and the diagnosis was confirmed in the fetus. The other was diagnosed too late in pregnancy; the baby has Tay-Sachs disease. Of those at risk that were diagnosed as normal, 14 have now been carried to term and the babies are normal. The rest have not yet been delivered.

The combination of heterozygote detection and prenatal diagnosis provides a powerful approach to the genetic control of this disease.

TREATMENT

There is no treatment. Skilled supportive care may provide real support for the family as well as prolonging life for the patient.

REFERENCES

1. Volk, B. W. (ed.): *Tay-Sachs Disease* (New York: Grune & Stratton, 1964).
2. Svennerholm, L.: Chemical structure of normal human brain and Tay-Sachs gangliosides, Biochem. Biophys. Res. Commun. 9:436, 1962.
3. Okada, S., and O'Brien, J. S.: Tay-Sachs disease: Generalized absence of a beta-D-N-acetylhexosaminidase component, Science 165:698, 1969.
4. O'Brien, J. S., Okada, S., Chen, A., and Fillerup, D. L.: Tay-Sachs disease, detection of heterozygotes and homozygotes by serum hexosaminidase assay, N. Engl. J. Med. 283:15, 1970.
5. O'Brien, J. S., Okada, S., Fillerup, D. L., Veath, M. L., Adornato, B., Brenner, P. H., and Leroy, J. G.: Tay-Sachs disease: Prenatal diagnosis, Science 172:61, 1971.

JUVENILE GM₂ GANGLIOSIDOSIS
Late Infantile Amaurotic Idiocy

CARDINAL CLINICAL FEATURES

Progressive intellectual and psychomotor degeneration, seizures, ataxia, accumulation of the GM_2 ganglioside, partial deficiency of hexosaminidase A.

CLINICAL PICTURE

Two patients with this disease were described in 1968 by Bernheimer and Seitelberger,[1] who demonstrated by thin layer chromatography that the material accumulated in the central nervous system was the GM_2 ganglioside. Patients are normal at birth. The onset of symptoms is between 2 and 6 years of age. All patients described have been non-Jewish, in sharp contrast to those with Tay-Sachs disease.

Patients develop progressive ataxia, with striking dysarthria and impaired gait. Spasticity is prominent. They also have tonic and clonic seizures as well as myoclonic seizures. Intellectual deterioration is progressive. Ultimately, patients lose the ability to stand, walk or crawl. They become unable to speak and are very spastic. The deep tendon reflexes are hyperactive and there is ankle clonus. Flexion contractures develop around the joints. Blindness is a late feature. Shortly before death, usually caused by aspiration pneumonia, patients become unresponsive to visual and auditory stimuli. There are no hepatosplenomegaly, megalencephaly, radiologic bone changes or cherry red spots. The bone marrow is normal.

Brain biopsy reveals the appearance of a neurolipidosis. Electron microscopy shows three types of cytoplasmic inclusion: (1) lamellated neuronal inclusions resembling the cytoplasmic inclusions seen in Tay-Sachs disease; (2) inclusions similar to the zebra bodies seen in mucopolysaccharidosis; and (3) inclusions typical of lipofuscin. Chemical analysis indicates that it is GM_2 that accumulates, and that accumulation is to a lesser degree than that seen in Tay-Sachs disease. GM_2 is N-acetylgalactosaminyl-(1→4)-[(2→3)-N-acetylneuraminyl]-galactosyl-(1→4)-glucosyl-(1→1)-[2-N-acyl]-sphingosine (see chapter on Tay-Sachs Disease). Levels of this ganglioside are 100–300 times normal in Tay-Sachs disease and 40–90 times normal in this disease.[2] Enzyme assay reveals that there is a partial deficiency of hexosaminidase A.[2] The deficiency has been documented in serum, fibroblasts, liver, spleen and brain.

DIFFERENTIAL DIAGNOSIS

Tay-Sachs disease is the better known variant of hexosaminidase A deficiency (see chapter, this section). This infantile form of GM_2 gangliosidosis is an autosomal recessive disorder in which motor development is normal until about 5 months of age, after which the patient develops progressive cerebral degeneration. Macrocephaly, cherry red macular spots, optic atrophy and blindness are characteristic. Death occurs between 2 and 4 years of age. In Tay-Sachs disease hexosaminidase A activity is absent. The Sandhoff disease (see chapter, this section) is the third disorder in which hexosaminidase A is deficient. In these patients both hexosaminidase A and B are absent. In the Sandhoff disease GM_2 ganglioside also accumulates. The clinical picture is quite similar to that of Tay-Sachs disease.

CASE REPORT*

K. L. was the first born infant of healthy, unrelated young parents of mixed non-Jewish European stock. She was the product of a normal full-term pregnancy and delivery. The birth weight was 7 lb, 6 oz. At 2½ years of age she was toilet trained; she could speak entire sentences and memorize nursery rhymes. She appeared perfectly normal and was in good health until the age of 34 months when she had a seizure (Fig. 1). Minor motor in type, the seizures increased in frequency up to 1–2 per day, and her speech became repetitive. She would repeat the same words or phrases over and over again. At 40 months she lost the ability to walk, apparently as a consequence of weakness in coordination and problems with balance. Mental and motor retardation were progressive. She was seen at 3⁹/₁₂ years by Dr. John Menkes.[3] Physical examination at that time revealed an alert, chatty girl with dysarthria. There was no hepatosplenomegaly, lymphadenopathy, bony changes or macrocephaly. She was spastic and ataxic (Fig. 2), and her feet toed in. Muscle tone was increased, deep tendon reflexes were hyperactive, and there was bilateral ankle clonus. Several months later she lost the ability to speak. By 5 years of age, her mental and motor retardation had progressed. She was mute and unable to walk, stand or crawl. She still recognized her parents, but this ability, too, was slowly being lost. She had an increased reaction to startle that her parents thought had been present for some time. She had spas-

---

*Case Report kindly provided by Dr. John S. O'Brien of the University of California San Diego and reprinted with permission from J. Pediatr.[2] and Arch Neurol.[3]

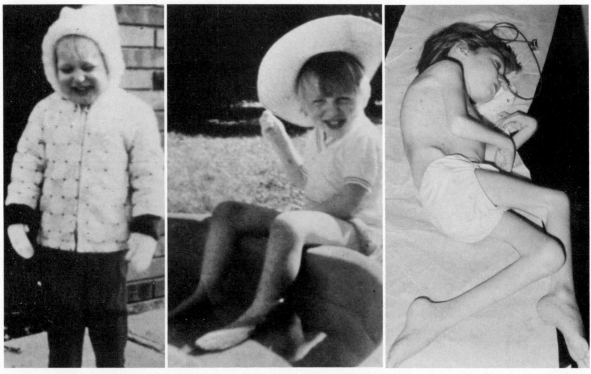

Fig. 1 (left). – K. L. (Case Report). A 3-year-old with juvenile GM$_2$ gangliosidosis. She had developed ataxia but appeared normal.

Fig. 2 (center). – K. L. At 4 years of age she was spastic and was beginning to go downhill.

Fig. 3 (right). – K. L. At 6 years there was virtually complete cerebral deterioration. She was rigid and had to be fed by tube. (Figures 1–3 courtesy of Dr. John S. O'Brien of the University of California San Diego.)

Fig. 4. – Histopathology of the brain showing ballooning of neurons. (Courtesy of Dr. John S. O'Brien of the University of California San Diego and reprinted with permission from Arch. Neurol.[3])

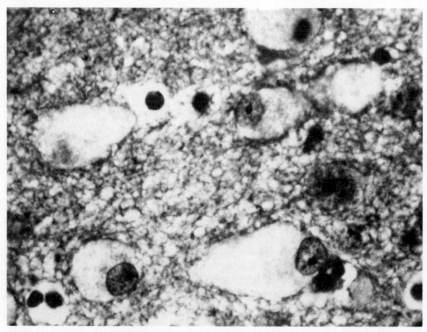

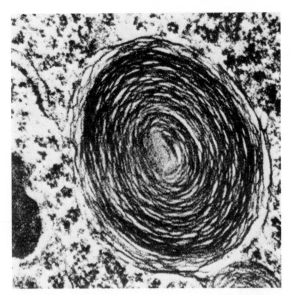

Fig. 5.—Electron microscopy of the brain showing intraneuronal lamellated inclusions resembling those found in Tay-Sachs disease. (Courtesy of Dr. John S. O'Brien of the University of California San Diego and reprinted with permission from Arch. Neurol.[3])

Fig. 6.—Cytoplasmic inclusions similar to the zebra bodies seen in mucopolysaccharidoses. (Courtesy of Dr. John S. O'Brien of the University of California San Diego and reprinted with permission from Arch. Neurol.[3])

tic paraplegia. There was an intention tremor bilaterally and some truncal titubation. Vision and hearing remained normal. Funduscopy revealed normal discs and retinas. She continued to deteriorate intellectually. By 6 years of age she was unresponsive and had contractures (Fig. 3). She died a few years later.

A brain biopsy at 5 years of age and autopsy revealed neuronal lipidosis and a massive accumulation of GM$_2$ ganglio-

side. This amounted, at age 5, to 40% of the total ganglioside. On microscopic examination the neurons of the cortex were enlarged and swollen (Fig. 4). By electron microscopy almost every neuron contained abnormal cytoplasmic inclusions (Figs. 5 and 6). Hexosaminidase A in fibroblasts was 12% of the total hexosaminidase activity, whereas in controls it averaged 49% and in patients with Tay-Sachs disease, 3.4%.

A younger brother and sister were well. Hexosaminidase A activity in the sister was normal (41%). The brother and both parents were shown to carry the gene. Their activities were 29%, 26% and 25%, respectively.

GENETICS

Juvenile GM$_2$ gangliosidosis is a genetic disorder transmitted as an autosomal recessive. Assay of the activity of hexosaminidase A in the serum or in fibroblasts cultured from skin biopsy can be used to detect the heterozygote. Values are intermediate between those of affected patients and control individuals. Prenatal diagnosis is possible but has not yet been reported.

The occurrence of related genes in different populations is interesting. This disease has been reported only in non-Jews, whereas the Tay-Sachs gene is found in one of 40 Ashkenazi Jews.

TREATMENT

None. The disease is always progressive to a fatal outcome.

REFERENCES

1. Bernheimer, H., and Seitelberger, F.: Über das Verhalten der Ganglioside im Gehirn bei 2 Fallen von spätinfantiler amaurotischer Idiotie, Wien. Klin. Wochenschr. 80:163, 1968.
2. Okada, S., Veath, M. L., and O'Brien, J. S.: Juvenile GM$_2$ gangliosidosis: Partial deficiency of hexosaminidase A, J. Pediatr. 77:1063, 1970.
3. Menkes, J. H., O'Brien, J. S., Okada, S., Grippo, J., Andrews, J. M., and Cancilla, P. A.: Juvenile GM$_2$ gangliosidosis biochemical and ultrastructural studies on a new variant of Tay Sachs disease, Arch. Neurol. 25:14, 1971.
4. Okada, S., Veath, M. L., LeRoy, J., and O'Brien, J. S.: Ganglioside GM$_2$ storage disease: Hexosaminidase deficiencies in cultured fibroblasts, Am. J. Hum. Genet. 23:55, 1971.
5. Suzuki, Y., and Suzuki, K.: Partial deficiency of hexosaminidase component A in juvenile GM$_2$ gangliosidosis, Neurology 20:848, 1970.
6. Volk, B. W., Adachi, M., Schneck, L., Saifer, A., and Kleinberg, W.: G$_5$ ganglioside variant of systemic late infantile lipidosis. Generalized gangliosidosis, Arch. Pathol. 87:393, 1969.
7. Okada, S., and O'Brien, J. S.: Tay Sachs disease: Generalized absence of a beta-D-N-acetylhexosaminidase component, Science 165:698, 1969.
8. Okada, S., McCrea, M., and O'Brien, J. S.: Sandhoff's disease (GM$_2$ gangliosidosis type 2): Clinical, chemical, and enzyme studies in five patients, Pediatr. Res. 6:606, 1972.

GENERALIZED GANGLIOSIDOSIS
GM$_1$ Gangliosidosis

CARDINAL CLINICAL FEATURES

Type I: Severe cerebral degeneration; coarse facial features and extreme Hurler-like skeletal deformities; hepatosplenomegaly; cherry red macular spot; accumulation of the ganglioside GM$_1$ in the brain and viscera and of mucopolysaccharides in viscera; deficiency of β-galactosidase.

CLINICAL PICTURE

Type I

Generalized gangliosidosis is a storage disease in which the GM$_1$ ganglioside accumulates in the brain and viscera.[1] The resultant cerebral degenerative disease is a devastating one, and involved patients usually die before 2 years of age.

The disease differs from most of the storage diseases in that abnormalities are present from birth. Psychomotor retardation is evident early in poor sucking and appetite and a failure to thrive. The infant is hypotonic and hypoactive and may appear dull. The facial features may be coarse, even very early. The expression is dull. There is frontal bossing, downy hirsutism over the forehead, a depressed nasal bridge, large low-set ears and an increased distance between the nose and upper lip (Figs. 1 and 2). Patients have hypertrophy of the gums and a large tongue.[1-3] A cherry red spot is visible bilaterally in about half of these patients.[4] Nystagmus may be present and the cornea may show mild clouding. Facial edema and pitting edema of the extremities are regular findings. Hepatomegaly is prominent and the spleen may be palpable.

By 8 months of age the infant may be able to hold up his head, but he cannot sit or crawl. He may follow objects with his eyes and even reach for them, but his grasp is poor. Movements are uncoordinated. The infant rarely smiles and appears uninterested in his environment. His cry is weak. Parents may say he is a "good baby," meaning that he sleeps a lot and is immobile much of the day. Hepatomegaly is uniformly found after 6 months, and splenomegaly in 80% of patients. Coarse features may be accentuated. Macrocephaly may develop, but it is less frequent and less prominent than in Tay-Sachs disease. Reflexes become hyperactive, and hyperacusis may develop. Muscle weakness progresses, and there may be a head lag on elevating the shoulders. After the first year, deterioration is rapid. Convulsions may develop. Swallowing is so poor that tube feeding is required. Recurrent pneumonias complicate the course. By 16 months the patient is blind and deaf and has decerebrate rigidity. Flexion contractures develop, and there is no response to stimuli. Death by 2 years of age usually follows pneumonia.

The skeletal manifestations of this disease also are progressive throughout the life of the patient. The fingers are short and stubby. The joints are stiff and limited in motion. There are flexion contractures of the fingers, especially the fifth. The wrists and ankles are enlarged. The dorsolumbar kyphosis may be a prominent gibbus. Flexion contractures also are seen at the elbows and knees.

Roentgenographic findings are those of an early and very severe dysostosis multiplex. These changes are similar to those of I cell disease and more severe than those of the Hurler syndrome.[1, 2, 5] The long tubular bones are shortened and widened in midshaft, tapering distally and proximally (Fig. 3). Subperiosteal new bone formation is characteristic of this syndrome. The consequent cloaking of the already widened bones, particularly evident in the humerus, and the pinching-off of the end of the bones are striking features. The distal ends of the radius and ulna tilt obliquely toward each other. All middle and proximal phalanges are widened. The metacarpals are short and broad and taper proximally. The carpal centers are hypoplastic. The lumbar vertebrae are hypoplastic and beaked anteriorly at the site of the kyphosis (Fig. 4). The ribs are thickened and spatula-like. The ilia are flared. The sella turcica is shallow and elongated, giving it a shoe-shaped appearance.

Neuronal lipidosis is prominent on histologic examination of the brain. The ballooned neurons look exactly like those of Tay-Sachs disease. Indeed, electron microscopy shows identical lamellar cytoplasmic inclusion bodies in the neurons. These membranous cytoplasmic bodies also may be seen in retinal ganglion cells.[4] In viscera such as the liver there are histiocytosis and vacuoles in the hepatocytes and histiocytes

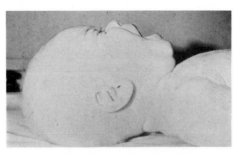

Fig. 2.–C. D. Lateral view. The coarse features are more clearly seen, as are the low-set ear and depressed nasal bridge. The prominent maxillary area is characteristic. (Courtesy of Dr. John S. O'Brien of the University of California San Diego.)

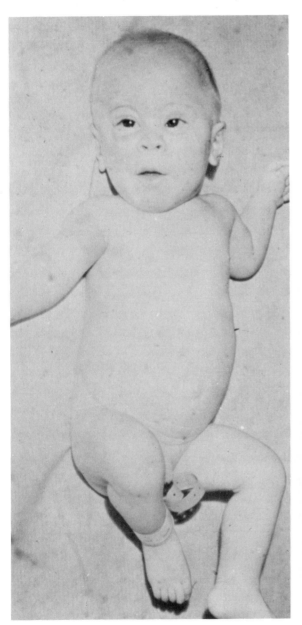

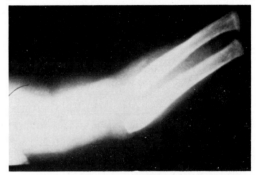

Fig. 3.–C. D. Roentgenogram of the arm at 7 months. The picture was that of dysostosis multiplex. The bones were short and had prominent midshaft thickening. The distal ends of the radius and ulna had begun to point toward each other. The distal end of the humerus showed the characteristic pinched-off appearance. The changes in the bones in this syndrome are just like those of the Hurler syndrome, but in this disease they are prominent much earlier in life. (Courtesy of Dr. John S. O'Brien of the University of California San Diego.)

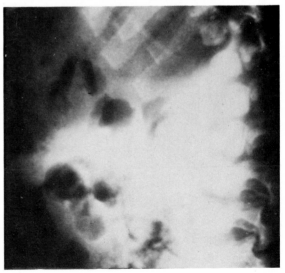

Fig. 1.–C. D. (Case Report). A 4-month-old with generalized GM₁ gangliosidosis. Features were coarse. There was frontal bossing, a depressed nasal bridge and low-set ears. (Courtesy of Dr. John S. O'Brien of the University of California San Diego.)

Fig. 4.–C. D. Roentgenogram of the spine at 7 months. The vertebral bodies were hypoplastic. There was anterior beaking of L1 and L2. (Courtesy of Dr. John S. O'Brien of the University of California San Diego.)

that stain with PAS. A characteristic lesion is the cytoplasmic vacuolation and ballooning of the renal glomerular epithelial cells[2] seen only in this disorder and in the Fabry disease. There are vacuolated lymphocytes in the peripheral blood and foamy histiocytes in the bone marrow (Fig. 5). Mucopolysacchariduria usually has not been detected, but keratan sulfate has been reported in the urine in excess in one patient.[6]

Figure 6 shows the thin layer chromatographic patterns of the gangliosides stored in the nervous system in five gangliosidoses. The ganglioside that accumulates in the brain and in the viscera in generalized gangliosidosis is the GM_1 ganglioside[1, 3, 7] (see Fig. 6). Its structure is galactosyl-(1→3)-N-acetylgalactosaminyl-(1→4)-[(2→3)-N-acetylneuraminyl]-galactosyl-(1→4)-glucosyl-(1→1)-[2-N-acyl]-sphingosine. Thus, it is a ceramide glycoside, a ceramide trihexoside with an N-acetylneuraminic acid side chain. It differs from the Tay-Sachs, or GM_2, ganglioside in the presence of a terminal galactose on GM_1. This ganglioside is present in 10 times normal abundance in the brain of patients with generalized gangliosidosis. Mucopolysaccharides also are stored in peripheral tissues.[7] The magnitude of the storage is similar to that in the Hurler disease but the mucopolysaccharides stored differ; they more closely resemble keratan sulfate. It appears reasonable that the bony abnormalities are due to the mucopolysaccharide storage and that cerebral degeneration is a consequence of the storage of ganglioside in the brain.

The enzyme defect is an absence of the lysosomal β-galactosidase.[8] The activity of this enzyme is 2–10% of normal in tissues of patients with this disease. It may be demonstrated using artificial substrate such as ρ-nitrophenylgalactoside. It also may be done using isolated GM_1 ganglioside. This enzyme normally splits the terminal galactose from this molecule, converting it to GM_2. It also normally cleaves the mucopolysaccharides that accumulate in the viscera of patients with this disease, releasing galactose.[1, 9] This galactosidase activity is of course also deficient in patients with the disease. Thus, the storage of both ganglioside and mucopolysaccharide may be seen as a direct consequence of the deficiency of the galactosidase.

Detection of the enzyme defect may be accomplished using tissues obtained at autopsy or biopsy, leukocytes[10] or cultured fibroblasts.[11]

Type II

GM_1 gangliosidosis type II, or juvenile GM_1 gangliosidosis, is another inborn error of ganglioside metabolism in which the GM_1 ganglioside accumulates in the brain.[12-14] Patients with this disease also have progressive cerebral deterioration but it begins later, between 6 and 24 months of age. Onset usually is with ataxia. There may be incoordination or frequent falling. Speech, if present, is lost. These patients also develop spasticity and rigidity as well as muscular weakness, and they have seizures. They usually die between 3 and 10 years of age in a state of decerebrate rigidity.

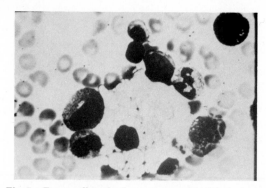

Fig. 5.—Foam cell in the bone marrow. This histiocyte was obtained from the marrow of a patient with juvenile GM_1 gangliosidosis.

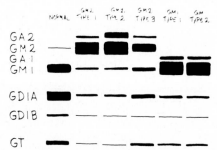

Fig. 6.—Thin layer chromatographic patterns of the gangliosides stored in the nervous system in five gangliosidoses: GM_1 type I is generalized gangliosidosis; GM_1 type II is the juvenile form; GM_2 type I is Tay-Sachs disease; GM_2 type II is the Sandhoff disease; and GM_2 type III is characterized by a partial deficiency of hexosaminidase A.

The facial appearance is normal. There are no funduscopic findings, hepatosplenomegaly or skeletal deformities, as in type I. Patients have vacuolated cells in the bone marrow and membranous cytoplasmic bodies in the neuronal cytoplasm and accumulate large amounts of GM_1 ganglioside in the brain.[12] Keratosulfaturia has been described.[12] These patients also have a deficiency of β-galactosidase.[12] The degree of deficiency in tissues is so great that it has been difficult to distinguish type II from type I patients by enzymatic assay. However, evidence that there are in fact two different genetically determined variants has been reported[15] in studies on fibroblasts in which the activity of the enzyme in type I was less than 1%, whereas that in type II was 3–4%, of normal. Furthermore, there were qualitative differences in pH optima and in thermolability between the two variants.

DIFFERENTIAL DIAGNOSIS

The differential diagnosis of gangliosidosis type I also includes the Hurler syndrome (see section on the Mucopolysaccharidoses), Tay-Sachs disease (see chapter, this section) and Niemann-Pick disease. Patients with Niemann-Pick disease and Tay-Sachs disease do not have skeletal deformities.

CASE REPORT*

C. D., a 6-month-old white male was studied at the Los Angeles Children's Hospital. He had shown slow development and failed to thrive from the start. He sucked ineffectively and was fed by gavage. A large head and coarse features (see Figs. 1 and 2), as well as hypotonia and bilateral talipes equinovarus, were noted at 3 months.

Physical examination at 6 months revealed hepatomegaly and rigidity of all of the extremities. There were bilateral rhonchi, and roentgenograms revealed diffuse pulmonary infiltrates. There was a dorsolumbar kyphosis and inferior beaking of L2 (see Fig. 4). Peripheral blood lymphocytes contained cytoplasmic vacuoles. Liver obtained at biopsy showed vacuolation of Kupffer's cells. The bone marrow contained vacuolated histiocytes such as those of Niemann-Pick disease. The patient died at 8 months of age, and there was histochemical evidence of accumulation of glycolipid in cerebral neurons, hepatic histiocytes and renal glomerular epithelium. The glycolipid was isolated from the brain and other tissues, purified and characterized chemically as GM_1. It comprised approximately 84% of the gangliosides of the gray matter.

GENETICS

GM_1 gangliosidosis type I is transmitted as an autosomal recessive trait. A risk of recurrence of 25% in siblings is consistent with observed data.[1] The rate of consanguinity in reported families has been high. The enzyme is detectable in cultured fibroblasts or leukocytes.[10] Detection of the heterozygous carrier should be feasible, as should intrauterine diagnosis by assay of β-galactosidase. The enzyme is active in cultured normal amniotic fluid cells.

GM_1 gangliosidosis type II appears to represent a different mutant gene and to be transmitted also as an autosomal recessive. Only one form or the other, never both, is seen in a given sibship.

TREATMENT

None.

REFERENCES

1. O'Brien, J.: Generalized gangliosidosis, J. Pediatr. 75:167, 1969.
2. Landing, B. H., Silverman, F. N., Craig, J. M., Jacoby, M. D., Lahey, M. E., and Chadwick, D. L.: Familial neurovisceral lipidosis, Am. J. Dis. Child. 108:503, 1964.
3. O'Brien, J. S., Stern, M. B., Landing, B. H., O'Brien, J. K., and Donnell, G. N.: Generalized gangliosidosis, Am. J. Dis. Child. 109:338, 1965.
4. Emery, J. M., Green, W. R., Wyllie, R. G., and Howell, R. R.: GM_1-gangliosidosis, Arch. Ophthalmol. 85:177, 1971.
5. Caffey, J.: Gargoylism (Hunter-Hurler disease, dysostosis multiplex, lipochondrodystrophy); prenatal and neonatal bone lesions and their early postnatal evolution, Hosp. Joint Dis. Bull. 12:38, 1951.
6. Severi, F., Magrini, U., Tettamanti, G., Bianchi, E., and Lanzi, G.: Infantile GM_1 gangliosidosis. Histochemical, ultrastructural and biochemical studies, Helv. Paediatr. Acta 26:192, 1971.
7. Suzuki, K.: Cerebral GM_1-gangliosidosis: Chemical pathology of visceral organs, Science 159:1471, 1968.
8. Okada, S., and O'Brien, J. S.: Generalized gangliosidosis: Beta-galactosidase deficiency, Science 160:1002, 1968.
9. MacBrinn, M. C., Okada, S., Ho, M. W., Hu, C. C., and O'Brien, J. S.: Generalized gangliosidosis: Impaired cleavage of galactose from a mucopolysaccharide and a glycoprotein, Science 163:946, 1969.
10. Singer, H. S., Nankervis, G. A., and Schafer, I. A.: Leukocyte beta-galactosidase activity in the diagnosis of generalized GM_1 gangliosidosis, Pediatrics 49:352, 1972.
11. Sloan, H. R., Uhlendorf, W., Jacobson, C. B., and Fredrickson, D. S.: β-Galactosidase in tissue culture derived from human skin and bone marrow: Enzyme defect in GM_1 gangliosidosis, Pediatr. Res. 3:532, 1969.
12. Wolfe, L. S., Callahan, J., Fawcett, J. S., Andermann, F., and Scriver, C. R.: GM_1-gangliosidosis without chondrodystrophy or visceromegaly, Neurology 20:23, 1970.
13. Patton, V. M., and Dekaban, A. S.: GM_1-gangliosidosis and juvenile cerebral lipidosis, Arch. Neurol. 24:529, 1971.
14. Singer, H. S., and Schafer, I. A.: Clinical and enzymatic variations in GM_1 generalized gangliosidosis, Am. J. Hum. Genet. 24:454, 1972.
15. Pinsky, L., and Powell, E.: GM_1-gangliosidosis types 1 and 2: Enzymatic differences in cultured fibroblasts, Nature 228:1093, 1970.

---

*This patient is the one from whose tissues O'Brien and colleagues[3] first isolated the GM_1 ganglioside, establishing the disorder as a generalized gangliosidosis. Clinical details are reported through the courtesy of Dr. John S. O'Brien of the University of California San Diego and reprinted with permission from the Am. J. Dis. Child.[3]

FABRY DISEASE
Angiokeratoma Corporis Diffusum Universale

CARDINAL CLINICAL FEATURES

Red, papular angiectatic skin lesions, crises of fever and excruciating pains in the extremities, nephropathy, coronary and cerebral vascular disease, deficiency of ceramide trihexosidase.

CLINICAL PICTURE

The Fabry disease is an inborn error of glycolipid metabolism, transmitted as an X-linked recessive character.[1,2,5,6] Pain, which may be the initial symptom, is often excruciating and has a burning quality. It is most often noted in the fingers and toes. Narcotics may not provide relief. Crises of pain are self-limited, disappearing spontaneously only to return later. These patients also have recurrent episodes of fever without infection or other obvious explanation.

The skin lesions develop slowly as clusters of dark red angiectases in the superficial layer of the skin (Figs. 1 and 2). They do not blanch with pressure. There is some tendency for bilateral symmetry. Areas of most common involvement are the hips, back, thighs, buttocks and scrotum. The oral mucosa and conjunctiva also are frequently involved.

Ocular lesions may involve the conjunctiva, cornea or retina. There is aneurysmal dilatation of thin ven-

ules. Corneal opacities are found in involved males and in some heterozygous female carriers. Proteinuria develops followed by signs of renal impairment. Examination of the urine reveals red cells, casts and birefringent lipid globules within and outside cells. Renal function gradually deteriorates, culminating in severe renal failure. Patients develop cardiomegaly and hypertension. Vascular disease is the rule. Myocardial ischemia or infarction, or cerebral vascular disease is common and often occurs before the age of 25 years. Patients may have seizures, hemiplegia, aphasia or other signs of cerebral vascular accident. Death usually results from uremia or from vascular disease of the heart or brain. Congestive heart failure and arrhythmias are common.

Lipid accumulates in the tissues of patients with the Fabry disease.[4] It has been characterized as a glycolipid and is known as ceramide trihexoside. The metabolic defect in the disease is in the catabolism of this glycolipid. In involved males there is a deficiency of the enzyme that catalyzes hydrolysis of the terminal galactose molecule of ceramide trihexoside.[3] The reaction is as follows:

$$\text{ceramide trihexoside} + H_2O \xrightarrow{\text{ceramide trihexosidase}}$$
$$\text{ceramide lactoside} + \text{galactose.}$$

The diagnosis can be made clinically. It has been established by determining the level of ceramide tri-

Fig. 1 (left).—Angiokeratomas of the skin. These red-purple macules or maculopapules feel hyperkeratotic. They are prominent over the hips, buttocks and scrotum.

Fig. 2 (right).—Closer view of angiokeratomas of the skin.

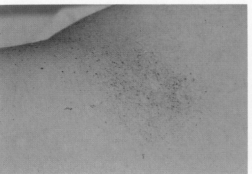

FABRY'S DISEASE
W. Kindred – La Jolla, 1970

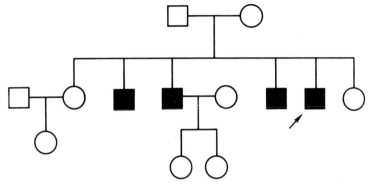

Fig. 3. —(Case Report.) Pedigree of the family. The pattern is consistent with that of an X-linked recessive character. *Arrow* indicates the patient, E. W.

hexosidase activity in biopsy material from the small intestinal mucosa or the kidney. The glycolipid has been found to accumulate in cultured fibroblasts obtained by skin biopsy.

CASE REPORT

E. W., a 31-year-old white male, was admitted to University Hospital with complaints of fever, headache and a problem

Fig. 4. – E. W. (Case Report). Fine telangiectases were present in the skin as well as in the conjunctiva, where tortuous dilated vessels are commonly seen. Slit lamp examination may reveal corneal opacities or a pathognomonic cataract of the posterior capsule of the lens.

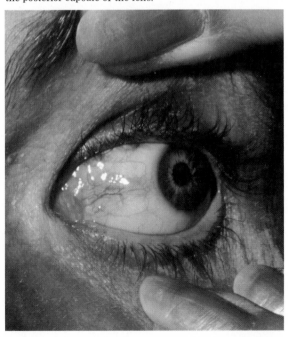

with vision. He had had recurrent attacks of fever, vomiting and chills and severe pains in his fingers and toes since the age of 18 months. He gave a history of repeated cerebral vascular accidents with residual involvement, which included a loss of 80% of his hearing. He had had blurring of vision for a number of years.

Family history revealed two older brothers with the Fabry disease, chronic heart failure and myocardial ischemia as well as a brother who had died with a diagnosis of multiple sclerosis but a history of recurrent fever and of skin lesions (Fig. 3). The mother had hypertension and raised red spots that bled easily. Physical examination revealed a young adult male with a marked hearing loss who appeared chronically ill.

In the skin there were dark red angiectatic lesions (see Fig. 1) over the buttocks, thighs and scrotum. He had dilated, tortuous vessels in his conjunctiva (Fig. 4) and retina. He also had corneal opacities, visible by slit lamp examination. Vision was 20/25 in both eyes.

The blood pressure was 130/60. Examination of the heart was not remarkable; however, the ECG revealed left ventricular hypertrophy. There were ischemic changes.

Laboratory examination showed a BUN of 12 mg/100 ml and an erythrocyte sedimentation rate of 28–50 mm/hour. Routine urinalysis was normal.

GENETICS

The Fabry disease is transmitted as an X-linked genetic defect. The male hemizygous patient has a total absence of the enzyme ceramide trihexosidase in his tissues. Female carriers of the condition have intermediate levels of ceramide trihexosidase activity.

TREATMENT

There is no specific treatment. Coronary or cerebral vascular disease should be treated in the usual fashion. Renal failure may be subject to treatment by hemodialysis or renal transplantation.

REFERENCES

1. Sweeley, C. C., and Klionsky, B.: Glycolipidosis: Fabry's Disease, in Stanbury, J. B., Wyngaarden, J. B., and Frederickson, D. S. (eds.): *The Metabolic Basis of Inherited Disease* (New York: McGraw-Hill Book Co., 1966), p. 618.

2. Johnston, A. W., Weller, S. D. V., and Warland, B. J.: Angiokeratoma corporis diffusum, Arch. Dis. Child. 43:73, 1968.

3. Brady, R. O., Gal, A. E., Bradley, R. M., Martensson, E., War-shaw, A. L., and Laster, L.: Enzymatic defect in Fabry's disease—Ceramidetrihexosidase deficiency, N. Engl. J. Med. 276:1163, 1967.

4. Schibanoff, J. M., Kamoshita, S., and O'Brien, J. S.: Tissue distribution of glycosphingolipids in a case of Fabry's disease, J. Lipid. Res. 10:515, 1969.

5. Wallace, R. D., and Cooper, W. J., Angiokeratoma corporis diffusum universale (Fabry), Am. J. Med. 39:656, 1965.

6. Frost, P., Spaeth, G. L., and Tanaka, Y.: Fabry's disease: Glycolipid lipidosis, Arch. Intern. Med. 117:440, 1966.

I CELL DISEASE
Leroy Syndrome, Mucolipidosis II

CARDINAL CLINICAL FEATURES

Hurler-like coarse features, severe retardation of growth and mental development, limitation of motion of the joints, advanced dysostosis multiplex, cytoplasmic inclusions in fibroblasts.

CLINICAL FEATURES

I cell disease was first described by Leroy and De Mars[1] who discovered that fibroblasts in cell culture derived from the skin of two patients showed striking and abundant cytoplasmic inclusions by phase contrast microscopy. These patients resembled those with the Hurler syndrome, but they did not have cloudy corneas and they did not have increased urinary excretion of mucopolysaccharides. Leroy et al.[2] gave the disorder the name I cell disease, in which I stands for inclusions. The term mucolipidosis reflects the abnormalities in mucopolysaccharide and glycolipid.

The syndrome resembles that of the Hurler syndrome, but it is more severe.[2-5] A severe degree of psychomotor retardation is evident from birth and the IQ is very low. Involved patients do not learn to sit, walk, roll over or speak. They also have severe retardation of growth (Figs. 1 and 2). Most patients with the Hurler syndrome reach a final height of 39–42 inches, whereas patients with I cell disease reach a maximum height of 29–30 inches by 2 years of age.[2] The features are coarse, and this abnormality is progressive. There often is a high, narrow forehead with a metopic ridge, prominent epicanthal folds and puffy eyelids. The bridge of the nose is flat, and the tip of the nose is wide and the nostrils anteverted. The distance between the nose and the upper lip is increased. The cornea usually is clear. Slit lamp examination may reveal a fine granularity.[6] There may even be corneal opacity.[7] These patients have gingival hypertrophy, and the voice is hoarse. The ears are thick and firm. The skin is thick and waxy and, especially in early infancy, may be smooth, firm and tight; it cannot be pinched anywhere.

Patients with this syndrome also have a limitation of joint motion, with contractures at the hips, knees, shoulders, elbows and fingers. They also have a claw-hand deformity with ulnar deviation of both hands. The wrists are broad. There is a marked lumbar gibbus with dorsolumbar kyphosis, a protuberant abdomen, diastasis recti and an umbilical hernia. In males inguinal hernia is common. Hepatomegaly is minimal and splenomegaly absent or slight. Patients usually have a history of frequent nasal discharge, respiratory infections and ear infections. Most patients die between 2 and 8 years of age, usually with pneumonia

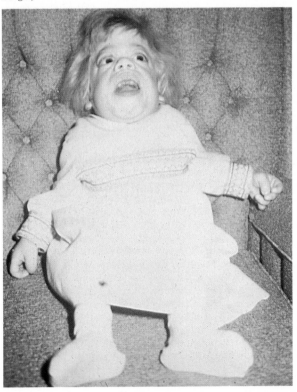

Fig. 1.—A girl with I cell disease (original patient of Dr. Jules Leroy) shown at age 7½. The features were markedly Hurler-like, but the corneas were clear. The patient had a prominent forehead, depressed nasal bridge and epicanthal folds. There were flexion contractures of fingers, hips and knees. Height was 30 inches, height age 1 year. (Courtesy of Dr. John S. O'Brien of the University of California San Diego.)

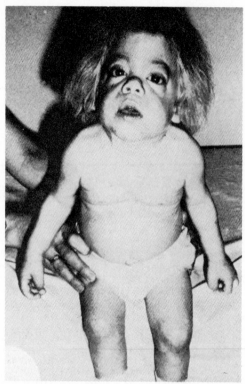

Fig. 2.—Patient shown in Figure 1, at 5 years of age. The coarse features and retarded physical development are evident. (Reprinted with permission from J. Pediatr.[3])

a high activity of acid phosphatase and low activity of multiple lysosomal enzymes.[3, 11, 12] These include β-glucuronidase, β-galactosidase, α-mannosidase, α-fucosidase, N-acetyl-β-D-galactosaminidase and aryl-sulphatase-A. Wiesmann and colleagues[12] found that lysosomal enzymes were high in the medium surrounding cultured fibroblasts. They postulated that there were leaky lysosomes. High levels of activity of these lysosomal enzymes also were found in the serum. These included hexosaminidase A and B and α-galactosidase. These patients do not have excessive excretion of urinary mucopolysaccharides.

A recent report has suggested that in I cell disease there is a defect in the uptake and intercellular localization of lysosomal enzymes.[13] I cells will not correct the abnormal $^{35}SO_4$ accumulation of Hunter and Hurler cells. It appears that lysosomal enzymes such as hexosaminidase A are taken up into cells by absorptive pinocytosis much more efficiently than are other proteins. This seems to be related to carbohydrates on these glycoprotein enzymes, for oxidation of the enzyme by sodium metaperiodate rapidly diminishes this uptake without affecting catalytic activity. I cell disease could reflect a deficiency in a carbohydrate recognition acceptor that might be common to

Fig. 3.—Same patient. Roentgenogram of the arm and hand at age 4 years. The findings reveal an advanced degree of dysostosis multiplex. (Reprinted with permission from J. Pediatr.[3])

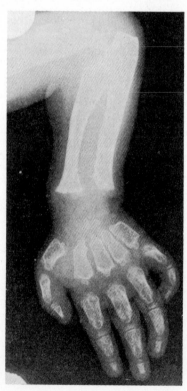

and congestive failure. At necropsy there may be thickening of the endocardium and myocardium and of the coronary arteries and aorta.

Roentgenographic findings are those of dysostosis multiplex, as seen in the Hurler syndrome (see section on Mucopolysaccharidoses), but they are more severe and appear at an earlier age. They are similar to those of GM_1 gangliosidosis (see chapter, this section). Skeletal changes are present at birth and may be extreme, even in infancy. In early infancy there may be extensive periosteal new bone function, with cloaking of the long tubular bones.[3] These changes are progressive and produce long bones that are short, wide and thick. The distal radius and ulna tilt toward each other. There are bullet-shaped proximal phalanges and proximal pointing of the metacarpals, with widening of their distal ends (Fig. 3). The ribs are broad and spatulate. The vertebral bodies are short and rounded, and there is anterior inferior beaking at L1 and D12. The proximal tibia and fibula may be deeply notched.[6] The calcaneus may have two ossification centers.

The cytoplastic inclusions (Fig. 4) in skin fibroblasts are large dark granules which fill the cytoplasm.[1, 7-9] Vacuolated lymphocytes may be seen in the blood or bone marrow,[10] and vacuolated hepatocytes are found.[9] The inclusions are sudanophilic. They contain

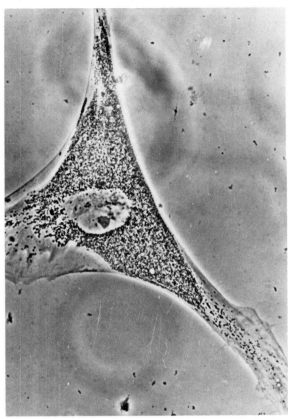

Fig. 4.—I cell in culture showing the characteristic cytoplasmic inclusions. (Courtesy of Dr. Jules Leroy of the State University of Antwerp, Belgium.)

a group of lysosomal enzymes. A-galactosyl transferase might be an example of such a protein. Consistent with this hypothesis, I cells have been found not to be leaky in that they retain added purified lysosomal enzyme—if anything, better than other cells. On the other hand, acid hydrolases derived from I cells are not effectively taken up from the medium by any cells.[14] Thus, it appears that these enzymes are normally secreted into the medium, after which they must be specifically recognized and taken up by cells. The I cell mutation could interfere with the recognition site on a number of hydrolases.

GENETICS

The syndrome is transmitted as an autosomal recessive. Multiple sibs, both male and female, have been reported with normal parents. Consanguinity between the parents also has been reported.[6] Abnormal inclusions have been found in the fibroblasts of some phenotypically normal parents.

TREATMENT

None. The outcome is uniformly fatal.

REFERENCES

1. Leroy, J. G., and De Mars, R. I.: Mutant enzymatic and cytological phenotype in culture human fibroblasts, Science 157:804, 1967.
2. Leroy, J. G., De Mars, R. I., and Opitz, J. M.: "I Cell" Disease, in *Proceedings of the First Conference on Clinical Delineation of Birth Defects*, Birth Defects Original Article Series, Vol. V, No. 4 (New York: The National Foundation-March of Dimes, 1969), p. 174.
3. Leroy, J. G., Spranger, J. W. Feingold, M., Opitz, J. M., and Crocker, A. C.: I-cell disease: A clinical picture, J. Pediatr. 79:360, 1971.
4. Leroy, J. G., and Spranger, J. W.: I cell disease, N. Engl. J. Med. 283:598, 1970.
5. Luchinger, N., Buhler, E. M., Mehes, K., and Hirt, H. R.: I cell disease, N. Engl. J. Med. 282:1374, 1970.
6. Blank, E., and Linder, D.: I-cell disease (mucolipidosis II): A lysosomopathy, Pediatrics 54:797, 1974.
7. de Montis, G., Garnier, P., Thomassin, N., Job, J.-C., and Rossier, A.: La mucolipidose type II (maladie des cellules à inclusions). Étude d'un cas et revue de la littérature, Ann. Pediatr. 19:369, 1972.
8. Hanai, J., Leroy, J. G., and O'Brien, J. S.: Ultrastructure of cultured fibroblasts in I-cell disease, Am. J. Dis. Child. 122:34, 1971.
9. Kenyon, K. R., Sensenbrenner, J. A., and Wyllie, R. G.: Hepatic ultrastructure and histochemistry in mucolipidosis II (I-cell disease), Pediatr. Res. 7:560, 1973.
10. Rapola, J., Autio, S., Aula, P., and Nanto, V.: Lymphocytic inclusions in I-cell disease, J. Pediatr. 85:88, 1974.
11. Gilbert, E. F., Dawson, G., ZuRhein, G. M., Opitz, J. M., and Spranger, J. W.: I-cell disease, mucolipidosis II. Pathological, histochemical, ultrastructural and biochemical observations in four cases, Z. Kinderheilkd. 114:259, 1973.
12. Wiesmann, U. N., Lightbody, J., Vassella, F., and Herschkowitz, N. N.: Multiple lysosomal enzyme deficiency due to enzyme leakage?, N. Engl. J. Med. 284:109, 1971.
13. Spranger, J. W., and Wiedeman, H. R.: The genetic mucolipidoses and diagnosis and differential diagnosis, Hum. Genet. 9:113, 1970.
14. Hickman, S., and Neufeld, E. F.: A hypothesis for I-cell disease: Defective hydrolases that do not enter lysosomes, Biochem. Biophys. Res. Commun. 49:992, 1972.

Other Disorders of Metabolism and Disorders of Transport

HYPOPHOSPHATASIA

CARDINAL CLINICAL FEATURES

Irregular, incomplete ossification of the cartilage of the bones; bowing, shortening and multiple fractures of long bones; cutaneous dimples; defective ventilation and respiratory problems; bulging fontanel or craniostenosis; low activity of alkaline phosphatase; elevated urinary excretion of phosphoethanolamine.

CLINICAL PICTURE

Hypophosphatasia was first recognized as a distinct clinical entity by Rathbun in 1948.[1] There is a spectrum of involvement, the most severely affected presenting manifestations at birth. A few patients have been diagnosed in utero. In the most severe prenatal or neonatal form, the cranium is soft and globular, giving the picture of a boneless skull. Skull films reveal a well-calcified base and marked lack of calcification of the other bones. There may be patchy ossification of the frontal bones and small plaques of parietal or occipital bone.

Deformities of the extremities are pronounced. The extremities are short. Dimples are characteristically present, whether or not there is bowing.[2] Roentgenographic examination[3-5] reveals generalized rarefaction of the skeleton. The roentgenographic abnormalities of this disease distinguish it from all other disorders of bone.[4] The long bones have irregular and incomplete ossification in the metaphyseal areas, with deep segmental defects. They also are bowed and there are often associated overlying skin dimples. Multiple fractures are the rule. Thoracic cage defects and short ribs may lead to deficient pulmonary ventilation.

These patients may be stillborn or die within hours of birth of respiratory difficulty, or shortly later of pulmonary infection. They fail to thrive; they may have vomiting, fever or convulsions. Loud cries, as if in pain, have been observed even in very young infants.

A moderately severe or infantile form may present within the first months of life. There have been no survivals in patients with hypophosphatasia presenting with clinical manifestations prior to the end of the first 6 months. These infants also have generalized skeletal deformities. The cranial sutures are wide, and a bulging anterior fontanel and prominent scalp veins usually develop. A rachitic type of thoracic rosary may be present.[1]

In patients with hypophosphatasia first seen after the seventh month of life, the disease may be less severe. It is still characterized by generalized skeletal deformities. Craniostenosis is a common sequel in these infants. They regularly develop dental problems, which include premature loss of deciduous teeth, dental hypoplasia and marked dental caries. Walking often is delayed and awkward. The lower extremities may be bowed or may show genu valgum. In some patients the disease may present in adult life. These and other patients may have bone pains, and occasional fractures remain a part of life.

The major criterion for diagnosis is an absent or extremely low activity of alkaline phosphatase of the serum.[6] Alkaline phosphatase also is low in tissues. There is no observable relationship between the degree of alkaline phosphatase abnormality and the clin-

Fig. 1.—Phosphoethanolamine excretion, as quantitated using the Automatic Acid Analyzer.

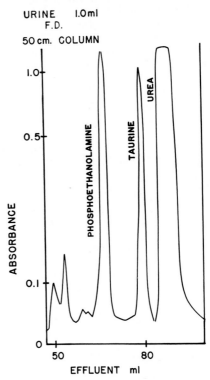

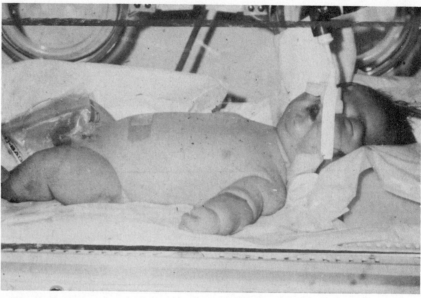

Fig. 2. — J. G. (Case Report). Neonatal hypophosphatasia. The baby had short, deformed limbs. The legs were bowed and had prominent dimples of the skin.

ical expression of the disease. Phosphoethanolamine ($H_2NCH_2CH_2OPO_3H_2$) excretion in the urine[7] is regularly increased (Fig. 1). This amino acid is not found in normal urine using the usual screening methods of chromatography on paper or thin layer plates. The concentrations of inorganic pyrophosphate in plasma and urine[8] have been reported to be increased. In the newborn type the serum calcium may be elevated.[1] Proteinuria, casts and impaired renal function are common in the very young.

The microscopic appearance of the bones[1, 5] resembles rickets. There are wide, irregular zones of proliferative cartilage and a lack of calcification of the osteoid.

CASE REPORT

J. G. was transported by helicopter to the University Hospital neonatal intensive care unit because of respiratory distress at birth. Multiple fractures were noted. He was the product of a fifth pregnancy. The first child, a 6-year-old girl, and the fourth, a 16-month-old male, were well. The second and third were stillborn and had multiple deformities such as those of the patient. There was no consanguinity. Gestation was of 9 months' duration but was complicated by polyhydramnios. Roentgenograms prior to delivery showed multiple fractures and incomplete ossification of the skull and long bones. A diagnosis of osteogenesis imperfecta was suspected. Birth weight was 8 lb, 13 oz. The baby was cyanotic at birth and breathed poorly. He was intubated and transported.

Physical examination revealed a cyanotic child with a frail chest and poor aeration. The head was globular and felt boneless. The sclera were white. The extremities were short, bowed and deformed. There were multiple dimples in the skin overlying the bowed long bones (Figs. 2 and 3). There was a grade II/VI systolic cardiac murmur.

Roentgenographic survey of the skeleton (Figs. 4-7) revealed multiple fractures, most marked in the metaphyseal areas. The skull showed very poor ossification, with only scattered islands of bone present (see Fig. 4). The vertebral bodies were poorly ossified.

The alkaline phosphatase activity of the blood was initially 13 I.U. (100 I.U. is a normal value equivalent to 20 Bodansky units). A repeat determination revealed 5 I.U. Phosphoethanolamine was found in the urine. Its excretion was quantitated at 1.25 mg/mg of creatinine (normal individuals excrete less than 0.04 mg). There was a 3+ proteinuria. The serum calcium was approximately 10.5 mg/100 ml.

The infant survived in the newborn special care unit for 11 weeks. He had persistent severe hypoventilation, which led to tracheostomy and artificial ventilation. Even at this age he demonstrated panic whenever the respirator was removed. He had several episodes of pneumonia and sepsis. Prior to death he developed seizures, cardiomegaly, and heart failure on the right side.

Autopsy revealed multiple fractures and bilateral pneumo-

Fig. 3. — J. G. The arm, like the other extremities, was short and rather puffy in appearance. Prominent dimples indicated underlying fractures.

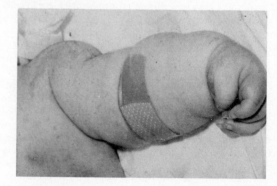

nia. The bones cut with a consistency of hyaline cartilage. Cross-sections showed a normal amount of marrow.

GENETICS

Hypophosphatasia is a rare autosomal recessive disease. Its incidence has been estimated at one in 100,000 live births. The heterozygous state usually can be demonstrated by measuring the serum activity of alkaline phosphatase. Persons with levels of alkaline phosphatase below 2 Bodansky units are considered to be carriers.[6] However, it often is impossible to distinguish carriers from patients on the basis of activity of alkaline phosphatase. Phosphoethanolamine excretion is increased in the urine of carriers. Variability and errors in the assay of alkaline phospha-

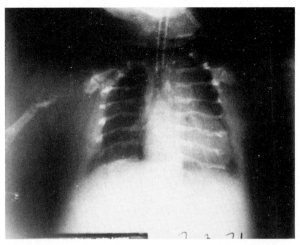

Fig. 6.—J. G. Roentgenogram of the chest. The ribs were ribbon-like, delicate and short. There was poor ossification and multiple fractures.

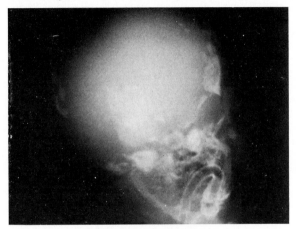

Fig. 4.—J. G. Roentgenogram of the skull. There was very poor ossification. Only scattered islands of bone were present.

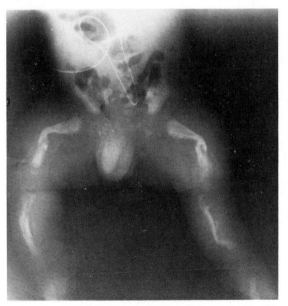

Fig. 7.—J. G. Roentgenogram of the lower extremities. The bones were demineralized and ribbon-like. Marked angulation of both femora was interpreted to represent old fractures. The metaphyses were irregular, and ossification was markedly deficient. The tibiae were bowed and showed multiple fractures.

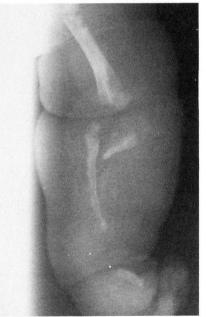

Fig. 5.—J. G. Film of the upper extremities revealed incredibly poor ossification and irregularity of the metaphyses. There were multiple fractures of the humerus, ulna and radius.

tase are so common that excretion of phosphoethanol-amine is probably a more reliable method for the study of families. It also is true that all heterozygotes studied have not had increased phosphoethanolamine excretions, even in families in which some do,[7] but it is possible that with more sensitive methods most heterozygotes can be detectable.[9]

The clinical picture of hypophosphatasia is characterized by heterogeneity. The uniformity of the age at onset and course in affected members of a single family suggests that we are dealing with a number of different genetic diseases due either to variation in alleles at a single locus or involvement of a number of loci.[4]

TREATMENT

A number of treatments have been tried without convincing benefit. Phosphate supplementation may be worth trying.[10] It may be necessary to correct craniosynostosis surgically. The usual measures are employed for the management of fractures and skeletal deformities. In patients surviving the critical periods of early infancy, rates of calcification and clinical improvement may be striking.

REFERENCES

1. Rathbun, J. C.: Hypophosphatasia, Am. J. Dis. Child, 75:822, 1948.
2. Weller, S. D. V.: Hypophosphatasia with congenital dimples, Proc. Roy. Soc. Med. 52:637, 1959.
3. Gwinn, J. L., and Lee, F. A.: Radiological case of the month, Am. J. Dis. Child. 122:151, 1971.
4. Currarino, G., Neuhauser, E. B. D., Reyersbach, G. C., and Sobel, E. H.: Hypophosphatasia, Am. J. Roentgenol. Radium Ther. Nucl. Med. 78:392, 1957.
5. Fraser, D.: Hypophosphatasia, Am. J. Med. 22:730, 1957.
6. Rathbun, J. C., MacDonald, J. W., Robinson, H. M. C., and Wanklin, J. M.: Hypophosphatasia: A genetic study, Arch. Dis. Child. 36:540, 1961.
7. Harris, H., and Robson, E. B.: A genetical study of ethanolamine phosphate excretion in hypophosphatasia, Ann. Hum. Genet. 23:421, 1958.
8. Russell, R. G. G., Bisaz, S., Donath, A., Morgan, D. B., and Fleisch, H.: Inorganic pyrophosphate in plasma in normal persons and in patients with hypophosphatasia, osteogenesis imperfecta, and other disorders of bone, J. Clin. Invest. 50:961, 1971.
9. Pimstone, B., Eisenberg, E., and Silverman, S.: Hypophosphatasia: Genetic and dental studies, Ann. Intern. Med. 65:722, 1966.
10. Bongiovanni, A. M., Album, M. M., Root, A. W., and Hope, J. W.: Metabolic studies in hypophosphatasia and response to high phosphate intake, (Abstr.) Pediatr. Res. 1:314, 1967.

CYSTINOSIS
Cystine Storage Disease, Syndrome of Lignac, Fanconi, Debré and De Toni

CARDINAL CLINICAL FEATURES

Nephropathy and renal failure, glycosuria, phosphaturia, generalized aminoaciduria, acidosis, rickets, growth retardation, retinopathy, refractile corneal bodies, cystine crystals in bone marrow and other tissues, increase in intracellular concentration of cystine.

CLINICAL PICTURE[1, 2]

In classic cystinosis the patient usually appears normal at birth. He usually presents at 6–10 months of age with symptoms generated by renal tubular dysfunction. These patients have a classic renal Fanconi syndrome in which there is renal tubular deficiency in reabsorption of glucose, phosphate, amino acids and other organic acids, as well as of sodium, potassium, bicarbonate and water. The aminoaciduria is a generalized aminoaciduria in which those amino acids usually found in the urine in moderate quantity, such as glycine, are excreted in very large amounts; those usually found in very small amounts, such as leucine, are excreted in moderate amounts; and the imino acids, such as proline, which are not normally found in urine, are excreted. Renal losses of phosphate lead to hypophosphatemia and rickets. This is a form of vitamin D-resistant rickets in the sense that it develops in spite of, and will not respond to, the usual antirachitic doses of vitamin D. Patients may have tremendous polyuria and polydipsia. This makes these children highly vulnerable to dehydration, especially in the presence of otherwise minor infectious disease. Losses of sodium, bicarbonate, and potassium lead to chronic acidosis and hypokalemia. Growth retardation is extreme. Renal damage also involves the glomeruli. Uremia and renal failure are progressive and lead usually to death within the first 10 years of life.

The pathology of the disease reflects the extensive deposition of cystine crystals throughout the tissues of the body. The disorder may be readily diagnosed by examination of aspirated bone marrow for cystine crystals (Fig. 1). Conjunctival biopsy also has been used for this purpose. Crystals also may be demonstrated in the lymph nodes. Cystine deposition in the kidney leads to the nephropathy that characterizes the disease.

Refractile crystalline bodies also may be demonstrated by slit lamp examination of the cornea (Fig. 2). These birefringent refractile bodies are sufficiently typical to be diagnostic of cystinosis. However, the nature of these bodies has not yet been defined. Photophobia is a prominent and disturbing symptom to the patient. Patients with cystinosis also have a characteristic retinopathy. The peripheral retina has a patchy pattern of hyperpigmentation and depigmentation, often in regular distribution, varying from about a tenth of a disc in diameter to a fine, pepper size. Retinal changes (Fig. 3) that usually are more marked temporally may be the earliest clinical manifestations of the disease.

Patients generally have a fair complexion and very fair hair (Fig. 4).

A molecular explanation for cystinosis has not yet been elucidated. The site of the defect has been indicated by the accumulation of large amounts of cystine intracellularly. Within the cell the lysosomes appear to be the locus for this unusual cystine storage, which can be as high as 100 times the normal cellular content. The biochemical determination of the content of cystine in isolated leukocytes or in fibroblasts in cell culture provides an elegant definitive diagnosis of the disease. Rapid cytologic diagnosis in cell culture can be made by the induction of vacuolation by exposure to the mixed disulfide of penicillamine and cysteine.[3]

Chemical definition of the disease in this way has permitted the understanding of clinical heterogeneity in cystinosis, which is increasingly being defined. Types of cystinosis other than the classic form are being found that have contents of intracellular cystine, which are intermediate between that of the usual patient and of control individuals. One form has been described under the heading of benign cystinosis, or adult type cystinosis. Involved patients have typical corneal manifestations of cystinosis but none of the renal features of the disease, no retinopathy and no photophobia. In an increasing number of intermediate

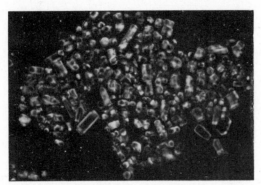

Fig. 1.—Cystine crystals aspirated from the marrow of a patient with cystinosis. (Courtesy of Dr. Jerry Schneider of the University of California San Diego.)

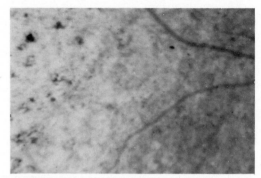

Fig. 3.—The peripheral retinopathy of cystinosis. (Courtesy of Dr. Jerry Schneider of the University of California San Diego. Reprinted with permission from Ann. Intern. Med.[2])

or late-onset types of cystinosis, the nephropathic expression is attenuated and permits survival into the second or third decade. It seems likely that increasing genetic heterogeneity will be recognized.

Case Report

D. I. M.[4] was admitted to the hospital at 9 months of age because of fever and dehydration. He had been thought to be normal until about 6 months of age, at which time he was first admitted with a temperature of 104° F and moderate dehydration. A diagnosis of bilateral otitis media was made, and he was treated with penicillin and fluids given intravenously. At 8 months of age he was again admitted, with a temperature of 104.5 F and severe dehydration. He again responded well to intravenous therapy. A diagnosis of tonsillitis was made at that time, and he was given a 10-day course of penicillin.

There was no family history of genetic disorder. The patient had two normal older brothers. His parents were second cousins.

Physical examination revealed a weight of 14 lb and a height of 25.2 inches. The pulse was 130, and respirations 48/minute. Blood pressure was 105/65. He appeared to be an emaciated white male who was approximately 5% dehydrat-

Fig. 2.—Slit lamp examination of the cornea. The refractile bodies are seen on the left and in the central cornea and on the right in the peripheral cornea. (Courtesy of Dr. Jerry Schneider of the University of California San Diego. Reprinted with permission from Ann. Intern. Med.[2])

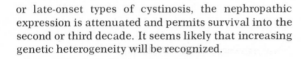

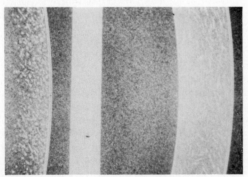

ed. He had very fine, dry hair and extremely long eyelashes. Four teeth had a yellowish color, and the enamel was poor. The wrists seemed somewhat thickened. Otherwise the extremities appeared normal. The skin was very dry and fair.

Laboratory examination revealed a hemoglobin of 10.5 gm/100 ml, WBC, 12,000/cu mm and platelet count, 150,000/cu mm. The ESR was 45. Serum electrolyte concentrations were sodium, 140; potassium, 2.8; chloride, 115; CO_2, 8 mEq/L. Calcium was 9, phosphorus 3.4 and BUN 20 mg/100 ml.

This form of early presentation is obviously not as readily interpreted as typical of cystinosis and points up the difficulties of early diagnosis. Cystinosis was diagnosed by the demonstration of crystals in bone marrow aspirate at 20 months of age. The patient has since had a sister with cystinosis.

Genetics

Cystinosis is transmitted by a single, autosomal recessive gene. Heterozygotes can be detected by analysis of the intracellular content of cystine in the leukocyte or fibroblast. In classic cystinosis the heterozygous cell contains a four- to five-fold greater content of cystine than do control cells. Variant forms of cystinosis are each transmitted by separate recessive genes. If a given form of cystinosis occurs in a family, other involved individuals will have the same type of disease.

Since the content of cystine in the cell can be used to diagnose the disease, intrauterine diagnosis has been possible. Pulse labeling with ^{35}S-cystine markedly increases the sensitivity of the assay.[6]

Treatment

There is as yet no definitive treatment for cystinosis. Dietary therapy has received extensive trial, as has penicillamine, but no therapeutic effect has been observed. Supportive therapy involves the correction of renal tubular losses with fluid, potassium and alkali. Vitamin D doses in the range of 5,000–10,000 units along with replacement therapy are sufficient for

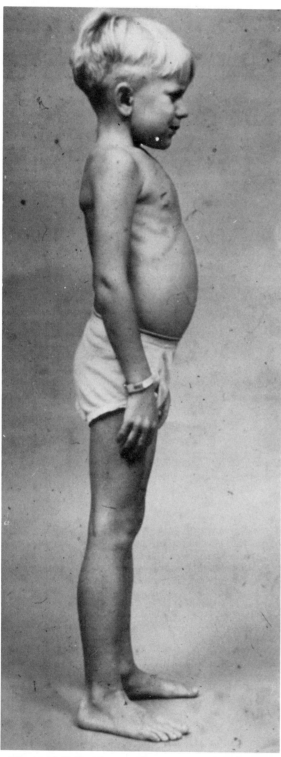

management of the rickets. Cleland's reagent, which reduces disulfides to sulfhydryls, has been used in tissue culture to remove cystine from fibroblasts and experimentally in vivo. It is not yet clear whether this form of treatment is useful.

In classic cystinosis some patients with advanced renal failure have now been treated with renal transplantation. This is rather heroic therapy and probably not generally available, but it is increasingly clear that it is effective and that the transplanted kidney does not, at least within the few years of current experience, develop cystine nephropathy. Hair color has been observed to darken following transplantation.

Benign cystinosis requires no treatment.

REFERENCES

1. Schneider, J. A., and Seegmiller, J. E.: Cystinosis and the Fanconi Syndrome, in Stanbury, J. B., Wyngaarden, J. B., and Fredrickson, D. S. (eds.): *Metabolic Basis of Inherited Diseases* (3d ed.; New York: McGraw-Hill Book Co., 1972), p. 1581.
2. Seegmiller, J. E., Friedmann, T., Harrison, H. E., Wong, V., and Schneider, J. A.: Cystinosis, Combined Clinical Staff Conferences at the National Institutes of Health, Ann. Intern. Med. 68:883, 1968.
3. Schulman, J. D., and Bradley, K. H.: Cystinosis: Selective induction of vacuolation in fibroblasts by L-cysteine-D-penicillamine disulfide, Science 169:595, 1970.
4. Crawhall, J. C., Lietman, P. S., Schneider, J. A., and Seegmiller, J. E.: Cystinosis: Plasma cystine and cysteine concentrations and the effect of D-penicillamine and dietary treatment, Am. J. Med. 44:330, 1968.
5. Schulman, J. D., Fujimoto, W. Y., Bradley, K. H., and Seegmiller, J. E.: Identification of heterozygous genotype for cystinosis in utero by a new pulse-labeling technique: Preliminary report, J. Pediatr. 77:468, 1970.
6. Schneider, J. A.: Cystinosis, in Nyhan, W. L. (ed.): *Heritable Disorders of Amino Acid Metabolism* (New York: John Wiley & Sons, Inc., 1974).

Fig. 4.—W. G. R.[4] A 7-year-old boy with advanced cystinosis had the characteristic very blond hair and pot belly. He was 42.5 inches tall. (Courtesy of Dr. Jerry Schneider of the University of California San Diego.)

RENAL TUBULAR ACIDOSIS

CARDINAL CLINICAL FEATURES

Failure to thrive, shortness of stature, rickets, osteomalacia, hyperchloremic acidosis with low serum bicarbonate, proportionately high urinary pH, nephrocalcinosis.

CLINICAL PICTURE

Renal tubular acidosis (RTA) represents a fundamental defect in renal tubular function in which the normal ability to establish a pH gradient between blood and urine is interfered with. Infants with RTA present with failure to thrive. It may be 6 months to 2 years before they are brought to the attention of a physician who makes the diagnosis; most are seen at about 1 year of age. Most have had loss of appetite for some time, and many have vomited more than the average infant, leading to intestinal interpretation of their failure to thrive. Linear growth is inhibited. Many patients, but not all, become irritable or apathetic. The infant seen at this stage usually shows little on physical examination that adds to the picture (Figs. 1 and 2).

Roentgenographic examination of the bones may reveal findings that are diagnostic of rickets (Fig. 3). However, it should be emphasized that a severely acidotic infant who is not growing may show no signs of rickets. Patients with persistent untreated disease or patients who develop RTA later in infancy or childhood invariably manifest rickets in spite of the administration of vitamin D in the usual doses. Such patients ultimately develop nephrocalcinosis and urinary tract stones. Onset in adult life leads to osteomalacia, often with linear transverse areas of demineralization, or pseudofractures.

A lowered serum concentration of bicarbonate is essential to the diagnosis. In the absence of treatment, patients seldom have a CO_2 content of over 19 mEq/L. On the other hand, the values usually are over 10 mEq/L. The pH of the serum generally is normal. Relative alkalinity of urinary pH is the rule in these patients and critical to diagnosis. It is important to remember, however, that in an acidotic patient a urinary pH of 6 may be relative alkalinity. A pH much over 7 is not often found in patients with RTA. They cannot establish a pH gradient in which the urine:plasma H^+ ratio exceeds 80.1, which is what one obtains when the

pHs of the urine and plasma are 6 and 7.38, respectively; normally, this ratio easily exceeds 800.1, as when the urine pH is 5 and the plasma 7.38.

In the presence of a defect in acidification of the urine, increased amounts of sodium are excreted, much as in acetazolamide-induced diuresis. Osmotic diuresis and polyuria result. There is tubular exchange of sodium for potassium, and thus potassium also is lost. In defense against acidosis, bone resorption takes place, leading to hypercalciuria. The serum calcium is normal. Phosphaturia also is increased, and serum phosphorus may be decreased. These findings lead to changes in the bones. When there is active bone disease, the serum activity of alkaline phosphatase is increased. Hypercalciuria, along with a neutral or less acid urine, and the low urinary citrate also found in this condition, leads to nephrocalcinosis and nephrolithiasis.

Renal tubular acidosis occurs in a number of conditions both genetic and acquired, e.g., in the Fanconi syndrome (see chapter on Cystinosis). Even among patients with idiopathic and classic RTA, there is increasing evidence of heterogeneity. The disorder was first described by Lightwood[1] in 1935. He was struck by renal medullary calcification in six of 850 consecutive autopsies. Examination of the clinical records revealed a syndrome of manifestations that has since come to be recognized as typical of RTA. Lightwood[2] then described 35 infants in whom the full clinical syndrome was documented at approximately 1 year of age.

These observations established the clinical syndrome as seen in the infant. Significantly, the prognosis was excellent and the disorder proved to be transient in this series. Albright and colleagues[3, 4] studied a series of patients with late rickets and osteomalacia in older children and adults, respectively, and established the clinical picture of RTA in these age groups. The disease in these patients was not transient. These investigators clarified the status of what had been known as Milkman's disease—an idiopathic disorder of bone, with bone pains and pseudofractures—as simple osteomalacia with RTA. It is now customary to refer to the transient form of RTA as the Lightwood syndrome and to the permanent disorder as the Albright syndrome; this is the way in which Lightwood himself refers to these disorders. In this sense, the Lightwood syndrome has virtually disappeared. The patients we see even in infancy have a permanent,

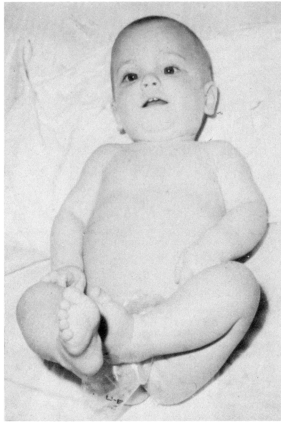

Fig. 1.–D. H. (Case 1), an 11-month-old boy with newly diagnosed renal tubular acidosis. Failure to thrive is evident in his appearance, which is much younger than his stated age. His abdomen was protuberant.

apparently inborn tubular defect. Lightwood's very large experience with infants obviously reflected relatively epidemic use of some nephrotoxin, possibly mercury, at least in England. On the other hand, we cannot escape the feeling that Lightwood's initial series of autopsied patients probably represented the

Fig. 2.–D. H. Close-up of the face illustrating the bright eyes and short, sparse dry hair.

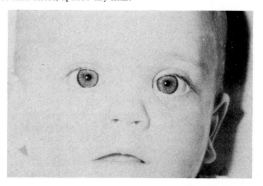

type of infant we see today with nontransient disease. As clinicians, we would more appropriately consider the clinical syndrome described by Lightwood in the infant as the Lightwood syndrome. In this view, the Albright syndrome would refer to the clinical syndrome he described in which there was late onset of disease in adulthood or later childhood.

Among patients with nontransient RTA, Soriano, Boichis and Edelmann[5] have provided evidence for subclassification on the basis of the renal threshold for bicarbonate. They have considered those patients who spill bicarbonate at less than the normal threshold as having proximal RTA. This is consistent with the fact that patients with the Fanconi syndrome who have other signs of proximal tubular defect have a lowered bicarbonate threshold. Morris[6] has considered the evidence for an anatomic site of the defect in patients with bicarbonate-losing to be unconvincing. He has proposed three types of classic RTA: (1) one conforming largely to the distal tubular type of Soriano et al.,[5] in that bicarbonate excretion is trivial at normal or subnormal plasma bicarbonate concentrations; (2) one having a lowered threshold; and (3) one in which bicarbonate is spilled at normal as well as subnormal plasma bicarbonate concentrations. In any case it is clear that there is heterogeneity. Experience with additional patients and genetic analysis should help to clarify the situation.

In the workup of a patient with suspected RTA, the diagnosis often can be made from the clinical picture, a diminished plasma concentration of bicarbonate and a relatively alkaline urine. On the other hand, it is not always easy to distinguish borderline patients from normal. In this case a loading test with ammonium chloride for 3–5 days may be very useful. For these purposes the urine should be collected for a measured period of time and assayed for titratable acidity, NH_4^+ and HCO_3^-.[7] The excretion of H^+ can be calculated from the excretions of titratable acidity and ammonia minus bicarbonate as follows:

$$UV_{H^+} = UV_{TA} + UV_{NH_4^+} - UV_{HCO_3^-}.$$

These data can then be considered in terms of other factors such as the plasma HCO_3^- concentration. Studies of bicarbonate reabsorption and threshold[5, 8] are carried out by the infusion of bicarbonate at a rate that produces a progressive increase in the serum concentration of bicarbonate while monitoring the serum concentration and urinary excretion.

Case Reports

CASE. 1.–D. H. was an 11-month-old white male referred for evaluation of failure to thrive and hypocalcemia. He was the product of a term pregnancy and uncomplicated delivery. Birth weight was 6 lb, 2 oz. The history was of poor appetite, failure to thrive and apathy for as long as the mother could remember. He particularly refused liquids and milk. He sat alone at 8 months but remained too weak to sit long. Two weeks prior to admission to University Hospital, he was hospitalized elsewhere for pneumonia. At that time he was found to

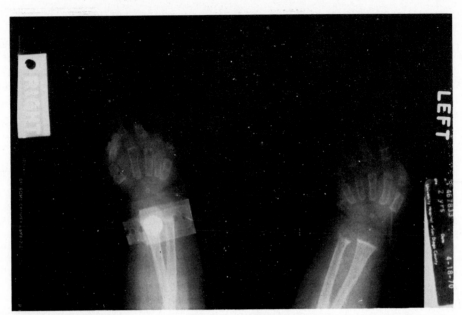

Fig. 3.—D. H. Roentgenographic examinations of the bones showing early rickets at the wrist.

have severe growth retardation and hypocalcemia. He was treated with oral calcium and referred for further study.

There was no similar condition in the family. Two siblings appeared well. A 2-year-old sister was short, but a bone survey and electrolytes were normal.

Physical examination revealed a height of 25 inches and a weight of 12 lb, 13 oz. Height age was that of a 4 month old, and weight age that of a 6 month old. The patient appeared to be poorly developed, irritable and querulous (see Figs. 1 and 2). Examination was otherwise normal. Ophthalmologic examination revealed no crystal deposits in the cornea.

Laboratory data included a hemoglobin of 12.6 gm/100 ml and a hematocrit of 40. The WBC count was 8,800. Electrolyte concentrations were CO_2, 15; chloride, 108; sodium, 136; potassium, 5.1 mEq/L. The calcium was 7 mg/100 ml and phosphorus 4.4. Alkaline phosphatase was 35 Bodansky units. Urinary pH was tested several times and found to be 8.1, 7.9, 8.0 and 6.0. Specific gravity of the urine was 1.014, 1.007 and 1.022. Creatinine clearance was 59 ml/minute, which is normal. There was no glycosuria and no aminoaciduria. A 6.5-hour water deprivation test was normal, indicating no defect in renal concentrating mechanisms. An ammonium chloride load was given (NH_4Cl, 100 mEq/1.73 M^2 S.A./24 hours in three to four doses orally). There was no increase in titratable acidity over a 4-day period of study. Calcium excretion was 2 mg/24 hours, which is normal, whereas 187 mg of phosphate was excreted in 24 hours, indicating a distinct phosphaturia. Tubular reabsorption of phosphorus was 79% (normal is over 90%). Glucose excretion was only 26 mg/24 hours. A bicarbonate loading study was done using 40 mEq/200 cc. The serum bicarbonate rose from 14 to as high as 22.4 mEq/L. The urinary bicarbonate revealed increased excretion of bicarbonate, whereas the serum concentration was as low as 18.5 mEq/L. These data indicate that the child had a low bicarbonate threshold, which is consistent with proximal RTA.

Roentgenographic views of the spine, skull and lower extremities demonstrated evidence of fraying and cupping of metaphyseal ends of the bones. The findings were diagnostic of rickets, but the extent of the disease was mild (see Fig. 3).

There was widespread demineralization of bony structures. A bone age of 4 months was at the lower limits of 2 s.d. There was no nephrocalcinosis. An intravenous pyelogram showed a normal collecting system and good function.

The infant was treated with Shohl's solution: citric acid: 140 gm, sodium citrate: 98 gm in 1,000 ml of water. While receiving a dose of 15 ml b.i.d., his CO_2 returned to normal. He was given vitamin D, 400 units daily. One month after discharge his weight had increased from 12 lb, 13 oz to 15 lb, 10 oz, and his height had increased from 25 to 26 1/2 inches. Meanwhile, he had a complete personality change. He became friendly, outgoing and pleasant to be with. He ate hungrily and drank milk. Muscle tone improved.

CASE 2.—T. K., a 20-month-old boy of Portuguese-Hawaiian descent, was seen with a referral diagnosis of RTA, rickets and nephrocalcinosis. He had been hospitalized elsewhere 1 month previously for bronchiolitis and was found to have hyperchloremic acidosis, hypokalemia, hypocalcemia, hypophosphatemia and an alkaline urine. On admission, electrolyte concentrations were sodium, 144; potassium, 2.5; chloride, 106; and CO_2, 5.8 mEq/L. The pH was 7.16 and BUN 6.6 mg/100 ml. Three days later the chloride concentration was 117 mEq/L. The calcium was 7.5 and phosphorus 2.2 mg/100 ml. The urinary pH was 8.

The patient was the product of an uncomplicated pregnancy and delivery. Birth weight was 5 lb, 3 oz. He began refusing solids early in infancy and demonstrated excessive thirst and polyuria. The most striking feature was a failure to thrive. At the age of 5 months the infant weighed 6–8 lb; at 19 months, 10 lb and at 17 months, 11 lb. He sat at 8 months, stood at 10 months and walked at 18 months.

There was no family history of renal problems. The mother, father and one sibling were living and in good health.

Significant physical findings were a height of 26 1/2 inches (height age 6 months and certainly less than the 3d percentile) and a weight of 15 1/2 lb (less than the 3d percentile). There were six deciduous teeth. There was a diastasis recti and slight bowing of both lower extremities.

X-rays revealed rickets (Fig. 4). The distal ends of the long

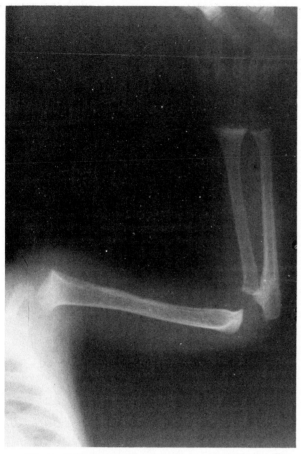

Fig. 4.—T. K. (Case 2). Roentgenograms of the bones, illustrating more advanced rickets.

bones revealed fraying and cupping at the metaphyseal ends and poor calcification of the provisional zone of calcification. There was slight inward flaring of the metaphyseal areas of the weight-bearing parts of the femora and upper tibiae. There was no evidence of nephrocalcinosis. An intravenous pyelogram was normal.

The patient was treated initially with a solution of 50 gm of sodium and 50 gm of potassium citrate in 500 cc of water. He also was given 2,000 units of vitamin D daily. The solution was later changed to sodium citrate, 45 gm; potassium citrate, 45 gm; and citric acid, 140 gm and the vitamin D was stopped. As soon as treatment was started, his excessive thirst and urination disappeared. He grew from his original weight of 15 lb and height of 25 inches to a weight of 26 lb and a height of 35 inches at 3 years of age. The electrolytes returned to normal and the rickets cleared.

GENETICS

It has begun to be apparent that RTA is heterogeneous clinically and physiologically. Therefore, it is likely that it also is heterogeneous genetically. Information is not available to assess the genetics of RTA with renal bicarbonate wasting. In most patients with RTA of any sort, the pattern has been sporadic. However, a few important families have been described in which a number of patients have been observed. From these it is clear that RTA can be transmitted as an autosomal dominant. It may be that other mechanisms apply in other families.

TREATMENT

Treatment of RTA is fundamentally the treatment of acidosis. Shohl's solution of citric acid and sodium citrate is very useful for this purpose. This solution is given in doses sufficient to keep plasma CO_2 and chloride in the normal range. Some patients may require supplemental potassium, especially early in therapy. Extra quantities of calcium or vitamin D are not required. Phosphaturia subsides with treatment of the acidosis. Rickets and osteomalacia heal and linear growth is initiated. Patients usually begin to eat, and they promptly lose their irritability or apathy. The response may be quite dramatic.

REFERENCES

1. Lightwood, R.: Calcium infarction of kidneys in infants, Arch. Dis. Child. 10:205, 1935.
2. Lightwood, R., Payne, W. W., and Black, J. A.: Infantile renal acidosis, Pediatrics 12:628, 1953.
3. Albright, F., Consolazio, W. V., Coombs, F. S., Sulkowitch, H. W., and Talbott, J. H.: Metabolic studies and therapy in a case of nephrocalcinosis with rickets and dwarfism, Johns Hopkins Med. J. 66:7, 1940.
4. Albright, F., Burnett, C. H., Parson, W., Reifenstein, E. C., and Roos, A.: Osteomalacia and late rickets, Medicine 25:399, 1946.
5. Soriano, J. R., Biochis, H., and Edelmann, C. M.: Bicarbonate reabsorption and hydrogen ion excretion in children with renal tubular acidosis, J. Pediatr. 71:802, 1967.
6. Morris, R. C.: Renal tubular acidosis. Mechanisms, classification, and implications, N. Engl. J. Med. 281:1405, 1969.
7. Elkinton, J. R., Huth, D. J., Webster, G. D., and McCance, R. A.: The renal excretion of hydrogen ion in renal tubular acidosis, Am. J. Med. 29:554, 1960.
8. Edelmann, C. M., Soriano, J. R., Biochis, H., Gruskin, A. B., and Acosta, M. I.: Renal bicarbonate reabsorption and hydrogen ion excretion in normal infants, J. Clin. Invest. 46:1309, 1967.

HYPOPHOSPHATEMIC VITAMIN D-RESISTANT RICKETS
Vitamin D-Resistant Rickets, X-Linked Hypophosphatemic Rickets, Familial (Hereditary) Hypophosphatemia

Sherwood A. Libit and William J. Lewis

CARDINAL CLINICAL FEATURES

Short stature, active rickets, hypophosphatemia and phosphaturia, unresponsiveness to usual doses of vitamin D in absence of renal failure or multiple renal tubular defects.

CLINICAL PICTURE

Hypophosphatemic vitamin D-resistant rickets is an X-linked dominant inborn abnormality of renal tubular reabsorption of phosphate.[1, 2] The earliest definitive description was that of Albright, Butler and Bloomberg[3] in 1937. Patients with this syndrome have a marked phosphaturia that is unresponsive to vitamin D therapy. In children the values for alkaline phosphatase in the serum are always elevated. The concentration of calcium is normal or only slightly reduced. The primary defect appears to be a decrease in proximal renal tubular reabsorption of phosphate. Other suggested defects have been defective gastrointestinal absorption of calcium, abnormal metabolism of vitamin D or a diminished rate of new bone formation. Studies of gastrointestinal calcium absorption have been normal. There is no evidence of secondary hyperparathyroidism. Vitamin D appears to have a prolonged half-life in these patients.[4] However, use of the active metabolite of vitamin D (1,25-dihydroxycholecalciferol) in physiologic amounts has not corrected the biochemical findings in this disease.[5] The diagnosis is generally made by demonstrating that serum concentration of phosphate is low for the patient's age and that there is active rickets that does not respond to the usual dose of vitamin D.

Infants with this disease are normal at birth. The roentgenographic signs of rickets appear in the middle of the first year, and there is progressive physical deformity during infancy. Particularly prominent is bowing of the lower extremities when weight-bearing is initiated. The patient may have a waddling gait. Genu valgum also may be seen. The roentgenographic findings of rickets include coarsened bony trabecular pattern, cystic-appearing areas in the metaphysis and epiphysis and shortening and broadening of long bones. Coxa vara is common. The growth rate in these

Fig. 1. — Vitamin D and its metabolism.

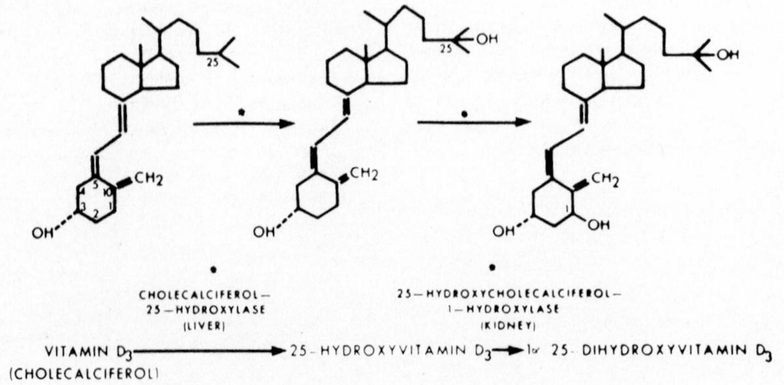

patients is slow once signs of rickets appear.[6] The head may be dolichocephalic. Craniosynostosis may develop, and convulsions during infancy have been noted in a few children. In adults pseudofractures are common and abnormal bony protuberances develop at the sites of muscle attachments.[7] Osteomalacia and pseudofractures may cause bone pain, and the bony protuberances at the site of muscle attachments may lead to limitation of joint mobility. Degenerative ar-

Fig. 2.–C. G. (Case Report) and her 21-year-old mother. Both had vitamin D-resistant rickets.

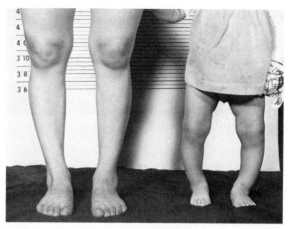

Fig. 3.–C. G. View of the legs illustrates the bowing, much worse in the little girl.

thritis may develop. Spinal cord compression has been reported in untreated adults.[8]

The teeth have large pulp chambers and there is enamel hypoplasia.[9] Eruption of the teeth may be slow. Gingival and periapical infections are common. The rachitic deformities do not affect intellectual function or life expectancy.

Fig. 4.–C. G. Roentgenogram of the lower extremities, illustrating florid rickets. There is enlargement and fraying and cupping of the ends of the shafts of the long bones.

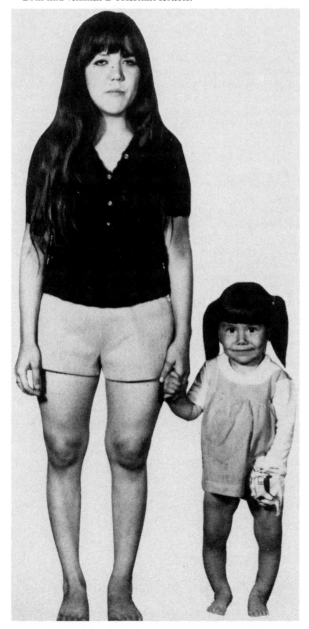

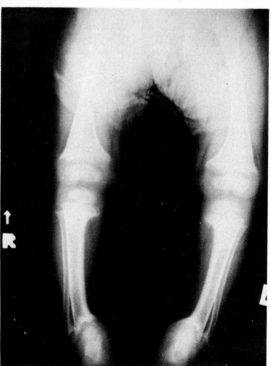

Differential Diagnosis

It is still true in 1975 that most patients we see with rickets have vitamin D deficiency. This is true even in Florida and southern California. Such patients usually have a modest aminoaciduria, in contrast to patients with vitamin D-resistant rickets. For this reason it is useful to treat infants with rickets first with the usual therapeutic doses of vitamin D. A regimen in which 1,000–1,500 I.U. is given daily usually restores the serum phosphate to normal within a week. Roentgenographic improvement occurs over 2–6 weeks. The use of larger doses, especially on the assumption that vitamin D deficiency has disappeared, may cloud the picture.

A variety of renal tubular disorders are accompanied by rickets. These include the Fanconi syndrome and cystinosis (see chapter, this section), the Lowe syndrome and renal tubular acidosis (see chapter, this section). Rickets also is seen in tyrosinosis. It may be seen in patients with steatorrhea, malabsorption or celiac disease, or in patients with hepatic biliary obstruction.

Another form of refractory rickets was described in 1961 by Prader, Illig and Heierli[10] as hereditary pseudo-deficiency rickets. In this disorder rickets occurs during the first year. There is little hypophosphatemia and a stubborn hypocalcemia. Enamel defects involve the permanent teeth. There may be hyperaminoaciduria. Remineralization and return of the serum calcium to normal is observed with doses of vitamin D from 50,000–100,000 I.U./day. The disorder appears to be transmitted by an autosomal gene. Inheritance is recessive.[11] This disorder is thought to result from an inborn error of vitamin D metabolism in which the conversion of 25-hydroxyvitamin D to 1α, 25-dihydroxyvitamin D (Fig. 1) by an enzyme in the kidney is defective. The argument for this hypothesis is that, whereas massive doses of vitamin D or 25-hydroxyvitamin D are required to heal the rachitic lesions, tiny, presumably physiologic doses of 1α, 25-dihydroxyvitamin D accomplish the same effect.[11, 12]

Case Report

C. G. was a 1¼-year-old Mexican-American female who presented to the Naval Regional Medical Center, San Diego, because of marked bowing of her lower extremities (Fig. 2). She was the 7 lb product of a normal pregnancy and full-term delivery to a 20-year-old primigravida mother. The child's general health had been good and she had received vitamin D-fortified formula or milk up to the time of examination. Intellectual development was normal, as was tooth eruption.

The child's mother was 56 inches tall. She had obvious bowing of the lower extremities (Figs. 2 and 3). The father was 67 inches tall and had normal leg structure. The maternal grandfather was reported to be of short stature.

The child was 31 inches long. This was less than the third percentile. Her weight was 21½ lb, which was in the 10th percentile. Head size was normal. She had mild frontal bossing. There was lateral bowing of both femora and tibiae (see Fig. 3). There was palpable widening of the distal epiphyses of tibiae and radii.

Serum calcium was 9 mg/100 ml, phosphorus 3.2 mg/100 ml and alkaline phosphatase 500 mU/ml. Tubular reabsorption of phosphate was calculated at 60%. Normal values for serum electrolytes, urea nitrogen, creatinine, creatinine clearance, urinary amino acids, urinary acidification and concentration excluded other significant associated renal defects. Roentgenograms of the long bones showed active rickets (Fig. 4). The child was treated with oral phosphate solution and vitamin D 50,000 I.U. daily. Healing of the rickets was observed.

Genetics

Hypophosphatemic vitamin D-resistant rickets is an X-linked dominant disorder.[1] Males are characteristically more severely involved than females. Individuals, usually female, may be identified in whom there is hypophosphatemia but no significant bone disease or growth retardation.

Treatment

The use of vitamin D in large doses has been the usual approach to therapy in this disease. However, it remains controversial. The usual recommended dosage has been 50,000–150,000 I.U./day. The danger of vitamin D in high dosage is, of course, the occurrence of nephrocalcinosis and renal damage. McNair and Stickler[13] have observed the regular occurrence of episodes of hypercalcemia with therapy and little over-all benefit in ultimate height. The use of an oral phosphate supplement, along with lower doses of vitamin D, appears to offer an attractive alternative method of treatment.[14] This treatment will heal active rickets in a child and the bony complications seen in the adult. However, these are still substantial doses of vitamin D, in excess of 50,000 I.U./day, and they carry a continued risk of hypercalcemia and nephrocalcinosis. The newer metabolites of vitamin D appear to have the same side-effects, and they are required in this disease in amounts in excess of the equivalent of 25,000 I.U. of vitamin D daily. Frequent monitoring of serum and urinary calcium is a recommended part of the treatment program for these children. Adults with this disorder do not have active rickets and so the argument that this treatment may be dangerous is cogent. They may require orthopedic correction for bony deformity of the lower extremities.

REFERENCES

1. Winters, R. W., Graham, J. B., William, T. F., McFalls, V. W., and Burnett, C. H.: A genetic study of familial hypophosphatemia and vitamin D resistant rickets with a review of the literature, Medicine 37:97, 1958.

2. Fraser, D., and Salter, R. B.: The diagnosis and management of the various types of rickets, Pediatr. Clin. North Am. 5:417, 1958.

3. Albright, F., Butler, A. M., and Bloomberg, E.: Rickets resistant to vitamin D therapy, Am. J. Dis. Child. 54:529, 1937.

4. DeLuca, H. F., Lund, J., Rosenbloom, A., and Lobech, C. C.: Metabolism of tritiated vitamin D_3 in familial vitamin-D-resistant rickets with hypophosphatemia, J. Pediatr. 70: 828, 1967.

5. Brickman, A. S., Coburn, J. W., Jurokawa, K., Bethune, J. E., Harrison, N. E., and Norman, A. W.: Action of 1,25-dihydroxycholecalciferol in patients with hypophosphatemic, vitamin-D-resistant rickets, N. Engl. J. Med. 289:495, 1974.

6. Harrison, H. E., Harrison, H. C., Lifshitz, F., and Johnson, A. D.: Growth disturbance in hereditary hypophosphatemia, Am. J. Dis. Child. 112:290, 1966.

7. Williams, T. F., and Winters, R. W.: Familial (Hereditary) Vitamin D-Resistant Rickets with Hypophosphatemia, in Stanbury, J. B., Wyngaarden, J. B., and Fredrickson, D. S. (eds.): *The Metabolic Basis of Inherited Disease* (3d ed.; New York: McGraw-Hill Book Co., 1972), pp. 1465–1485.

8. Yoshikama, S., Shiba, M., and Susuki, A.: Spinal-cord compression in untreated adult cases of vitamin-D-resistant rickets, J. Bone Joint Surg. [Am] 50:743, 1968.

9. Archard, H. O., and Witkop, C. J., Jr.: Hereditary hypophosphatemia (vitamin D-resistant rickets) presenting primary dental manifestations, Oral Surg. 22:184, 1966.

10. Prader, A., Illig, R., and Heierli, E.: Eine besondere Form der primären Vitamin-D-resistenten Rachitis mit Hypocalcämie und autosomaldominantem Erbang: Die hereditäre Pseudo-mangelrachitis, Helv. Paediatr. Acta 16: 452, 1961.

11. Fraser, D., Kooh, S. W., Kind, P., Holick, M. F., Tanaka, Y., and DeLuca, H. F.: Pathogenesis of hereditary vitamin-D-dependent rickets. An inborn error of vitamin D metabolism involving defective conversion of 25-hydroxyvitamin D to 1α,25-dihydroxyvitamin D, N. Engl. J. Med. 289:817, 1973.

12. Prader, A.: Pseudo vitamin D deficiency (vitamin D dependency), Proceedings of XIV International Congress of Pediatrics, No. 10. Genetics. Metabolism, 1973, Buenos Aires, p. 106.

13. McNair, S. L., and Stickler, G. B.: Growth in familial hypophosphatemic vitamin-D-resistant rickets, N. Engl. J. Med. 281:511, 1969.

14. Glorieux, F. H., Scriver, C. R., Reade, T. M., Goldman, H., and Roseborough, A.: Use of phosphate and vitamin D to prevent dwarfism and rickets in X-linked hypophosphatemia, N. Engl. J. Med. 287:481, 1972.

Cytogenetic Syndromes

INTRODUCTION
Syndromes Associated with Chromosomal Abnormalities

Uta Francke

Human cytogenetics is a fairly new discipline. For less than 20 years, it has been possible to study human chromosomes in an easily accessible tissue, the peripheral blood lymphocyte. The correlation of distinct clinical syndromes with recognizable changes in the chromosomes has led to definition of the now classic chromosomal disorders presented in the following chapters.

In more recent years, differential staining methods have been developed that produce patterns of bright and dark bands that are characteristic for each chromosome. These transverse banding patterns serve to identify each individual chromosome pair and provide landmarks along the chromosomal strand. It is now possible to distinguish more than 300 regions in the haploid human genome and this has greatly enhanced the resolution of studies of human chromosomes.

In current terminology, Q-, G-, R- and T-banding are used for chromosome identification.[1]

Q-banding, obtained by staining the chromosomes with quinacrine or quinacrine mustard, and G-banding, produced by Giemsa-staining after various pretreatments, result in basically identical banding patterns with exception of the heterochromatic regions discussed below.

Heat denaturation technics producing R-banding[1] or T-banding[2] result in a reverse banding pattern. The chromosome regions that are stained intensely with Q and G, stain weakly with R or T, and vice versa. The Q- and G-banding methods are most widely used for routine chromosome identification. The R and T technics have the advantage of better defining chromosome ends and, thus, are useful in detecting terminal deletions.

Q-banding is particularly useful in studying the Y chromosome, as it produces brilliant fluorescence in the heterochromatic region on the distal long arm of the Y chromosome. This remains identifiable as the brilliant Y-body in the nucleus of interphase cells. This property permits the sexing of cells from buccal smears, hair roots and amniotic fluid. The Y-body identifies the male and is complementary to testing for the Barr body, which identifies the female.

C-banding and Giemsa II staining distinguish certain heterochromatic, often polymorphic, regions on a number of chromosomes.[3] Common and less common variations in the size and position of these C-band regions are heritable characteristics of the individual chromosomes and are generally not associated with any clinical manifestation.[1]

Applying one or more of the new banding methods, the specific chromosome involved in each of the classic chromosomal syndromes has been identified unequivocally. The chromosome that is trisomic in the Down syndrome has been named chromosome 21. The new identifications have been correlated with the results of radioautographic studies in other classic syndromes, such as the 4p−, 5p−, 13q− or 13 ring, trisomy 13, trisomy 18, 18p−, 18q− and 18 ring syndromes.

With the increased resolution of the new banding technics, previously unidentifiable chromosomal changes are now being recognized. These include exchanges of nearly equal-sized pieces between chromosomes or small deletions from an end or from the middle of a chromosome. An increasing number of clinical syndromes associated with autosomal partial trisomies and deletions are now being described. Among the best characterized so far are trisomy for the short arms of chromosome 4 (trisomy 4p[4]) and of chromosome 9 (trisomy 9p[5]). Further distinct clinical syndromes begin to emerge as results of trisomies for parts of the long arms of chromosomes 7 (trisomy 7q[6]) and 10 (trisomy 10q[7]).

REFERENCES

1. *Standardization in Human Cytogenetics. Paris Conference*, Birth Defects: Original Article Series, Vol. VII, No. 7 (New York: The National Foundation-March of Dimes, 1971).
2. Dutrillaux, B.: Nouveau système de marquage chromosomique: Les bandes T, Chromosoma 41:395, 1973.
3. Bobrow, M., Madan, K., and Pearson, P. L.: Staining of some specific regions of human chromosomes, particularly the secondary constriction of No. 9, Nature 238:122, 1972.
4. Rethoré, M. O., Dutrillaux, B., Giovannelli, G., Forabosco,

A., Dallapiccola, B., and Lejeune, J.: La trisomie 4p, Ann. Genet. (Paris) 17:125, 1974.

5. Rethoré, M. O., Hoehn, H., Rott, H., Couturier, J., Dutrillaux, B., and Lejeune, J.: Analyse de la trisomie 9p par dénaturation ménagée, Humangenetik 18:129, 1973.

6. Vogel, W., Siebers, J., and Reinwein, H.: Partial trisomy 7q, Ann. Genet. (Paris) 16:277, 1973.

7. Yunis, J., and Sanchez, O.: A new syndrome resulting from partial trisomy for the distal long arm of chromosome 10, J. Pediatr. 84:567, 1974.

TRISOMY 18
Trisomy E, Edwards Syndrome

Cardinal Clinical Features

Multiple anomalies including micrognathia, low-set, malformed ears; characteristically a failure of fisting in which the index fingers overlie the third fingers; hypoplasia of the nails, especially of the fifth fingers and toes; rocker-bottom feet. Also, low birth weight for gestational age and severe degree of retardation of growth and mental development, usually with early demise; congenital heart disease; narrow pelvis with limited abduction of the hip; low-arch dermal ridge pattern on the fingertips.

Clinical Picture

The 18 trisomy syndrome was first described by Edwards and colleagues[1] in 1960. Smith, Patau and colleagues[2] in the same year provided a fuller clinical description of two patients with the syndrome and characterized it as due to trisomy of chromosome 18 rather than 17. Most patients have a characteristic over-all appearance that permits their immediate diagnosis[2-4] (Figs. 1 and 2). The first appearance is of a baby with a prominent occiput and elongation of the skull in its anterioposterior diameter, as well as a very narrow pelvis. A preponderance of these patients is female. Closer inspection of the face, which may appear round, reveals a narrow forehead with frontal bossing, hypertelorism and prominent epicanthal folds. Ptosis is common. The palpebral fissures are narrow, small and often antimongoloid, and there may be microphthalmus, corneal or lenticular opacities or colobomas of the optic disc or uvea. Congenital glaucoma or optic atrophy is rarely seen. The ears are low set and malformed. Deafness is common among these children. The nasal bridge is depressed, but the nose has been described as Grecian. These infants have micrognathia, a high-arched palate and a small mouth. Clefts of the lip or palate are sometimes seen. The neck is short, and there may be redundant skin at the nape of the neck or an appearance of webbing.

Other characteristic features include a short sternum and a narrow chest. Patients almost always have congenital heart disease, usually a ventricular septal defect or a patent ductus arteriosus.[5] Atrial septal defect, dextrocardia and pulmonic stenosis also have

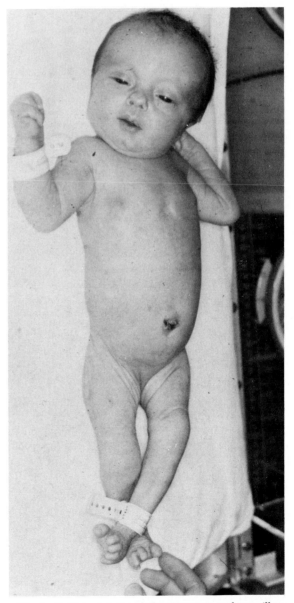

Fig. 1.—G. H. An infant with the 18 trisomy syndrome, illustrating the over-all appearance, with typical hands, face and feet, and a very narrow pelvis.

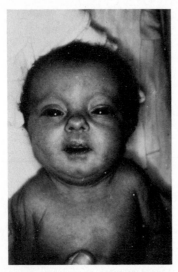

Fig. 2.—G. H. The tip of the nose was prominent and coarse, somewhat reminiscent of the peasants in the paintings of Brueghel. Otherwise, the facial features in this syndrome are rather fine.

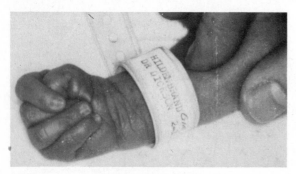

Fig. 3.—G. H. The hand.

been reported. Patients may have renal anomalies, the most common of which is the horseshoe kidney. Cystic kidneys, or duplications of the ureter or kidney, are common. Pyloric stenosis has been described, as has tracheoesophageal fistula. Inguinal and umbilical hernias are common. Imperforate anus has been reported[6] and esophageal atresia without fistula.[6, 8] The pelvis is characteristically very narrow, and the hips are frequently dislocated. Genitalia often are abnormal, in that in males there are undescended testes and a microphallus or, in females, hypoplastic labia major and a hypertrophic clitoris.

The typical position of the hand is clenched, with flexion contractures of the fingers and underfolded thumb (Fig. 3). The index and fifth fingers characteristically override the third and fourth fingers. There is ulnar deviation of the wrist. The thumbs are proximally placed. They may be rudimentary, or even absent. The thumb may overlap the next digit. The thenar eminence and its underlying muscles are hypoplastic, as are the nails.

Dermatoglyphic analysis (see section on Dermatoglyphics) reveals a greatly simplified fingertip pattern with a predominance of arches on the fingertips. This abnormality may be associated with absent or rudimentary flexion creases on the fingers. Simian creases are common, and there are distally placed axial triradii.[9, 10] Some patients have exhibited an unusual so-called "surrender posture" with adduction at the shoulders and flexion of the elbows.[11] There often is syndactyly of the second and third toes, and the great toe may be unusually short, broad and dorsiflexed. A rocker-bottom foot in which there is calcaneovalgus deformity or a prominent calcaneus is characteristic.

Generally these babies have low birth weights for their gestational age. They continue to have severe growth retardation and mental retardation. They have feeding difficulties, a poor sucking reflex and a high-pitched cry. They often have attacks of apnea. Most of these babies die in infancy, before 6 months of age. Exceptional cases have been reported in two patients who survived until the ages of 11 and 15 years, respectively.[12, 13] The first of these patients survived a Wilms' tumor. The physical characteristics of patients tend to become less striking as they grow older.

CASE REPORT

S. E., a 2-week-old female, was admitted to University Hospital for evaluation of congenital heart disease and multiple congenital anomalies. She had a poor sucking reflex and was difficult to feed. She was the product of a normal pregnancy of 38 weeks. Her mother was 20 years old during pregnancy. There was no maternal history of illness, drugs or exposure to x-ray during pregnancy.

On examination the child weighed 4 lb, 1 oz and was 18 inches in length. She had a high-pitched cry and was hypertonic. She had prominent stigmata of the 18 trisomy syndrome (Figs. 4–7). X-rays revealed a smooth-bordered density in the right lower lobe area. There was cardiomegaly, with an upturned apex suggestive of right ventricular hypertrophy.

While in the hospital, the child had repeated spells of apnea and cyanosis. She required gavage feeding and gained weight poorly. Two weeks after admission the baby had a cardiac arrest following an apneic episode and died. Autopsy revealed congenital heart disease. There was a patent foramen ovale, patent ductus arteriosus, subpulmonic infundibular stenosis, a high interventricular septal defect with a bicuspid pulmonic valve and right ventricular hypertrophy. The baby also had Meckel's diverticulum and eventration of the diaphragm.

Chromosome study revealed 18 trisomy (Fig. 8).

GENETICS

18 Trisomy usually occurs sporadically and probably is the result of nondisjunction in the first or second meiotic division during the formation of the gamete. That the extra chromosome is an 18 may be confirmed by autoradiography, as chromosome no. 18 is more heavily labeled than is no. 17. Staining with quinacrine mustard or Giemsa stain also can provide posi-

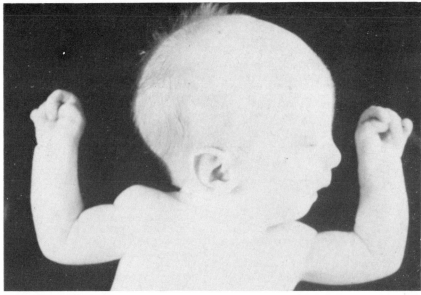

Fig. 4.—S. E. (Case Report). Stigmata of the 18 trisomy syndrome. The infant was tiny and frail and had a feeble cry. The hands were clenched and revealed abnormal fisting. The occiput was prominent, the mandible small and the ears low set.

Fig. 5.—S. E. The hand shows the classic overriding of the index finger on the next finger, and the adducted thumb. The fifth finger also overrides the ring and third finger. The low-set, malformed ear is visible.

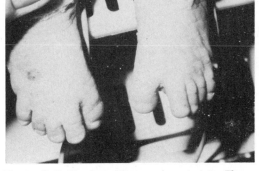

Fig. 6.—S. E. The feet also were characteristic. The great toe was short. It usually was held in dorsiflexion, although not in this picture. There was some syndactyly of the second, third and fourth toes. The fifth toe also was proximally placed.

Fig. 8.—S. E. Karyotype. (Courtesy of Dr. Uta Francke of the University of California San Diego.)

Fig. 7.—S. E. The neck. The cutis laxa, producing redundant skin at the nape of the neck in this syndrome, illustrates clearly that the Turner syndrome is not the only cause of this manifestation. This finding in a young infant always is an indication for cytogenetic study.

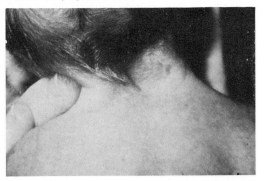

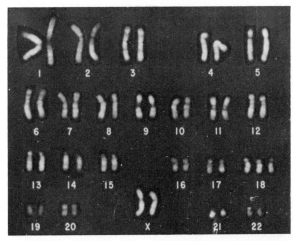

tive identification of chromosome 18. An occasional patient is trisomic for more than one chromosome, such as 18 and 13 or XXX. Increased maternal age is characteristic of the 18 trisomy syndrome.

TREATMENT

None.

REFERENCES

1. Edwards, J. H., Harnden, D. G., Cameron, A. H., Crosse, V. M. and Wolff, O. H.: A new trisomic syndrome, Lancet 1: 787, 1960.
2. Smith, D. W., Patau, K., Therman, E., and Inhorn, S. L.: A new autosomal trisomy syndrome: Multiple congenital anomalies caused by an extra chromosome, J. Pediatr. 57: 338, 1960.
3. Taylor, A. I.: Autosomal trisomy syndromes: A detailed study of 27 cases of Edwards' syndrome and 27 cases of Patau's syndrome, J. Med. Genet. 5:227, 1968.
4. James, A. E., Belcourt, C. L., Atkins, L., and Janower, M. L.: Trisomy 18, Radiology 92:37, 1969.
5. Warkany, J., Passarge, E., and Smith, L. B.: Congenital malformations in autosomal trisomy syndromes, Am. J. Dis. Child. 112:502, 1966.
6. Grosfeld, J. L., Kontras, S. B., and Sommer, A.: Chromosomal abnormalities of the E group and surgical anomalies in neonates, Surgery 69:451, 1971.
7. Weber, F. M., and Sparkes, R. S.: Trisomy E (18) Syndrome: Clinical spectrum in 12 new cases, including chromosome autoradiography in 4, J. Med. Genet. 7:363, 1970.
8. Benady, S. G., and Harris, R. J.: Trisomy 17–18. A study of five cases, three of whom were associated with oesophageal atresia, Acta Paediatr. Scand. 58:445, 1969.
9. Taylor, A. I., and Polani, P. E.: Autosomal trisomy syndromes, excluding Down's, Guys Hosp. Rep. 13:231, 1964.
10. Penrose, L. S.: Dermatoglyphics in trisomy 17 or 18, J. Ment. Defic. Res. 13:44, 1969.
11. Butler, L. J., Snodgrass, G. J. A. I., France, N. E., Sinclair, L., and Russell, A.: E (16–18) trisomy syndrome: Analysis of 13 cases, Arch. Dis. Child. 40:600, 1965.
12. Surana, R. B., Bain, H. W., and Conen, P. E.: 18-Trisomy in a 15-year-old girl, Am. J. Dis. Child. 123:75, 1972.
13. Geiser, C. F., and Schindler, A. M.: Long survival in a male with 18-trisomy syndrome and Wilms' tumor, Pediatrics 44:111, 1969.

TRISOMY 8
Trisomy 8 Mosaicism
Kenneth W. Dumars and Pamela J. Reed

CARDINAL CLINICAL FEATURES

Absent patellae; mental retardation; facies notable for its anteverted nose, long philtrum, micrognathia and malformed ears; flexion deformities of the fingers or toes; deep plantar V-shaped cleft between the first and third interdigital web of foot, "pli capitonne"; trisomy for C-8.

CLINICAL PICTURE

Trisomy 8 has been recognized cytogenetically only since the advent of chromosomal banding. Prior to that time it was impossible to identify with certainty the pairs within the C group, 6–12. To date, more than 30 patients have been identified as having trisomy 8.[1-7]

Children with trisomy 8 have a striking resemblance to each other in early childhood. They tend to be short and slender with a very narrow pelvis. The face has characteristically an anteversion of the nose; a long philtrum; a large, protruding, or everted, rather petulant-appearing lower lip; and a shallow nasal bridge with a relatively broad root, giving the eyes a wide-set appearance. A cleft of the soft palate frequently is present. The ears are malformed and usually large; an absence of the normal convolutions gives a sail or floppy appearance. They may be low set.[1] The mandible is small. Most children have unilateral or bilateral strabismus. Corneal opacity is frequent.

The skeletal abnormalities in this syndrome are unusual and their combination appears to be distinct. Absence of the patellae is one of the most interesting. Another is the deep plantar furrow or cleft between the first and third interdigital web of the foot, which has been called the pli capitonne and is associated with a large retroflexed great toe.[7] Abnormally deep furrows also may be seen in the palm. Joint mobility is limited in the fingers and toes, usually in flexion. In our patient (Case Report), limitation of motion, particularly of the metacarpal-phalangeal and interphalangeal joints, was striking. Articular mobility also may be limited in the large joints such as the knee. The articular abnormalities may be progressive. Limited supination at the elbow has been reported. Club foot or pes planus may be present. There is clinodactyly of the fifth finger. Anomalies of the vertebral column are common and include spina bifida occulta in unusual locations such as C-1 or L-1.[2] Patients have been reported with six lumbar vertebrae.[2] There may be a Sprengel's deformity.

Urinary tract anomalies are sufficiently common in

Fig. 1.—D. A. F. (Case Report). A patient with trisomy 8, at age 3½ years. The corneal opacity, ptosis and esotropia are evident on the left side of the face, as well as the long philtrum, slender pelvis and absent patellae.

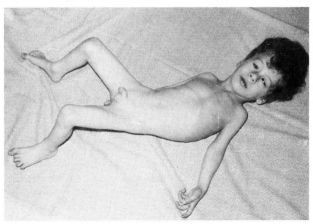

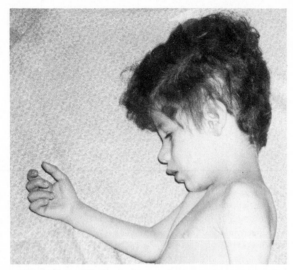

Fig. 2.—D. A. F. The ears appeared large, and there was anteversion of the nose, a long philtrum and retrognathia.

this syndrome that its diagnosis merits study of the urinary tract. Hypospadias is frequent, as is cryptorchidism. Radiographic examination of the urinary tract often reveals reflux and uretero-pelvi-calyceal dilatation, or hydronephrosis. Most patients have congenital heart disease. Atrial and ventricular septal defects have been described. Anomalies of the great vessels also have been reported, including an aortic arch with vascular ring on the right side. The linear growth of children affected with C trisomy, in contrast to many of the chromosomal abnormalities, is not strikingly delayed, i.e., below the third percentile. To date the life expectancy of these children is unknown. Cerebral anomalies such as absence of the corpus callosum and hydrocephalus may be present. Patients with this syndrome may have a mild mental retarda-

tion in the 70 IQ range, characterized by problems with language conceptualization and visual-motor coordination,[2] or mental retardation may be severe (Case Report).

Dermatoglyphic analysis usually reveals a distal axial triradius and an increased number of arches.[2, 6] Simian creases have been reported in about half the patients.

It is important to distinguish this condition from the nail patella syndrome (see section on Syndromes with Skeletal Dysmorphogenesis), particularly because of the different implications for counseling in an autosomal dominant condition. In these patients there are no iliac horns. The nails may be short or, rarely, dystrophic.[5]

CASE REPORT

D. A. F. was seen at age 3 years, 1 month. He had been born 3 weeks prematurely, weighing 6 lb. The father was 36 years of age, the mother 34. This was their first live-born infant; the parents had previously tried for 8 years to conceive, and a miscarriage followed this birth.

Physical examination revealed a thin, scrawny, white male weighing 25 lb, 9 oz. Length was 39 inches. There were flattening of the occiput, ptosis and strabismus on the left, with 15 prism diopters of esotropia (Fig. 1). A corneal opacity was present on the left. The fundi appeared normal. The philtrum was rather long, the mandible small (Fig. 2). The child had a high-arched palate and a submucous cleft. The ears lacked the normal convolutions and appeared rather floppy. They were not low set. There was a grade III/IV pansystolic cardiac murmur along the left sternal border transmitted to the axilla and back. A testis was palpable on the right but not on the left.

The patellae were absent bilaterally (see Fig. 1). The palms had accentuated mounds at the bases of the fingers, producing deep grooves between these mounds (Fig. 3). In the feet there was a deep V-shaped cleft between the first and third interdigital web (pli capitonne) (Fig. 4). There were syndactyly of the

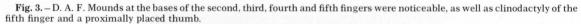

Fig. 3.—D. A. F. Mounds at the bases of the second, third, fourth and fifth fingers were noticeable, as well as clinodactyly of the fifth finger and a proximally placed thumb.

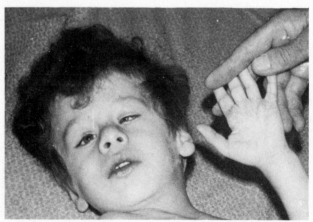

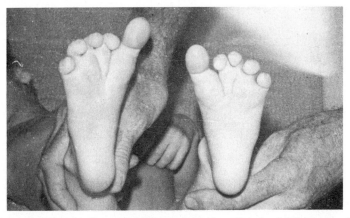

Fig. 4. – D. A. F. The deep V-shaped groove between the first and third interdigital web (pli capitonne) and the long, large great toe are illustrated.

second and third toes and a long great toe (Fig. 5). The child had been treated with casting for bilateral club feet.

Cardiac catheterization revealed a ventricular septal defect. ECG revealed right ventricular hypertrophy. An intravenous pyelogram revealed bilateral vesicoureteral reflux and uretero-pelvi-calyceal dilatation on the right. Roentgenograms revealed Sprengel's deformity on the right. There were mild flexion contractures involving the proximal interphalangeal joints and the ankles.

Developmental evaluation at a chronologic age of 1 year 4 months gave a mental age of 5 months for an IQ of 31. The child was severely retarded and resided in a long-term residential facility.

Chromosomal analysis of both blood and skin revealed a karyotype of 46XY/47XYC-8+. The additional C chromosome was identified as C-8 with the use of ASG-banding (Fig. 6).

GENETICS

Among the reported patients with trisomy 8, mosaicism has been frequent. This was true of 10 of 14 pa-

tients reported. In two children with no evidence of mosaicism, only blood was examined, whereas two had both blood and skin examined. The C group chromosomes are sufficiently similar in appearance that banding is required for the diagnosis of 8 trisomy.

Recurrence risks are unknown, but advanced parental age leads us to suspect that recurrence or occurrence risk may increase with increasing parental age. This disorder should be detectable using amniocentesis and chromosomal karyotyping for prenatal diagnosis.

C group trisomy also has been found in the bone marrow cells of patients with a variety of hematologic disorders, including malignant diseases such as acute myelocytic leukemia. The extra chromosome has been

Fig. 5. – D. A. F. Large great toe and flexion contractures of the interphalangeal joints.

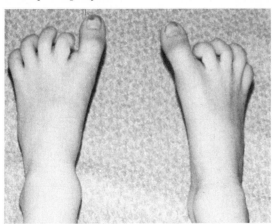

Fig. 6. – D. A. F. Karyotype reveals trisomy for C-8.

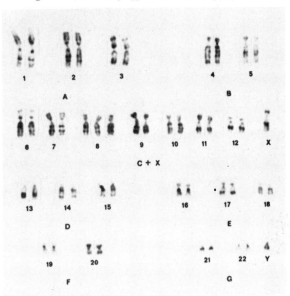

identified as no. 8 in two patients: one with pancyto-penia and one with granulocytopenia and thrombocy-topenia.[8] Karyotype of the peripheral blood in these patients was normal, and there were no other pheno-typic abnormalities.

TREATMENT

No specific treatment is known. However, these pa-tients may have a good prognosis and therefore should not be stereotyped as being developmentally delayed because of chromosome abnormality. All measures to inform and support the family with early infant devel-opmental intervention are indicated.

REFERENCES

1. Lejeune, J., Dutrillaux, B., Rethoré, M. O., Berger, R., De-bray, H., Veron, P., Gorce, F., et Grossiord, A.: Sur trois cas de trisomie C, Ann. Genet. (Paris) 12:28, 1969.

2. Riccardi, V. M., Atkins, L., and Holmes, L. B.: Absent patel-lae, mild mental retardation, skeletal and genito-urinary anomalies and C group autosomal mosaicism, J. Pediatr. 77:664, 1970.

3. Caspersson, T., Lindsten, J., Zech, L., Buckton, K. E., and Price, H.: Four patients with trisomy 8 identified by flu-orescence and giemsa banding techniques, J. Med. Genet. 9:1, 1972.

4. Kakati, S., Nihill, M., and Sinha, A. K.: An attempt to estab-lish trisomy 8 syndrome, Humangenetik 19:293, 1973.

5. Tuncbilek, E., Halicioglu, C., and Say, B.: Trisomy 8 syn-drome, Humangenetik 23:23, 1974.

6. Bijlsma, J. B., Wijffels, J. C. H. M., and Tegelaers, W. H. H.: C8 trisomy mosaicism syndrome, Helv. Paediatr. Acta 27: 281, 1972.

7. Laurent, C., Robert, J. M., Grambert, J., and Dutrillaux, B.: Observations cliniques et cytogénétiques de deux adultes trisomiques C en mosaïque, Lyon Med. 226:827, 1971.

8. Chapelle, de la, A., Schroder, J., and Vuopio, P.: 8 Trisomy in the bone marrow. Report of two cases, Clin. Genet. 3:470, 1972.

TRISOMY 13
Patau Syndrome, Bartholin-Patau Syndrome

CARDINAL CLINICAL FEATURES

Multiple congenital anomalies, including severe mental retardation; holoprosencephaly; major ocular anomalies such as anophthalmia, microphthalmia or colobomas; clefts of lip and palate; polydactyly; congenital heart disease; cutaneous defects of the scalp.

CLINICAL PICTURE

The syndrome was first described by Patau and colleagues[1] in 1960 with the report of an infant in whom an extra autosome of the D group was associated with major anomalies of the heart and brain, anophthalmia, cleft lip and palate and polydactyly. The description of subsequent cases has revealed a considerable similarity, so that this syndrome is readily recognized clinically.[2, 3] The syndrome appears to be uncommon, less common for instance than the 18 trisomy syndrome.

The characteristic phenotype includes a number of

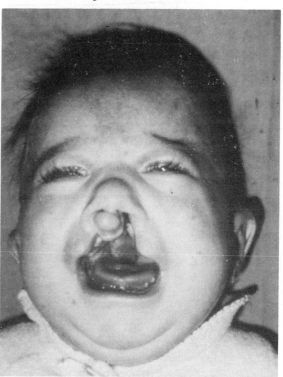

Fig. 2.—A baby with 13 trisomy. She had the typical bilateral cleft of the lip and a cleft palate. The head was small, the forehead sloped backward and the ears were low set.

Fig. 1.—B. G. An infant with 13 trisomy with the usual bilateral cleft. This patient had microphthalmia. She also had hydrocephalus with a very special appearance of the skull. This is a rare concomitant of 13 trisomy.

Fig. 3.—A close-up of the cleft of patient shown in Figure 2.

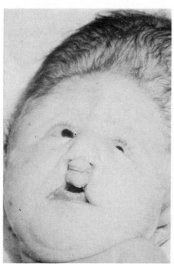

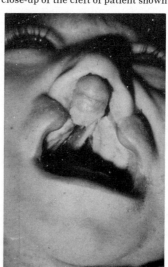

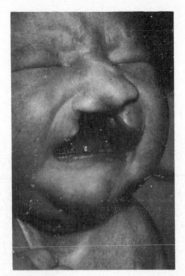

Fig. 4.—M. S. A male infant with 13 trisomy. The cleft was huge.

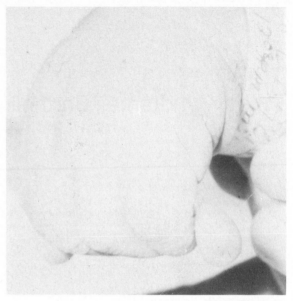

Fig. 6.—M. S. Polydactyly of the hand.

anomalies of the face. Severe anomalies of the eye are cardinal features of the syndrome, ranging from anophthalmia or microphthalmia (Fig. 1) to colobomas of the irides or retina. Cataracts, cloudy corneas or glaucoma is seen. The retina may be dysplastic, and retinal folds may be visible. Facial clefts usually are large (Figs. 2–4) and bilateral, and the palatal cleft usually is complete. The nose often appears flat and short, squashed or bulbous (see Fig. 4). The head usually is small, with a sloping forehead and prominent occiput. There may be trigonocephaly and premature closure of the metopic suture. Ocular hypotelorism is common. The ears are small, low set or malformed. The skin defects (Fig. 5) in the parietal-occipital area are diagnostic. Polydactyly (Figs. 6 and 7) and flexion contractures of the fingers are characteristic.

Patients with 13 trisomy usually have a somewhat low birth weight for gestational age. They have severe retardation of postnatal growth as well as mental retardation. Their cry is weak. A failure to respond to sound is common.[3] Studies of the temporal bone have

revealed bony and membranous anomalies of the inner ear.[4] Soon after birth these infants develop spells of apnea and cyanotic attacks that are difficult to distinguish from seizures. They also regularly develop seizures, usually myoclonic, and have prominent EEG abnormalities. Life expectancy is very short. Most patients die within the first days or months of life. A few exceptions have been reported to live into childhood.

The cerebral defect in these patients is associated with major structural malformations. The most characteristic pathologic finding is arhinencephaly or holoprosencephaly. There is variably incomplete development of the forebrain, often with absence of the olfactory nerve and bulb and absence of the corpus callosum. Fusion of the frontal lobes and the single ventricle is suggestive of cyclopia. Arhinencephaly is typical of 13 trisomy, but it may be seen in patients with a normal karyotype. In all patients with holoprosencephaly the appearance of the face is highly predictive of the cerebral defect.[5] Congenital heart disease is a prominent feature of this syndrome. In these patients the

Fig. 5.—M. S. Characteristic punched-out lesion of the posterior scalp.

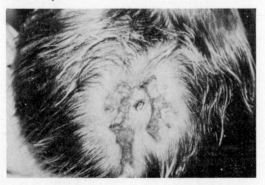

Fig. 7.—H. W. Polydactyly of the foot.

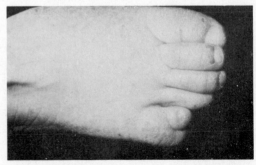

Fig. 8. – The thumb may be proximally placed. The broad flattened hand had a Sydney line.

most common cardiac defects are ventricular septal defect, patent ductus arteriosus and atrial septal defect. Gastrointestinal anomalies include malrotations and Meckel's diverticulum. There may be renal anomalies, particularly cystic kidneys or unilateral renal agenesis. Girls with this syndrome often have bicornuate uteri and may have abnormal fallopian tubes or hypoplastic ovaries. Male patients frequently have cryptorchidism. Scrotal skin may extend to the tip of the penis. Inguinal and umbilical hernias are common, as are accessory spleens; some patients have an ectopic spleen, especially in the pancreas. Sound spectrograms of the cry of an infant with this syndrome have been published[6] and may be characteristic.

A capillary hemangioma on the forehead is a frequent finding. The chin usually is small and the neck short, with loose skin folds and a low posterior hairline. The polydactyly is on the ulnar and fibular sides of the extremities (see Figs. 6 and 7). Flexion contractures of the fingers are prominent and there may be a so-called "trigger thumb," in which clicks are palpable at the metacarpal-phalangeal joint on passive extension. The thumbs may be retroflexible and proximally placed (Fig. 8). The nails are narrow and hyperconvex. The thumb and index finger have a tendency to overlap the third finger. The fifth finger may overlap the fourth. There may be marked posterior prominence of the heels, although not as frequently as in trisomy 18 and talipes equinovarus. The roentgenographic features of this syndrome are not unique, but the occurrence of hypotelorism and small, poorly formed orbits with a sloping forehead, cleft palate and polydactyly usually is diagnostic.[7] The bones of the skull may be poorly ossified and the ribs may be ribbon-like.

On dermatoglyphic analysis (see section on Dermatoglyphics) these patients have distal axial triradii and increased atd angles. Arch patterns on the fingertips are increased. An S-shaped modification of a hallucal arch fibular A pattern may be unique to this syndrome[3, 8] and has been called an A^f-S pattern. Tibial loops may be seen in the proximal thenar region. Simian creases are regularly seen.

Some interesting hematologic abnormalities are present in this syndrome. There may be persistent or unusually high levels of fetal and embryonic hemoglobins.[9] These include hemoglobin F, hemoglobin Bart's γ^4 and hemoglobin Gower 2. Nuclear projections or appendages in the polymorphonuclear leukocytes[10] may be unique. This is most readily recognized by counting the number of nuclear projections.

CASE REPORT

Baby S., a female, was referred at 6 hours of life to University Hospital for evaluation of multiple congenital anomalies. She was her mother's first child. Pregnancy and delivery were normal, and during pregnancy there was no exposure to illness, x-rays or drugs. The mother was 19 years old. Birth weight was 4 lb, 5 oz.

Physical examination revealed a microcephalic baby with an ulcerative, punched-out skin defect at the posterior part of the skull that was approximately 1 cm in diameter. There was hypertelorism and colobomas of the irides. The ears were low set. The nose was flat and there were clefts of the lip and palate. There was bilateral polydactyly of the feet. Karyotype revealed the presence of an extra group D chromosome.

The patient had a weak cry. She experienced repeated attacks of apnea, cyanosis and seizures. During one of these attacks, on the eighth day of life, she died. Autopsy revealed hypoplasia of the anterior cerebral fossa, absence of the olfactory nerve and lobe (arhinencephaly) and absence of the corpus callosum. There were a patent ductus arteriosus, a patent foramen ovale and a ventricular septal defect. The lungs were abnormally lobulated. Multiple accessory spleens were found. The uterus was bicornuate.

GENETICS

Trisomy of chromosomes in the D group usually is the result of nondisjunction, most often in meiosis. Forty-seven chromosomes are evident in the karyotype (Fig. 9), and identification of the one involved in the trisomy is accomplished by Giemsa or quinacrine mustard banding analysis. These methods have revealed

Fig. 9. – Karyotype of a patient with 13 trisomy.

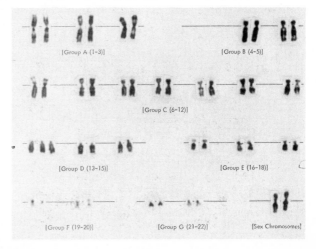

that the extra chromosome in this syndrome is chromosome 13.[11] Chromosome 13 is the easiest of the D group chromosomes to identify in this way. Autoradiographic technics have provided the same result.

The syndrome also may be caused by a translocation, usually D/D,[12] due to centric or near centric fusion, to which this acrocentric chromosome appears to be particularly susceptible. The karyotype in this event shows 46 chromosomes, with only five instead of six in group D, and a chromosome that appears to be an extra A chromosome but is really two D group chromosomes whose long arms are joined at their centromeres. Noncentric translocation of an extra D group chromosome to another D group chromosome also may occur, as may translocation to a chromosome of another group. Most cases associated with translocations are sporadic, although a few have been shown to be the result of inheritance from a mother with a balanced translocation.[13] The relative frequency of translocation is much greater in 13 trisomy than it is in the Down syndrome.[13] Mosaicism also is more common in 13 trisomy than in the Down syndrome. Increased maternal age is associated with trisomy 13,[13, 14] as it is with other types of nondisjunction. The maternal ages of mothers of trisomic patients show a bimodal distribution. This is consistent with the possibility that there are two groups of trisomies, one age-dependent and the other age-independent. A karyotype should be done on any patient with this syndrome. In the presence of a translocation, the mother also should be tested because of the high risk of recurrence for a translocation carrier.

TREATMENT

None.

REFERENCES

1. Patau, K., Smith, D. W., Therman, E., Inhorn, S. L., and Wagner, H. P.: Multiple congenital anomaly caused by an extra autosome, Lancet 1:790, 1960.
2. Warkany, J., Passarge, E., and Smith, L. B.: Congenital malformations in autosomal trisomy syndromes, Am. J. Dis. Child. 112:502, 1966.
3. Smith, D. W., Patau, K., Therman, E., Inhorn, S. L., and DeMars, R. I.: The D₁ trisomy syndrome, J. Pediatr. 62:326, 1963.
4. Maniglia, A. J., Wolff, D., and Herques, A. J.: Congenital deafness in 13–15 trisomy syndrome, Arch. Otolaryngol. 92:181, 1970.
5. De Myer, W., Zeman, W., and Palmer, C. G.: The face predicts the brain: Diagnostic significance of median facial anomalies for holoprosencephaly (arhinencephaly), Pediatrics 34:256, 1964.
6. Ostwald, P., Peltzman, P., Greenberg, M., and Meyer, J.: Cries of a trisomy 13–15 infant, Dev. Med. Child Neurol. 12:472, 1970.
7. James, A. E., Jr., Belcourt, C. L., Atkins, L., and Janower, M. L.: Trisomy 13–15, Radiology 92:44, 1969.
8. Uchida, I. A., Patau, K., and Smith, D. W.: Dermal patterns of 18 and D₁ trisomics, Am. J. Hum. Genet. 19:345, 1962.
9. Walzer, S., Gerald, P. S., Breau, G., O'Neill, D., and Diamond, L. K.: Hematologic changes in the D₁ trisomy syndrome, Pediatrics 38:419, 1966.
10. Huehns, E. R., Lutzner, M., and Hecht, F.: Nuclear abnormalities of the neutrophils in a D₁ (13–15) trisomy syndrome, Lancet 1:589, 1964.
11. Miller, D. A., Allderdice, P. W., Miller, O. J., and Breg, W. R.: Quinacrine fluorescence patterns of human D group chromosomes, Nature 232:24, 1971.
12. Apple, D. J., Holden, J. D., and Stallworth, B.: Ocular pathology of Patau's syndrome with an unbalanced D/D translocation, Am. J. Ophthalmol. 70:383, 1970.
13. Taylor, M. B., Juberg, R. C., Jones, B., and Johnson, W. A.: Chromosomal variability in the D₁ trisomy syndrome, Am. J. Dis. Child. 120:374, 1970.
14. Magenis, R. E., Hecht, F., and Milham, Jr., S.: Trisomy 13 (D₁) syndrome: Studies on parental age, sex ratio, and survival, J. Pediatr. 73:222, 1968.
15. Taylor, A. I.: Autosomal trisomy syndromes: A detailed study of 27 cases of Edwards' syndrome and 27 cases of Patau's syndrome, J. Med. Genet. 5:227, 1968.
16. Taylor, A. I., and Polani, P. E.: Autosomal trisomy syndromes, excluding Down's, Guys Hosp. Rep. 13:231, 1964.
17. Smith, D. W.: Autosomal abnormalities, Am. J. Obstet. Gynecol. 90:1055, 1961.
18. Warburg, M., and Andersen, S. R.: Ocular changes in simple trisomy and a few cases of partial trisomy, Acta Ophthalmol. (Kbh.) 46:372, 1968.

DOWN SYNDROME
Trisomy 21

CARDINAL CLINICAL FEATURES

Mental retardation; hypotonia; upward slanting palpebral fissures, epicanthal folds, flat, round facies, flat nasal bridge, flattening of the occiput; hyperextensible joints; dysplastic pelvis; clinodactyly, simian creases.

CLINICAL PICTURE

In 1866 Down,[1] a British physician, described a group of mentally retarded patients among whom he recognized a great resemblance. He also considered that these patients had an Oriental appearance, probably because of the eyes, and coined the term mongolian idiocy. Parents and others have objected to the term mongolism, and it is being abandoned in favor of the Down syndrome. The Down syndrome is the most common malformation syndrome, occurring once in every 660 live births.[2] The association of this disorder with a chromosomal aberration by Lejeune and colleagues[3] in 1959 was a landmark in the understanding of mental retardation and human malformations. This was the first chromosomal aneuploidy recognized in man. The extra, small, acrocentric chromosome seemed to fit into the 21–22 group; and so they called this abnormality 21 trisomy.

Patients with the Down syndrome are strikingly similar to one another in appearance (Figs. 1–4). They often can be recognized at a glance. However, there may be considerable variation.[4] The face is characteristically round and flat. The occiput is flat and the over-all appearance of the skull brachycephalic, or shortened in its anterior-posterior diameter. The shape of the head is likely to be round in the newborn infant, becoming brachycephalic with growth. The nasal bridge is flat. The palpebral fissures are narrow, delicate and upward slanting. Epicanthal folds are prominent, but they may not be evident in the newborn infant. Brushfield's spots (Fig. 5) are seen in 85% of these patients: these are tiny white spots that form a ring in the midzone of the iris. They are seen most often in blue eyes. Brushfield's spots also are seen in 25% of normal individuals. Patients with the Down syndrome also may have hypoplasia of the periphery

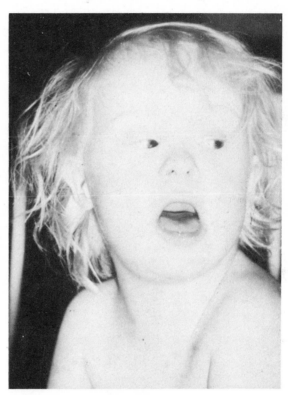

Fig. 1.—A girl with the classic appearance of the Down syndrome. She had prominent forehead, upward slant of the palpebral fissures and epicanthal folds. The nasal bridge was depressed. The mouth was small and open, the tongue protruding.

of the iris. Refractive errors are common. The mouth is often held open and the tongue protrudes, not because of macroglossia but because of the small size of the mouth. The arch of the palate often is high. The tongue is characteristically fissured or furrowed in adult patients. The ears may be low set. The auricle usually is small, the helix folded and the lobe small or absent. With age, the ears may become protruding. The neck is short and, in the newborn period, there is extra, loose skin over the nape.

The hands are short and broad. Clinodactyly of the

117

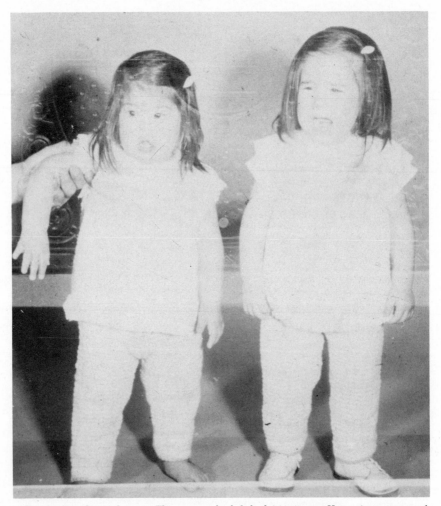

Fig. 2. – Nonidentical twins. The one on the left had 21 trisomy. Her twin was normal. The patient was shorter and had the typical facies, as described in Figure 1. Clinodactyly of the fifth finger is evident.

fifth finger (Fig. 6) results from hypoplasia of the middle phalanx. Instead of the two normal flexion creases on the fifth finger, these patients may have only one. The thumbs are proximally placed (Fig. 7). A simian crease (Fig. 8) often is present unilaterally or bilaterally. An increased distance between the first and second toes is characteristic.

Fingertip dermatoglyphs often feature ulnar loops on all 10 fingers. Whorl and arch patterns are seen rarely. Radial loops are sometimes found on the fourth and fifth fingers. A pattern is sometimes seen in the third interdigital space. The palmar triradius is normally found between the thenar and hypothenar areas and the flexion crease at the wrist; in these patients, however, it is displaced distally and located in the center of the palm. Consequently the atd angle (formed by lines drawn from the triradius at the base of the second finger and the triradius at the base of the fifth finger to the palmar triradius) is greater than the normal 45 degree angle. A tibial arch present on the hallucal area of the sole is highly typical of these patients. It is seen in only 1% of the normal population. A distal loop in the fourth interdigital area of the foot is seen twice as frequently in the Down syndrome as in normal individuals.

These patients are floppy and hypotonic, especially in infancy. The newborn infant may lack the Moro reflex. Muscle tone improves with age. The joints are lax and hyperextensible. They are also hyperflexible, especially at the hips. The skin in the Down syndrome is characterized in the newborn period by acrocyanosis and cutis marmorata. Thereafter the skin may appear pale, pasty or waxy. In older patients the skin is rough, dry and prematurely wrinkled (Fig. 9). In the Down syndrome the hair is fine and sparse.

Patients with the syndrome often have a low birth

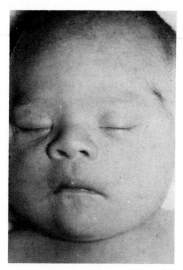

Fig. 3. – A newborn infant with 21 trisomy. This picture of the face illustrates that visual diagnosis is not always easy. The baby did have a round face and a depressed nasal bridge. The eyelids were delicate, and the palpebral fissures were narrow and slanted upward. The epicanthus became evident when the eyes opened.

weight for their gestational age, although normal weight also may be seen. Postnatal growth usually is relatively slow and skeletal maturation delayed. Ultimate height is most often reached by 15 years of age.

Mental retardation in this syndrome usually is severe. In infancy these babies are very quiet and sleep a great deal. Developmental milestones are delayed. When they do talk, they learn only a few words and short sentences. Their attention span is very short. IQs generally range between 25 and 50, with rare instances of IQs of 70–80. Most patients are considered trainable but not educable. Developmental progress tends to slow down with age. As many as 23% of patients under 3 years of age may have an IQ over 50, whereas virtually none over 3 years has an IQ over 50. The mean IQ for older patients is 24. These children generally are very cheerful and pleasant. Their social performance is much better than the IQ. Children with the Down syndrome often live at home, and the social adjustment of the patient and family may be very good.

Serious congenital anomalies are commonly associated with the syndrome. Congenital heart disease is seen in at least 40% of these patients. Septal defects are the most common. Upper gastrointestinal obstructions such as duodenal atresia or annular pancreas occur frequently. Hirschsprung's disease also is seen with greater than normal frequency. Imperforate anus has been reported in the Down syndrome. Tracheo-esophageal fistula may be associated.

Various other disorders may be associated with the Down syndrome. Males may have undescended testicles and are always infertile. Girls with the syndrome

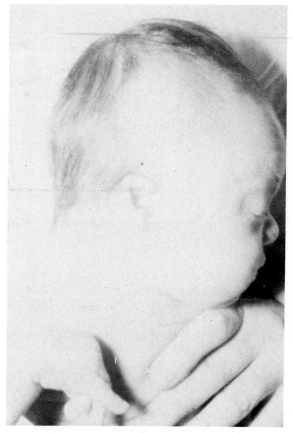

Fig. 4. – Another view of the baby shown in Figure 3 illustrates the upward slant of the eye. The flat occiput is striking. The anomalous ear also is evident.

have late puberty and early menopause. There are several reports of fertility in women with the syndrome. Other endocrine disorders associated with the Down syndrome include hyperthyroidism or hypothyroidism and sexual precocity. Acute leukemia is seen with increased incidence in patients with the syndrome. These patients are highly susceptible to infec-

Fig. 5. – Brushfield's spots.

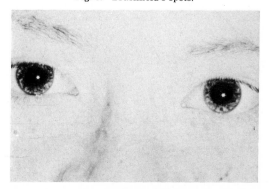

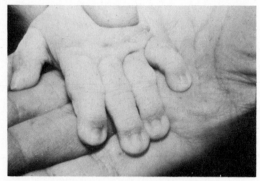

Fig. 6. – Clinodactyly of the fifth finger.

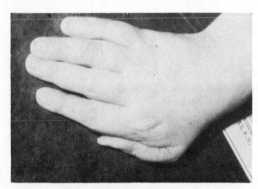

Fig. 7. – Proximally placed thumb.

Fig. 8. – The simian crease.

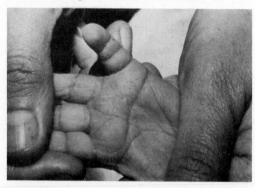

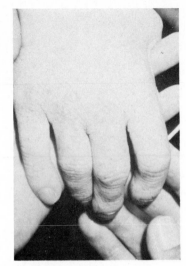

Fig. 9. – Prematurely aged skin in the hand of a 19-year-old with the Down syndrome. Cyanosis also is evident in the nail beds.

pubic bones, as they form the inferior margin of the acetabulum. A line is then drawn along the acetabular roof, and the angle of intersection of the two lines is determined on each side. The iliac angle uses the same horizontal and a line drawn along the lateral ilium from the top of the acetabulum to the anterior iliac spine. This angle too is narrow in the Down syndrome. The iliac index is the sum of the acetabular angle and the iliac angle divided by two.

One third of patients have 11 ribs. Multiple centers of ossification in the manubrium of the sternum are diagnostic. The middle phalanx of the fifth finger is dysplastic.

GENETICS

The Down syndrome is a cytogenetic disease. The fundamental abnormality is an extra quantity of chromosomal material of chromosome 21. In most patients there are 47 chromosomes and a trisomy of chromosome no. 21 (Fig. 10). Using older methods, it was very difficult to distinguish chromosomes 21, 22 and the Y, which led to some confusion early in the phenotype associated with trisomy 21. Autoradiography has been used[6] to distinguish the Y chromosome from other small acrocentric chromosomes. The Y has a distinctive late-replicating DNA pattern. With the advent of Giemsa and fluorescent staining technics,[7] it has become possible to identify each chromosome during metaphase. In patients with the Down syndrome there are five small acrocentric chromosomes, or six including the Y in males. Fluorescent technic readily identified the Y chromosome with its very bright, long arms. In addition, in the patient with the Down syndrome there are three bright, small acrocentric chromosomes

tion. Hepatitis is common in institutions, and Australia antigen was discovered in patients with the Down syndrome.

These patients age prematurely. This is most readily seen in the skin (see Fig. 9), but it may be appreciable in the blood vessels. Alzheimer's senile dementia has been reported.

Roentgenographic findings[5] in the Down syndrome include a flat acetabulum and lateral flaring of the iliac wings. The acetabular angles are narrow. To assess this angle, a horizontal line is drawn across both

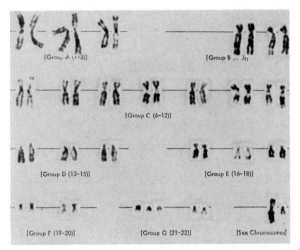

Fig. 10. — Karyotype of a patient with 21 trisomy. (Courtesy of Dr. O. W. Jones of the University of California San Diego.)

(the no. 21 chromosomes) and two very faint ones (the 22 chromosome pair).

Trisomy 21 in the Down syndrome has been thought from the beginning[8] to be due to nondisjunction. Nondisjunction is associated with increased maternal age. The risk of occurrence of the Down syndrome for a woman under 25 years old is one in 1,600 births, and from age 25–29 it is one in 1,350. In women between 30 and 34 the risk is one in 800; 35–39 years, one in 260; 40 to 44 years, one in 100; and 45–49, one in 50.[6] The incidence of the Down syndrome in the general population is one in 700 live births. The peak maternal age in the normal population is 24 years, whereas among mothers of children with the Down syndrome, it is 41.

The cytogenetic defect is due to 21 trisomy in 94% of all patients with the Down syndrome. This results from a defective separation of the chromosomes, or nondisjunction, during meiosis in the formation of the ovum. A major cause of nondisjunction is increased maternal age. It is not known what the conditions are in growing older that influence the production of nondisjunction. It is clear that they are environmental in the sense of the environment of the ovum rather than genetic. Factors of this sort or genetic factors could be responsible for a clustering of nondisjunction in a family in which more than one offspring have trisomy 21 or in which one offspring has trisomy 21 and another the Klinefelter syndrome (XXY), trisomy 18 or the Turner syndrome (XO). These associations occur rarely. Actually, trisomy 21 has been reported along with the Klinefelter syndrome in the same patient on several occasions.

Three per cent of patients with the Down syndrome have a translocation. These are predominantly of the D/G (Fig. 11) or G/G (Fig. 12) type. Phenotypically these patients are identical to those with trisomy 21. Increased maternal age is not a factor. Patients with

the Down syndrome due to a translocation have karyotypes that reveal a normal total of 46 chromosomes, but one of them is a translocation chromosome in which all or a large part of a 21 chromosome is translocated to a chromosome of the D group (D/G translocation) or there is a 21/21 or 21/22 translocation. The extra, translocated material in this syndrome always is that of a 21 chromosome. In patients reported with a D/G translocation the D group chromosome almost always has been a no. 14.

Translocation may be sporadic or inherited in the sense that a phenotypically normal parent of the patient may carry the translocation. In such a carrier the translocation is balanced by the corresponding dele-

Fig. 11. — Karyotype of a patient with a D/G translocation. (Courtesy of Dr. O. W. Jones of the University of California San Diego.)

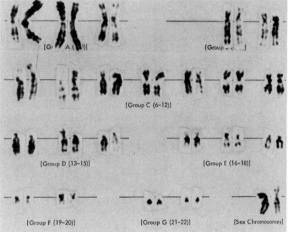

Fig. 12. — Karyotype of a patient with a G/G translocation. The Giemsa-banding technic clearly shows the nature of the translocation. (Courtesy of Dr. O. W. Jones of the University of California San Diego.)

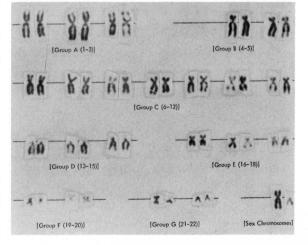

tion. A translocation in the case of all or most of a small chromosome may leave the carrier with only 45 chromosomes. The carrier may be either parent.

Genetic counseling in families of patients with the Down syndrome depends on the etiology. Therefore each patient with this syndrome should be studied cytogenetically. In patients with trisomy 21, the risk of recurrence can be determined by referral to the figures on maternal age. In women over 30 the risks are greater for those who have already had an affected infant than for others in those high risk age groups. The possibility that a pregnancy can be monitored for cytogenetic abnormality by amniocentesis has markedly influenced the nature of genetic counseling in this situation. In many communities it is possible to carry out amniocentesis and monitor for trisomy 21 in any pregnant woman over 35. Certainly this would be recommended in anyone over 30 who had had an infant with the Down syndrome. In the young mother with an infant with trisomy 21, the risk of recurrence in a subsequent pregnancy is one in 290. This is enough to recommend the monitoring of the pregnancy by amniocentesis, and the procedure can be particularly reassuring in such a family.

In the counseling of the family of a patient found to have a translocation, the first step is to determine whether either parent carries it. The risk of recurrence among sporadic translocations is small. Among patients with the Down syndrome due to D/G translocation, at least 50% of cases are sporadic, new events that occur during meiosis in one of the gametes. Among those cases inherited from a carrier parent, it usually is from the mother. Theoretically, the D/G translocation carrier could produce four types of gametes: (1) A normal gamete. (2) A gamete that would lack the 21 chromosome and upon fertilization produce a zygote monosomic for 21. (3) A gamete that would lack one free D and one free G chromosome but have instead an additional, large D/G translocation chromosome; fertilization would produce an individual with a balanced translocation, chromosomally and phenotypically like the parent (45,D–,G–,t(DqGq)+). (4) A gamete that would have the 21 chromosome as well as the D/G translocation chromosome and, upon fertilization, would lead to an offspring with the Down syndrome (46,D–,t(DqGq)+). Thus, it would appear that one fourth of the offspring of a parent with a balanced D/G translocation would be at risk for having the Down syndrome. If one assumed that monosomy was lethal, the theoretical risk would become one in three. In actuality, only 10% of the offspring of a carrier mother have the Down syndrome.[10] The risk for a family in which the father is the carrier is very small, estimated at about one fifth of the risk observed in the female carrier.

In patients with this syndrome who have a G/G translocation, 94% of cases are sporadic. If the carrier parent has 45 chromosomes and a 21/21 translocation, the risk of the Down syndrome in the offspring will be 100%.[11] The two possible gametes formed in this situation are (1) one lacking a 21 chromosome, which, fertilized, is assumed to lead to a lethal monosomy; and (2) one with the translocated 21/21 chromosome, which, fertilized, would lead to a child with the Down syndrome. If the carrier parent has a 21/22 translocation, again there are four possible gametes: (1) one normal; (2) one lacking a 21; (3) one lacking a 21 but containing the 21/22 translocation chromosome; and (4) one containing a 21 chromosome as well as the 21/22 translocation chromosome. One might calculate a theoretical incidence of the Down syndrome of 25% or 33%, depending on whether monosomy 21 was considered viable or nonviable. It turns out that the incidence of the syndrome is 9% when the mother has a 21/22 translocation. When the father is a 21/22 translocation carrier, the numbers are small and, as in D/G translocations, less reliable than in the case of the female. The current estimate for the male carrier of a 21/22 translocation is 4%, but the true figure may be lower. In any given family the risk for both females and males may be even smaller.[12]

In 1–3% of patients with the Down syndrome, there is mosaicism. There are two cell lines—one normal and one with trisomy 21. The diagnosis of mosaicism depends on demonstration of the presence of different cell lines either within a single tissue or in various tissues. In some cases a leukocyte culture may have only cells with a normal karyotype, whereas a fibroblast culture may reveal two cell lines. Thus, increasing the numbers of tissues analyzed increases the chances of discovering a mosaic pattern. There is wide phenotypic variation among these patients. Some are typical patients with the Down syndrome. Among patients with this phenotype the discovery of mosaicism is no basis for optimism in the prognosis for intelligence level.[13] On the other hand, some patients are completely normal clinically and are apt to be discovered only when they produce offspring with the syndrome. Their risk for producing an infant with trisomy 21 is higher than normal, especially when the mosaic pattern is present in the gonads. A mosaic mother may produce two types of gametes—one normal and one with an extra 21 chromosome. Consequently their offspring either may be normal or trisomic for chromosome no. 21. The age of the mosaic mother of a child with the Down syndrome is not a factor. However, it may be found that the maternal grandmother was of advanced age at the time of the birth of the mosaic mother.[14]

Female patients with the Down syndrome may be fertile. Such a patient should produce normal and trisomic gametes in equal numbers. Fertilized by normal spermatozoa, they should result in trisomic and normal zygotes in a ratio of 1:1. The observed ratio in a series of 21 reported is 1.6:1 non-Down to Down syndromes.[15] This number does not deviate significantly from 1:1 statistically, but it appears to be increasing as the numbers of pregnancies reported increase, suggesting a tendency toward selection for euploid ova.

TREATMENT

There is no treatment that can be directed at the primary process, although programs of prevention can be mounted on the basis of the information in the Genetics discussion in this chapter.

Medical attention usually is directed to the complications of the Down syndrome, including infections, cardiac defects, gastrointestinal anomalies and leukemia, all of which are common in these patients.

Counseling parents is important, especially regarding the nature of the disorder, the limitations of the child and the potential for development. Counseling is not easy. In our view it is best not to be overly directive.

REFERENCES

1. Down, J. L. H.: Observations on an ethnic classification of idiots, Clin. Lecture Rep. London Hosp. 3:259, 1866.
2. Penrose, L. S., and Smith, G. F.: *Down's Anomaly,* (Boston: Little, Brown and Co., 1966).
3. Lejeune, J., Gauthier, M., and Turpin, R.: Les chromosomes humains en culture de tissus, Acad. Sci. (Paris) 248:602, 1959.
4. Hall, B.: Mongolism in newborns, a clinical and cytogenetic study, Acta Paediatr. 154:1, 1964.
5. Caffey, J.: *Pediatric X-Ray Diagnosis* (4th ed.; Chicago: Year Book Medical Publishers, 1961), p. 680.
6. Laxova, R., McKeown, J. A., Paldana, P., and Timothy, J. A. D.: A case of XYY Down's syndrome confirmed by autoradiography, J. Med. Genet. 8:215, 1971.
7. Francke, U.: Quinacrine mustard fluorescence of human chromosomes: Characterization of unusual translocations, Am. J. Hum. Genet. 24:189, 1972.
8. Lejeune, J., Gautier, M., and Turpin, R.: Etude des chromosomes somatiques de neuf enfants mongoliens, Acad. Sci. (Paris) 248:1721, 1959.
9. Carter, C. O., and Evans, K. A.: Risk of parents who have had one child with Down's syndrome (mongolism) having another child similarly affected, Lancet 2:785, 1961.
10. Mikkelsen, M., and Stene, J.: Genetic counselling in Down's syndrome, Hum. Hered. 20:457, 1970.
11. Hecht, F., Delay, M., Seeley, J. R., and Stoddard, G. R.: Meiotic evidence in Down's syndrome for 21/21 chromosome translocation or isochromosome, J. Pediatr. 76:298, 1970.
12. Yang, S.-J., and Rosenberg, H. S.: 21/22 Translocation Down's syndrome: A family with unusual segregating patterns, Am. J. Hum. Genet. 21:248, 1969.
13. Kohn, G., Taysi, K., Atkins, T. E., and Mellman, W. J.: Mosaic mongolism. I. Clinical correlations, J. Pediatr. 76:874, 1970.
14. Aarskog, D.: Down's syndrome transmitted through maternal mosaicism, Acta Paediatr. Scand. 58:609, 1969.
15. Reiss, J. A., Lovrien, E. W., and Hecht, F.: A mother with Down's syndrome and her chromosomally normal infant, Ann. Genet. (Paris) 14:225, 1971.

4p– SYNDROME
Partial Deletion of the Short Arm of No. 4 Chromosome, the Wolf-Hirschhorn Syndrome

CARDINAL CLINICAL FEATURES

Unusual facies with prominent glabella, hypertelorism, antimongoloid slant, misshapen nose, defects of midline fusion, low-set ears; undescended testes and hypospadias; seizures; profound retardation of growth and mental development.

CLINICAL PICTURE

Wolf[1] described a patient in 1965 in whom a deletion of the short arm of a group B chromosome was asso-

ciated with a phenotype that he believed was distinct from the B group deletion that had been described in 1963 by Lejeune as the cri-du-chat syndrome (see chapter, this section). It has subsequently been established that Wolf's patient represented a distinct syndrome due to a partial deletion of the short arm of chromosome 4, whereas the deletion in the cri-du-chat syndrome is of chromosome 5. Hirschhorn and colleagues,[2] also in 1965, described a child with multiple congenital anomalies who was very similar to Wolf's patient and had identical cytogenetic findings.

Patients with this syndrome are low in birth weight for their gestational age, and they continue to have severe failure to thrive. Retardation of growth and mental development is profound. Bone age also is retarded. Tonic and clonic seizures are common in these patients. Death in or following seizures has been re-

Fig. 1.– R. R. (Case Report). A 4-month-old with the 4p– syndrome. He had a striking face with a prominent misshapen nose that seemed to flow in smooth lines along a broad nasal bridge arching into high, prominent glabellae. The forehead was long and prominent. The infant had hypertelorism, strabismus, an antimongoloid slant and prominent epicanthal folds. The curves of the mouth turned down. (Courtesy of Dr. Doralys Arias of Orlando, Florida.)

Fig. 2.– K. K. Another patient with the 4p– syndrome. She illustrated the same features around the nose and forehead as described in Figure 3. The eyes also were identical but, in addition, she had ptosis more severely on the right.

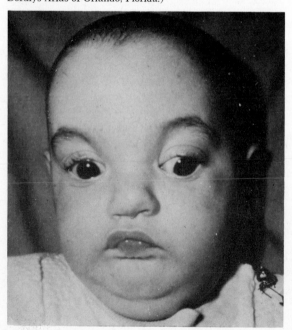

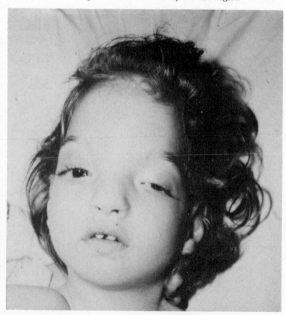

ported. Mothers of these patients have noted weak fetal activity during pregnancy. These babies have very feeble cries, which do not sound like the infant with the cri-du-chat syndrome. These patients do not smile or respond to social stimuli. Most can roll over, but they seldom learn to sit, stand, walk or talk.

Patients may be recognized as having this syndrome from the appearance of the face alone. (Figs. 1 and 2). The skull is microcephalic. The eyes have antimongoloid slants, hypertelorism and epicanthal folds. There also may be colobomas of the irides (Fig. 3), and these may be distinguishing features. Cataracts, ptosis, strabismus or nystagmus may be additional eye findings. The glabella is prominent. There may be a midline defect of the skull and in infancy an overlying midline scalp defect is sometimes present. The ears are low set and prominent. They also may be simple (Fig. 4). The ear canal is narrow, and there may be a preauricular tag or sinus. The nasal bridge is broad and the nose itself often is asymmetrical, flattened and misshapen. The philtrum is short, giving an appearance of a full upper lip, pulled up in the center and curled down at the corners. This gives the mouth a carp-like appearance. The mandible is hypoplastic. There may be a cleft of the lip or palate, or both. The uvula may be bifid. Hemangiomas may be present on the nape of the neck or on the brows.

These specific details do not describe the characteristic appearance nearly as well as a photograph of the face (see Figs. 1 and 2). If an artist were to dissect the lines of the face, one mass would be composed of the nose and cranium, with the lateral margin of the nose ever widening to form the curved eyebrows with the skull sitting on top. Below the curved eyebrows the rest of the face is recessed, often bisected by the cleft of the lip. The medial portion of the eyebrows is so sparse that the brow may appear to begin over the center of the eye. The entire configuration has been described as a look of being startled (see Fig. 1).[3]

Congenital heart disease is a leading cause of death in these patients. Genital anomalies include hypospadias (Fig. 5) and cryptorchidism in the male. A sacral dimple may be seen in some infants. Anomalies of the extremities include simian creases, clinodactyly, hypo-

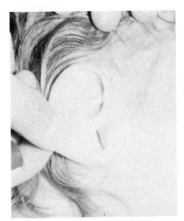

Fig. 4. – K. K. The ear was relatively simple. The narrow ear canal is characteristic of the syndrome.

plastic dermal ridges and low total ridge counts. There may be valgus or other deformities of the foot. Oblique ridges may be seen in the nails.

DIFFERENTIAL DIAGNOSIS

This syndrome can be differentiated from the cri-du-chat syndrome not only by the absence of the characteristic cry but also by the absence of syndactyly and premature graying and by the presence of the promi-

Fig. 5. – R. R. Hypospadias. The testes were undersized and the scrotum hypoplastic. (Courtesy of Dr. Doralys Arias of Orlando, Florida.)

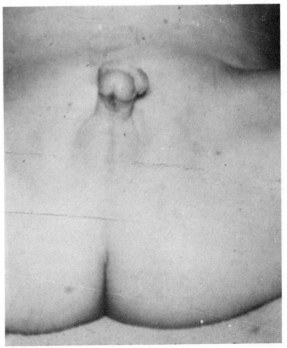

Fig. 3. – K. K. Coloboma of the iris.

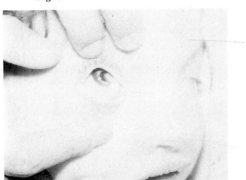

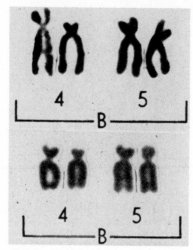

Fig. 6.—Partial karyotypes of two patients with the 4p— syndrome, demonstrating the deletion of the short arm of a chromosome Y in each. (Reproduced with permission from J. Pediatr.[4])

nent features of the 4p— syndrome. These, especially, might include the profound degree of retardation, low birth weight and higher incidence of seizures. This syndrome also should be differentiated from 13 trisomy (see chapter, this section), in which too there

Fig. 7.—Autoradiograph of the group B chromosomes illustrating the deleted chromosome to be one of the pair with the more dense, relatively late labeling. (Reproduced with permission from J. Pediatr.[4])

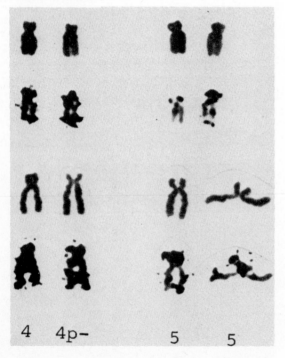

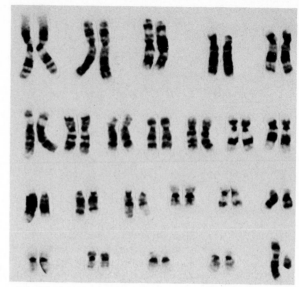

Fig. 8.—Giemsa banding patterns of the chromosomes of a patient with the 4p— syndrome. (Courtesy of Michael Gene Brown of the Cytogenetics Laboratory, Children's Hospital and Health Center, San Diego.)

is a midline scalp defect. Polydactyly, characteristic of trisomy 13, was consistently absent in the 4p— syndrome.

CASE REPORT

R. R., a white male infant previously reported by Arias and colleagues[4] was born after a 40-week gestation. His 23-year-old mother and 28-year-old father were well and unrelated. The first pregnancy had resulted in a normal male infant. This second pregnancy and delivery were normal, although the infant had shown little intrauterine movement.

The patient weighed 4½ lb at birth, measured 17½ inches in length and had a head circumference of 30.5 cm. A cleft palate and hypospadias (see Fig. 5) were noted at the time of birth. At 2–3 months of age, the infant had failed to grow, was hypotonic, had no head control and was generally unresponsive.

He was noted to have hypertelorism, epicanthal folds, a misshapen nose, low-set ears, microcephaly, micrognathia and a prominent glabella (see Fig. 3).

Roentgenographic studies of the urinary tract revealed delayed excretion of dye bilaterally and, during cystography, there was reflux of dye up both ureters. He had difficulty initiating a cry, and, when he did, the sound was weak and plaintive but not cat-like. He was thought to be blind, but electroretinography showed his vision to be intact. At 9 months of age he had a tonic-clonic seizure with low grade fever. He went on to have sporadic seizures in spite of anticonvulsant therapy. A pneumoencephalogram showed a mild degree of internal hydrocephalus. At 11 months of age he weighed 13½ lb, was 25½ inches long and had a head circumference of 38 cm. Motor retardation was severe. He still lacked head control and the ability to turn over or sit up. The karyotype was 4p— (Fig. 6).

GENETICS

All reported cases of the 4p− syndrome have been sporadic in origin. No proved balanced translocation carriers among parents have as yet been reported. However, it seems likely that instances will be encountered; in genetic counseling the karyotypes of the parents should regularly be examined for this possibility. If the parents' karyotypes are normal, the risk of recurrence is negligible. Maternal age appears not to be a factor in this syndrome.

The diagnosis of the syndrome at birth is possible on clinical grounds alone. Using older cytogenetic methods, it is very difficult to distinguish between chromosomes no. 4 and 5. By autoradiography (Fig. 7) chromosome 4 is late replicating, whereas the long arm of no. 5 is an early replicant. Chromosome no. 4 is statistically longer, but this is very hard to see without the most careful measurement, and the means are too close and the variation too great to permit distinction in this way. On the other hand, using the fluorescent or Giemsa (Fig. 8) staining technics, the identification of these chromosomes has become easy.

TREATMENT

No treatment is available. These patients are very profoundly retarded and are almost always institutionalized.

REFERENCES

1. Wolf, U., Reinwein, H., Porsch, R., Schroter, R., and Baitsch, H.: Defizienz an der kurzen Armen eines Chromosoms Nr. 4, Humangenetik 1:397, 1965.
2. Hirschhorn, K., Cooper, H. L., and Firschein, I. L.: Deletion of short arms of chromosome 4–5 in a child with defects of midline fusion, Humangenetik 1:479, 1965.
3. Miller, O. J., Breg, W. R., Warburton, D., Miller, D. A., de-Capoa, A., Allderdice, P. W., Davis, J., Klinger, H. P., McGilvray, E., and Allen, F. H.: Partial deletion of the short arm of chromosome no. 4 (4p−): Clinical studies in five unrelated patients, J. Pediatr. 77:792, 1970.
4. Arias, D., Passarge, E., Engle, M. A., and German, J.: Human chromosomal deletion: Two patients with the 4p− syndrome, J. Pediatr. 76:82, 1970.
5. Miller, O. J., Warburton, D., and Breg, W. R.: *Deletions of Group B Chromosomes*, Birth Defects Original Articles Series, Vol. V, No. 5 (New York: The National Foundation-March of Dimes, 1969).
6. Coffin, G. S., and Wilson, M. G.: Wolf-Hirschhorn syndrome, Am. J. Dis. Child. 121:265, 1971.
7. Guthrie, R. D., Aase, J. M., Asper, A. C., and Smith, D. W.: The 4p− syndrome, a clinically recognizable chromosomal deletion syndrome, Am. J. Dis. Child. 122:421, 1971.

5p– SYNDROME
The Cat Cry Syndrome, Cri-du-Chat Syndrome

Cardinal Clinical Features

A characteristic cry in infancy reminiscent of the mewing of a cat; microcephaly and characteristic facies with epicanthi, hypertelorism and antimongoloid slant; profound mental retardation.

Clinical Picture

Lejeune and colleagues[1] in 1963 described a syndrome in which a deletion of the short arm of one of

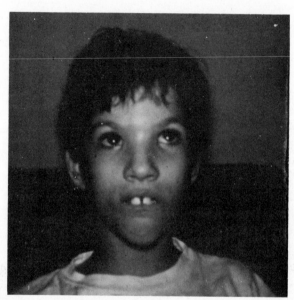

Fig. 2.—J. W. A 5-year-old girl with the cri-du-chat syndrome. The bridge of the nose was wide, and she had hypertelorism, epicanthi and a face that was flattened from side to side. (Courtesy of Dr. Doralys Arias of Orlando, Florida.)

Fig. 1.—N. W. A 6-year-old girl with the cat cry syndrome. She was severely retarded and had hypertelorism and an antimongoloid slant. The base of the nose was broad, the mouth enormous. The face was characteristically long and thin and was asymmetrical, although this was better shown when she demonstrated more emotion. (Permission to photograph this patient granted by Drs. William Morris and William Clover of Pacific State Hospital in Pomona.)

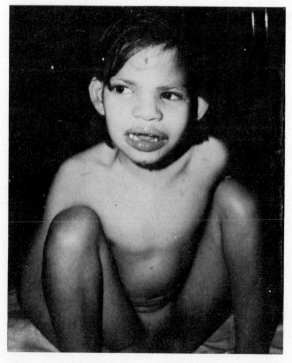

the B group chromosomes was associated with a characteristic cry that he felt was cat-like. The name cri-du-chat used in the original report has stuck. Lejeune commented on the similarity of chromosomes 4 and 5 but judged that the deletion was on 5. In 1964 German and colleagues[2] identified the chromosome involved in the deletion as no. 5 using radioautographic technic.

Infants with this syndrome are characteristically low in birth weight for their gestational age and, as infants, fail to thrive. Growth retardation is a continuing part of the picture. They are microcephalic and have profound retardation of intellectual and motor development. The IQ is usually less than 20. The characteristic cry is high pitched and is clearly distinctive. As one listens, especially knowing what to listen for, it does resemble somewhat the mewing of a cat. The sound is due to a small, narrow, hypoplastic larynx.

The facies are characteristic. In infancy the face is

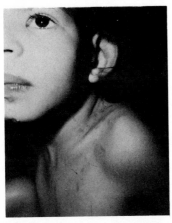

Fig. 3.—N. W. The large and prominent ear, as well as the epicanthal fold, is shown.

round or moon-shaped,[3] but soon it becomes long and thin[4] (Figs. 1 and 2). There usually is some asymmetry. Concomitantly the hypertelorism tends to disappear. These patients have epicanthal folds (Fig. 3) and an antimongoloid slant. Strabismus, which may be divergent, is common. The nasal bridge is broad, the philtrum usually short. The ears are low set, and there may be a preauricular skin tag. The ear canals may be narrow. There also is a high-arched palate. Patients may have micrognathia (Fig. 4), with overbite or prognathism.[5] Malocclusion is common among older chil-

Fig. 4.—J. W. Lateral view illustrates a slight micrognathia. (Courtesy of Dr. Doralys Arias of Orlando, Florida.)

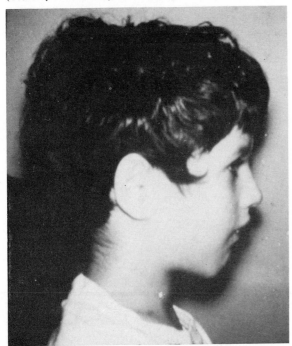

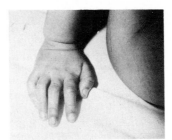

Fig. 5.—N. W. The metacarpals were short. Note clinodactyly of the fifth finger and proximally placed thumbs.

dren. The characteristic cry disappears with age: it may disappear as early as 2 weeks of age,[5] and it rarely is heard after the first year or 2 of life. Patients seldom develop speech. They have premature graying of the hair.

Congenital heart disease is common. Patients may also have diastasis recti. Renal anomalies rarely are seen.

The upper extremities usually show some anomalies, generally minor. The metacarpals most often are short (Fig. 5). Clinodactyly and simian creases are common. Dermatoglyphic analysis[6] reveals a decreased number of digital ulnar loops and an increased number of whorls. There is a distal axial triradius. The total ridge count is normal. A thenar pattern is common, with a fourth interdigital loop. Occasionally there are partial syndactylies of fingers and toes. Short metatarsals and pes planus may be features of the lower extremities.

Patients with this syndrome are so severely retarded that they learn to walk very late, but most eventually learn this skill. The gait is wide based and unsteady. Hypotonia is prominent in early childhood but disap-

Fig. 6.—S. L. (Case Report). At 3 years of age. (Courtesy of Dr. Doralys Arias of Orlando, Florida.)

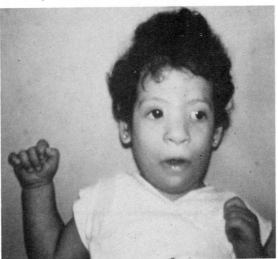

pears later,[4] occasionally to be replaced by hypertonia.[7]

Roentgenographic findings are consistent with the microcephaly and hypertelorism. Scoliosis is common. The wings of the ilia may be small. Pneumonencephalogram may reveal dilated ventricles and very thin cortex.[1, 8, 9]

The cat cry syndrome should be differentiated from the 4p− syndrome (see chapter on Partial Deletion of No. 4 Chromosome, this section). Both syndromes share the hypotonia, profound mental retardation and antimongoloid slant. Patients with the 4p− syndrome, however, characteristically have colobomas of the iris and midline defects. Males with 4p− have hypospadias and undescended testes. Seizure disorders are common. Of course, the cat-like cry is unique to the 5p− syndrome.

CASE REPORT*

S. L., a 2-month-old white female of Colombian ancestry, was referred because of a heart murmur. She had been born after 38 weeks' gestation weighing 4 lb, 3 oz. She did not breathe spontaneously but responded to resuscitative measures and oxygen. She fed poorly, appearing to tire and stop sucking after 1–1.5 oz of formula. She had a weak, high-pitched cry that was reminiscent of that of patients with the cri-du-chat syndrome.

Examination revealed a 7 lb, 14 oz infant who was 21 inches long. She had hypertelorism, an antimongoloid slant and prominent epicanthal folds. There was micrognathia. The head circumference was 34.3 cm. Weight, height and head circumference were below the third percentile. There were bilateral simian creases. The cardiac rhythm was normal and the rate 170. There was a grade III–IV holosystolic murmur, maximal along the left sternal border but heard all over the chest. There was a faint thrill. The heart was enlarged and the liver was palpated 3.5 cm below the right costal margin. There was an abnormal implantation of the fourth toe on each foot and abnormal separation of the first and second toes. Deep tendon reflexes were 4+, and there was ankle clonus on the left.

Laboratory examination revealed a partial deletion of chromosome no. 5. Roentgenograms showed cardiomegaly and an ectopic right kidney. ECG revealed right ventricular hypertrophy.

The patient was treated with digitalis, and the heart diminished in size. Figure 6 shows the patient at age 3. By 6½ years of age she was walking but did not speak.

GENETICS

Patients with the cat cry syndrome have a deletion of the short arm of chromosome no. 5. Their chromosome constitution is 46, 5p−. Chromosome 5 is very similar to chromosome no. 4 and difficult to distinguish from it with older methods. Autoradiography[2] has been used to distinguish the two: chromosome no. 4 replicates late compared to the relatively early completion of DNA synthesis in the long arm of no. 5. The length of the long and even the short arms of chromosome no. 4 is greater than that of chromosome no. 5,[9] but this is not easy to detect. The new technics of fluorescent (Fig. 7) and Giemsa banding make identification of chromosomes 4 and 5 easy; they have made all

Fig. 7.—N. W. Karyotype. The fluorescent technic clearly distinguishes between chromosomes 4 and 5. The deletion is in the short arm of chromosome 5 on the left. (Courtesy of Dr. Uta Francke of the University of California San Diego.)

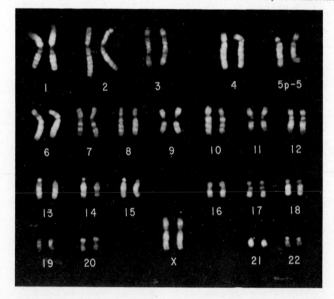

---

*We are indebted to Dr. Doralys Arias of Orlando, Florida, for details of this patient.

of the other approaches to this problem obsolete. The chromosomal anomaly of the cat cry syndrome usually is sporadic. Several cases have been reported in which there was familial transmission, resulting from a balanced translocation in one of the parents.[10] The mother of two offspring with a 5p— chromosomal constitution was found to be a mosaic in whom a small number of cells were 5p—. Presumably, in an individual of this sort, an occasional ovum turns out to be 5p—.

TREATMENT

None. Institutionalization has been required in most patients.

REFERENCES

1. Lejeune, J., Lafourcade, J., Berger, R., Vialatte, J., Boeswillwald, M., Seringe, P., and Turpin, R.: Trois cas de délétion partielle du bras court d'un chromosome 5, C. R. Acad. Sci. [D] (Paris) 257:3098, 1963.
2. German, J., Lejeune, J., Macintyre, M. N., and de Grouchy, J.: Chromosomal autoradiography in the *cri du chat* syndrome, Cytogenetics 3:347, 1964.
3. Gellis, S. S., and Feingold, M.: Picture of the month. Denouement and discussion: Cri du chat syndrome, Am. J. Dis. Child. 117:699, 1969.
4. Miller, O. J., Wartburton, D., and Breg, W. R.: *Deletions of Group B Chromosomes*, Birth Defects: Original Articles Series, Vol. V, No. 5 (New York: The National Foundation-March of Dimes, 1969), p. 100.
5. Breg, W. R., Steele, M. W., Miller, O. J., Warburton, D., deCapoa, A., and Allderdice, P. W.: The cri du chat syndrome in adolescents and adults: Clinical findings in 13 older patients with partial deletion of the short arm of chromosome No. 5 (5p—), J. Pediatr. 77:782, 1970.
6. Warburton, D., and Miller, O. J.: Dermatoglyphic features of patients with a partial short arm deletion of a B-group chromosome, Ann. Hum. Genet. 31:189, 1967.
7. Platt, M., and Holmes, L. B.: Hypertonia in older patients with the 5p— syndrome, Lancet 2:1429, 1971.
8. James, A. E., Atkins, L., Feingold, M., and Janower, M. L.: The cri du chat syndrome, Radiology 92:50, 1969.
9. James, A. E., Merz, T., Janower, M. L., and Dorst, J. P.: Radiological features of the most common autosomal disorders: Trisomy 21–22 (mongolism or Down's syndrome), trisomy 18, trisomy 13–15, and the cri du chat syndrome, Clin. Radiol. 22:417, 1971.
10. Warburton, D., Miller, D. A., Miller, O. J., Breg, W. R., deCapoa, A., and Shaw, M. W.: Distinction between chromosome 4 and chromosome 5 by replication pattern and length of long and short arms, Am. J. Hum. Genet. 19:399, 1967.
11. deCapoa, A., Warburton, D., Breg, W. R., Miller, D. A., and Miller, O. J.: Translocation heterozygosis: A cause of five cases of the *cri du chat* syndrome and two cases with a duplication of chromosome number five in three families, Am. J. Hum. Genet. 19:586, 1967.
12. Philip, J., Brandt, N. J., Friis-Hansen, B., Mikkelsen, M., and Tygstrup, I.: A deleted B chromosome in a mosaic mother and her cri du chat progeny, J. Med. Genet. 7:33, 1970.

13q– SYNDROME
Ring D Chromosome

CARDINAL CLINICAL FEATURES

Absent thumbs, severe cerebral defect, bifid scrotum with ambiguous genitalia in the male,[1] ocular abnormalities, congenital cardiac disease, characteristic dysmorphic facies.

CLINICAL PICTURE

Some 50 patients have been reported in which a ring chromosome of the D group has been the common denominator. Considerable variation has been observed in phenotypic appearance. The formation of a ring-shaped chromosome results from the partial deletion of chromatid material at the ends of the long and short arms of the chromosome and the subsequent sticking together of the ends to form a circle or ring[1] (Fig. 1). Variability in the congenital abnormalities observed could be due to variation in the amount of chromatid material lost. It also could reflect a heterogeneity resulting from deletions of different D group chromosomes in distinct syndromes. The banding technics should resolve these possibilities.

The most characteristic features of patients with

Fig. 1.–Karyotype (46,XY). The ring was a D group chromosome, no. 13.

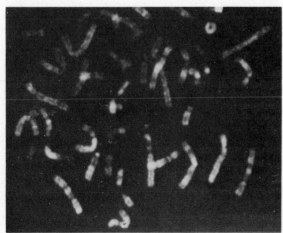

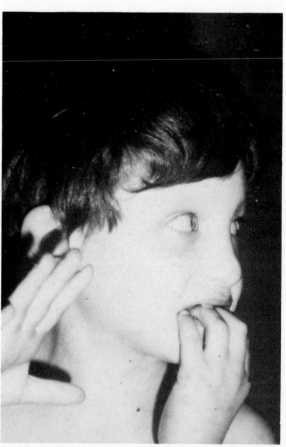

Fig. 2.–R. U. This 7-year-old boy was tiny and severely retarded. He had low-set ears and a high-arched palate. The fingers were long and tapering and he had clinodactyly of the fifth finger. (We are grateful to Dr. Charles Fish of Fairview State Hospital for the opportunity to study this patient.)

ring D chromosomes that are recognizable as a syndrome[2-4] are mental retardation, absent thumbs and bifid scrotum. The mental defect is severe (Fig. 2; see Fig. 3 for karyotype of patient shown in Fig. 2). There often is a major underlying anomaly of the central nervous system. This frequently takes the form of a single large cerebral ventricle, a holistic prosen-

132

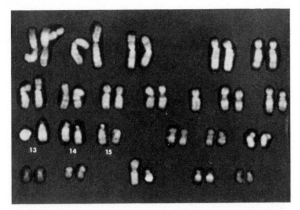

Fig. 3.—R. U. Karyotype illustrating the ring 13. One chromosome 10 is lying across short arms of one chromosome 1. (Courtesy of Dr. Uta Francke of the University of California San Diego.)

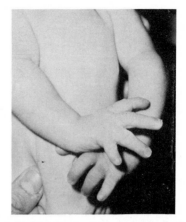

Fig. 5.—J. V. The hands. Each had four fingers and no thumb.

cephalon or arhinencephaly and markedly reduced cerebral substance. There is an associated microcephaly (Fig. 4) or trigonocephaly and usually a flat occiput. These infants frequently have a weak or high-pitched cry. They often feed poorly and fail to thrive. Birth weight may be low. They may be stillborn or die early in life.

The hands in this syndrome may be pathognomonic. Characteristically they have absent thumbs (Fig. 5) or the thumb may be rudimentary. The first metacarpal also is absent. The radii are present, but bilateral dislocation of the radial head has been observed.[3]

The dermatoglyphics are consistent with the absence of the thumbs. Bilateral simian creases are common. The fifth fingers are characteristically short and incurved and have a single flexion crease. The middle phalanx of the fifth digit may be absent. There may be fusion of the fourth and fifth metacarpals. Other skeletal abnormalities include hemivertebrae

and dysplasia of the hips. The feet may mirror the hands, with absent (Fig. 6) or rudimentary great toes and simian creases. The fourth and fifth toes may be fused. Talipes equinovarus has been reported.

Ocular anomalies also may be striking. These include an antimongoloid slant, epicanthal folds, ptosis and narrow palpebral fissures. Microphthalmia may be unilateral or bilateral (Fig. 7). Colobomas of the irides are characteristic. There may be colobomas of the retina as well.

Facial asymmetry may be impressive. The bridge of the nose is depressed and wide. The palate is high arched. The ears may be low set, malformed or malrotated. Micrognathia may be present (see Fig. 7). With growth and development there may be a posterior cervical cutis laxa, as in the Turner syndrome. The nipples may be hypoplastic.

Serious congenital heart disease has been the rule. Interventricular septal defect and an overriding aorta

Fig. 6.—J. V. The feet. There were four toes on each foot. The toe on the tibial side appeared hallucal, but it still seemed likely that the great toe was missing and that this was the second toe.

Fig. 4.—J. V. A male infant with a ring chromosome 13. He had a low birth weight for gestational age and a weak cry. He was microcephalic and had abnormalities of the hands and feet.

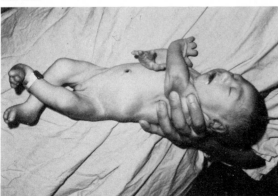

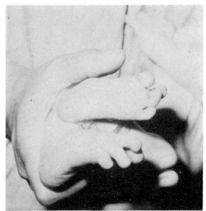

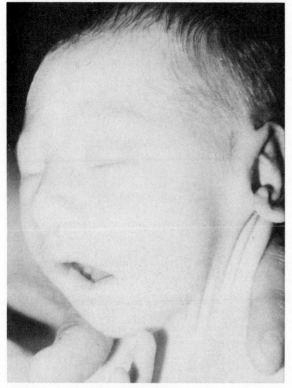

Fig. 7.–J. V. Head and face were small. The forehead sloped back, as did the micrognathic chin. The eyes appeared particularly small and the palpebral fissures were very narrow. The ears were large and malformed. There was redundant skin on the neck.

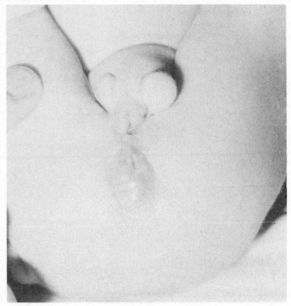

Fig. 8.–J. V. The genitalia were striking. The scrotum was bifid to such a degree that there really was no scrotum but, rather, a persistence of the labioscrotal swellings above. The phallus was very small and hypospadiac.

have been reported[2]–and observed in our patient (J. V.).

A striking genital malformation may be a distinct feature in the male with this syndrome. In two females reported,[2, 4] the only genital abnormality was a bicornuate uterus.[2] In our patient (J. V.) and in the patient of Sparkes and colleagues,[3] the scrotum was bifid and the testes descended no further than the inguinal canal. A small phallic structure with a urethral opening at the base completed the ambiguous appearance of the genitalia (Fig. 8). Posterior to the urethra was an anteriorly placed anus or a cloacal opening. A bifid scrotum with palpable testes, perineal hypospadias and imperforate anus with anoperineal fistula also has been reported,[5] as well as a hypoplastic bifid scrotum and undescended testes.[6] A female patient[7] had an imperforate anus and an anteriorly placed orifice through which stool was passed, as well as large, redundant anterior labia.

Many of the same abnormalities have been observed in patients with a deletion of the long arm of chromosome 13 (13q–) without formation of a ring.[5, 6] Thus, it appears that this syndrome results from the deletion and is a monosomy for the distal segment of chromosome 13. One of the patients,[5] with the 13q– had, in addition to the other features of the syndrome, bilateral radioulnar synostoses. Thumb anomalies have been prominent in patients who had the 13q– as well as those with ring 13.[5] The differential diagnosis of absent or rudimentary thumbs is given in the section on Thumb and Radial Dysplasias (Holt-Oram syndrome).

Other phenotypes have been reported in patients with a ring chromosome D. Some with milder malformations could represent a deletion of less chromosomal material than in patients who have defined the syndrome.

REFERENCES

1. Wang, H. C., Melnyk, J., McDonald, L. T., Uchida, I. A., Carr, H. D., and Goldberg, B.: Ring chromosomes in human beings, Nature 195:733, 1962.
2. Bain, A. D., and Gauld, I. K.: Multiple congenital abnormalities associated with ring chromosome, Lancet 2:304, 1963.
3. Sparkes, R. S., Carrel, R. E., and Wright, S. W.: Absent thumbs with a ring D2 chromosome: A new deletion syndrome, Am. J. Hum. Genet. 19:644, 1967.
4. Juberg, R. C., Adams, M. S., Venema, W. J., and Hart, M. G.: Multiple congenital anomalies associated with a ring-D chromosome, J. Med. Genet. 6:314, 1969.
5. Allderdice, P. W., Davis, J. G., Miller, O. J., Klinger, H. P., Warburton, D., Miller, D. A., Allen, F. H., Jr., Abrams, C. A. L., and McGilvray, E.: The 13q– syndrome, Am. J. Hum. Genet. 21:499, 1969.

6. Grace, E., Drennan, J., Colver, D., and Gordon, R. R.: The 13q— deletion syndrome, J. Med. Genet. 8:351, 1971.

7. Bilchik, R. C., Zackai, E. H., Smith, M. E., and Williams, J. D.: Anomalies with ring D chromosome, Am. J. Ophthalmol. 73:83, 1972.

8. Biles, A. R., Luers, T., and Sperling, K.: D₁ ring chromosome in newborn with peculiar face, polydactyly, imperforate anus, arrhinencephaly, and other malformations, J. Med. Genet. 7:399, 1970.

9. Holmes, L. B., Moser, H. W., Halldorsson, S., Mack, C., Pant, S. S., and Matzilevich, B.: *Mental Retardation, an Atlas of Diseases with Associated Physical Abnormalities* (New York: MacMillan Co., 1972).

18q— SYNDROME
Deletion of the Long Arm of Chromosome 18

CARDINAL CLINICAL FEATURES

Mental retardation; microcephaly; shortness of stature; midfacial dysplasia, carp-mouth and large, malformed, low-set ears.

CLINICAL PICTURE

In 1964 de Grouchy and colleagues[1] described a mentally retarded girl who had microcephaly, deafness and a partial deletion of the long arm of chromosome no. 18. Lejeune et al.[2] described two other microcephalic children with similar clinical features in 1966 and suggested that the deletion of the long arm of 18 was responsible for a recognizable syndrome.

Patients with this syndrome may have normal or low birth weight for gestational age. There regularly is a subsequent slowness of growth and shortness of stature. The patients are microcephalic and profoundly retarded. IQs range from 20 to 30. Hypotonia may be extreme, and lack of coordination is the rule. Nevertheless, developmental progress may be expected. Patients may have seizures and an abnormal EEG. They have normal life expectancy.

Facial features are characteristic.[1-7] They include frontal bossing, hypoplasia of the supraorbital ridges with deep-seated eyes, hypertelorism and prominent epicanthal folds. The ears are large, malformed and low set. There is a prominent anthelix and/or antetragus, and occasionally patients have narrow or atretic external auditory canals and deafness. Underdevelopment of the middle portion of the face and a hypoplastic maxilla are striking. The nose is small and upturned and the nasal bridge is depressed. The mouth is large, usually open, with downward-slanted corners so that the angles are well below the midpoint of the lower lip; this causes the typical fish- or carp-mouth appearance. In profile the upper lip appears to be rolled up, the lower lip rolled down. The upper lip may be cleft, and the columella of that lip may be absent. The hair of these patients is sparse. They may have congenital nystagmus. Funduscopy may reveal anomalies such as posterior staphyloma or a tilted optic disc or optic atrophy. The mandible usually is normal but may be small or mildly prognathic.

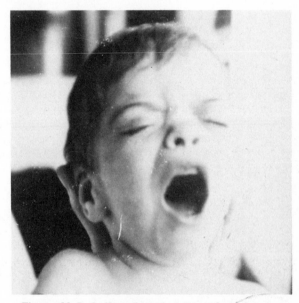

Fig. 1.—M. P. C. (Case Report). A 7-month-old female infant with a deletion of the long arm of chromosome 18. The large fish- or carp-mouth is illustrated, as well as the midface retraction. (Courtesy of Dr. Primarosa de Chieri of Buenos Aires.)

The palate may be high arched, cleft or pigmented. The uvula may be bifid.

The nipples are widely spaced. Occasionally there is congenital heart disease. Horseshoe kidneys have been observed. Spina bifida, hemivertebrae and an extra vertebra have been observed,[4] as well as vertebral fusion. There may be 13 ribs. Male patients may have undescended testicles or a microphallus. Hypospadias may be present, and the scrotum may be simple or bifid. Girls may have an underdeveloped clitoris and absent or hypoplastic labia minora.

The fingers are long and tapered or fusiform. Dimples are present over the knuckles and also may be seen over the elbows. The thumbs are proximally placed. These patients may have an equinovarus deformity of the foot, and there may be an abnormal implantation of the toes. In one patient there was ankylosis of the knees in extension. The dermatoglyphs of

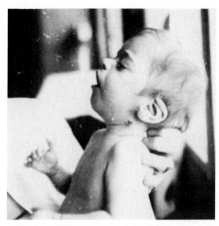

Fig. 2.—M. P. C. The carp-mouth is clearly seen, as well as the large, malformed, low-set ear. (Reprinted with permission from J. Genet. Hum.[12])

Fig. 3.—M. P. C. Unusual hand. The position of the thumb was simian and it was small, incurved and dysplastic.

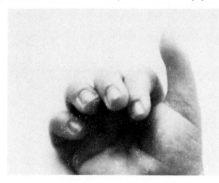

these patients contain an increased number of whorls on the fingers. Simian lines are common.

IgA deficiency has been reported[3, 8, 9] in association with partial deletion of the long arm of chromosome 18 or a ring 18 chromosome. Deletion of the *short* arm of chromosome 18 (18p—) also has been reported,[10, 11] but there is no recognizable syndrome. Involved pa-

tients with 18p— have some stigmata in common with the 18q— syndrome, e.g., mental retardation; low birth weight; large, floppy, low-set ears; epicanthal folds; hypertelorism and micrognathia.[12]

Case Report

M. P. C.[13] was born to a 32-year-old mother and a 34-year-old father who had previously had three healthy children. There was no consanguinity and the family history was negative. There was no history of drugs, infections or radiation during pregnancy. The birth weight was 4 lb, 12 oz.

Physical examination revealed a syndrome of anomalies that persisted. The length was 18.9 inches, the head circumference 32 cm. There was a prominent depression of the bridge of the nose and the midface (Figs. 1 and 2). There was microcephaly, a flattened nose and a carp-mouth. The patient had ocular hypertelorism and hypoplasia of the supraorbital ridges, and the eyes seemed deep seated. She had a small mandible and a high-arched palate. Nystagmus was observed. The ears were low set, oddly formed and large (see Fig. 2). There was a simian line on the left palm. The thumb was placed in a simian position and was unusual (Fig. 3). The patient was hypotonic and flaccid (Fig. 4), especially in the lower extremities. Deep tendon reflexes could not be obtained, although she would withdraw.

The fifth toes were abnormally placed (Fig. 5). There was a mild syndactyly of the second, third and fourth toes, as well as talipes equinovarus. There was a small umbilical hernia. The genitalia were very striking in that the labia minora were absent.

Roentgenographic examination revealed fusion of the first and second thoracic vertebrae. Dermatoglyphs revealed whorls on each of the fingers.

Growth and development were markedly slow. At 2 years of age the patient could do nothing but fix the examiner with her eyes. She was markedly hypotonic. The legs were flaccid and areflexic. On the other hand, she appeared to have some hypersensitivity to cutaneous stimulation, as for instance with alcohol or cotton. She gained only 2 lb, 3 oz between 1 and 2 years of age.

The karyotype (Fig. 6) revealed 46 chromosomes. One 18 was replaced with an acrocentric chromosome whose arms were about the length of the G group. It was considered to be an 18q—. In each of 32 karyotypes, the short arms also appeared smaller than the normal 18. Studies with ^{3}H-thymidine

Fig. 4.—M. P. C. at 2 years of age. She was hypotonic and flaccid. The configuration of the midface and mouth remained striking, as was the hypertelorism.

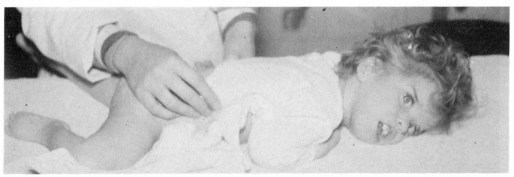

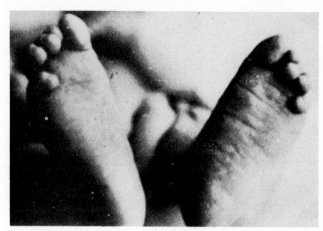

Fig. 5.—M. P. C. Abnormal placement of the fifth toes.

revealed the deleted chromosome to be late labeling. The patient was considered to have an 18q— deletion, with an apparent deletion of the short arms of the same chromosome.

GENETICS

Most of the reported cases of this syndrome have been sporadic. The size of the deletion on the long arm

Fig. 6.—M. P. C. Karyotype, illustrating deletion of the long arm of chromosome 18. In this patient there also was a deletion of the short arm (18q—, 18p—). (Reprinted with permission from J. Genet. Hum.[12].)

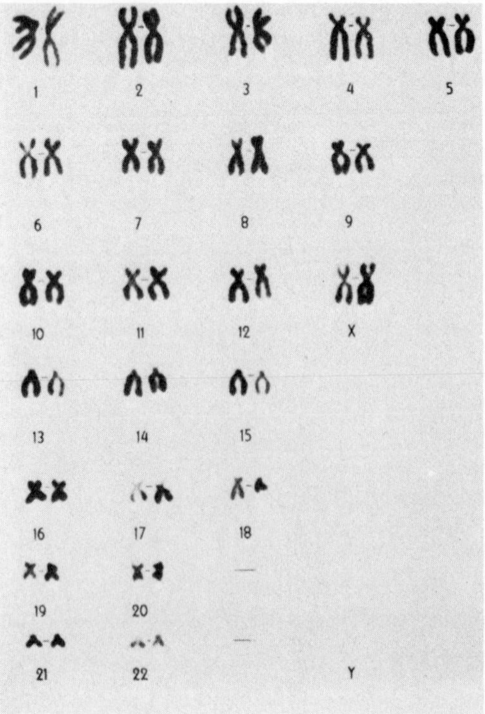

of the no. 18 chromosome (18q—) varies from one fourth to one third of the length of the long arm. Some patients have had a ring 18 chromosome. The diagnosis may be confirmed by autoradiography, as chromosome 18 replicates relatively late[6, 8, 13, 14] and is heavily labeled, whereas no. 17 replicates early and is little labeled. Staining with quinacrine mustard produces a characteristic fluorescent banding pattern by which all of the human metaphase chromosomes can be identified, including no. 18.[7, 15] It is particularly easy to recognize deletions using the banding technics.

Familial occurrence has been reported in which two of three siblings had deletions of the long arm of chromosome 18. In this case it was discovered that the mother's karyotype showed a translocation of part of the long arm of chromosome 18 to the short arm of a G group chromosome.[16] Another child with the syndrome has been reported whose father had a balanced translocation involving material of chromosome 18. In one family[3] the father and two siblings of the patient had a metacentric chromosome the size of a normal 18 that appeared to replace one member of the 18 pair. This was interpreted as a pericentric inversion [inv(18p + q—)]. The syndrome also may result from inheritance from a mosaic parent.[17]

TREATMENT

None.

REFERENCES

1. de Grouchy, J., Royer, P., Salmon, C., and Lamy, M.: Délétion partielle des bras longs de chromosome 18, Pathol. Microbiol. 12:579, 1964.
2. Lejeune, J., Berger, R., Lafourcade, J., and Réthoré, M.O.: La délétion partielle de bras long du chromosome 18. Individualisation d'un nouvel état morbide, Ann. Genet. (Paris) 9:32, 1966.
3. Wertelecki, W., and Gerald, P. S.: Clinical and chromosomal studies of the 18q— syndrome, J. Pediatr. 78:44, 1971.

4. Insley, J.: Syndrome associated with a deficiency of part of the long arm of chromosome no. 18, Arch. Dis. Child, 42: 140, 1967.

5. Destiné, M. L., Punnett, H. H., Thovichit, S., DiGeorge, A. M., and Weiss, L.: La délétion partielle du bras long du chromosome 18 (syndrome 18q–). Rapport de deux cas, Ann. Genet. (Paris) 10:65, 1967.

6. Curran, J. P., Al-Salihi, F. L., and Allderdice, P. W.: Partial deletion of the long arm of chromosome E– 18, Pediatrics 46:721, 1970.

7. Parker, C. E., Mavalwala, J., Koch, R., Hatashita, A., and Derencsenyi, A.: The syndrome associated with the partial deletion of the long arms of chromosome 18 (18q–), Calif. Med. 117:65, 1972.

8. Feingold, M., Schwartz, R. S., Atkins, L., Anderson, R., Bartsocas, C. S., Page, D. L., and Littlefield, J. W.: IgA deficiency associated with partial deletion of chromosome 18, Am. J. Dis. Child. 117:129, 1969.

9. Stewart, J. M., Go, S., Ellis, E., and Robinson, A.: Absent IgA and deletions of chromosome 18, J. Med. Genet. 7:11, 1970.

10. de Grouchy J.: Chromosome 18: A topologic approach, J. Pediatr. 66:414, 1965.

11. de Grouchy, J., Lamy, M. Thieffry, S., Arthuis, M., and Salmon, C.: Dysmorphie complexe avec oligophrenie: Délétion des bras courts d'un chromosome 17– 18, C. R. Acad. Sci. [D] (Paris) 256:1028, 1963.

12. de Chieri, P. R., Cedrato, A., and Albores, J. M.: Possible 46, XX, 18q–, 18p– syndrome, J. Genet. Hum. 19:127, 1971.

13. de Grouchy, J.: The 18p–, 18q–, and 18r Syndromes, in Bergsma, D. (ed.): *The First Conference on the Clinical Delineation of Birth Defects, Part V: Phenotypic Aspects of Chromosomal Aberrations* (New York: The National Foundation-March of Dimes, 1969), p. 74.

14. Schmid, W.: DNA replicant pattern of human chromosomes, Cytogenetics 2:175, 1963.

15. Francke, U.: Quinacrine mustard fluorescence of human chromosomes, Am. J. Hum. Genet. 24:189, 1972.

16. Law, E. M., and Masterson, J. G.: Partial deletion of chromosome 18, Lancet 2:1137, 1966.

17. Day, E. J., Marshall, R., Macdonald, P. A. C., and Davidson, W. M.: Deleted chromosome 18 with paternal mosaicism, Lancet 2:1307, 1967.

CAT EYE SYNDROME

Cardinal Clinical Features

Coloboma of the iris; anal atresia; renal anomalies; extra, acrocentric chromosome.

Clinical Picture

The syndrome of coloboma of the iris, anal atresia and renal anomalies first reported by Haak in 1878 was associated with the presence of an extra, acrocentric chromosome by Schachenmann and colleagues[1] in 1965. They described three patients with this syndrome and also commented on their facial appearance. The extra chromosome was smaller than the chromosomes of group G. The mother of the third patient also had colobomas of the irides and renal anomalies, an IQ of 76 and a karyotype identical to that of her daughter. The patient's brother, maternal uncle and maternal grandmother, all phenotypically normal, had mosaicism in which a small proportion of their cells displayed the abnormal karyotype.

These patients have colobomas of the iris that usually are vertically placed in the lower iris and account for a feline appearance of the eye. Coloboma of the choroid also may be present. Microphthalmia[2] is relatively common and there may be epicanthal folds, an antimongoloid slant and strabismus. A coarse nystagmus may be present. Patients also have a mild hypertelorism. The ears may be low set. Preauricular skin tags are regularly observed and there may be a preauricular fistula.

Anal atresia is a characteristic feature and usually is associated with a rectovestibular, rectovaginal or rectoperineal fistula. Renal anomalies are common and may include hypoplasia or absence of a kidney, hydronephrosis, ureteral reflux or stenosis in the urinary tract. There has been a preponderance of females in reported cases, but the disorder also occurs in the male.[3] Cryptorchidism may be present. Dislocation of the hip has been reported. Mental development may be normal, or there may be moderate to severe retardation.

Dermatoglyphic analysis may reveal absent axial triradii in association with arches in the hypothenar area[4] or an unusually distal t triradius with a large atd angle.[5]

There may be considerable variation in the pheno-

type of patients with the cat eye syndrome. This may relate to the presence of mosaicism and the relative proportion of the normal and abnormal cell lines. The syndrome has been reported in a patient with normal chromosomes,[6] possibly reflecting a transient developmental or otherwise subtle mosaicism. It also is possible, of course, for a small chromosomal fragment to come from a wide variety of chromosomes, and this could result in widely different phenotypes.[7] The cat eye syndrome should be differentiated from the VATER anomaly (Say syndrome)[8] (see section on Thumb and Radial Dysplasias), in which imperforate anus is associated with preaxial polydactyly or hypo-

Fig. 1.– F. G. (Case Report). A 48-day-old boy with the cat eye syndrome. Visible are the small misshapen helix, antimongoloid slant of the eyes and hypertelorism. He had mild micrognathia. (Courtesy of Dr. Primarosa de Chieri of Buenos Aires.)

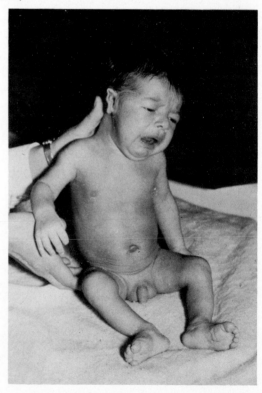

plasia of the thumb, vertebral anomalies and renal anomalies. Coloboma of the iris also may be seen in 13 trisomy (see chapter, this section).

CASE REPORT*

F. G. was seen at 48 days of age. He had been born of healthy, unrelated parents and weighed 6 lb, 3 oz. There had been no other pregnancies. The mother was 24 and the father 21 years of age. The maternal karyotype was normal.

Examination revealed a male infant weighing 8 lb. The small, unusually formed ear (Fig. 1) had a normal implantation. The patient had hypertelorism and an antimongoloid slant. There were bilateral colobomas involving the iris, cho-

Fig. 2.—F. G. Karyotype. The small chromosomal fragment is shown between the 22 and the Y. (Courtesy of Dr. Primarosa de Chieri of Buenos Aires.)

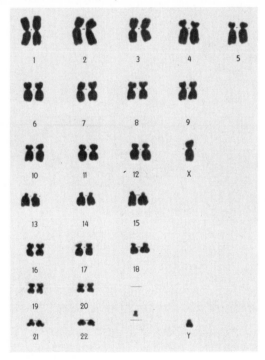

roid and retina. The mandible was small. Also present were an atrial and a ventricular septal defect. There was a simian crease on the right hand. The right hip was subluxed. The examination was otherwise normal. The patient died shortly after he was seen. Details of the terminal illness are not known and autopsy was not performed. The patient's karyotype showed a small chromosomal fragment between the 22 and the Y (Fig. 2).

GENETICS

The cat eye syndrome is characterized by the presence of extra chromosomal material in the form of a small, acrocentric fragment. This fragment is half the size or smaller than the G group chromosomes. Increased maternal age has been reported. Transmission from mother to daughter over three generations has been reported[1] in a family in which mosaicism was documented in the grandmother.

TREATMENT

None.

REFERENCES

1. Schachenmann, G., Schmid, W., Fraccaro, M., Mannini, A., Tiepolo, L., Perona, G. P., and Sartori, E.: Chromosomes in coloboma and anal atresia, Lancet 2:290, 1965.
2. Weber, F. M., Dooley, R. R., and Sparkes, R. S.: Anal atresia, eye anomalies, and an additional small abnormal acrocentric chromosome (47,XX,mar+): Report of a case, J. Pediatr. 76:594, 1970.
3. Fryns, J. P., Eggermont, E., Verresen, H., and van den Berghe, H.: A newborn with the cat-eye syndrome, Humangenetik 15:242, 1972.
4. Pfeiffer, R. A., Heimann, K., and Heiming, E.: Extra chromosome in "cat-eye" syndrome, Lancet 2:97, 1970.
5. Darby, C. W., and Hughes, D. T.: Dermatoglyphics and chromosomes in cat-eye syndrome, Br. Med. J. 3:47,1971.
6. Neu, R. L., Assemany, S. R., and Gardner, L. I.: "Cat-eye" syndrome with normal chromosomes, Lancet 1:949, 1970.
7. Borgaonkar, S., Schimke, R. N., and Thomas, H.: Report of five unrelated patients with a small, metacentric, extra chromosome or fragment, J. Genet. Hum. 19:207, 1971.
8. Say, B., Balci, S., Pirnar, T., and Hicsonmez, A.: Imperforate anus/polydactyly/vertebral anomalies syndrome. A hereditary trait?, J. Pediatr. 79:1033, 1971.
9. Buhler, E. M., Mehes, K., Muller, H., and Stalder, G. R.: Cat eye syndrome, a partial trisomy 22, Humangenetik 15:50, 1972.

*We are indebted to Dr. Primarosa de Chieri of the Fundación de Genética Humana in Buenos Aires for the description of this patient.

TURNER SYNDROME
45/XO Syndrome, Gonadal Dysgenesis

CARDINAL CLINICAL FEATURES

Shortness of stature, ovarian dysgenesis, sexual infantilism, webbed neck, shield chest, cubitus valgus, short fourth metacarpals, hypoplastic nails, multiple pigmented nevi.

CLINICAL PICTURE

In 1938 Turner[1] described seven patients 16–23 years of age who presented with sexual infantilism, short stature, webbed neck, a wide carrying angle at the elbow and primary amenorrhea. Wilkins and Fleischmann[2] later showed that a critical pathologic feature was the absence of normal ovaries, which were usually replaced by thin white streaks of stroma without follicles. The genetic importance of this syndrome was recognized with the demonstration by Ford and colleagues[3] in 1959 that a patient with the disorder had a 45/XO chromosomal constitution. This had been predicted in 1954 by Polani et al.[4] in their interpretation of the earlier finding that these patients were sex chromatin negative.

Fundamental features of the disease are shortness of stature, infantilism and ovarian dysgenesis. A group of associated malformations is so characteristic that they have been referred to as Turner stigmata.[5] The stature may be seen as short very early in life, even at birth. The average adult height of patients with the Turner syndrome is 54–56 inches. Other stigmata include webbing of the neck (Fig. 1), cubitus valgus, short fourth metacarpals, hypoplastic nails, a shield chest, multiple pigmented nevi and coarctation of the aorta. Peripheral edema in the newborn infant, especially over the dorsa of the feet (Fig. 2) in the Bonnevie-Ullrich pattern, also is so characteristic that it may be considered one of the stigmata.

Patients with the Turner syndrome may have very redundant skin at the nape of the neck, especially in the newborn period (see Fig. 2), and a low posterior hairline. They also may have epicanthal folds and occasionally ptosis and strabismus. The nasal bridge may be broad and the palate high arched, and there may be micrognathia. The ears may be low set and have prominent pinnae. Recurrent otitis media and

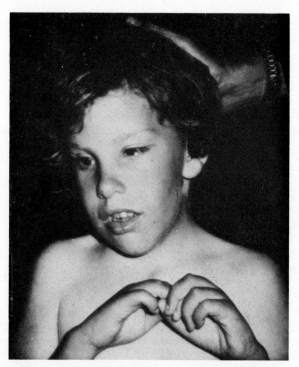

Fig. 1.–D. C. A child with XO Turner syndrome. She had shortness of stature and a web neck.

hearing loss are common. In the shield chest the nipples appear to be widely spaced. They are occasionally hypoplastic or inverted. Pectus excavatum may be present. In addition to the wide carrying angle at the elbow and the short fourth metacarpals, the bone age usually is delayed, and there may be clinodactyly. The hypoplastic nails are characteristic of the newborn period. The nails later become hyperconvex. Rarely dyskeratotic nails have developed. Alopecia has been described in a few of these patients. The multiple pigmented nevi develop with puberty, and there is a tendency to formation of keloids.

Congenital heart disease is frequent among these patients, the most common lesion being coarctation of the aorta. Hypertension reported in patients with this syndrome usually is on this basis, but essential hyper-

142

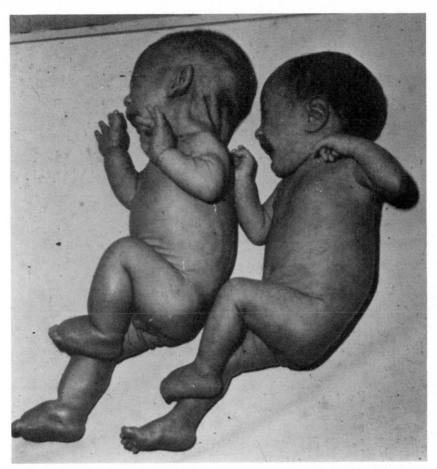

Fig. 2.—Dizygotic twins. The one on the left had the Turner syndrome. This baby was an XO/XX mosaic. The redundant skin or cutis laxa is localized to the neck in this syndrome. The edema of the feet and legs was prominent.

tension has rarely been seen. Other defects include aortic stenosis and aneurysm of the aorta. There may be cystic medial necrosis of the aorta. A variety of renal anomalies (Fig. 3) commonly associated with the syndrome include horseshoe and ectopic kidneys, ureteropelvic obstruction, a double collecting system, a bifid renal pelvis and hydronephrosis.[6, 7] Gastrointestinal hemorrhage has been reported in some patients with the Turner syndrome. This has been associated with multiple telangiectases. Other vascular anomalies that may bleed include hemangioma of the gastrointestinal tract and dilated, tortuous serosal veins, with a bag-of-worms appearance.

Most patients with the Turner syndrome present first because of their short stature.[8] They may present somewhat later because of failure of sexual development or amenorrhea. The breasts do not develop. The external genitalia remain infantile, the labia minora underdeveloped. The vaginal mucosa is red and infantile because of the lack of estrogen. The uterus usually is present but may be too tiny to be felt. The ovaries are

merely streaks of connective tissue, lacking in germinal cells. Slight hypertrophy of the clitoris has been reported in a few patients with the syndrome. Rarely an XO individual has normal menses; this is more likely to signify an XO/XX mosaicism.[8] A few cases have been reported[9] in which women with an XO type of Turner syndrome have given birth to normal children. The possibility of an unrecognized mosaicism, with an XX cell population in these patients or an XY population in those masculinized, always remains. Plasma levels of follicle-stimulating hormone (FSH) and luteinizing hormone (LH) are elevated in patients with gonadal dysgenesis.[10, 11] FSH concentrations are higher than normal even in patients under 10 years of age, whereas concentrations of LH become higher than normal after the age of 11. Elevated quantities of urinary gonadotropin are found in these patients.

Other diseases commonly associated with the Turner syndrome are the Hashimoto thyroiditis[12] and primary nongoitrous hypothyroidism. Other possibly autoimmune diseases seen with this syndrome include

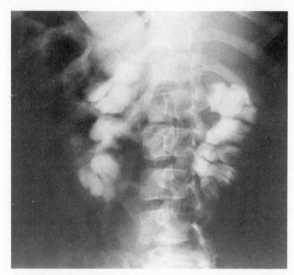

Fig. 3.—Intravenous urogram in a patient with the Turner syndrome and repeated urinary tract infection. There was considerable bilateral caliectasis. There was no accompanying pyelectasis, and peristalsis proceeded normally through both pelves and ureters, which were normal in caliber. There was no evidence of obstruction.

the Addison disease, ulcerative colitis and regional ileitis. There is a high incidence of diabetes mellitus in adult patients. Osteoporosis also may be seen in the Turner syndrome.[8]

Dermatoglyphic abnormalities found in the syndrome are not as striking as those seen in the autosomal trisomies. They include elevated total finger ridge counts and a–b ridge counts.[13, 14] Ulnar loops are frequent on the first digit and whorls on the fifth digit. The distal axial triradius is relatively distally placed.

Patients with the Turner syndrome generally have normal intelligence. However, their performance IQ as assessed in the Wechsler test usually is significantly lower than their verbal IQ. Money and colleagues[15-17] have reported a series of studies that indicate that these girls have a specific pattern of performance that is demonstrable on psychologic testing. They have problems in space-form perception, as evidenced by their unusual distortions in drawings[15] of people and figures, that has been called a developmental dysgnosia, or possibly a dyspraxia. They also have a poor right-left directional sense[16] and difficulty with numbers and counting (dyscalculia) that may reflect a relative distractability. One patient had such poor directional sense that she frequently became lost. These observations are interesting in that they provide a link between chromosomal structure and cognitive function.

Differential Diagnosis

The Turner syndrome must be differentiated from the Noonan syndrome (see section on Distinct Syndromes with Very Short Stature). Patients with this syndrome have a phenotype similar to that of the Turner syndrome, but have normal karyotypes. They may be male as well as female. The female patient has normal ovaries. Patients with pure gonadal dysgenesis have bilateral streak ovaries and normal karyotypes 46, XX.[18, 19] These patients are phenotypic females and have normal stature. They do not have the Turner stigmata. They do not develop sexually and often present for primary amenorrhea. Virilization has been reported in some of these patients. The mechanism appears to be a consequence of a deficiency of estrogen and the attendant elevation in gonadotropin, which may stimulate the production of excess androgen by the gonadal streak tissue. There also are patients with pure gonadal dysgenesis who are chromosomal males. Patients with this condition, referred to as XY gonadal dysgenesis,[20] are phenotypic females with streak gonads. There may be varying degrees of masculinization or genital ambiguity, but these patients usually are reared as girls. The occurrence of gonadoblastoma is more frequent in such patients,[20, 21] although this may occur in XX gonadal dysgenesis or even in the Turner syndrome. Gonadoblastoma is particularly common in mixed gonadal dysgenesis, where there is an XO and XY cell line (see chapter on XO/XY Mosaicism, this section).

Case Reports

Case 1.—C. O., a 16-year-old Mexican girl, was seen at University Hospital because of primary amenorrhea and lack of development of secondary sex characteristics. She was the product of a normal pregnancy and delivery. The mother was 17 years old at the time of this birth. C. O. was small at birth and had puffiness of the hands and feet along with redundancy of the skin of the neck and upper back. She remained small throughout childhood. She was always the shortest member of her class.

At 16 years of age she was 51¼ inches tall and weighed 69½ lb. Blood pressure was 118/68. She had a short, webbed neck and a low hairline. There was an increased number of pigmented nevi (Fig. 4). The nipples appeared widely spaced, and the breasts showed no sign of development. There was a grade

Fig. 4.—C. O. (Case 1). A 16-year-old patient with XO Turner syndrome, illustrating the increased number of pigmented nevi seen in this syndrome.

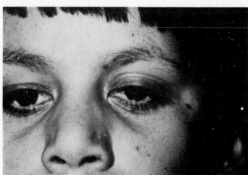

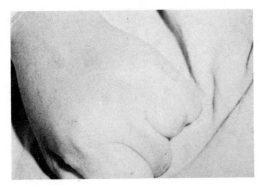

Fig. 5. – C. O. The fourth metacarpal is classically short in this syndrome.

III/VI systolic precordial ejection murmur. There was no axillary or pubic hair, and the vaginal mucosa was pink. No masses could be palpated rectally. There was a wide carrying angle at the elbow. The fourth metacarpals were short (Fig. 5), and there was clinodactyly.

A buccal smear revealed only 2–4% Barr bodies. The karyotype was 45,XO. An intravenous pyelogram was normal. Plasma testosterone was normal but plasma FSH and LH concentrations were elevated. Roentgenogram of the hands revealed short fourth metacarpals. There was generalized bony demineralization. The bone age was 11½ years.

CASE 2. – J. F., a 5-week-old white female, was admitted to University Hospital for pneumonia. She was the product of a normal pregnancy and delivery and had a birth weight of 5 lb, 9 oz. Her mother was 15 years old. The infant was found at birth to have multiple congenital anomalies suggestive of the Turner syndrome. A karyotype showed the characteristic 45,XO constitution.

On physical examination her height was 19 inches, and she weighed 6 lb, 3 oz. She had a triangular face, prominent epicanthal folds and striking redundant skin at the nape (Fig. 6). The posterior line was low, the chin small and the palate high arched. Ears were large and low set, the neck short (Fig. 7). The nipples appeared widely spaced. The clitoris was 0.5 cm in length. There were a wide carrying angle at the elbow and striking pitting edema over the dorsa of both hands and feet (Fig. 8). The edema extended up the lower leg. An intravenous pyelogram showed ureteropelvic obstruction on the left and a slightly enlarged left kidney.

CASE 3.* – This patient was one of the original patients diagnosed by Dr. H. Turner in 1934.[1] She was a 57-year-old white female who was born prematurely, weighing 4 lb, 8 oz. During childhood and adolescence her parents noted that she had short stature, no breast development and no menses. She saw Dr. Turner when she was 19 years old because of primary amenorrhea. He treated her with pituitary extract. In 1938 she had several episodes of menstrual flow.

At 57 years of age, she was 4 feet, 11 inches tall. She had a short, webbed neck (Fig. 9) and a low hairline. There was a wide carrying angle at the elbow. She had minimal breast development and scanty pubic and axillary hair. A blowing diastolic murmur was present at the cardiac apex, and a systolic ejection murmur along the sternum. Pelvic examination revealed essentially normal findings but a marked lack of evi-

*History and illustrations of this patient kindly provided by her present physician, Dr. Samuel Yen, of the University of California, San Diego.

dence of estrogen stimulation. Roentgenograms revealed short fourth metacarpals in both hands, and the epiphyses of the iliac crests were unfused. Demineralization was evident in the spine. Intravenous pyelogram revealed a renal cyst on the right kidney, which was successfully aspirated. Gonadotropins were elevated for both LH and FSH, although both were lower than those seen in younger Turner patients. Chromosome analysis was carried out for the first time in her life by Dr. Kurt Benirschke and revealed a 45,XO pattern (Fig. 10).

GENETICS

Most patients with the Turner syndrome have a karyotype in which there is a 45,XO pattern. The buccal smear for Barr bodies is negative (Fig. 11). Studies of the X chromosomal marker color blindness also have uniformly indicated that these patients have only one X chromosome. Monosomy for the X chromosome may result from nondisjunction during meiosis. This would yield gametes with 2 X chromosomes and others with no X chromosome. Fertilization between a normal X-bearing gamete and the latter would yield an XO gamete. This type of a zygote has been found to be viable in a variety of animal species.

Nondisjunction, which produces the Turner syndrome, can occur during spermatogenesis or oogenesis.[2] There is no evidence of increased maternal age at the time of birth in the Turner syndrome,[5] as has been found in the Down syndrome (21 trisomy) and the Klinefelter syndrome (XXY), in both of which the mechanism appears to be one of nondisjunction. In a majority of patients with the Turner syndrome, the X chromosome present is that of the mother (X^mO). There could be a familial tendency to undergo chromosomal alteration in the formation of the gametes; there have been several reports of multiple siblings in one family in which there was 21 trisomy in one child and the Turner syndrome in another.[22] A family was reported with two siblings with the XO Turner syndrome,[23] and another was recorded in which monozygotic twins with the XO Turner syndrome had an identical phenotype.[24] Experience with twinning, in which monozygotic twinning is rather common in the Turner syndrome and in which the usual case is for mosaicism with dissimilarity between the twins (Fig. 12), suggests that nondisjunction in the zygote is the primary event in the pathogenesis of most patients with the Turner syndrome.

Some patients with the Turner syndrome have readily demonstrated mosaicism. Chromosomal mosaicism is more common in patients with the Turner syndrome than it is in any other chromosomal aberration. It has been estimated to occur in as many as 30% of individuals with the Turner syndrome.[25] On the other hand, among patients having spontaneous abortions, some 20% have chromosomal anomalies, and the most frequent of these is XO.[25, 26] Mosaicism for XO/XX (see Fig. 12) was not found in these fetuses, nor were structural defects of the X chromosome.[26] Approximately 96–98% of all XO conceptuses die in utero. These ob-

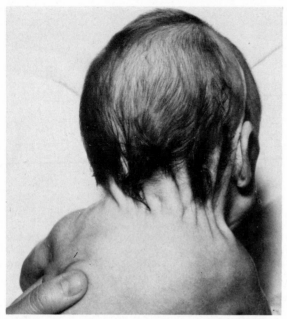

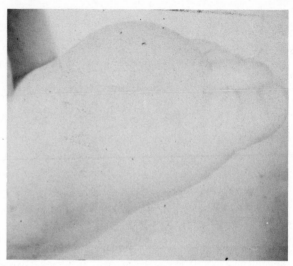

Fig. 8. – J. F. The edema of the feet and legs was enormous. It did pit, as illustrated. Dysplasia of the nails also is evident.

Fig. 6. – J. F. (Case 2). A 5-week-old baby with an XO karyotype. The redundant skin is very striking.

servations indicate that the XO constitution is highly lethal in the fetus and that mosaicism provides an increased chance of viability.

Types of mosaicism associated with this syndrome include XO/XXX and XO/XX/XXX, as well as XO/XX.[27, 28] The clinical picture in XO/XX mosaics generally is milder than in XO patients. Shortness of

Fig. 9. – Case 3, a 58-year-old woman with the Turner syndrome. This patient was one of the original patients reported by Turner in 1938. (Case 2 in that series). The web neck was still prominent, as were the shortness of stature, widely spaced nipples and wide carrying angle of the arms. (Courtesy of Dr. Samuel Yen of the University of California San Diego.)

Fig. 7. – J. F. Epicanthal folds. The ears were large and low set. The short, webbed neck is visible.

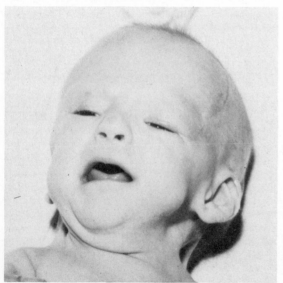

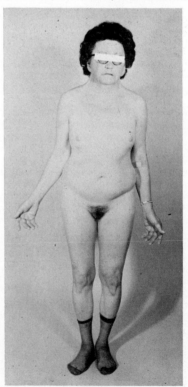

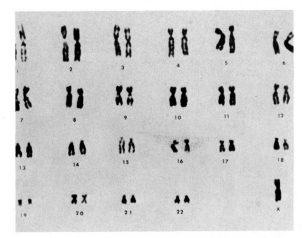

Fig. 10.—(Case 3). Karyotype of patient in Figure 9. She had an XO chromosomal constitution. (Courtesy of Dr. Samuel Yen and Dr. Kurt Benirschke of the University of California San Diego.)

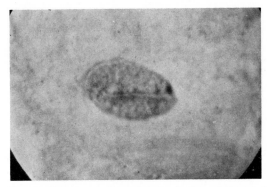

Fig. 11.—Sex chromatin pattern in the buccal smear of a normal female patient. The patient with the Turner syndrome lacks the chromatin body at the rim of the nucleus in the 5 o'clock position.

Fig. 12.—Karyotype of twin on left in Figure 2, illustrating XX/XO mosaicism. The upper two cells represented had 46,XX and the lower two 45,XO constitutions. (Courtesy of Dr. Samuel Yen and Dr. Kurt Benirschke of the University of California San Diego.)

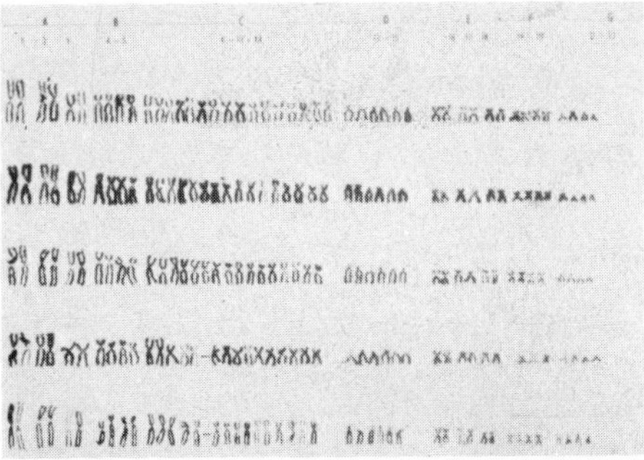

stature is not so completely uniform.[5] Phallic hypertrophy is slightly more common. Cardiac anomalies are much less frequent in patients with mosaicism.[8] Normal pubertal development may be seen in XO/XX mosaicism.[27]

Structural abnormalities of the X chromosome also have been reported in patients with the Turner syndrome. These include deletions and iso-X chromosomes. The patient with an isochromosome component may be more likely to have mental retardation,[8] and many patients with thyroiditis have had duplications or deletions.[12] It has been concluded that absence of the short arm of the X chromosome is necessary for the development of the typical stigmata of the Turner syndrome.[5, 29] Dysgenesis of the ovaries may be seen, with deletion of the long arms as well as that of the short arms. The frequent occurrence of structural abnormalities suggests that the presence of an abnormal chromosome predisposes it to nondisjunction and particularly to mosaicism. This appears to be true also of the Y and other chromosomes. It also has been postulated that, to explain the occurrence of multiple siblings with mosaicism involving the X and Y chromosomes in a single family, there might be an autosomal recessive gene that promotes mitotic instability.[32] There is some precedent for this in maize genetics in which a recessive gene, called sticky, leads to mitotic nondisjunction in the homozygote.[33]

Treatment

Counseling is a primary consideration in the management of patients with the Turner syndrome. The prognosis in this condition is considerably better than in most cytogenetic abnormalities. However, shortness of stature and failure of sexual development are both matters that require sensitive help. Feminization can be achieved using estrogen treatment. Johanson, Brasel and Blizzard[34] have recommended preceding this with androgen treatment in order to gain a few inches in height.

REFERENCES

1. Turner, H. H.: A syndrome of infantilism, congenital webbed neck, and cubitus valgus, Endocrinology 23:566, 1938.
2. Wilkins, L., and Fleischmann, W.: Ovarian agenesis: Pathology, associated clinical symptoms and the bearing on the theories of sex differentiation, J. Clin. Endocrinol. 4:357, 1944.
3. Ford, C. E., Jones, K. W., Polani, P. E., de Almeida, J. C., and Briggs, J. H.: A sex-chromosome anomaly in a case of gonadal dysgenesis (Turner's syndrome), Lancet 1:711, 1959.
4. Polani, P. E., Hunter, W. F., and Lennox, B.: Chromosomal sex in Turner's syndrome with coarctation of the aorta, Lancet 2:120, 1954.
5. Ferguson-Smith, M. A.: Karyotype-phenotype correlations

in gonadal dysgenesis and their bearing on the pathogenesis of malformations, J. Med. Genet. 2:142, 1965.

6. Matthies, F., MacDiarmid, W. D., Rallison, M. L., and Tyler, F. H.: Renal anomalies in Turner's syndrome. Types and suggested embryogenesis, Clin. Pediatr. 10:561, 1971.

7. Persky, L., and Owens, R.: Genitourinary tract abnormalities in Turner's syndrome (gonadal dysgenesis), J. Urol. 105:309, 1971.

8. Goldberg, M. B., Scully, A. L., Solomon, I. L., and Steinbach, H. L.: Gonadal dysgenesis in phenotypic female subjects. A review of eighty-seven cases, with cytogenetic studies in fifty-three, Am. J. Med. 45:529, 1968.

9. Nakashima, I., and Robinson, A.: Fertility in a 45,X female, Pediatrics 47:770, 1971.

10. Penny, R., Guyda, H. J., Baghdassarian, A., Johanson, A. J., and Blizzard, R. M.: Correlation of serum follicular stimulating hormone (FSH) and luteinizing hormone (LH) as measured by radioimmunoassay in disorders of sexual development, J. Clin. Invest. 49:1847, 1970.

11. Guyda, H. J., Johanson, A. J., Migeon, C. J., and Blizzard, R. M.: Determination of serum luteinizing hormone (SLH) by radioimmunoassay in disorders of adolescent sexual development, Pediatr. Res. 3:538, 1969.

12. Melin, K., and Samuelson, G.: Gonadal dysgenesis with lymphocytic thyroiditis and deletion of the long arm of the X chromosome, Acta Paediatr. Scand. 58:625, 1969.

13. Holt, S. B., and Lindsten, J.: Dermatoglyphic anomalies in Turner's syndrome, Ann. Hum. Genet. 28:87, 1964.

14. Borgaonkar, D. S., and Mules, E.: Comments on patients with sex chromosome aneuploidy: Dermatoglyphs, parental ages, Xg[a] blood group, J. Med. Genet. 7:345, 1970.

15. Money, J., Alexander, D., and Ehrhardt, A.: Visual-constructional deficit in Turner's syndrome, J. Pediatr. 69:126, 1966.

16. Alexander, D., Walker, Jr., H. T., and Money, J.: Studies in direction sense, Arch. Gen. Psychiatry 10:337, 1964.

17. Money, J., and Alexander, D.: Turner's syndrome: Further demonstration of the presence of specific cognitional deficiencies, J. Med. Genet. 3:47, 1966.

18. Judd, H. L., Scully, R. E., Atkins, L., Neer, R. M., and Kliman, B.: Pure gonadal dysgenesis with progressive hirsutism. Demonstration of testosterone production by gonadal streaks. N. Eng. J. Med. 282:881, 1970.

19. Nielsen, J., and Friedrich, U.: Pure gonadal dysgenesis, Clin. Genet. 3:52, 1971.

20. Chemke, J., Carmichael, R., Stewart, J. M., Geer, R. H., and

Robinson, A.: Familial XY gonadal dysgenesis, J. Med. Genet. 7:105, 1970.

21. Fischer, P., Golob, E., and Holzner, H.: XY gonadal dysgenesis and malignancy, Lancet 2:110, 1969.

22. Casteels-Van Daele, M., Proesmans, W., Van den Berghe, H., and Verresen, H.: Down's anomaly (21 trisomy) and Turner's syndrome (46, XXqi) in the same sibship, Helv. Paediatr. Acta 25:412, 1970.

23. Dunlap, D. B., Aubry, R., and Louro, J. M.: The occurrence of the 45, X Turner's syndrome in sisters, J. Clin. Endocrinol. Metab. 34:491, 1972.

24. Riekhof, P. L., Horton, W. A., Harris, D. J., and Schimke, R. N.: Monozygotic twins with the Turner syndrome, Am. J. Obstet. Gynecol. 112:59, 1972.

25. Hecht, F., and MacFarlane, J. P.: Mosaicism in Turner's syndrome reflects the lethality of XO, Lancet 2:1197, 1969.

26. Carr, D. H.: Lethal Chromosome Errors, in Benirschke, K. (ed.): Comparative Mammalian Cytogenetics (New York: Springer-Verlag New York, Inc., 1969), p. 68.

27. Francks, R. C., and Engel, E.: Turner's syndrome with XO/XX mosaicism and normal puberty, J. Pediatr. 79:1035, 1971.

28. Hsu, L. Y. F., and Hirschhorn, K.: Unusual Turner mosaicism (45,X47,XXX; 45,X/46, XXqi; 45,X/46,XXr): Detection through deceleration from normal linear growth or secondary amenorrhea, J. Pediatr. 79:276, 1971.

29. Bachman, R., de la Cruz, F. F., Al-Aish, M., and Santell, F.: Short arm deletion of an X chromosome in a 19-year-old girl, Am. J. Ment. Defic. 75:435, 1971.

30. Hecht, F., Jones, D. L., Delay, M., and Klevit, H.: Xq– Turner's syndrome: Reconsideration of hypothesis that Xp– causes somatic features in Turner's syndrome, J. Med. Genet. 7:1, 1970.

31. Finley, W. H., Taylor, Jr., P. T., Rosecrans, C. J., Finley, S. C., Waldrop, E. G., and Pittman, C. S.: Gonadal dysgenesis associated with a deleted X chromosome, Am. J. Ment. Defic. 74:553, 1970.

32. Hsu, L. Y. F., Hirschhorn, K., Goldstein, A., and Barcinski, M. A.: Familial chromosomal mosaicism, genetic aspects, Ann. Hum. Genet. 33:343, 1970.

33. Beadle, G. W.: A gene for sticky chromosomes in Zea mays, Indu. Abstammungs. Vererbungslehre 63:195, 1932.

34. Johanson, A. J., Brasel, J. A., and Blizzard, R. M.: Growth in patients with gonadal dysgenesis receiving fluoxymesterone, J. Pediatr. 75:1015, 1969.

XO/XY MOSAICISM
Mixed Gonadal Dysgenesis

CARDINAL CLINICAL FEATURES

Ambiguous genitalia, short stature with variable Turner stigmata, female or müllerian internal genital development, gonadoblastoma.

CLINICAL PICTURE

Patients with mixed gonadal dysgenesis usually have a phenotype that is generally female. Most are reared as girls. They usually appear first in the physician's office around the age of puberty because of amenorrhea. Clitoral hypertrophy may be present from birth; it also may increase with development. There may be posterior fusion of the labioscrotal folds and a urogenital sinus. There is generally a lack of female secondary sexual characteristics. A few patients raised as boys may present because of hypospadias, bifid scrotum or undescended testes. Rarely a patient with an entirely male external phenotype may present with a unilateral testis or an inguinal hernia in which an abnormal gonad is found. A uterus or other müllerian elements may be found in the hernia.

Most patients of either type have some Turner stigmata. The typical short stature is common but not invariable in this syndrome, as are the webbed neck, low hairline, shield chest, cubitus valgus and dysplastic nails. The span often is greater than the height, consistent with the fact that short legs are characteristic of X chromosomal monosomy, and longer than normal legs with an extra X in the XXY syndrome. An occasional simian crease is seen. Numerous nevi frequently develop. Mental retardation is seen in a small proportion of patients, but more often than in the Turner XO syndrome.

Laparotomy in patients with mixed gonadal dysgenesis may reveal asymmetrical gonadal differentiation. The streak gonads typical of the Turner syndrome may be seen on one or both sides. The usual presentation is with a streak on one side and what looks like a testis on the other. However, there is considerable variation. The streaks may contain primitive testicular tissue or ovarian precursors. A testicular or ovarian structure may be present, and testicular and ovarian elements may be found in the same gonadal structure.

There may be a testis on one side and no evidence of a gonad on the other (Case 2). The histology of the gonads present usually is primitive, as in other patients with gonadal dysgenesis. Usually it is testicular structure that is best developed. Only the histologic Sertoli cell picture may be seen. Evidence of spermatogenesis or oogenesis is conspicuously absent.

The internal genital development always is müllerian. There are a uterus and an upper vagina as well as fallopian tubes.[1, 2] Cytogenetic examination reveals the patient to be a mosaic in which there are XO and XY cell lines. Mosaicism may be evident in the peripheral blood, or it may be necessary to culture cells from other tissues.[3]

The presence of two sex chromosomes, XX in the female and XY in the male, is considered necessary for normal, complete development of the gonads. Patients with mixed gonadal dysgenesis usually have dysgenic testes.[1, 2] It appears that they fail to produce müllerian inhibitory factor, which is normally made early in fetal life around the eighth week of gestation, and consequently the müllerian duct persists and develops into the uterus and other structures just as it would in the normal female. These dysgenic testes may produce a variable amount of testosterone, and this may result in varying degrees of masculinization. This may involve the external genitalia and the secondary sex characteristics. Some patients may experience virilization with puberty. Concentrations of testosterone may be as high as those of the male. Elevated concentration of follicle stimulating hormone (FSH) and luteinizing hormone are common in these patients, presumably reflecting the response of the pituitary to gonadal dysgenesis.[3]

The development of gonadoblastoma is common in patients with mixed gonadal dysgenesis. These tumors tend to develop in any patient with gonadal dysgenesis, but especially in those in whom there is an XY cell line.[4] Histologically these tumors are characterized by embryonal gonadal cells, malignant germ cells that stain intensively with PAS and with stains for alkaline phosphatase. They may occur in nests in fibrous stroma or alongside primitive testicular or ovarian tissue.[4, 5] Other tumors also may occur in the gonads, including dysgerminoma, teratoma, arrhenoblastoma, mesoblastoma and choriocarcinoma.[5] Histologic dis-

149

tinction is important because simple excision is all that is required in the management of a gonadoblastoma. More extensive procedures are required for these other neoplasms, which have highly invasive and metastatic propensities. Excision always should be bilateral. Abdominal pain may be a presenting symptom of a gonadoblastoma, or there may be virilization or an increase in virilization.[3] Fibroblasts derived from patients with XY gonadal dysgenesis have shown an increased propensity for transformation with simian papovavirus (SV40), like those of patients with the Down and Klinefelter syndromes and Fanconi anemia.[6]

CASE REPORTS

CASE 1.*—S. L. was a 19-year-old white female seen for primary amenorrhea. (Fig. 1). She was found to have ambiguous genitalia. She had been born with an imperforate anus, which was corrected in the first week of life. She had otherwise been healthy all her life, although she was always the shortest one in her class. She was 53 inches tall and weighed 105 lb. The neck was short and had mild webbing and a low hairline. The breasts were lacking in development and the nipples appeared

Fig. 1.—S. L. (Case 1). A 19-year-old who presented with primary amenorrhea. She had short stature and sexual infantilism. There was no breast development. She also had Turner stigmata including short neck, widely placed nipples and wide carrying angle at the elbows. (Courtesy of Drs. Donna L. Brooks and David Allan of Mercy Hospital, San Diego.)

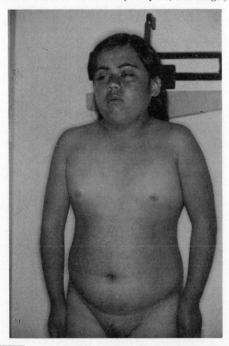

*Reported through the courtesy of Drs. Donna L. Brooks of San Diego, David Allan of Mercy Hospital, San Diego, and Kurt Benirschke of the University of California, San Diego.

Fig. 2.—S. L. (Case 1). Ambiguous genitalia. There were clitoromegaly and partial fusion. (Courtesy of Drs. Donna L. Brooks and David Allan of Mercy Hospital, San Diego.)

widely spaced. There was a large clitoris 2–3 cm in length and mild labial fusion (Fig. 2). The vagina appeared initially to be absent. There was a urogenital sinus.

Karyotype showed a mosaic pattern in which one cell line was 45,XO (Fig. 3a) and the other 46,XY (Fig. 3b). Concentrations of FSH in the plasma were increased. An intravenous pyelogram and cystogram were normal. Contrast roentgenography revealed a tiny vagina, a normal cervix and uterine cavity and a right fallopian tube. Laparotomy revealed a streak ovary on one side (Fig. 4A) and a streak testis (Fig. 4B) on the other. Both streaks were removed and the labial fusion separated. The patient was started on cyclic therapy and had a good feminization response, with development of breasts and withdrawal bleeding.

CASE 2*—S. P. was a 3-year-old white male who presented at the U.S. Naval Hospital, San Diego, for evaluation of ambiguous genitalia. He was the product of a normal pregnancy. The mother, who was 25 years old at the time of birth, had received no hormonal therapy during pregnancy. Delivery was 2 weeks early. Birth weight was 4 lb, 13 oz. Ambiguous genitalia were detected at birth and consisted of hypospadias and a bifid scrotum with a gonad in the right labioscrotal fold but none on the left. Investigation in another hospital revealed the 24-hour urinary excretion of 17-ketosteroids and pregnanetriol to be normal. The karyotype revealed the presence of two cell lines, one XO and one XY. An x-ray examination was said to suggest female anatomy.

Physical examination at 3 years of age revealed a height of 36 inches, for a height age of 2 and 5/12 years. Weight was 26 lb, 4 oz. The patient appeared alert and of normal intelligence. There were no stigmata of the Turner syndrome. The external genitalia were indeed ambiguous (Fig. 5). The scrotum was bifid. The right side was larger than the left and contained a 3 x 3 cm mass, which felt like a testis. The left side was empty. The phallus was small and poorly developed and had severe chordee. A hypospadias or urogenital sinus was represented by an opening at the junction of the phallus and the fused labioscrotal folds. On roentgenography injected contrast medium revealed that the urogenital sinus opened into a vagina and refluxed as far as a fallopian tube.

Biopsy of the gonad established that it was a testis. Exploratory laparotomy revealed no evidence of a gonad on the left side, not even a streak. On the right was the testis with a fallopian tube connected to it (Fig. 6). Both were removed. A uter-

*Reported through the courtesy of Drs. Patrick Walsh and John E. Wimmer of the U. S. Naval Hospital, San Diego.

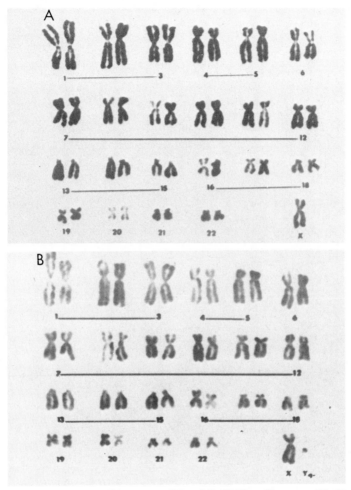

Fig. 3.—S. L. (Case 1). Karyotype, with two cell lines: *A,* 45 XO and *B,* 46 XYq–. (Courtesy of Dr. Kurt Benirschke of the University of California San Diego.)

us, vagina and left fallopian tube also were removed. Karyotypes of lymphocytes and the testis revealed two cell lines in both, one XO and one XY. The Y chromosome in this cell line completely lacked the brightly fluorescent ends of the long arms (Fig. 7). The presence in the testis of a nonfluorescent Y cell line supports the hypothesis that maleness is determined by areas of the Y chromosome other than the distal long arms.

GENETICS

Patients with mixed gonadal dysgenesis are chromatin negative; no Barr bodies are seen. The karyotype is that of a mosaic pattern of XO and XY cell lines.[7] It seems likely that in the genesis of the XO cell line the Y chromosome is lost in mitosis early in the development of the mesoderm. This could reflect nondisjunction or simple loss due to a lag in anaphase. Among the patients described, a certain number of

abnormalities have been present in the Y chromosomes. These have included deletion of the short arm of the Y (p–), isochromosome for the Y chromosome, and a very small, acrocentric Y with deletions on both the long and the short arms.[8] Rarely, a karyotype is seen in this syndrome in which there is XO/XX/XY, XO/XXXY or XO/XYYY mosaicism.[9] It is thought that genes determining maleness are located on the short arms of the Y and on the portions of the long arms closest to the centromere, whereas the strikingly fluorescent parts of the Y at the distal ends of the long arms are heterochromic.[10]

TREATMENT

Bilateral gonadectomy is an important part of therapy in these patients because of the high incidence of gonadoblastoma. The discovery of a Y chromosome in

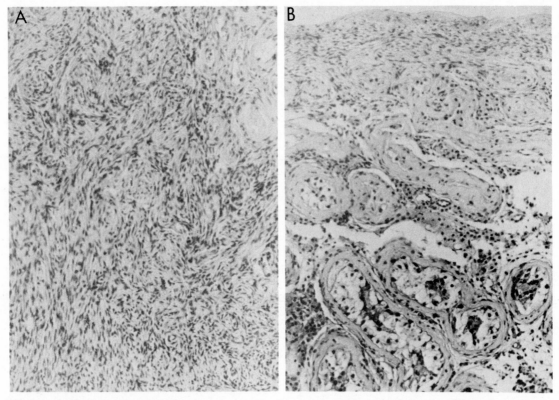

Fig. 4.—S. L. (Case 1). *A,* histology of the streak ovary from one side. *B,* histology of the testis on the other side, illustrating hypoplastic seminiferous tubules. (Courtesy of Dr. Kurt Benirschke of the University of California San Diego.)

any cell line in a patient with gonadal dysgenesis is an indication for prompt gonadectomy.

The presence of ambiguous genitalia in the newborn infant should be an alerting signal[11, 12] for chromosomal analysis, as well as for tests for adrenal hyperpla-

Fig. 5.—S. P. (Case 2), a 2½ year old with mixed gonadal dysgenesis. This patient presented as a male pseudohermaphrodite. He had no Turner stigmata. Below the small phallus was a urogenital sinus. On the right there was a scrotal swelling containing a testis. No gonad was palpable on the left. (Courtesy of Drs. Patrick Walsh and John E. Wimmer of the U. S. Naval Hospital, San Diego.)

sia. Once the diagnosis is clear, it is important to evaluate the phallus critically. It is generally thought that unless the phallus will be clearly functional, the child is better reared as a girl, regardless of the karyotype. Among patients raised as girls, cyclic estrogen therapy is indicated, following which they will usually develop breasts and withdrawal bleeding.

REFERENCES

1. Federman, D. D.: *Abnormal Sexual Development. A Genetic and Endocrine Approach to Differential Diagnosis* (Philadelphia and London: W. B. Saunders Co., 1968).

Fig. 6.—S. P. (Case 2). Testis, with an attached fallopian tube. (Courtesy of Drs. Patrick Walsh and John E. Wimmer of the U.S. Naval Hospital, San Diego.)

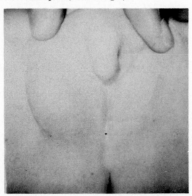

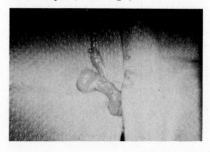

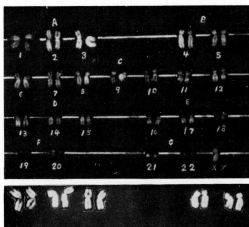
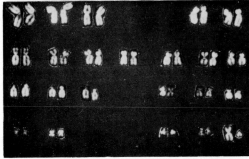

Fig. 7.—S. P. (Case 2). Karyotype, with two cell lines: **Top,** 45 XO and **bottom,** 46 XY, in which the typical bright fluorescence of the distal part of the long arm of the Y is lacking. (Courtesy of Dr. O. W. Jones of the University of California San Diego.)

2. Blizzard, R. M.: Sexual Development, Normal and Abnormal, in *The Clinical Delineation of Birth Defects, Part X, The Endocrine System*, Birth Defects Original Article Series Vol. III, No. 6 (Baltimore, The Williams & Wilkins Company, 1971), p. 108.

3. Josso, N., Nezelof, C., Picon, R., de Grouchy, J., Dray, F., and Rappaport, R.: Gonadoblastoma in gonadal dysgenesis: A report of two cases with 46, XY/45,X mosaicism, J. Pediatr. 74:425, 1969.

4. Marquez-Monter, H., Armendares, S., Buentello, L., and Villegas, J.: Histopathologic study with cytogenetic correlation in 20 cases of gonadal dysgenesis, Am. J. Clin. Pathol. 57:449, 1972.

5. Schellhas, H. F., Trujillo, J. M., Rutledge, F. N., and Cork, A.: Germ cell tumors associated with XY gonadal dysgenesis, Am. J. Obstet. Gynecol. 109:1197, 1971.

6. Mukerjee, D., Bowen, J. M., Trujillo, J. M., and Cork, A.: Increased susceptibility of cells from cancer patients with XY-gonadal dysgenesis to simian papovavirus 40 transformation, Cancer Res. 32:1518, 1972.

7. Hirschhorn, K., Decker, W. H., and Cooper, H. L.: Human intersex with chromosome mosaicism of type XY/XO, N. Engl. J. Med. 263:1044, 1960.

8. Ferguson-Smith, M. A., Boyd, E., Ferguson-Smith, M. E., Pritchard, J. G., Yusuf, A. F. M., and Gray, B.: Isochromosome for long arm of Y chromosome in patient with Turner's syndrome and sex chromosome mosaicism (45,X/46,XYqi), J. Med. Genet. 6:422, 1969.

9. Hortling, H., de la Chapelle, A., Teppo, L., and Kovioja, O.: Mixed gonadal dysgenesis, a case with male phenotype and 45,X/46,XY mosaicism, Acta Endocrinol. 65:229, 1970.

10. Severi, F., Tiepolo, L., and Scappaticci, S.: Identification of the Y chromosome by the fluorescence technique in an XY/XO gonadal dysgenesis, Acta Paediatr. Scand. 60:716, 1971.

11. Davidoff, F., and Federman, D. D.: Mixed gonadal dysgenesis, Pediatrics 52:725, 1973.

12. Hung, W., Verghese, K. P., Picciano, D., Jacobson, C. B., and Chandra, R.: Mixed gonadal dysgenesis with XO/XY mosaicism in multiple tissues, Obstet. Gynecol. 36:373, 1970.

13. Hansen, I.: XO/XY-Mosaik bei einem Knaben mit ovariell-testikulärer Dysgenesie, Dtsch. Med. Wochenschr. 95: 1599, 1970.

14. Sohval, A. R.: Hermaphroditism with "atypical" or "mixed" gonadal dysgenesis. Relationship to gonadal neoplasm, Am. J. Med. 36:281, 1964.

KLINEFELTER SYNDROME
XXY SYNDROME

Cardinal Clinical Features

In the preadolescent patient, slim, eunuchoid body proportions with relatively longer legs, small testes, behavior problems, dull mentality; later, gynecomastia, underdeveloped secondary sexual characteristics, elevated gonadotropins, hyalinization and atrophy of seminiferous tubules; sex chromatin-positive buccal smear, 47 XXY karyotype.

Clinical Picture

The syndrome was first described by Klinefelter and colleagues in 1942[1] in a series of nine men with gynecomastia, small testes, azoospermia and elevated levels of follicle stimulating hormone (FSH). In 1956 Plunkett and Barr discovered that the buccal smear in patients with this syndrome regularly displayed the nuclear sex chromatin body[2] previously seen only in female cells. Jacobs and Strong[3] described the abnormal XXY karyotype in 1959. The syndrome is relatively common; its incidence is 1.7 in a thousand live-born male infants.[4, 5]

The patient appears normal at birth. Diagnosis usually is not made until puberty or later, when advice is sought for sexual infantilism or sterility. Of course, a number of patients have been turned up in the course of programs in which screening has been carried out in the newborn period by chromosome analysis or by examination of the amnion for chromatin bodies.[5] Other patients have been discovered by chromosome analysis of patients with mental retardation. The syndrome is recognizable in childhood.[6] Clues may be small external genitalia and, in particular, small testes,[7] or a tall, slim, eunuchoid appearance in which the legs are long and the upper/lower ratio is low. The span usually exceeds the height.[1] The hips may have a broad, female configuration (Fig. 1). These patients may have grossly recognizable mental retardation. Mental retardation is not always present, but patients usually have at best a borderline intelligence. They frequently have behavior problems, especially in school. There may be limited supination of the elbow, and roentgenograms may reveal radioulnar synostosis or dislocation. Even in the absence of radioulnar syn-

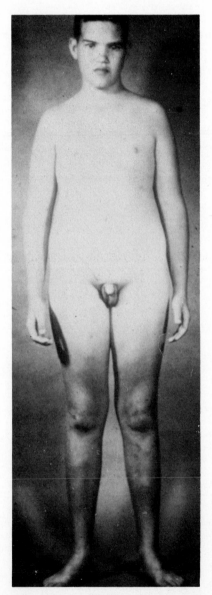

Fig. 1.—A boy with the Klinefelter syndrome. He was tall and had a female habitus and hair distribution. At this age he did not have gynecomastia.

154

ostosis or dislocation, patients have a significant cubitus varus or reduction of the normal carrying angle. Patients with this syndrome have taurodont teeth.

Gynecomastia characteristically develops following puberty, which generally begins between ages 12 and 14 in these patients.[1] It always is bilateral, and it is present in about 40% of cases. The breasts resemble those of the adolescent female. Secondary male sexual characteristics remain underdeveloped. Beard growth usually is scanty, as is axillary and body hair. Pubic hair takes a female distribution. Body fat also assumes a female distribution.[2] Temporal baldness may not develop. The testes are very small but usually of normal firmness. Undescended testes probably are more frequent in this syndrome than in normal individuals.[8] The patients are characteristically sterile. A few reported patients who have been fertile have been mosaics in whom the cell lines were XY and XXY.

The Klinefelter syndrome is the most common cause of primary hypogonadism in the male. An association between abnormal behavior and the XYY syndrome is generally recognized. It is probably also true that behavioral abnormality is a product of the XXY chromosomal constitution.[5] Chromatin-positive males seem to be found more frequently among populations of mentally subnormal men incarcerated for sexual offenses than among populations of retarded persons with no criminal record. Some of these may of course be XXYY, but it is clear that the XXY individual also is prone to behavioral disorders. Schizophrenia has been associated with the Klinefelter syndrome.[9] It also is clear that an XXY chromosomal constitution is compatible with a normal social adjustment.[8] The association of breast cancer in the male and a 47,XXY karyotype is well known.[10-12] Interstitial tumor of the testes[11] and other tumors[12] also have been described in this syndrome, suggesting that the karyotype or the resultant abnormal endocrine situation predisposes to neoplasia.

Dermatoglyphic analysis reveals an increased number of digital arches. The digital patterns are small and the total ridge count is decreased.[13-15] Among series of patients with increasing numbers of X chromosomes, the ridge count tends to decrease as the number of chromosomes increases.[13] Thus an XXY individual has a lower ridge count than an XY individual, and those with XXXY and XXXXY karyotypes have even lower counts.

Elevated gonadotropin levels always are seen in the Klinefelter syndrome. FSH is elevated, whereas testosterone is low or at the lower range of normal. Azoospermia is the rule. Aspermia may be present. Biopsy of the testis in the puberal patient reveals a decreased number of germ cells. In the adult there is hyalinization, fibrosis and atrophy of the seminiferous tubules. There are no germ cells, but there are abundant Leydig cells. In the buccal smear the cells have one sex chromatin body.[2] The karyotype is that of a 47,XXY chromosomal constitution.[3]

The Klinefelter syndrome also is seen in patients whose karyotype is 48,XXXY.[16] These patients tend to be more retarded than the 47,XXY patient. The buccal smear reveals two Barr bodies. Patients with a 48,XXYY chromosomal constitution are taller.[17] Their dermatoglyphics may be identical to those with 47,XXY, but they also may have a hypothenar pattern.[14, 15]

CASE REPORT*

W. S. was a 50-year-old man (Fig. 2) who had been in state hospitals since the age of 12 and in Fairview State Hospital, Costa Mesa, since he was 37 years of age. He was the third child of healthy unrelated parents. His mother was 29 years old when he was born. Two sisters were normal. Pregnancy, delivery and early development were normal. He walked at 14 months.

He entered school at 8 years but only stayed part of a year. He had been in an automobile accident at 6 years of age in which he was thrown through a windshield, injuring his temple. He had also had rickets and rheumatic fever with chorea. An IQ of 60 was determined at 12 years of age. It was recorded in his institutional record when he was 16 that he showed "marked and incorrigible sex tendencies" and was "utterly

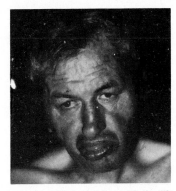

Fig. 2.—W. S. (Case Report). Man with the Klinefelter syndrome. He was mentally retarded. There was some prognathism.

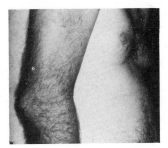

Fig. 3.—W. S. Gynecomastia.

*We are indebted to Dr. Charles Fish of Fairview State Hospital, Costa Mesa, California, for the details of this patient's history and for permission to report him here.

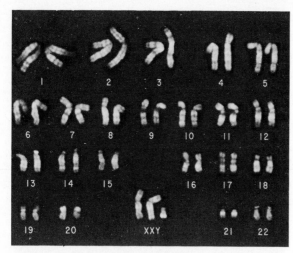

Fig. 4.—Karyotype of a patient with the Klinefelter syndrome. The XXY chromosomal constitution is clearly shown in this fluorescent preparation. (Courtesy of Dr. Uta Francke of the University of California San Diego.)

untrustworthy in this regard." A positive Wassermann was found at about that time, and he had rectal gonorrhea at 33 years. Operations included an appendectomy at 9 and a herniorrhaphy at 10 years of age.

Physical examination revealed a mildly obese man whose height was 5 feet, 8½ inches. Blood pressure was 110/70. He spoke fairly well. His head was brachycephalic and the occiput flattened. He had bilateral ptosis. He had large, fatty breasts (Fig. 3). The testicles were small, the penis of normal proportions. IQ was 52. EEG showed a generalized dysrhythmia. Cytogenetic analysis revealed an XXY karyotype (Fig. 4).

GENETICS

The extra X chromosome in the Klinefelter syndrome is thought to arise through nondisjunction during oogenesis or spermatogenesis. A proportion of patients are mosaics. There may be two cell lines, e.g., XY/XXY, XX/XXY, XXY/XXXY or XXY/XXYY, or three cell lines, e.g., XX/XY/XXY, XY/XXY/XXYY or XX/XXY/XXXY.[5] In these individuals mitotic nondisjunction probably takes place at an early cleavage division of the fertilized ovum. Maternal age at the time of birth of a 47,XXY child with the Klinefelter syndrome is higher than that of the general population. The maternal or paternal origin of the extra X chromosome has been studied using the Xg blood group.[18, 19] The data indicate that in about 65% of XXY males, both Xs are of maternal origin.[5, 18, 19] A proneness to nondisjunction is illustrated by a 42-year-old woman who had a son with both the Klinefelter syndrome and the Down syndrome, with 21 trisomy as well as an XXY constitution.[20] The discovery of the XXY syndrome was important in human developmental biology, for it demonstrated for the first time that masculine development is determined by the Y chromosome.

TREATMENT

Patients with the Klinefelter syndrome usually have a normal life span. Counseling, special classes or institutionalization may be required for behavior problems. Treatment with testosterone at puberty[6] and later may lead to more normal development of sexual characteristics. Gynecomastia is sometimes treated by mastectomy.

REFERENCES

1. Klinefelter, H. F., Jr., Reifenstein, E. C., Jr., and Albright, F.: Syndrome characterized by gynecomastia, aspermatogenesis without A-Leydigism, and increased excretion of follicle-stimulating hormone, J. Clin. Endocrinol. 2:615, 1942.
2. Plunkett, E. R., and Barr, M. L.: Testicular dysgenesis affecting the seminiferous tubules principally, with chromatin-positive nuclei, Lancet 2:853, 1956.
3. Jacobs, P. A., and Strong, J. A.: A case of human intersexuality having a possible XXY sex-determining mechanism, Nature 183:302, 1959.
4. Maclean, N., Harnden, D. G., Court Brown, W. M., Bond, J., and Mantle, D. J.: Sex-chromosome abnormalities in newborn babies, Lancet 1:286, 1964.
5. Court Brown, W. M.: Sex chromosome aneuploidy in man and its frequency, with special reference to mental subnormality and criminal behavior. Int. Rev. Exp. Pathol. 7: 31, 1969.
6. Caldwell, P. D., and Smith, D. W.: The XXY (Klinefelter's) syndrome in childhood: Detection and treatment, J. Pediatr. 80:250, 1972.
7. Laron, Z., and Hochman, I. H.: Small testes in prepubertal boys with Klinefelter's syndrome, J. Clin. Endocrinol. 32: 671, 1971.
8. Laron, Z.: Klinefelter's syndrome: Early diagnosis and social aspects, Hosp. Practice 7:135, 1972.
9. Sperber, M. A., Salomon, L., Collins, M. H., and Stambler, M.: Childhood schizophrenia and 47,XXY Klinefelter's syndrome, Am. J. Psychiatry 128:1400, 1972.
10. Harnden, D. G., Maclean, N., and Langlands, A. O.: Carcinoma of the breast and Klinefelter's syndrome, J. Med. Genet. 8:460, 1971.
11. Dodge, O. G., Path, M. C., Jackson, A. W., and Mudal, S.: Breast cancer and interstitial-cell tumor in a patient with Klinefelter's syndrome, Cancer 24:1027, 1969.
12. Coley, G. M., Otis, R. D., and Clark, W. E., II: Multiple primary tumors including bilateral breast cancers in a man with Klinefelter's syndrome, Cancer 27:1476, 1971.
13. Penrose, L. S.: Finger-print pattern and the sex chromosomes, Lancet 1:298, 1967.
14. Hunter, H.: Finger and palm prints in chromatin-positive males, J. Med. Genet. 5:112, 1968.
15. Borgaonkar, D. S., and Mules, E.: Comments on patients with sex chromosome aneuploidy: Dermatoglyphs, parental ages, Xg[a] blood group, J. Med. Genet. 7:345, 1970.
16. Ferguson-Smith, M. A., Johnston, A. W., and Handmaker, S. D.: Primary amentia and micro-orchidism associated with an XXXY sex-chromosome constitution, Lancet 2: 184, 1960.
17. Ferrier, P. E., and Ferrier, S. A.: XXYY Klinefelter's syndrome: Case report and a study of the Y chromosomes'

DNA replication pattern, Ann. Genet. (Paris) 11:145, 1968.

18. Frøland, A., Sanger, R., and Race, R. R.: Xg blood groups of 78 patients with Klinefelter's syndrome and of some of their parents, J. Med. Genet. 5:161, 1968.

19. Soltan, H. C.: Genetic characteristics of families of XO and XXY patients, including evidence of source of X chro-mosomes in 7 aneuploid patients, J. Med. Genet. 5:173, 1968.

20. Ford, C. E., Jones, K. W., Miller, O. J., Mittwoch, U., Pen-rose, L. S., Ridler, M., and Shapiro, A.: The chromosomes in a patient showing both mongolism and the Klinefelter syndrome, Lancet 1:709, 1959.

XXXXY SYNDROME

Cardinal Clinical Features

Severe mental retardation; hypoplastic, ambiguous genitalia; skeletal anomalies including proximal synostosis of the radius and ulna and clinodactyly of the fifth finger; one to three chromatin bodies; four X chromosomes.

Clinical Picture

Fraccaro and colleagues[1] in 1960 reported a male with 49 chromosomes. Since then more than 60 similar patients have been reported,[2] and it is clear that the syndrome reflects an XXXXY chromosomal constitution. Sexual development in this syndrome is interesting because when male patients with one X chromosome are compared with those with two, three or four, there seems to be a dosage effect that is operative in early development.

The scrotum in these patients is underdeveloped and usually cleft. It may be very poorly rugated. The penis most often is very small, although occasionally it can be normal in size, and it usually arises behind the anterior ridge of the bifid scrotum, which tends to surround it as labia do the clitoris. A large mons pubis may further obscure the small genitalia.[2] Hypospadias may be present. The testes are small and often are undescended. Testicular biopsy reveals very few tubules and no germ cells or Leydig cells (Fig. 1). It is thought that tubular atrophy occurs, so that the older patient has fewer tubules and more stroma than the younger patient. In some patients the ambiguous genital malformation may not be present, but these patients too have gross and microscopic testicular hypoplasia. Renal abnormalities occasionally seen include dilatation of the renal pelvis and ureters.[1]

Mental retardation is present in all of these patients and usually is severe.[3] The IQ is less than 60 and most often less than 35.[3] One patient had a normal Gesell Adaptive Maturity Level at 15 months,[4] but by 41 months he had clear-cut mental retardation[5]; thus, the rate of mental development may decelerate in this syndrome after the first year of life.

The facies may be characteristic. It usually is rounded and flat. The eyes tend to be widely spaced and there are prominent epicanthal folds. Strabismus occurs occasionally. A slight mongoloid slant may be seen, and the facial features have been said to suggest the Down syndrome. The nose may be wide and the bridge flat. Cleft palate has been described. Mild prognathia may occur. The ears may be low set or malformed. The occiput may be flat. Some patients have had a webbed neck.[3]

The most characteristic of the skeletal malformations is at the elbow where radioulnar synostosis or dislocation may lead to an abnormal carrying angle and limitation of motion, especially of supination at the elbow.[3, 6, 7] The proximal ulnae may be expanded or club shaped. The long bones in general tend to be shapeless and to lack modeling. There may be missing ossification centers, especially at the wrist and hand.[1] Supernumerary centers or pseudoepiphyses occur, notably among the metacarpals. Similar changes may be seen in the metatarsals.[7] Clinodactyly of the fifth finger is seen in about 90% of patients. In the lower extremities coxa valga and pes planus are common. The iliac wings often are narrow. Intrauterine growth retardation may be present in this syndrome[8] and short stature may be a persistent feature.

Dermatoglyphic analysis reveals an appreciable decrease in the total ridge count.[9] There is a tendency for a progressive decrease in this value with increasing numbers of sex chromosomes. Thus, the value of 50 for the XXXXY syndrome was the lowest in the series recorded at the Galton Laboratory.[9] The total ridge count in the normal XY individual was 145. The fingertip pattern is one of an increase in the number of arches. An arch configuration has no triradius, and thus it scores a zero in ridge counts. In the XXXXY

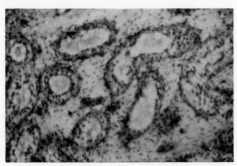

Fig. 1.—Histologic appearance of the testis. Multiple small seminiferous tubules are seen but no spermatogonia or Leydig cells.

158

syndrome an increased number of arches leads to a low ridge count. Simian creases occasionally are seen.

Buccal smears prepared from patient reveal a mixture of cells with one, two and three chromatin bodies.[10-13] Cytogenetic evaluation reveals 49 chromosomes. The Y chromosome is normal. There are three extra X chromosomes.[3, 10]

The presence of three chromatin bodies indicates that all of the extra X chromosomes are inactivated. This is consistent with the Lyon hypothesis.[13, 14] Yet examination of the syndromes that occur with extra X chromosomes in males reveals considerable evidence for a dosage effect, indicating that these extra X chromosomes are active in early development. The Klinefelter syndrome in which the karyotype is XXY is at one end and the XXXXY syndrome at the other. In the male the degree of mental retardation increases considerably as the number of X chromosomes is increased. The same is true of radioulnar synostosis, which occurs rarely in the XXY individual, more commonly in the XXXY individual[15] and commonly in those with the XXXXY constitution. The progression in genital anomaly is very striking.[2] In the Klinefelter syndrome testicular hypoplasia may not be evident until adolescence, and there are abundant Leydig cells. The XXXY patient is intermediate in these respects and may have small or normal genitalia. The XXXXY individual often has a small penis and may have the full-blown syndrome of rather ambiguous genitalia. It is generally accepted that the Y chromosome determines masculine differentiation. This is progressively suppressed in some fashion by a progressive increase in numbers of X chromosomes.

CASE REPORTS

CASE 1.*—The patient was admitted at 7 months of age to the U. S. Naval Hospital, San Diego, because of ambiguous genitalia. The infant had been delivered after 36 weeks of an uneventful pregnancy with a birth weight of 5 lb. The length

Fig. 2.—Case 1. In this view the genitalia appeared ambiguous.

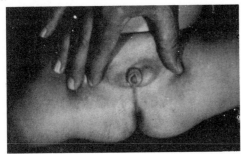

———————

*We are indebted to Dr. Douglas Cunningham of the U.S. Naval Hospital, San Diego, for information on this patient and permission to report him. Data on this patient have been published.[2]

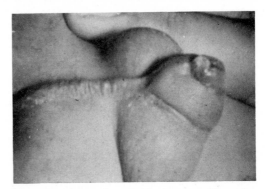

Fig. 3.—Case 1. Very short phallus, heavy median raphe and hypoplastic scrotum.

was 17½ inches. The Apgar score at 5 minutes was 4, and there was meconium staining.

Physical examination at 7 months revealed a small infant weighing 15 lb. Height was 24½ inches, height age 3½ months. He had a rounded, flat face. The eyes appeared widely spaced, and there were prominent epicanthal folds. A grade II/VI systolic ejection murmur was heard at the base of the heart. The genitalia were to some extent obscured by a large overhanging mons pubis. The penis was small—1 cm in length. The meatus was posteriorly placed. The scrotum was hypoplastic and cleaved by a hypertrophic median raphe (Figs. 2 and 3). There were two small pea-sized testes. The patient had clinodactyly of the fifth fingers.

The buccal smear contained cells with one, two and three chromatin bodies. Cytogenetic analysis of cultured leukocytes and skin fibroblasts revealed an XXXXY karyotype.

By 22 months of age the height had decreased to 4 s.D. below the third percentile. A Denver Developmental Test gave a developmental age of 9–10 months for gross motor and 16–18 months for fine motor activity. A testicular biopsy showed small seminiferous tubules surrounded by hyaline

Fig. 4.—T. D. (Case 2). One-year-old boy with the XXXXY syndrome. He had a rounded face, epicanthal folds and a left strabismus.

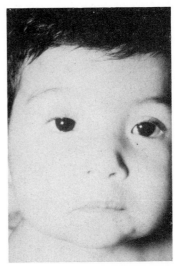

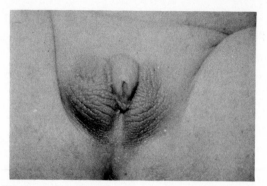

Fig. 5.—T. D. The scrotum was bifid and the penis very small.

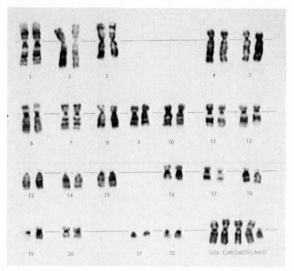

Fig. 7.—Giemsa preparation of chromosomes of a patient with the XXXXY constitution. (Courtesy of Dr. Uta Francke of the University of California San Diego.)

material. Sertoli cells were present but spermatogonia and Leydig cells were absent.

CASE 2.*—T. D. was first seen by Dr. Joyner at 1 year of age because of ambiguous genitalia. At that time a karyotype revealed an XXXXY constitution, and the child was referred to University Hospital. He was a Mexican boy in a foster home and his past and family histories were unknown. Over the past 6 months his development had been slow. He could not crawl or roll over, though he could sit with support and say, "Mama."

Physical examination revealed a baby small for his age. He was below the third percentile for both height and weight. He had a round face, prominent epicanthal folds, left strabismus (Fig. 4) and a slight antimongoloid slant. He also had a high-arched palate. His nipples were widely spaced. Abnormalities of the genitalia included a bifid scrotum and micropenis (Fig.

5) with first degree hypospadias. The testes were both palpable in the scrotum but small (1 cm/0.05 cm). He had clinodactyly of the fifth fingers. He was hypotonic and obviously retarded. A buccal smear revealed three Barr bodies in many cells. Cytogenetic analysis of fibroblasts derived from skin showed a 49,XXXXY karyotype (Fig. 6).

Fig. 6.—T. D. Karyotype, illustrating the XXXXY chromosomal constitution. (Courtesy of Dr. O. W. Jones of the University of California San Diego.)

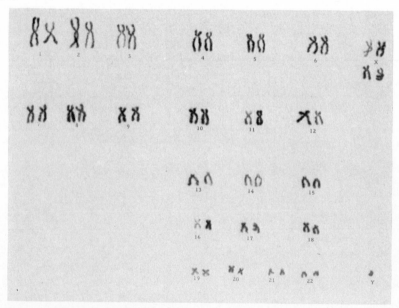

―――――――――

*We are indebted to Dr. Joseph Joyner of San Diego for allowing us to evaluate this patient and for permission to include this report.

GENETICS

The karyotype of patients with this syndrome is an XXXXY pattern (Fig. 7). Studies using the Xg antigen have indicated that all of the X chromosomes are maternal. Thus, it is likely that the four chromosomes result from two nondisjunctional events during gametogenesis of the ovum. Fertilization of the resultant XXXX ovum by a Y-bearing sperm would yield an XXXXY zygote. From this mechanism one would predict that there would also be females with 5 X chromosomes, and this situation has been described.

A few patients have been described in whom an XXXXY karyotype has occurred with mosaicism, with an XXXY cell line[16] and with XXY as well as XXXY cell lines.[17] The data in such a patient may more resemble the Klinefelter than the XXXXY syndrome. In these patients the mechanism probably involves mitotic nondisjunction. A predisposition to nondisjunction appears to be the case in certain families such as the one reported by Miller et al.[18] in which an XXXXY male had two 21 trisomic sisters with the Down syndrome and a father with chronic leukemia.

TREATMENT

None.

REFERENCES

1. Fraccaro, M., Kaijser, K., and Lindsten, J.: A child with 49 chromosomes, Lancet 2:899, 1960.
2. Cunningham, M. D., and Ragsdale, J. L.: Genital anomalies of an XXXXY male subject, J. Urol. 107:872, 1972.
3. Zaleski, W. A., Houston, C. S., Pozsonyi, J., and Ying, K. K.: The XXXXY chromosome anomaly: Report of three new cases and review of 30 cases from the literature, Can. Med. Assoc. 94:1143, 1966.
4. Shapiro, L. R., Hsu, L. Y. F., Calvin, M. E. and Hirschhorn, K.: XXXXY boy: A 15-month-old with normal intellectual development, Am. J. Dis. Child. 119:79, 1970.
5. Shapiro, L. R., Brill, C. B., Hsu, L. Y. F., Calvin, M. E., and Hirschhorn, K.: Deceleration of intellectual development in a XXXXY child, Am. J. Dis. Child. 122:163, 1971.
6. Fraser, J. H., Boyd, E., Lennox, B., and Dennison, W. M.: A case of XXXXY Klinefelter's syndrome, Lancet 2:1064, 1961.
7. Houston, C. S.: Roentgen findings in the XXXXY chromosome anomaly, J. Can. Assoc. Radiol. 18:258, 1967.
8. Joseph, M. C., Anders, J. M., and Taylor, H. I.: A boy with XXXXY sex chromosomes, J. Med. Genet. 1:95, 1964.
9. Penrose, L. S.; Fingerprint pattern and the sex chromosomes, Lancet 1:298, 1967.
10. Barr, M. L., Carr, D. H., Pozsonyi, J., Wilson, R. A., Dunn, H. G., Jacobson, T. S., Miller, J. R., Lewis, M., and Chown, B.: The XXXXY chromosome abnormality, Can. Med. Assoc. J. 87:891, 1962.
11. Fraccaro, M., Klinger, H. B., and Schutt, W.: A male with XXXXY sex chromosomes, Cytogenetics 1:52, 1962.
12. Atkins, L., and Connelly, J. R.: XXXXY sex chromosome abnormality, Am. J. Dis. Child. 106:514, 1963.
13. Lyon, M. F.: Sex chromatin and gene action in the mammalian X chromosome, Am. J. Hum. Genet. 14:135, 1962.
14. Galindo, J., and Baar, H. S.: The XXXXY sex chromosome abnormality, Arch. Dis. Child. 41:82, 1966.
15. Ferguson-Smith, M. A., Mack, W. S., Ellis, P. M., Dickson, M., Sanger, R., and Race, R. R.: Parental age and the source of the X chromosomes in XXY Klinefelter's syndrome, Lancet 1:46, 1964.
16. Turner, B., den Dulk, G. M, and Watkins, G.: The XXXXY syndrome, Med. J.. Aust. 2:715, 1963.
17. Kardon, N. B., Beratis, N. G., Hsu, L. Y. F., Moloshok, R. E., and Hirschhorn, K.: 47,XXY/48,XXXY/49,XXXXY mosaicism in a 4-year-old child, Am. J. Dis. Child. 122:160, 1971.
18. Miller, O. J., Berg, W. R., Schmickel, R. D., and Tretter, W.: A family with an XXXXY male, a leukaemic male, and two 21-trisomic mongoloid females, Lancet 2:78, 1961.

THE XXX FEMALE
The Triple X Syndrome

CARDINAL CLINICAL FEATURES

Normal female phenotype; two sex chromatin bodies; 47, XXX karyotype.

CLINICAL PICTURE

Jacobs and colleagues[1] in 1959 first found a female with three X chromosomes. The buccal smear showed two Barr bodies. Since then several cases have been described.[2]

The incidence of triple X females among newborn infant girls is 1.2/thousand.[3] However, the incidence among mentally retarded females is 4/thousand.[4] The risk of mental illness in the XXX female also appears to be increased.[4] Most XXX individuals have normal physical development.[2] They do not have a characteristic set of associated syndromic features. The first patient was a young woman with secondary amenorrhea, underdeveloped breasts and genitalia, and ovaries with deficient follicular development.[1] A number of others have had ovarian dysfunction.[4] However, many of these patients are endocrinologically normal and fertile.

Mental retardation is a variable feature,[2, 4] not always seen. Many of the patients reported have been retarded, reflecting the fact that they were found in programs in which retarded persons were screened. There seems to be a relationship between the incidence and severity of retardation with the number of additional X chromosomes in the female, as is so in the male. Patients with an XXXX constitution are more retarded than those with an XXX, and those with five Xs usually are profoundly retarded. Brachycephaly, high-arched palate, clinodactyly and transverse palmar creases have been described.[5] Radioulnar synostosis has been reported in the XXXX syndrome.[6] In a recent prospective study of XXX girls diagnosed first by unselected examination of the amnion, development did not differ greatly from normal, as represented by siblings or a group of XXX mosaics. However, in the XXX group about a third of the patients had a delay in early motor development and speech, a mild intellectual deficiency and disturbance in interpersonal relationships.[7]

Dermatoglyphic analysis may reveal a reduced total finger ridge count.[6] The buccal smear shows two sex chromatin bodies in a sizable proportion of cells. The peripheral smear does not reveal the number of drumstick polymorphonuclear leukocytes to be increased,[4] but occasional cells with double drumstick appendages are seen. The karyotype is 47,XXX (Fig. 1).

CASE REPORT*

G. P. was a 47-year-old Mexican female seen at the Fairview State Hospital, Costa Mesa. At age 17 she had been admitted to institution at Pacific State Hospital because of mental retardation, and was described by Day, Larson and Wright in 1964.[8] She was the first-born child of a healthy couple. The mother was 19 years old at the time of her birth, the father 32. Birth weight was 5 lb. She had six other siblings, all normal. The patient had been retarded since early childhood. She walked at

Fig. 1.—Karyotype of a patient with 3 X chromosomes. (Courtesy of Dr. Uta Francke of the University of California San Diego.)

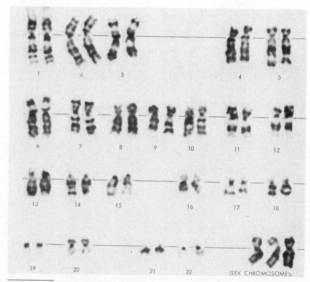

*We are indebted to Dr. Charles Fish of Fairview State Hospital in Costa Mesa, California, for permission to study and photograph the patient.

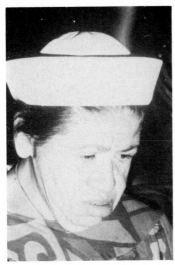

Fig. 2.—G. P. (Case Report). A 47-year-old woman with an XXX chromosomal constitution. She was retarded and institutionalized. She had short stature and the unusual facies included very prominent lips. Her physical characteristics were those of her associated cretinism.

5 years and spoke her first words at 7. She had been diagnosed as having athyreotic cretinism and was treated with thyroid sporadically during childhood. At age 17 her IQ was 30. Radioactive iodine uptake was 0.8% in 24 hours. She was treated with 2 grains of desiccated thyroid daily.

Physical examination revealed a short, obese, mentally retarded female. Height was 52.6 inches. Her upper/lower ratio was 1.11, indicating the short limb dwarfism characteristic of cretinism. She had features typical of cretinism, prognathism and a short neck (Figs. 2 and 3). Figure 4 shows the whorls present on all 10 fingers. The thyroid gland was not palpable. The buccal smear showed two chromatin bodies. A karyotype was 47,XXX.

Fig. 3.—G. P. Lateral view.

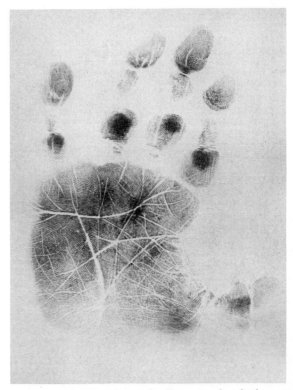

Fig. 4.—G. P. Print of the palm illustrating the whorls present on all 10 fingers.

GENETICS

The abnormal 47,XXX chromosomal constitution in the triple X female may be the result of nondisjunction in oogenesis or spermatogenesis in either of the meiotic divisions. It also may follow an error in mitosis in an early cleavage division of a fertilized ovum. Maternal age at the time of the birth of a triple X female is higher than that of the average population.[4,8] Some patients are mosaics with XO and XXX, or XX and XXX cell lines.[2]

The offspring of XXX females usually have been found not to have chromosome aneuploidy.[9] A triple X female would be expected to have two gametes, one ovum with one X and the other unbalanced with XX chromosomes. However, most of the children reported have had normal karyotypes. The exceptions include XXY and XX,XXX mosaics.[4]

TREATMENT.

None.

REFERENCES

1. Jacobs, P. A., Baikie, A. G., Court Brown, W. M., Mac-Cregor, T. N., Maclean, N., and Harnden, D. G.: Evidence

for the existence of the human "super female," Lancet 2: 423, 1959.

2. Court Brown, W. M.: Sex chromosome aneuploidy in man and its frequency, with special reference to mental subnormality and criminal behavior, Int. Rev. Exp. Pathol. 7: 31, 1969.

3. Maclean, N., Harnden, D. G., Court Brown, W. M., Bond, J., and Mantle, D. J.: Sex-chromosome abnormalities in newborn babies, Lancet 1:286, 1964.

4. Barr, M. L., Sergovich, F. R., Carr, D. M., and Shaver, E. L.: The triple-X female: An appraisal based on a study of 12 cases and a review of the literature, Can. Med. Assoc. J. 101:247, 1969.

5. Johnston, A. W., Ferguson-Smith, M. A., Handmaker, S. D., Jones, H. W., and Jones, G. S.: The triple-X syndrome. Clinical, pathological, and chromosomal studies in three mentally retarded cases, Br. Med. J. 2:1046, 1961.

6. Telfer, M. A., Richardson, C. E., Helmken, J., and Smith, G. F.: Divergent phenotypes among 48,XXXX and 47,XXX females, Am. J. Hum. Genet. 22:326, 1970.

7. Tennes, K., Puck, M., Bryant, K., Frankenburg, W., and Robinson, A.: A developmental study of girls with Trisomy X, Am. J. Hum. Genet. 27:71, 1975.

8. Day, R. W., Larson, W., and Wright, S. W.: Clinical and cytogenetic studies on a group of females with XXX sex chromosome complements, J. Pediatr. 64:24, 1964.

9. Kohn, G., Winter, J. S. D., and Mellman, W. J.: Trisomy X in three children, J. Pediatr. 72:248, 1968.

10. Borgaonkar, D. S., and Leger, H.: The Triple-X Syndrome, in *The First Conference on the Clinical Delineation of Birth Defects*, Birth Defects: Original Article Series, Part V (New York: The National Foundation-March of Dimes), p. 138.

THE XYY MALE

CARDINAL CLINICAL FEATURES

Tall stature, radioulnar synostosis, behavioral abnormalities; a phenotype that may be normal, extra fluorescent Y body, XYY chromosomal constitution.

CLINICAL PICTURE

The XYY syndrome is controversial and exciting because of the possibility that this chromosomal aneuploidy may lead to aberrant, aggressive behavior. The information on the Y chromosome also is controversial. It has been conceded for some time that XY individuals are males, but it was not clear whether maleness represented the absence of an X or the presence of Y. This has now become clear through the study of individuals with deviations from chromosomal normality. An XXY individual is a phenotypic male, and an XO individual is a female. Thus, it is clear that the presence of a Y chromosome determines the presence of testes. There is suggestive evidence that hairy ears may be determined by a gene on the Y chromosome. This has not been conclusively demonstrated, and it is not known what other information is carried on the Y chromosome.

A relationship between the Y chromosome and behavior was first made by Jacobs and colleagues.[1] About 1% of the males in institutions for the mentally retarded are chromatin positive, and the majority of them are XXY. In 1963 Forssman and Hambert[2] surveyed the nuclear sex chromatin of 760 male patients in three Swedish institutions for males of subnormal intelligence who were either criminals or hard to manage: the unusually high incidence of 2% was found. Casey et al.[3] then made a similar study of some 942 males in England and found a similar incidence of 2.2%. He found that most of these patients had a chromosomal complement of XXYY or that they were mosaics with XXYY as the major chromosomal complement. Less than 1% of over 10,000 retarded males who did not have unusual behavior had an extra sex chromatin body, and one would predict from experience that most of these were XXY. Jacobs and colleagues[1] surveyed mentally subnormal male patients in an institution where the dangerous or violent, or those with criminal tendencies, were handled under conditions of special security. The data indicated that patients institutionalized for violent crimes such as rape, arson and assault who were mentally subnormal had a 3.5% incidence of an extra Y chromosome. The majority were of the XYY chromosomal complement, and one of 197 was XXYY. By contrast, in two control groups of newborn or adult males, each of over 200 patients, there were no examples of an extra Y chromosome, and Jacobs and colleagues reviewed the records of some 1,500 patients studied up to that time in the course of operation of their cytogenetic laboratory and found only one patient who had an extra Y.

The patients have since been carefully studied from a medical point of view. Unlike patients with an XXY chromosomal constitution, they usually do not look different from other people. Statistically, they are significantly taller than controls, and we have seen some patients who are XYY who have been strikingly tall (Figs. 1, 2 and 3).

Patients with this syndrome may have abnormalities around the elbow.[4] This may take the form of radioulnar synostosis or radioulnar dislocation. A scholarly study of the carrying angle at the elbow (angle of intersection of the long axis of the upper arm with the long axis of the supinated forearm when the elbow is fully supinated) was conducted by Baughman and colleagues.[5] The angle is consistently greater in women than in men. Furthermore, it is increased in the Turner (XO) syndrome, in which cubitus valgus is the rule. The XYY male is at the other end of the spectrum. The elbow is in cubitus varus or cubitus rectus even in the absence of radioulnar synostosis.

Mental subnormality certainly occurs in this condition. On the other hand, patients may have normal intelligence. Some patients have had prognathism.[6] Tooth size is larger in XYY males than in XY males.[7]

Hypogonadism, hypospadias and ambiguous genitalia have been seen in patients with an XYY karyotype, but usually the genitalia are normal.[8] There are oligospermia, a maturation arrest of germ cells and increase of low fertility.[8]

The most actively discussed aspect of the XYY syndrome is its relationship to aggressive behavior.[8, 9] There is a relationship. The incidence of XYY males in prisons and mental hospitals where inmates have been selected for being dangerous or violent has ranged from 1.8–12%, whereas in the general newborn population XYY males constitute 0.14–0.38%.[8] Furthermore, psychologic evaluation of a series of XYY men indicated them to be passive and incapable of controlling their impulses. It is possible that the

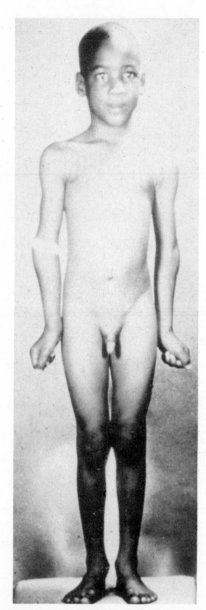

Fig. 1.—M. W. (Case Report). An 8-year-old boy with the XYY syndrome. He was tall and had bilateral abnormalities of the elbows.

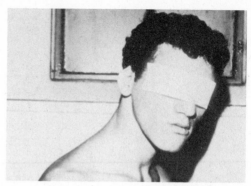

Fig. 2.—A 19-year-old Argentine man with the XYY syndrome. He was 6 ft, 2¾ in. tall and had an odd, pointed helix. He had been imprisoned for repeated sexual assaults. (Courtesy of Dr. Primarosa de Chieri of Buenos Aires.)

Pediatr.[4]) had a history of tantrums, fighting, defying authority, cursing his teachers and leaving school. He was in the second grade but was considered to be a slow learner. Development was reported to have been normal. The child had been difficult to control from early life. A deformity of both elbows had been present from birth. There was a family history of schizophrenia.

Physical examination revealed a tall, slender Negro boy with bilateral deformities of the elbows (see Fig. 1). There was almost complete inability to pronate and supinate the forearms bilaterally. The external genitalia were normal. The height at age 5⅓ years was 45 inches and at age 8¼ years, 54¼ inches—each in the 75th percentile. At the age of 8½ years the following measurements were recorded: span, 52 inches; upper segment, 24¼ inches; lower segment, 27 inches; ratio (upper segment/lower segment), 0.92.

Excretion of 17 ketosteroids was 1.7 mg/24 hours. Bone age was not accelerated. Roentgenographic examination of the elbows revealed bilateral symmetrical fusion of the proximal radius and ulna just distal to the joint (Fig. 4). An EEG was normal. The karyotype showed two Y chromosomes (Fig. 5).

The results of psychologic evaluation using the WISC were: full scale, 75; verbal; 87; performance, 76.

Fig. 3.—E. G. A tall, massive, retarded man with the XYY syndrome.

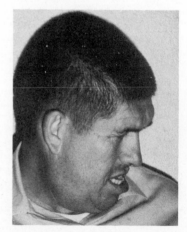

characteristic of the XYY male is increased impulsiveness rather than increased aggressiveness.[9] An extra Y chromosome does not predict antisocial behavior, but the risk is certainly increased.

CASE REPORT

M. W. was brought to the outpatient department of the Jackson Memorial Hospital, Miami, Florida, at the age of 8 years, because of behavioral problems. This patient (reported in J.

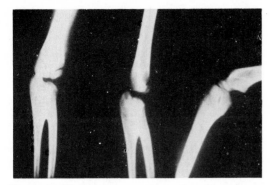

Fig. 4.—M. W. Roentgenogram reveals radioulnar synostosis.

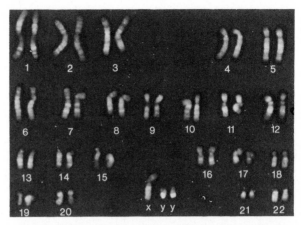

Fig. 6.—Karyotype of a patient with the XYY syndrome. (Courtesy of Dr. Uta Francke of the University of California San Diego.)

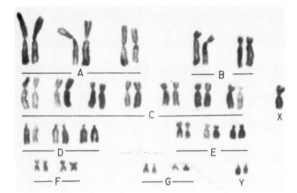

Fig. 5.—Karyotype of M. W. illustrating the two Y chromosomes.

Fig. 7.—Fluorescent Y bodies in a patient with the XYY syndrome.

GENETICS

The essential feature of this syndrome is an extra Y chromosome. The karyotype is XYY (Fig. 6) but an XXYY karyotype may lead to a similar phenotype. The XYY situation is the result of nondisjunction in the male. It appears to occur as frequently as one in 700 newborn males.[10] The Y chromosome is readily recognized by its bright fluorescence in chromosome preparations stained with quinacrine or quinacrine mustard.[11] The extra Y chromosome may be seen in whole cell preparations in which there are 2 fluorescent Y bodies (Fig. 7).

TREATMENT

None. The recognition of an extra Y chromosome in a newborn or a fetus raises an ethical dilemma. The response requires wisdom and individualization.

REFERENCES

1. Jacobs, P., Brunton, M., and Melville, M. M.: Aggressive behavior, mental subnormality and the XYY male, Nature 202:1351, 1965.
2. Forssman, H., and Hambert, G.: Chromosomes and antisocial behavior, Lancet 2:282, 1966.
3. Casey, M. D., Segall, L. J., Street, D. R. K., and Blank, C. E.: Sex chromosome abnormalities in two state hospitals for patients requiring special security, Nature 209:641, 1966.
4. Cleveland, W. W., Arias, D., and Smith, G. F.: Radioulnar synostosis, behavioral disturbance, and XYY chromosomes, J. Pediatr. 74:103, 1969.
5. Baughman, F. A., Jr., Higgins, J. V., Wadsworth, T. G., and Demaray, M. J.: The carrying angle in sex chromosome anomalies, J.A.M.A. 230:718, 1974.
6. Parker, C. E., Melnyk, J., and Fish, C. H.: The XYY syndrome, Am. J. Med. 47:801, 1969.
7. Alvesalo, L., Osborne, R. H., and Kari, M.: The 47,XYY Male, Y chromosome, and tooth size, Am. J. Hum. Genet. 27:53, 1975.
8. Gardner, L. I., and Neu, R. L.: Evidence linking an extra Y chromosome to sociopathic behavior, Arch. Gen. Psychiatry 26:220, 1972.

9. Hook, E. B.: Behavioral implications of the human XYY genotype, Science 179:139, 1973.

10. Lynch, D. A., Neu, R. L., and Gardner, L. I.: Nondisjunction in males: Commoner than suspected?, J.A.M.A. 222: 1311, 1972.

11. Wilson, M. G., Towner, J. W., Lipshin, J., and Fleisher, A.: Identification of an unusual Y chromosome in YY mosaicism by quinacrine fluorescence, Nature 231:388, 1971.

Disorders of Connective Tissue

PSEUDOXANTHOMA ELASTICUM
Groenblad-Strandberg Syndrome

CARDINAL CLINICAL FEATURES

Pseudoxanthomas of the skin, angioid streaks in the fundus, poor vision, weak peripheral pulses, hemorrhages from gastrointestinal tract.

CLINICAL MANIFESTATIONS

Pseudoxanthoma elasticum is a generalized inherited disorder of connective tissue. The first description

Fig. 1.–L. G. (Case Report). The changes in elasticity of the skin are well demonstrated in the neck, as are the raised pseudoxanthomatous lesions.

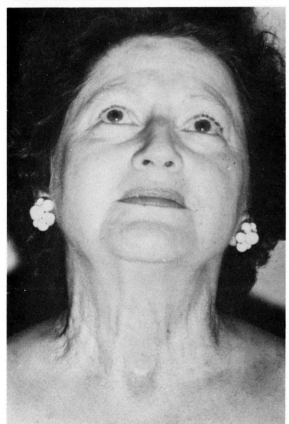

of the raised yellow lesions or pseudoxanthomas of the skin was made by Rigal[1] in 1881. The disorder involves primarily the skin, the eye and the cardiovascular system.

SKIN AND MUCOUS MEMBRANES.–Pseudoxanthoma elasticum usually is not considered a pediatric disorder. However, the syndrome may appear in childhood, and typical skin lesions have been noted at birth. The changes in the skin often are not recognized before the second decade of life or later. The lesions are most often seen on the neck (Fig. 1) and axillae. They also are prominent in the inguinal folds, antecubital area (Fig. 2) and periumbilical area. Nails are soft (Fig. 3). The lesions also are readily seen in the mucosa of the soft palate, the inner aspects of the lip (Fig. 4) and elsewhere in the buccal mucosa. Vaginal and rectal mucosae also have been involved. The skin is lax, redundant and at the same time inelastic. It is grooved, and between the grooves are thickened yellow patches, the pseudoxanthomas. There often are calcifications in the middle and deeper layers of the dermis, which can be identified radiographically.

THE EYE.–The characteristic changes consist of

Fig. 2.–L. G. Close-up of the skin of the antecubital space.

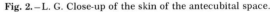

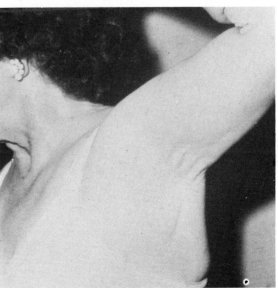

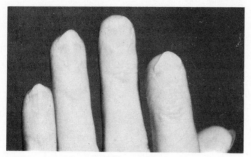

Fig. 3.—L. G. The nails were soft.

angioid streaks in the fundus.[2] These streaks are red, brownish or gray and are four or five times wider than the veins, but they resemble vessels in the manner in which they course over the fundus. Like the skin changes, they usually develop in the second decade or later. Macular involvement frequently results in severe diminution of visual acuity. Retinal hemorrhage can occur.

CARDIOVASCULAR SYSTEM.—Arterial involvement is expressed in pulse changes and symptoms of arterial insufficiency in the extremities. There is roentgenographic evidence of premature medial calcification of peripheral arteries. Patients also have symptoms of coronary insufficiency. They may develop hemorrhages in one or many areas. Hypertension may be the presenting complaint and has been observed at 6 and 10 years of age.[3] Weakness or absence of pulses in the extremities is a frequent finding. Calcifications of vessels have been seen as early as 9 years of age.[4]

Hemorrhages often constitute a major medical problem. Gastrointestinal hemorrhage is common and may be fatal. Gastrointestinal bleeding has been seen as

Fig. 4.—L. G. The lesions are particularly well seen in the mucous membranes.

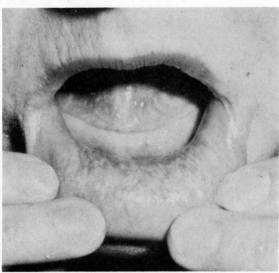

early as 6½ years of age.[1] Bleeding also has been seen in the subarachnoid, retinal, renal and uterine areas. Psychiatric disorders have been seen in many patients.

CASE REPORT

L. G., a 61-year-old white female, had been known to have pseudoxanthoma elasticum since diagnosis at 23 years of age. Skin lesions were progressive. She noted tarry stools on several occasions and complained at times of epigastric burning. She also developed severe muscle, tendon and joint pains at the age of 56 years. Intermittent paresthesias developed more recently. She had color blindness and had virtually complete loss of the central portion of her visual fields and partial loss of peripheral vision.

She had a brother and a sister, both of whom were affected.

Physical examination revealed the typical skin changes of pseudoxanthoma elasticum (see Figs. 1–3). These soft ivory yellow lesions were papular and arranged in a linear reticular pattern. Involved areas include the neck, shoulders, axillae and antecubital and periumbilical areas. The blood pressure was 138/65 and the pulse 72/minute and strong. Visual acuity was 20/400 in both eyes. She appeared essentially blind. Funduscopy revealed the characteristic angioid streaks.

X-ray examination showed intracranial calcification that appeared to be in the falx cerebri. There were ring calcifications in the region of the splenic artery.

Histologic pathology was observed following biopsy of the skin of the anterior axillary fold. There was fragmentation of the elastic fibers and calcium encrustation of some of these fibers. These changes were typical of pseudoxanthoma elasticum.

GENETICS

Pseudoxanthoma elasticum is transmitted as an autosomal recessive character. The frequency of the condition has been estimated at one in 160,000 persons. Persons presumed to be heterozygous for the gene may have an abnormally prominent choroidal vascular pattern.

TREATMENT

There is no definitive treatment for this disorder. Vitamin E (tocopherol) has been used, as have steroids. Plastic or vascular surgery may be indicated.

REFERENCES

1. McKusick, V. A., *Heritable Disorders of Connective Tissue* (3d ed.; St. Louis: C. V. Mosby Co., 1966), pp. 286–324.
2. Xatzkan, D. N.: Angioid streaks of the fundus, Am. J. Ophthalmol. 43:219, 1957.
3. Parker, J. C., Friedman-Kien, A. E., Levin, S., and Bartter, C.: Pseudoxanthoma elasticum and hypertension, N. Engl. J. Med. 271:1204, 1964.
4. Wolff, H. H., Stokes, J., and Schlesinger, B.: Vascular abnormalities associated with pseudoxanthoma elasticum, Arch. Dis. Child. 27:82, 1952.

EHLERS-DANLOS SYNDROME

CARDINAL CLINICAL FEATURES

Hyperextensible, fragile skin; cigarette paper scars; easy bruisability; hypermobility of joints; hernias; rupture of the great vessels.

CLINICAL PICTURE

The Ehlers-Danlos syndrome is one of the classic heritable disorders of connective tissue.[1] It was first reported in 1682 by a Dutch physician, Job van Meekeren, who described a Spanish patient who could pull his skin outward a considerable distance. For instance, he could pull the skin of his chin up to cover his eyes and, when released, the skin would resume its former smooth adherence to the structures below.[1] Ehlers[2] in 1901 described a patient with hyperextensible skin in whom other clinical features included hemorrhage into the skin and hypermobile, spontaneously subluxating joints. Danlos[3] in 1908 commented on the scars, the extraordinary hyperelasticity of the skin, the vulnerability of the skin to minor trauma and the molluscoid pseudotumors. These authors' descriptions indicated that they were dealing with a syndrome, one that has come to be known by their names.

The fundamental characteristic of the syndrome is an abnormality of connective tissue that leads to laxity of the ligaments around the joints, hernias and vascular problems, as well as dermal hyperextensibility. The skin is pale, thin and velvety. Because, when released, it immediately regains its former smoothness, it is hyperextensible but not lax. Patients with this disorder have worked in circuses and sideshows, often as "India rubber men." Skin fold thickness in these patients is significantly smaller than in control individuals.[4] The skin of the palms and soles may be redundant and furrowed (Fig. 1). Molluscoid or raisin-like tumors may appear in scarred areas. The progressive development of secondary creases may be well demonstrated by hand print analysis.[5] An increase in the number of secondary creases may be a useful sign of a connective tissue disorder.

Minor trauma may cause the skin to split.[6] Minor lacerations leave thin paper-like, gaping scars (cigarette paper scars), especially over the knees and other bony prominences. Poor healing of wounds is characteristic. Surgery is difficult because the skin is very friable and will not hold sutures well.[7] Cannulation of a vein may lead to a tear in the vessel and to subcutaneous hemorrhage. Angiography is especially hazardous.[8] Calcified spheroids often may be felt in the subcutaneous tissues of the forearms and shins. They may be demonstrated roentgenographically.[9]

About half of these patients display an unusual ability to touch the nose with the tip of the tongue[9] (Gorlin's sign). The ears are stretchable, and patients often have lop ears. Various hernias are seen in patients with the Ehlers-Danlos syndrome. Inguinal hernias are commonplace. Patients may actually present first to a physician with inguinal hernias, and recurrence after surgical repair may be the first clue to the diagnosis.[7] Diaphragmatic and hiatal hernias, as well as eventrations of the diaphragm, also have been observed. Intestinal diverticula are common, as is rectal prolapse in infants.

Ocular abnormalities[10] include blue sclerae, epicanthal folds, strabismus and myopia. Patients may have Metenier's sign, in which the extensibility of the skin permits easy eversion of the upper lids.[6] Redundant skin folds develop around the eye.[5] An appearance of hypertelorism probably is a telecanthus resulting from changes in the tissues alongside the nose. Beighton[10] described two siblings who had the characteristic features of the syndrome as well as extreme fragility of ocular tissues: one became blind following rupture of both globes after mild trauma; his sister had bilateral retinal detachments that led to blindness. Ectopia lentis has been reported. Patients with severe ocular complications may have a recessive disorder.[10]

Hypermobility of the joints is striking.[11, 12] Patients may have a disjointed gait and a limp handshake. Dislocations have been observed in the hips, patellae, shoulders, radii and clavicles. There may be a tendency to locking of the knees. Effusions may occur as a consequence of the trauma that occurs in an unstable joint. Early osteoarthritis is common. Major tendons may rupture. Genu recurvatum is regularly seen, and patients also may have pes planus.

Bleeding is a frequent problem. The severity varies greatly, but most common is a mild bruisability of the skin. Other patients may have frequent, prominent hematomas. Intestinal bleeding may consist of melena or massive hemorrhage.[13] Spontaneous perforation of the bowel has been reported. Coagulation studies are normal in patients with the syndrome. Fragility of the

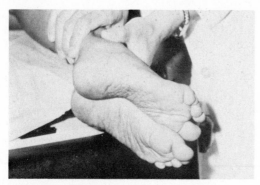

Fig. 1.—The soles of the feet are furrowed.

walls of even large arteries may lead to aneurysm, dissection or rupture. Varicosities of the veins are very common. Cerebrovascular accidents may reflect the formation of aneurysms or arteriovenous fistulas. Pregnancy may be particularly hazardous in patients with this syndrome.[14] Hemorrhage, as well as abdominal herniation and varicosities, may occur. Uterine rupture is a risk, especially in the woman with a previous cesarean section. The use of forceps has been reported to result in the extraction of the uterus, posterior wall of the bladder and part of a ureter, along with the baby.

Among groups of patients with the Ehlers-Danlos syndrome, it has long been known that there was clinical heterogeneity. Barabas[15] proposed a classification in 1967 in which he divided these patients into three types. Group 1, the classic type, included those with severe hyperextensibility of the skin and joints. These patients were often born prematurely after premature rupture of the fetal membranes. Group 2 included patients born at term having only mild skin and joint involvement, but often with severe problems with varicose veins. Group 3 included those with only minor extensibility of skin and joints, but severe problems of arterial rupture and bruising.

Beighton[6] made a case from a study of 100 patients for the presence of five distinct types: (1) *gravis type* (autosomal dominant): the most hyperextensible skin and most mobile joints, as well as gross splitting of the skin and moderate bruising. Prematurity occurs, and patients have varicose veins. Skeletal deformities may result from the problems with the joints; thoracic deformity is common.[12] (2) *Mitis type* (autosomal dominant): mild signs of the syndrome. Hypermobility of the joints may be confined to the hands and feet. (3) *Benign hypermobile type:* extreme hypermobility of the joints and variable but usually severe hyperextensibility of the skin; bruising and skin splitting generally mild. Varicose veins and prematurity usually are absent. (4) *Ecchymotic type:* bruising (the cardinal feature) so severe that minor trauma leads to gross ecchymosis. One patient developed a massive hemato-

ma when she dropped a loaf of bread on her shin.[6] There is dark, pigmented scarring of the bony prominences. Cardiovascular complications include arterial rupture and aortic dissection; gross gastrointestinal problems include intestinal perforation and hemorrhage. Melena may result from gastrointestinal diverticula. Hematuria, epistaxis and menorrhagia may be seen. Joint hypermobility is limited to the digits, and hyperextensibility of the skin is mild. Many of these patients die of arterial or hemorrhagic catastrophes.[15] (5) *X-linked type*[16]: prominent, generalized hyperextensible skin in which the skin fold is characteristically thick and does not give the sensation of coming away from the underlying tissues, as in the gravis form; joint hypermobility is mild, often limited to the digits. Bruising, splitting and scarring of the skin are minimal. In any one kindred, all affected members have only one type of the disease.

DIFFERENTIAL DIAGNOSIS

The differential diagnosis of the Ehlers-Danlos syndrome includes cutis laxa.[17-19] It is transmitted as an autosomal recessive and is characterized by loose, pendulous, bloodhound-like skin that, unlike that in the Ehlers-Danlos syndrome, does not have increased elasticity. The skin does not bruise easily or develop cigarette paper scars. Patients often have urinary tract and gastrointestinal diverticula. Emphysema rarely is seen in the Ehlers-Danlos syndrome but is often fatal in patients with cutis laxa.

A birth defect resembling the Ehlers-Danlos syndrome has been described in the female infant of a mother treated during pregnancy with penicillamine for cystinuria.[20] She had lax skin, hypermobility of the joints and impaired wound healing. Roentgenographic examination made because of severe vomiting led to pyloromyotomy. This was followed by a series of complications and then death.

Fig. 2.—D. P. (Case Report). One of identical Philippino twins with the Ehlers-Danlos syndrome. She had a subconjunctival hemorrhage in the right eye. Hemorrhage is common in this syndrome following mild trauma.

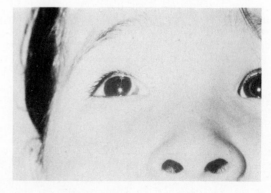

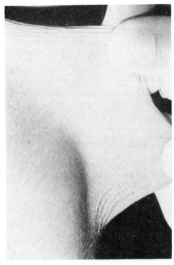

Fig. 3.—D. P. Hyperelasticity of the skin was striking.

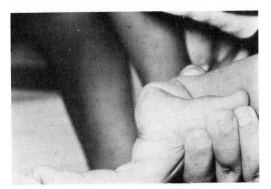

Fig. 5.—D. P. Hyperextensibility of the joints also is characteristic of the syndrome.

CASE REPORT

D. P. was an 8-year-old Philippine-American girl who came to University Hospital Emergency Room because of a subconjunctival hemorrhage (Fig. 2). She and her identical twin, L. P., had been diagnosed previously and had been reported at the age of 5 years as examples of the Ehlers-Danlos syndrome.[21] They were the products of a full-term pregnancy and delivered by cesarean section; the mother was then 29 years old, the father 37. Birth weights were low (D. P.: 3 lb, 5 oz; L. P.: 3 lb, 9 oz). Both had umbilical and inguinal hernias at birth, which were later repaired surgically. L. P. had a cutaneous hemangioma at birth. Both had histories of easy bruis-

Fig. 4.—D. P. The skin is easily damaged, and resulting scars take on a papyraceous or cigarette paper appearance.

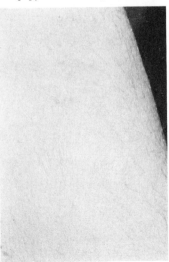

ing, frequent development of hematomas over the shins and occasional epistaxis.

On examination these girls had smooth, velvety, hyperextensible skin (Fig. 3), which was easily lacerated. Cigarette paper scars were present mainly on the forehead and shins (Fig. 4). Both children had strabismus and epicanthal folds. The joints were hypermobile (Fig. 5) and the ligaments lax.

The family history revealed no evidence of the Ehlers-Danlos syndrome. The father was of Philippine descent and the mother Hungarian. There was no consanguinity. A 10-year-old sibling was normal. In view of the age of the father at the time of conception, it appears likely that a spontaneous mutation had occured.

GENETICS

The Ehlers-Danlos syndrome is genetically heterogeneous, most forms being transmitted in an autosomal dominant fashion. In two families the disease was transmitted as an X-linked recessive trait.[5] The occurrence of the classic syndrome in two siblings who had no family history of the disease and normal parents first raised the possibility of an autosomal recessive type of the syndrome.[12] These patients had severe ophthalmologic complications.

Recent advances in collagen biochemistry have provided new understandings of the Ehlers-Danlos syndrome. The first development was the definition of a marked reduction in the content of hydroxylysine in the dermal collagen of two sisters with a form of the Ehlers-Danlos syndrome.[22] This disorder appears to be recessively transmitted and is referred to by some authors as type VI of the Ehlers-Danlos syndrome. These sisters had severe scoliosis, hypermobility of the joints and recurrent dislocations, blue sclerae and hyperextensible skin. Both were sufficiently hypotonic that a diagnosis of amyotonia congenita was entertained. They also had arachnodactyly and a positive Steinberg thumb sign.

A decrease in hydroxylysine in collagen would be expected to interfere with cross-linking and impair the stability of collagen. Lysine is hydroxylated in collagen. The lysylhydroxylase, or lysylprotocollagen hydroxylase, that catalyzes this conversion has been found to be low in fibroblasts derived from these patients.[23] The mother had an activity about 60% of normal, consistent with an autosomal recessive pattern. Hydroxylysine-deficient collagen also has been demonstrated in the family in which both siblings were blind following ocular catastrophes.[1]

In another subgroup of the Ehlers-Danlos syndrome, a defect has been found in the procollagen peptidase concerned with the conversion of procollagen to collagen.[24] Three unrelated patients with this defect had stretchable skin, short stature, hypermobile joints and multiple dislocations. These patients could represent a type VII of the Ehlers-Danlos syndrome.

TREATMENT

There is currently no treatment directed at the basic defect. Each of the complications of this disorder should be treated appropriately.

REFERENCES

1. McKusick, V. A.: *Heritable Disorders of Connective Tissue* (3d ed.; St. Louis: C. V. Mosby Co., 1966), p. 179.
2. Ehlers, E.: Cutis laxa, Neigung zu Haemorrhagien in der Haut, Lockergun mehrerer Artikulationen, Dermatologica 8:1973, 1901.
3. Danlos, M.: Un cas de cutis laxa avec tumeurs par contusion chronique de coudes et des genoux (xanthome juvenile pseudo-diabétique de MM. Hallopeau et Mace de Lepinay), Soc. Fr. Derm. Syph. Bull. 19:70, 1908.
4. Grahame, R., and Beighton, P.: Physical properties of the skin in the Ehlers-Danlos syndrome, Ann. Rheum. Dis. 28:246, 1969.
5. Goodman, R. M., Katznelson, M., and Frydman, M.: Evolution of palmar skin creases in the Ehlers-Danlos syndrome, Clin. Genet. 3:67, 1972.
6. Beighton, P., Price, A., Lord, J., and Dickson, E.: Variants of the Ehlers-Danlos syndrome. Clinical biochemical, haematological, and chromosomal features of 100 patients, Ann. Rheum. Dis. 28:228, 1969.
7. Woolley, M. M., Morgan, S., and Hays, D. M.: Heritable disorders of connective tissue. Surgical and anesthetic problems, J. Pediatr. Surg. 2:325, 1967.
8. Beighton, P., and Thomas, M. L.: The radiology of the Ehlers-Danlos syndrome, Clin. Radiol. 20:354, 1969.
9. Gorlin, R. J., and Pindborg, J. J.: *Syndromes of the Head and Neck*, (New York: McGraw-Hill Book Co., 1964), pp. 195–196.
10. Beighton, P.: Serious ophthalmological complications in the Ehlers-Danlos syndrome, Br. J. Ophthalmol. 54:263, 1970.
11. Seaton, D. G.: Bilateral recurrent dislocation of the patellas in the Ehlers-Danlos syndrome, Med. J. Aust. 1:737, 1969.
12. Beighton, P., and Horan, F.: Orthopaedic aspects of the Ehlers-Danlos syndrome, J. Bone Joint Surg. [Br.] 51:444, 1969.
13. Beighton, P., Murdoch, J., and Votteler, T.: Gastrointestinal complications of the Ehlers-Danlos syndrome, Gut 10:1004, 1969.
14. Beighton, P.: Obstetric aspects of the Ehlers-Danlos syndrome, J. Obstet. Gynaecol. Br. Commonw. 76:97, 1969.
15. Barabas, A. P.: Heterogeneity of the Ehlers-Danlos syndrome: Description of three clinical types and a hypothesis to explain the basic defect(s), Br. Med. J. 2:612, 1967.
16. Beighton, P.: X-linked recessive inheritance in the Ehlers-Danlos syndrome, Br. Med. J. 3:409, 1968.
17. Maxwell, E.: Cutis laxa, Am. J. Dis. Child. 117:479, 1969.
18. Schreiber, M. M., and Tilley, J. C.: Cutis laxa, Arch. Dermatol. 84:134, 1961.
19. Sesta, Z.: Ehlers-Danlos syndrome and cutis laxa: An account of families in the Oxford area, Ann. Hum. Genet. 25:313, 1962.
20. Mjølnerød, O. K., Dommerud, S. A., Rasmussen, K., and Gjeruldsen, S. T.: Congenital connective-tissue defect probably due to D-penicillamine treatment in pregnancy, Lancet 1:673, 1971.
21. Martin, G. I.: The Ehlers-Danlos syndrome. Report of two cases in monozygous Philippino-American twins, J. Chronic Dis. 23:197, 1970.
22. Pinnell, S. R., Krane, S. M., Kenzora, J. E., and Glimcher, M. J.: A heritable disorder of connective tissue. Hydroxylysine-deficient collagen disease, N. Engl. J. Med. 286:1013, 1972.
23. Krane, S. M., Pinnell, S. R., and Erbe, R. W.: Lysyl-protocollagen hydroxylase deficiency in fibroblasts from siblings with hydroxylysine-deficient collagen, Proc. Nat. Acad. Sci. USA 69:2899, 1972.
24. Lichtenstein, J. R., Martin, G. R., Kohn, L. D., Byers, P. H., and McKusick, V. A.: Defect in conversion of procollagen to collagen in a form of Ehlers-Danlos syndrome, Science 182:298, 1973.

MARFAN SYNDROME

CARDINAL CLINICAL FEATURES

Arachnodactyly; ectopia lentis; cardiovascular complications, especially aneurysm of the aorta.

CLINICAL PICTURE

The long, thin extremities of patients with the disorder that now bears his name were described in 1896 by Marfan,[1] who called it dolichostenomelia for that reason and characterized the legs as spider-like. Achard[2] first used the term arachnodactyly. Ectopia lentis was added as one of the major manifestations of the disease by Boerger[3] in 1914. Inheritance of the condition as an autosomal dominant (Fig. 1) was first recognized by Weve[4] in 1931. The integral relation of the major cardiovascular complications to the syndrome was first definitely made in 1943 by Baer, Taussig and Oppenheimer[5] and by Etter and Glover.[6] McKusick and Murdock et al.[7-10] have made major and continuing contributions to the understanding of the Marfan syndrome. McKusick has reported extensive studies of 257 patients and 74 families with the disorder and has classified it as a primary heritable disorder of connective tissue.

The skeletal features of the syndrome are characteristic. Stature of the patient is very tall. Furthermore, the major elongation is in the limbs. Thus, the upper to lower ratio of the patient with the Marfan syndrome is low (Fig. 2). A ratio of 0.85 in this syndrome may be compared to the normal adult Caucasian ratio of 0.93.[7] The arm span exceeds the height in patients with the syndrome. A useful clinical sign of arachnodactyly or the Marfan syndrome is the thumb sign, or the Steinberg sign,[11, 12] in which the thumb protrudes across and beyond the palm when the fist is clenched.

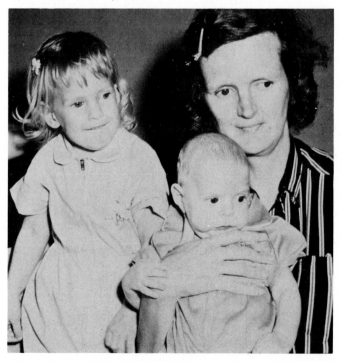

Fig. 1.—Mother and two children, all of whom had the Marfan syndrome, illustrating its dominant inheritance. Arachnodactyly is evident in the mother and baby. The sister also had iridodonesis.

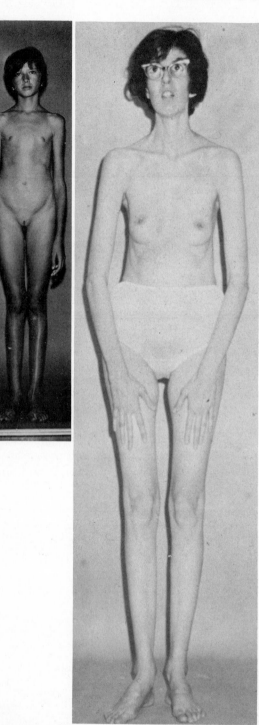

Feingold[12] has found that the Steinberg sign is positive in only 1.1% of a normal white population, in 2.7% of a normal black population and not present in patients with homocystinuria. In the so-called wrist sign, the patient's thumb and fifth finger overlap when clasped around the other wrist.[7] The fingers are typically very long and slender in these patients. The middle finger usually is one and a half times as long as the metacarpal of that finger. The ratio of the hand to the height usually is greater than 11% in these patients, and the ratio of the foot to the height usually is greater than 15%. The metacarpal index also has been used in the documentation of arachnodactyly.[7, 13] The length of the second, third, fourth and fifth metacarpals is measured in the roentgenogram of the right hand. The sum of these lengths is divided by the sum of the breadths measured at the midpoints. Data are available for normal adults[7] and normal infants, and in each case patients with the Marfan syndrome are greater than 2 s.d. from the normal mean.

The toes of these patients also are excessively long, particularly the great toes. Hallux valgus and contractures producing hammer toes are common. Contracture and camptodactyly of the fifth finger also occur frequently. In addition, the skeletal abnormalities of the disease regularly include kyphoscoliosis and deformity of the anterior chest. Pectus carinatum and pectus excavatum are commonly seen. The bones of the skull and face also participate in unusual longitudinal growth, producing a face that is long and narrow, a dolichocephalic skull, a high-arched palate and mild prognathism. Intelligence is normal. The joints are very lax and weak, which may lead to other skeletal abnormalities such as pes planus or genu recurvatum. Inguinal hernias are common, and femoral and diaphragmatic hernias occur.

Ocular abnormalities are frequently seen. Ectopia lentis is one of the hallmarks of the syndrome. It does not occur in 100% of patients, but it is present in most and is virtually always bilateral. The displacement may be upward, whereas in homocystinuria the subluxation tends to be downward. Iridodonesis, the tremor or dance of the iris, is an index of the presence of a dislocated lens. It may be particularly impressive in children in whom the dislocation may sometimes be difficult to detect. These patients usually are severely myopic. Retinal detachment is common. Patients may have nystagmus or glaucoma, and blue sclerae are occasionally observed.

Cardiovascular disease is an integral part of the syndrome.[5-9, 14] There is a structural weakness of the media of the aorta and other major vessels. This leads to dilatation of the ascending aorta or pulmonary artery, or a dissecting aneurysm. Aortic regurgitation is a regular complication of dilatation of the aortic ring. Mitral regurgitation also occurs.[7] Cardiovascular problems are progressive in this syndrome. A patient who develops angina pectoris or symptoms of left ventricular heart failure usually dies within a few years. Dis-

Fig. 2 (left).—Preadolescent girl with the Marfan syndrome. She was tall. Her arms were very long and her span measured greater than her height. Her lower segment was greater than the upper segment. She had subluxed lenses.

Fig. 3 (right).—N. M. (Case 1). Adult with the Marfan syndrome. The manifestations appear to be progressive. The patient had arachnodactyly and ectopia lentis.

secting aneurysm is a major cause of death. Certainly the Marfan syndrome is a leading concomitant of dissecting aneurysm in patients under 40 years of age.[7] Aortic dilatation and its complications account for 80% of the mortality in the Marfan syndrome.[9] Coarctation of the aorta may be seen. Atrial septal defects, predominantly patency of the foramen ovale, have been described,[8] as has bacterial endocarditis.

Pulmonary complications occur in association with the syndrome, repeated pneumothorax being the most frequent. Congenital cysts of the lung, as well as emphysema even in infants, have been reported.

Patients with the syndrome generally lack subcutaneous fat. There is a rare yet characteristic skin lesion called Miescher's elastoma, or elastoma intrapapillare performans verruciforme, which consists of small nodules or papules on the neck.[15] These are cysts containing whorls of material that stain like elastic fibers. Striae of the skin are frequently seen in the Marfan syndrome.

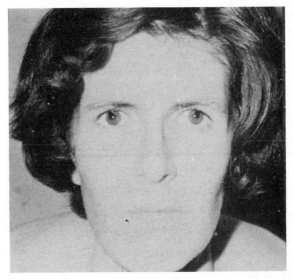

Fig. 5.– N. M. Close-up of the face, illustrating surgical colobomas of the iris and lensless eyes.

DIFFERENTIAL DIAGNOSIS

Homocystinuria (see section on Inborn Errors of Metabolism of Small Molecules) is the disease most frequently confused with the Marfan syndrome. Most of the patients we have seen with homocystinuria have not resembled very much the patients with the Marfan syndrome. They can be distinguished with certainty by the presence of homocystine in the urine. The enzyme cystathionine synthase is lacking in this disorder.

Congenital contractural arachnodactyly[16-18] is very similar to the Marfan syndrome. In fact it has been suggested[16] that Marfan's original patient did not have the Marfan syndrome but, rather, contractural arachnodactyly. Patients with this disorder have flexion contractures of several joints, especially of the fingers, elbows and knees, that are present at birth. The feet are long and slender and have elongated toes, and the fingers also are long and slender. Roentgenographically the cortices of the bones are thin; the ribs may

appear ribbon-like. Patients also have crumpled ears, absent or decreased subcutaneous fat, muscular hypoplasia, a high-arched palate and kyphoscoliosis. They do not have dislocated lenses, and the cardiovascular complications of the Marfan syndrome do not occur. Therefore, the prognosis is very different. Even the contractures tend to improve with time. The disorder is transmitted as an autosomal dominant.

Ectopia lentis is seen in about half of the patients with the Weil-Marchesani syndrome.[7] This syndrome is characterized by short stature, broad head, depressed nasal bridge and short fingers. It also may occur as an isolated inherited abnormality. It may be associated with aniridia and transmitted as a dominant. In association with ectopia of the pupils, ectopia lentis may be transmitted as either autosomal dominant or recessive. It may be associated with sulfocystinuria.[19]

CASE REPORTS

CASE 1.– N. M., a 28-year-old woman who had had problems with her eyes since early childhood, had been diagnosed only 5 years previously as having subluxed lenses. An operation was then undertaken for removal of the lenses, but she developed cardiac arrest; the lenses were removed later. She first showed cardiac symptoms, shortness of breath and anginal pain when pregnant. The child, the product of that pregnancy, was now 6½ years old. A second pregnancy was terminated by hysterectomy because of the cardiac condition. Her paternal grandparents were first cousins.

Physical examination revealed a very tall female with an appearance typical of that of the Marfan syndrome (Fig. 3). Her height was 71 inches, her span 72 inches. The lower segment was 36¾ inches. Her hand measured 8 inches and her foot 10 inches. There was pronounced arachnodactyly (Fig. 4).

Fig. 4.– N. M. The arachnodactyly was characteristic.

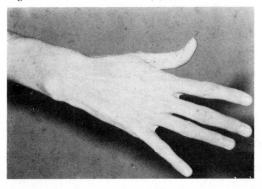

Examination of the eyes revealed surgical colobomas of each iris and no lenses (Fig. 5). Examination of the heart revealed a systolic ejection click and a grade II–III/VI systolic murmur.

CASE 2.—D. R. was the 6½-year-old son of N. M. He had grown 6 inches in the previous year, 4 of them during the previous 6 months. He had been squinting for about 2 years. He was a hyperactive child, especially so in school, and was being treated with Mellaril by a physician. He also had a problem with speech.

Physical examination revealed a tall, thin boy with definite arachnodactyly. His height was 52¾ inches, his span 54½ inches; the lower segment was 28 inches. Height age was 7½ years. The foot measured 8¾ inches. There was a precordial systolic ejection click and a grade II/IV systolic murmur.

GENETICS

The Marfan syndrome is transmitted as a simple mendelian autosomal dominant trait. There is a wide variability of expression even within the same family, but it is very rare that it does not express itself. Mild cases do go undiagnosed. Thorough examination of all family members is indicated when this disorder is diagnosed in any patient. There are, of course, a number of sporadic cases that are assumed to be the result of a new mutation. The proportion of sporadic cases is about 15% of all cases.[7] In Northern Ireland the mutation rate for this condition has been estimated at 5 per million genes/generation.[20] A figure for the prevalence of the Marfan syndrome is 1.5/hundred thousand population, and the gene frequency one half of this, or 0.73/hundred thousand.[7] Advanced paternal age at the time of conception has been documented for new mutations of the Marfan syndrome.[10, 20]

TREATMENT

There is no treatment for the basic defect. Treatment with propranolol or reserpine,[9] as agents that decrease myocardial contractility and diminish pulsatile flow, may decrease stress on a weakened aorta. Surgical treatment may be useful in the management of advanced aortic regurgitation. On the other hand, there is evidence that dissecting aneurysm of the aorta is better treated conservatively.[21] It may be useful, especially in girls, to inhibit excessive growth in height by hormonal induction of puberty.[7]

REFERENCES

1. Marfan, M. A.-B.: Un cas de déformation congénitale des quatre membres, plus prononcée aux extrémités, caractérisée par l'allongement des os avec un certain degré d'amincissement, Bull. Mem. Soc. Med. Hop. Paris 13:220, 1896.
2. Achard, C.: Arachnodactylie, Bull. Mem. Soc. Med. Hop. Paris 19:834, 1902.
3. Boerger, F.: Uber zwei Fälle von Arachnodaktylie, Z. Kinderheilkd. 12:161, 1914.
4. Weve, H.: Uber Arachnodaktylie. (Dystrophia mesodermalis congenita, Typus Marfan), Arch. Augenheilkd. 104:1, 1931.
5. Baer, R. W., Taussig, H. B., and Oppenheimer, E. H.: Congenital aneurysmal dilatation of the aorta associated with arachnodactyly, Johns Hopkins Med. J. 72:309, 1943.
6. Etter, L. E., and Glover, L. P.: Arachnodactyly complicated by dislocated lens and death from rupture of dissecting aneurysm of aorta, J.A.M.A. 123:88, 1943.
7. McKusick, V. A.: Heritable Disorders of Connective Tissue (4th ed.; St. Louis: C. V. Mosby Co., 1972), pp. 61–223.
8. McKusick, V. A.: The cardiovascular aspects of Marfan's syndrome: A heritable disorder of connective tissue, Circulation 11:321, 1955.
9. Murdoch, J. L., Walker, B. A., Halpern, B. L., Kuzma, J. W., and McKusick, V. A.: Life expectancy and causes of death in the Marfan syndrome, N. Engl. J. Med. 286:804, 1972.
10. Murdoch, J. L., Walker, B. A., and McKusick, V. A.: Parental age effects on the occurrence of new mutations for the Marfan syndrome, Ann. Hum. Genet. 35:331, 1972.
11. Steinberg, I.: A simple screening test for the Marfan syndrome, Am. J. Roentgenol. Radium Ther. Nucl. Med. 97:118, 1966.
12. Feingold, M.: The "thumb sign" in children, Clin. Pediatr. 7:423, 1968.
13. Joseph, M. C., and Meadow, S. R.: The metacarpal index of infants, Arch. Dis. Child. 44:515, 1969.
14. Symbas, P. N., Baldwin, B. J., Silverman, M. E., and Galambos, J. T.: Marfan's syndrome with aneurysm of ascending aorta and aortic regurgitation, Am. J. Cardiol. 25:483, 1970.
15. Haber, H.: Miescher's elastoma, Br. J. Dermatol. 71:85, 1959.
16. Beals, R. K., and Hecht, F.: Congenital contractural arachnodactyly, J. Bone Joint Surg. [Am.] 53:987, 1971.
17. Epstein, C. J., Graham, C. B., Hodgkin, W. E., Hecht, F., and Motulsky, A. G.: Hereditary dysplasia of bone with kyphoscoliosis, contractures, and abnormally shaped ears, J. Pediatr. 73:379, 1968.
18. MacLeod, P. M., and Fraser, F. C.: Congenital contractural arachnodactyly, Am. J. Dis. Child. 126:810, 1973.
19. Crawhall, J. C.: The Uncommon Disorders of Sulfur Amino Acid Metabolism, in Nyhan, W. L. (ed.): Heritable Disorders of Amino Acid Metabolism (New York: John Wiley & Sons, Inc., 1974), pp. 467–476.
20. Lynas, M. A.: Marfan's syndrome in Northern Ireland: An account of thirteen families, Ann. Human Genet. 22:289, 1958.
21. Wheat, M. W., Palmer, R. F., Bartley, T. D., and Seelman, R. C.: Treatment of dissecting aneurysms of the aorta without surgery, J. Thorac. Cardiovasc. Surg. 50:364, 1965.

OSTEOGENESIS IMPERFECTA

Cardinal Clinical Features

Brittle bones with numerous fractures, blue sclerae, deafness, dentinogenesis imperfecta.

Clinical Picture

Osteogenesis imperfecta probably is the most frequent of the heritable disorders of connective tissue.[1] It has clearly been present from antiquity. An authentic skeleton of an infant with the disease has been described in an Egyptian mummy dating from 1000 B.C.[2] Ekman[1] described the disease in 1788 in a family of three generations. The late form of the disease was described in 1835 by Lobstein,[3] and the congenital form in 1849 by Vrolik.[1, 4] The syndrome of brittle bones, blue sclerae and deafness was first put together by van der Hoeve and de Kleyn[5] in 1918. The terms osteogenesis imperfecta congenita and tarda were coined by Looser[6] in 1906.

Osteogenesis imperfecta is a generalized disorder of connective tissue. It involves classically the bones, sclerae and ear, but the ligaments, tendons and skin also are involved, as may be the heart. The disorder occurs in two basic forms. In the osteogenesis imperfecta congenita form, the disease is of such great severity that multiple fractures occur even in utero, and the infant either is stillborn or survives only a short time. At the other end of the spectrum, the mild tarda form may manifest itself only in the presence of blue sclerae, or fractures may occur first well into adult life. Unfortunately, from the point of view of easy classification, there is a full spectrum of gradation in between. It seems probable, as in the case of the mucopolysaccharidoses, that when the molecular nature of osteogenesis imperfecta is understood, there will be a considerable number of distinct varieties. At this time there are clinical and genetic advantages to distinguishing at least two forms of osteogenesis imperfecta.[1, 4, 7]

Osteogenesis Imperfecta Congenita

In the congenita form, the most striking feature is the brittleness of bones so that many fractures occur even in the fetus under the conditions of protection from trauma. The cranium is very soft and membranous. It may feel like a breech presentation to the examining obstetrician. In the newborn infant it may feel like the skull of a newborn with hypophosphatasia, but on x-ray it is apparent that there are bones throughout. The pattern is a mosaic with numerous wormian bones. Intracranial hemorrhage may occur during delivery. Very few of these infants survive the first 24 hours and hardly any the first year. Soft bones and multiple costal fractures may lead to a flail chest and respiratory embarrassment in the neonate. Pneumonia is a frequent cause of death. The extremities are short and bowed – short enough to be considered micromelia. The skin and soft tissues appear thick, and there are many extra folds. The limbs are often bowed on the chest and abdomen at birth. Prenatal bowing may be associated with cutaneous dimpling. Intrauterine diagnosis has been made by roentgenography. The bones of the newborn infant may be rather thick, reflecting the healing of fractures and the production of callus. Pseudoarthrosis of the tibia or other bones has been reported. There is a striking laxness and hypermobility of the joints. Hernias are frequent. Crepitation is commonly felt. The extremity of one of these patients has been described as feeling like a bag of walnuts.[4] Beading of the ribs with callus may be misinterpreted as a rachitic rosary. Progressive hydrocephalus often occurs in osteogenesis imperfecta congenita.

Most of these patients do not have impressively blue sclerae. Calcifications have been encountered in large peripheral arteries and in the pulmonary and cerebral arteries.[1] A preponderance of females among patients with this lethal form of osteogenesis imperfecta has suggested that males with the disease might have died even earlier in intrauterine life. The concentrations of calcium and phosphorus in the serum are normal. Alkaline phosphatase activity may be high because of healing fractures.

Osteogenesis Imperfecta Tarda

Osteogenesis imperfecta tarda is the usual form in which osteogenesis imperfecta is encountered. It has been estimated that approximately 90% of patients have the form of the disease in which there is a dominant mode of transmission, blue sclerae and other signs of mesenchymal dysplasia including easily fractured bones. There is wide range of expression of the syndrome. Some involved individuals have only blue sclerae. Nevertheless, the most important of the manifestations is the fragility of the bones, which leads to multiple fractures occurring after minor trauma. The nature of the trauma with which fractures have been seen in this syndrome is famous. One young man fractured both femurs when his fiancée sat on his lap.[4] The phalanges have been fractured by writing, and all of the metacarpals by wringing out a dish cloth. Apert[4] was quoted as referring to these patients as *les hommes de verre*, and one family has referred to their involved children as "china dolls."[8] Parallels with the story of Humpty Dumpty are unavoidable. Muscle pull occasionally is enough to shear off a segment of bone; for instance, the olecranon has been torn away by contraction of the triceps. The degree of deformity depends on the age at which fractures begin. Especially in childhood there is a tendency for greater involvement in the lower extremities. Some authors have distinguished a severe or gravis type and a milder or levis type. It is probable that in some patients with the former type, fracture may even occur in the neonatal period or before, but the survival pattern is much better than that of patients with osteogenesis imperfecta congenita. Among the milder forms, fractures may begin late in life or not at all. In general there is a tendency in patients for a decreased incidence of fractures after puberty, possibly with an increase again after menopause.

Callus formation in this disease is excellent. It may even be exuberant, and cases are recorded in which amputation was carried out because callus was thought to represent an osteogenic sarcoma. On the other hand, malignant disease of the bone does not appear to be a characteristic of this disease, as it is of the Paget disease, although a few patients have been described in whom there was metastatic osteogenic sarcoma. Pseudoarthrosis is common, especially of the tibia.

The bones generally have thin cortices, and in x-rays the long bones usually have slender shafts that widen toward the epiphyses. Occasionally the appearance of the bones may appear insufficient to distinguish the patient from normal. With fractures, callus and deformities, the bones may thicken in places or become cystic. Hourglass or codfish biconcave deformities of the vertebrae are the characteristic result of compression by pressure of the nucleus pulposus. Herniation of the nucleus into the vertebrae produces Schmorl's nodules. Growth is interfered with not only by fractures and deformity, especially in the legs, but by multiple microfractures in the epiphyses of the long bones. Patients may not only grow poorly but may become shorter with time.

The face may appear triangular because of a bulging cranium. The skull may have a bilateral temporal bulge and an overhanging occiput, frontal bossing and platybasia. It is as if a soft skull tended to flop in all directions like a soft cap; it has been referred to as "Tam-O'Shanter" skull.

The blue sclerae are the ocular hallmark of this syndrome and probably its most common manifestation.

Certainly in some families in which fragile bones and blue sclerae are found, there are members in whom the sclerae constitute the only sign of the disease. There is variation in the blueness of the sclerae in individuals with this disease, as there is in normal individuals, some of which is age related. The blue has been referred to as slate blue, marine blue, robin's egg blue and Wedgwood blue. The color is a reflection of the thinness of the sclerae. Rarely scleral perforation may occur. The area of sclera immediately adjacent to the cornea may be whiter, a so-called Saturn's ring, and there may be an opacity of the periphery of the cornea, called an arcus juvenilis or embryotoxon. Under the slit lamp the cornea is evidently thin as well, and keratoconus may be seen. Some patients have had ectopia lentis.[9]

Deafness is the least common of the major features of the disease. It seldom begins before the teens; it may begin in pregnancy. Its pattern is that of otosclerosis. It may involve the stapes or the cochlea, so that there may be conduction deafness or nerve deafness, or both.

The teeth are typically abnormal in this condition. They may be amber or yellowish brown. They usually are translucent or blue-gray in at least some portions and they wear down easily. On x-ray there may be no pulp canal. The dental condition has been referred to as dentinogenesis imperfecta, indicating that the problem lies in the dentin, not the enamel.

Laxity and hypermobility of the joints also is characteristic. This probably reflects the generalized nature of the defect in connective tissue. Hernias are common. Patients may be slow to walk as infants because of difficulty in fixing the joints. In one patient the Achilles tendon was described as having the diameter of a lead pencil.[4] Rupture of the patellar ligaments has been reported. The skin may be thin, delicate and translucent. Healing may be with unusually wide scars. Capillary fragility and subcutaneous hemorrhages may be seen. Cardiovascular manifestations occur late in this disorder and appear to reflect changes in the connective tissue of arterial walls or valves. Aortic regurgitation is the most common manifestation. Neurologic manifestations occur in the most severely involved patients and appear to result from compression of nerves or the spinal cord by platybasia, vertebral changes or specific fractures or, more rarely, intracranial hemorrhage at birth. Intelligence usually is normal.

The basic problem in this disease is thought to be in the maturation of collagen beyond the stage of reticulin fibers.[1, 10, 11] An absence of normal adult collagen has been seen histologically in the skin and the eye. Instead there are thin argyrophilic fibers with the appearance of reticulin.[10, 11] Fibroblasts of patients with osteogenesis imperfecta have been reported to have collagen that is more readily extractable than normal into salt and weak acid solutions, suggesting an abnormality in cross-linking, and to have abnormally low hydroxyproline formation from proline.[8] There may be an elevation in the total urinary excretion of proline, particularly in the peptide-bound fraction.[12] It seems likely that, with current advances in the understanding of collagen metabolism, the molecular nature of this disorder will be defined.

DIFFERENTIAL DIAGNOSIS

Osteogenesis imperfecta congenita should be distinguished from hypophosphatasia (see section on Other Disorders of Metabolism and Disorders of Transport). Multiple intrauterine fractures may occur in both conditions, producing short, bowed extremities with cutaneous dimpling. In hypophosphatasia the activity of alkaline phosphatase is very low and the urinary excretion of phosphoethanolamine is high. The skull films of the two conditions are quite distinct, with an almost complete absence of bone in the infant with hypophosphatasia and a complete, if very thin, complement of bone in osteogenesis imperfecta, with a mosaic of many wormian bones.

CASE REPORTS

CASE 1.—D. C. arrived at University Hospital via the transport team. She was born by breech delivery in a community hospital to a 22 year old in the 36th week of her first pregnancy. The Apgar score was 8–9. Anomalies were noted at birth,

Fig. 1.—D. C. (Case 1). Newborn girl with osteogenesis imperfecta congenita. The multiple intrauterine fractures are evident from the shortening and deformities of the lower extremities. The hips were held in a frogleg position and the thighs appeared abnormally large. There were no cutaneous dimples.

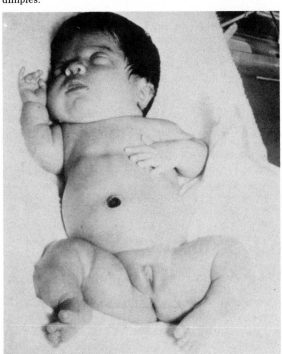

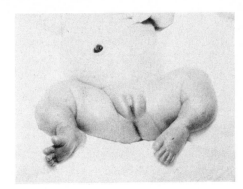

Fig. 2.—D. C. Closer view of the legs illustrates the bowing and deformity that is characteristic.

and the infant was transferred with a tentative diagnosis of achondroplasia. The family history was noncontributory. The father was 28 years old. There was no consanguinity.

The infant was small. She weighed 4 lb, 7 oz and had a length of 14 inches. Head circumference was 30 cm. The skull was soft and felt crepitant. The sutures were widely separated and the fontanels were wide. The sclerae appeared blue. The legs were grossly deformed (Figs. 1–3). There was evidence of multiple fractures and angulation as well as hyperextension at the knees and hips. There was a simian crease on the right hand. There were no spontaneous movements. The Moro reflex was absent, and the cry weak.

Roentgenographic evaluation (Figs. 4–6) revealed multiple fractures involving all of the long bones and the ribs. The skull was described as a membranous bag of wormian bones. The serum concentration of calcium was 9.2 mg/100 ml and the alkaline phosphatase 9.1 Bodansky units. The urine was negative for phosphoethanolamine.

The infant promptly became physiologically stable. She took feedings well and was discharged after 2 weeks to an extended care facility. She died there at the age of 4 months.

CASE 2.—M. O. (Fig. 7) was born at term of a 27-year-old unmarried woman in her first pregnancy. The presentation was breech and the delivery was by cesarean section. The infant moved very little and was quite blue. He was taken to the special care nursery. Family history was noncontributory.

The length was 17 inches and the head circumference 39.5 cm. The upper segment was 30 cm. The skull was very soft,

Fig. 3.—D. C. Position of the left leg at birth.

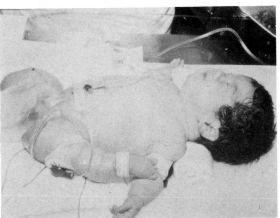

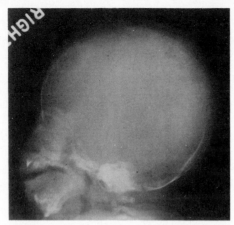

Fig. 4.—D. C. Roentgenogram of the skull illustrates retarded ossification in the parietal areas. This appearance is in
striking contrast to the skull of a patient with hypophosphatasia congenita.

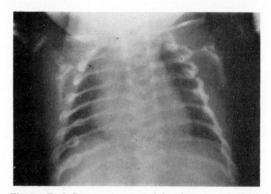

Fig. 5.—D. C. Roentgenogram of the chest reveals thin ribs,
with beading produced by multiple fractures.

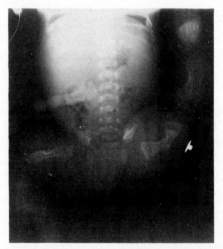

Fig. 6.—D. C. Roentgenogram of the lower extremities. The
bones were very poorly mineralized and lacking in cortex.
They were shortened and angulated as a consequence of
multiple fractures.

Fig. 7.—M. O. (Case 2). Another newborn infant with osteogenesis imperfecta congenita.

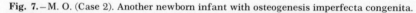

and most of the cranium felt membranous. The sclerae were not unusually blue for a newborn infant. There was a cleft of the palate. There appeared to be bilateral hip dislocations. The lower extremities were markedly bowed and shortened. There were pseudoarthroses of both tibiae.

Excretion of phosphoethanolamine in the urine was within normal limits. Roentgenographic examination (Fig. 8) revealed multiple fractures and multiple wormian bones in the skull. The serum calcium was 9.5 mg/100 ml and the alkaline phosphatase 126 I.U. The baby did well in the hospital and was discharged at 8 days of age. Bilateral inguinal hernias were found in the next few weeks.

GENETICS

Osteogenesis imperfecta tarda, the usual form, is transmitted as an autosomal dominant disorder with a wide range of expression. Skipped generations have been reported. Some authors have distinguished kindreds in which there were never any blue sclerae and in which transmission also was as a dominant. The classic or lethal form of osteogenesis imperfecta congenita appears to be recessive.[1, 7] The occurrence of dominant forms very early in life makes decisions as to counseling difficult. There is a common tendency to counsel the parents of an infant with osteogenesis imperfecta that, if a careful search of the parents and family reveals no stigmata of the disease, the patient probably represents a new mutation. In this case the possibility of giving birth to a second affected child would be negligible. We would not counsel the parents of an infant who died of osteogenesis imperfecta congenita in this way but, rather, would suggest that it is recessive unless otherwise proved and that the next sibling could be involved.[13]

TREATMENT

There is no known cure for osteogenesis imperfecta. Orthopedic management is essential. Pinning and plating of fractures are recommended to avoid immobilization. Deformities may be approached using osteotomies, bone grafts and what has been called a

Fig. 8.—M. O. Roentgenograms of the lower extremities.

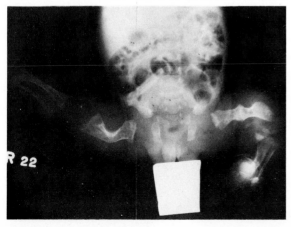

shishkebob operation in which a badly deformed long bone is fractured into many small pieces, which are then aligned on a rod through the medullary cavity.[14] The solid support of the rod also may reduce the number of fractures. Organizations of parents, such as the National Osteogenesis Imperfecta Foundation, may provide information and support to patients and their parents.

REFERENCES

1. McKusick, V. A.: *Heritable Disorders of Connective Tissue* (4th ed.; St. Louis: C. V. Mosby Co., 1972), p. 390.
2. Gray, P. H. K.: A case of osteogenesis imperfecta, associated with dentinogenesis imperfecta, dating from antiquity, Clin. Radiol. 20:106, 1969.
3. Lobstein, J. F.: *Lehrbuch der Pathologischen Anatomie*, (Stuttgart: Fr. Brodhag'sche Buchhandlung, 1835), Vol. 2, p. 179.
4. Seedorff, K. S.: *Osteogenesis Imperfecta—A Study of Clinical Features and Heredity Based on 55 Danish Families Comprising 180 Affected Persons* (Copenhagen: Ejnor Münksgaard, 1949).
5. van der Hoeve, J., and de Kleyn, A.: Blaue Sclera, Knochenbrüchigkeit und Schwerhörigkeit, Graefes Arch. Ophthalmol. 95:81, 1918.
6. Looser, E.: Zur Kenntnis der Osteogenesis imperfecta congenita und tarda (sogenannte idiopathische Osteopathyrosis), Mitt. Grenggeb. Med. Chir. 15:161, 1906.
7. Ibsen, K. H.: Distinct varieties of osteogenesis imperfecta, Clin. Orthop. 50:279, 1967.
8. Brown, D. M.: Collagen metabolism in fibroblasts from patients with osteogenesis imperfecta, (Abstr.) Pediatr. Res. 6:394, 1972.
9. Remigio, P. A., and Grinvalsky, H. T.: Osteogenesis imperfecta congenita. Association with conspicuous extraskeletal connective tissue dysplasia, Am. J. Dis. Child. 119:524, 1970.
10. Follis, R. H., Jr.: Histochemical studies on cartilage and bone. III. Osteogenesis imperfecta, Johns Hopkins Hosp. Med. J. 93:386, 1953.
11. Follis, R. H., Jr.: Maldevelopment of the corium in the osteogenesis imperfecta syndrome, Johns Hopkins Hosp. Med. J. 93:225, 1953.
12. Summer, G. K., and Patton, W. C.: Intravenous proline tolerance in osteogenesis imperfecta, Metabolism 17:46, 1968.
13. Goldfarb, A. A., and Ford, D.: Osteogenesis imperfecta congenita in consecutive siblings, J. Pediatr. 44:264, 1954.
14. Williams, P. F.: Fragmentation and rodding in osteogenesis imperfecta, J. Bone Joint Surg. [Br.] 47:23, 1965.
15. Freda, V. J., Vosburgh, G. J., and De Liberti, C.: Osteogenesis imperfecta congenita. A presentation of 16 cases and review of the literature, Obstet. Gynecol. 18:535, 1961.
16. Bock, J. E.: Osteogenesis imperfecta. A report of a case of the congenital form, Acta Obstet. Gynecol. Scand. 48:222, 1969.
17. Tan, K. L., and Tock, E. P. C.: Osteogenesis imperfecta congenita. Aust. Paediatr. J. 7:49, 1971.
18. Seriki, O.: Osteogenesis imperfecta congenita in one of twins, Acta Paediatr. Scand. 59:340, 1970.
19. Heckman, B. A., and Steinberg, I.: Congenital heart disease (mitral regurgitation) in osteogenesis imperfecta, Am. J. Roentgenol. Radium Ther. Nucl. Med. 103:601, 1968.

URBACH-WIETHE SYNDROME
Lipoid Proteinosis, Hyalinosis Cutis et Mucosae

CARDINAL CLINICAL FEATURES

Yellowish nodular infiltrations of the skin and mucous membranes, particularly around the mouth, eyes, elbows and knees; hoarseness; intracranial calcifications; seizures; parotid swelling.

CLINICAL PICTURE

Lipoid proteinosis is a rare disease first described by Urbach and Wiethe[1] in 1929. The unusual deposits that characterize the condition are found on the skin and mucous membranes. Lesions around the mouth and eyes are prominent. There are multiple bead-like, pin-sized, yellowish papules, which coalesce and grow larger. Involved areas become thick and leathery. In the eyes papular lesions of the lids often lead to itching. The eyelashes grow normally through the lids, but they are frequently lost. Microscopically the lesions show some acanthosis and lipid deposits. These are apparently largely phospholipid and are thought to be attached to protein. On funduscopic examination, numerous small, round, yellowish white lesions are seen scattered throughout the fundus. Lesions over the chin and at the angles of the mouth are common.

Most patients give histories of hoarseness that has been present since childhood but is nonprogressive.

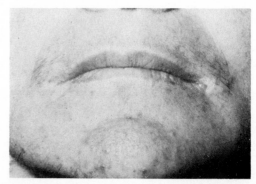

Fig. 2. – In the sister of V. S., lesions also were prominent at the angles of the mouth. These whitish lesions illustrate the problem of differential diagnosis from leukoplakia.

Direct laryngoscopy has revealed the false cords to be covered with a diffuse granular material similar to that seen on the other mucosal surfaces. Severe dyspnea may occur and laryngectomy may be necessary. There usually are whitish, plaque-like lesions on the inner surfaces of the lips. The tongue may become large and thickened, and the tonsils or the roof of the mouth may be diffusely infiltrated. From time to time, these patients may have a stenosis of Stensen's duct, with resultant enlargement of the parotid.[2] The denti-

Fig. 1. – V. S. (Case Report). Typical confluent papular lipoid lesions in a linear pattern across the chin and at the corners of the mouth.

Fig. 3. – V. S. The tongue was large and diffusely infiltrated. Thick deposits on the roof of the mouth were especially prominent.

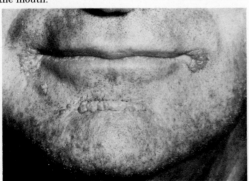

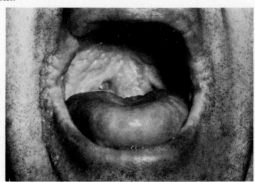

Fig. 4.—V. S. Very thick fissured area of infiltration over the elbow.

tion may be severely affected. Teeth may fail to develop or may be hypoplastic; they may erupt and then fall out. Diffuse nonscarring alopecia or thinning of the scalp hair may develop.

Intracranial calcifications have been found in most patients. Calcifications are located above the pituitary fossa and in the hippocampus, falx cerebri or temporal lobes. These lesions have been associated with the occurrence of convulsive seizures. Patients may have a positive tourniquet test.[3] Axillary hair may fall out, as may the hair of the head. Diabetes occurs in unusually high frequency.

CASE REPORT

V. S. was a 48-year-old white male Lebanese admitted to University Hospital for skin and mucosal lesions and hoarse-

Fig. 5.—V. S. Roentgenogram of the skull illustrating the intracranial calcifications, which appear like giant teeth projecting up into the hippocampal area.

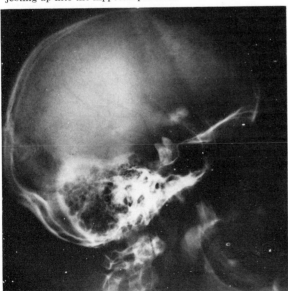

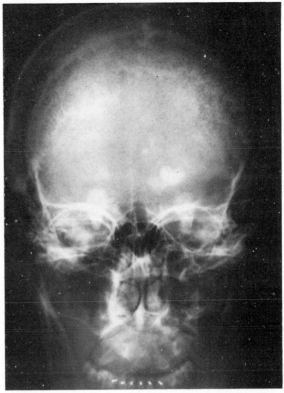

Fig. 6.—V. S. Roentgenogram of the skull in the anteroposterior view. Air studies revealed that calcification was highly selective, involving only the cardate and amygdaloid nucleus.

ness. These problems had been present since early childhood. He also had suffered from generalized seizures, for which he had been treated with Dilantin and Mysoline. His mother and father were second cousins. His 35-year-old sister had an identical disease and was also admitted for study. A deceased brother had had a similar illness, and four siblings were normal.

Examination revealed a healthy-looking man with prominent and unusual skin lesions. The edges of the eyelids were the sites of pinhead-sized papules, leading to general thickening of the lids. The lesions were yellowish brown. He also had bilateral pterygia. There were confluent lesions at the angles of the mouth (Figs. 1 and 2). The tongue was infiltrated and thick and had white plaques on the surface. The buccal surfaces, hard palate, pharynx and tonsillar areas were thick and diffusely infiltrated (Fig. 3). There was a linear arrangement of 3-mm nodules in the cleft of the chin at the upper part of the mandible (see Fig. 1). Soft, reddish brown, pebbly, verrucose lesions were present on the elbows (Fig. 4) and knees. The patient was quite hoarse, and lesions had been found on the larynx. Blood pressure was 190/150. He and his sister had lost all of their teeth during childhood. He had a very striking deficit in recent memory. The rest of the examination was unremarkable.

Roentgenographic examination of the skull revealed striking intracranial calcifications. These lesions (Figs. 5 and 6) were typical of lipoid proteinosis. Two well-circumscribed areas of calcification lay 2.5 cm from the midline and projected over the sella turcica in the area of the hippocampal gyrus. There were two other areas of irregular calcification in the basal ganglia.

Examination of the blood for sugar, creatinine, electrolytes, serum glutamic pyruvic transaminase, lactic dehydrogenase, triglycerides and cholesterol was normal.

Biopsies of the skin from the left elbow and chin as well as the posterior pharyngeal wall revealed hyperkeratosis and deposits below the mucosa of hyalinized eosinophilic material, which was PAS positive. This material also showed some metachromasia with the toluidine blue stain. These features are typical of lipoid proteinosis.

GENETICS

Lipoid proteinosis appears to be transmitted as an autosomal recessive trait. Familial patients have been observed with some frequency. Consanguinity among the parents of the affected individuals is common.[3] Direct transmission from an involved mother to three sons and a daughter[3] might suggest dominant transmission, but consanguinity makes an autosomal recessive transmission more likely.

TREATMENT

None.

REFERENCES

1. Urbach, E., and Wiethe, C.: Lipoidosis cutis et mucosae, Virchows Arch. [Pathol. Anat.] 273:285, 1929.
2. Sulzberger, M. B.: A case of lipoidosis cutis et mucosae—so-called "lipoid proteinosis," Urbach-Wiethe, Laryngoscope 52:286, 1942.
3. Rosenthal, A. R., and Duke, J. R.: Lipoid proteinosis, Am. J. Ophthalmol. 64:1120, 1967.
4. Burnett, J. W., and Marcy, S. M.: Lipoid proteinosis, Am. J. Dis. Child. 105:81, 1963.
5. Cowan, M. A., Alexander, S., Vickers, H. R., and Cowdell, R. H.: Case of lipoid proteinosis, Br. Med. J. 2:557, 1961.
6. Hewson, G. E.: Lipid proteinosis (Urbach-Wiethe syndrome), Br. J. Ophthalmol., 47:242, 1963.

Teratogenesis

FETAL ALCOHOL SYNDROME

Kenneth Lyons Jones, Jr.

CARDINAL CLINICAL FEATURES

Intrauterine growth retardation and retardation of postnatal growth and development, microcephaly, short palpebral fissures, maxillary hypoplasia, cardiac defects, joint anomalies.

CLINICAL PICTURE

The fetal alcohol syndrome was first described in 1973 by Jones, Smith, Ulleland and Streissguth,[1] and all patients described have been the offspring of women who were severe chronic alcoholics prior to and throughout pregnancy. The length at birth has been proportionately lower than birth weight. The postnatal rate of linear growth has averaged two thirds of normal and, by 1 year of age, the majority of patients have been markedly undersize and have been underweight even for their height age. The head circumference, below the third percentile at birth, has been below the third percentile for height age as well by 1 year of age. The IQ in these patients has ranged from below 50 to 83, with an average of 63. A fine motor dysfunction has consisted of poor eye-hand coordination and weak grasp. Tremulousness has been observed in the newborn period.

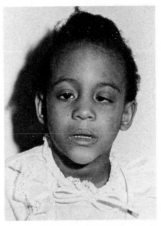

Fig. 2.—A 3¼-year-old girl with the fetal alcohol syndrome. She had short palpebral fissures, ptosis and strabismus on the left. (Reproduced with permission from Lancet.[1])

Cardiac defects most frequently observed have been ventricular septal defects. Joint anomalies have been variable. They have included abnormal joint positions, joint dislocations and inability to flex and/or extend completely. The palpebral fissures are characteristically short (Figs. 1–3). There also may be epicanthal folds, ptosis and strabismus (see Fig. 2). Other fea-

Fig. 1.—A 1-year-old American Indian girl with the syndrome. She had short palpebral fissures and maxillary hypoplasia. The upper lip appears thin. (Reproduced with permission from Lancet.[1])

Fig. 3.—A 2½-year-old boy with short palpebral fissures, maxillary hypoplasia and a thin upper lip. (Reproduced with permission from Lancet.[1])

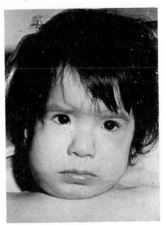

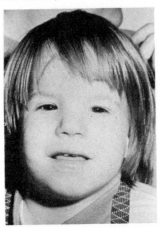

tures frequently seen are altered palmar crease patterns, anomalous external genitalia and capillary hemangiomas.

CASE REPORT

W. P., an American Indian boy, was born to a 30-year-old chronic alcoholic mother who had been drinking heavily for 6 years prior to his birth. During that time she had had three first trimester abortions. Her seven other children, all born prior to her becoming alcoholic, were normal. Delivery was at term from a breech presentation. The attending physician noted that the infant had "alcohol on his breath." Birth length was 17 inches (50th percentile for 32.5 weeks' gestation). The birth weight was 4½ lb (50th percentile for 34 weeks' gestation) and head circumference was 29 cm (below the 3d percentile). The child had problems with neonatal adaptation including respiratory distress and transient hypoglycemia and hypocalcemia. Marked tremulousness was unresponsive to medication.

Examination revealed pronounced hirsutism over the face (Fig. 4). The eyes were small and the palpebral fissures short, measuring 1.1 cm on the right and 1.2 cm on the left. A grade II/VI systolic murmur was heard along the left sternal border and was thought to represent a ventricular septal defect. There was a congenital dislocation of the left hip. There were bilateral simian creases.

At 10 months of age the height was 24½ inches (50th percentile for 4 months). Weight was 10 lb, 13 oz, and head circumference 39.5 cm (below the 3d percentile). Developmental age was approximately 5 months. The patient continued to be tremulous. He was hypotonic and had frequent breath-holding spells. An EEG showed generalized abnormalities.

GENETICS AND ETIOLOGY

The fetal alcohol syndrome appears to be nongenetic. The similarity in the over-all pattern of malformation among the children described suggests a singular

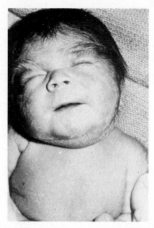

Fig. 4.—W. P. (Case Report). A 5-day-old infant with the fetal alcohol syndrome. The short palpebral fissures are clearly demonstrated, as is the hirsutism. The distance between nose and upper lip appears long, the upper lip very thin. (Reproduced with permission from Lancet.[2])

etiology. It is most likely related to the effect of maternal alcohol or one of its breakdown products.

TREATMENT

None.

REFERENCES

1. Jones, K. L., Smith, D. W., Ulleland, C. N., and Streissguth, A. P.: Pattern of malformation in offspring of chronic alcoholic mothers, Lancet 1:1267, 1973.
2. Jones, K. L., and Smith, D. W.: Recognition of the fetal alcohol syndrome in early infancy, Lancet 2:999, 1973.
3. Jones, K. L., Smith, D. W., Streissguth, A. P., and Myrianthopoulos, N. C.: Outcome in offspring of chronic alcoholic women, Lancet 1:1076, 1974.

DIPHENYLHYDANTOIN EMBRYOPATHY
Dilantin Embryopathy,
Fetal Hydantoin Syndrome

Kenneth Lyons Jones, Jr.

CARDINAL CLINICAL FEATURES

Prenatal onset deficiency of growth, mild to moderate mental deficiency, low nasal bridge, hypertelorism, hypoplasia of nails and distal phalanges.

CLINICAL PICTURE

Although an association between maternal ingestion of anticonvulsants during pregnancy and malformations in the offspring was first made in 1968 by Meadow,[1] data implicating diphenylhydantoin as a teratogen did not become available until the publication of epidemiologic studies by Fedrick[2] in 1973 and Monson et al.[3] in 1974. Hill et al.[4] in 1974 recognized a specific pattern of malformation in the offspring of women treated with diphenylhydantoin during pregnancy. Hanson and Smith[5] have referred to it as the fetal hydantoin syndrome.

The majority of affected children have mild to moderate mental deficiency. A slowed rate of growth usually is of prenatal onset. Postnatal growth rate has averaged 75% of normal. Frequently observed craniofacial abnormalities include microcephaly, a short nose with a low nasal bridge and eyes that appear widely set. The ears may be low set. The mouth often is wide and the lips prominent. Enlarged fontanels are characteristic in the newborn period. There may be gingival hypertrophy or a widened alveolar ridge. Limb defects have been variable; the most typical is hypoplasia of nails and distal phalanges. There may be a finger-like thumb. There is an increased incidence of low-arch dermal ridge patterns. The hypoplasia of the nail and distal digit usually involves the ulnar or fibular aspect of the limb most severely and often becomes less noticeable with advancing age. Other features frequently seen are a short or webbed neck with a low hairline and anomalies of the rib, sternum or spine. The nipples may be hypoplastic and appear widely spaced. Umbilical or inguinal hernias are seen, as are genital anomalies and pilonidal sinus. Less frequent anomalies are cleft lip and/or palate, cardiovascular anomalies, renal defects, positional deformities of a limb and diaphragmatic hernia.

DIFFERENTIAL DIAGNOSIS

Because of the hypoplastic nails and growth retardation, patients with this disorder have been confused with those with the Coffin-Siris syndrome and with the children reported by Senior.[6] The over-all pattern of malformation is reminiscent of the Noonan syndrome. One affected child was mistakenly diagnosed as having hypohidrotic ectodermal dysplasia.

Fig. 1. – R. H. (Case Report.) An infant with the fetal diphenylhydantoin syndrome.

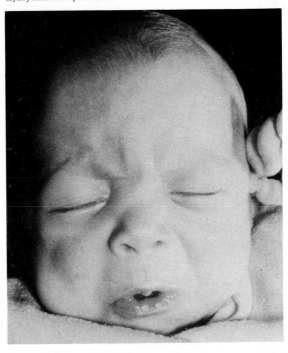

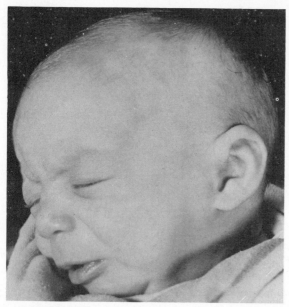

Fig. 2.—R. H. The nose was short and the nasal bridge low. There were epicanthal folds.

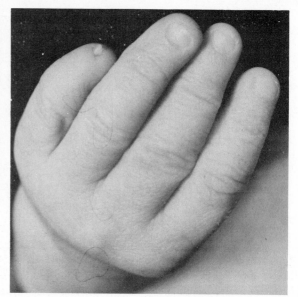

Fig. 3.—R. H. Hypoplasia of nails.

CASE REPORT

R. H. (Fig. 1) was born to an 18-year-old white primigravida female who was initially diagnosed as epileptic 3 years prior to the birth of her affected child. Throughout pregnancy she took Dilantin, 100 mg, b.i.d., phenobarbital, 30 mg, t.i.d., and mysoline, 250 mg, t.i.d. Her last seizure occurred 1 month prior to delivery. Birth was by breech presentation at 40 weeks of gestation. Birth weight was 5 lb, 11 oz; birth length 19 inches and head circumference 32.5 cm. The child was initially lethargic and had a weak cry. The anterior fontanel was enlarged. The nasal bridge was depressed (Fig. 2). Hypertelorism was present. A grade II/VI systolic murmur present along the left sternal border was subsequently diagnosed to represent mild pulmonic stenosis. Limb anomalies included marked hypoplasia of the fingernails (Fig. 3) and toenails bilaterally, increasing in severity from digits 1 to 5, a digitalized thumb on the right, a single upper palmar crease on the left hand and 10 of 10 low-arch dermal ridge patterns. A pilonidal sinus was present.

At 4 weeks of age the child was hospitalized for poor weight gain and episodes of tonic-clonic movements of all of the limbs, lasting approximately 30 seconds, associated with rolling back of the eyes and occurring twice a day. At that time he weighed 6 lb and was 19½ inches in length. Head circumference was 33.5 cm. EEG was within normal limits.

At 9 months of age poor weight gain continued. Developmental delay was evident.

GENETICS AND ETIOLOGY

Many of the infants with this syndrome have been born to women taking more than one anticonvulsant during pregnancy, but the over-all pattern of malformation has been seen only in children whose mothers took diphenylhydantoin anticonvulsants. There appears to be a two- to threefold greater chance that an infant exposed during the early months of gestation to daily diphenylhydantoin will have a significant malformation. Combination with barbiturates may increase the risk to the fetus. No "safe" dosage of hydantoins has yet been established.

TREATMENT

None.

REFERENCES

1. Meadow, S. R.: Anticonvulsant drugs and congenital abnormalities, Lancet 2:1296, 1968.
2. Fedrick, J.: Epilepsy and pregnancy: A report from the Oxford record linkage study, Br. Med. J. 2:442, 1973.
3. Monson, R. R., Rosenberg, L., Hartz, S. C., Shapiro, S., Heinonen, O. P., and Slone, D.: Diphenylhydantoin and selected congenital malformations, N. Engl. J. Med. 289:1049, 1974.
4. Hill, R. M., Verniand, W. M., Horning, M. G., McCulley, L. B., and Morgan, N. F.: Infants exposed *in utero* to antiepileptic drugs, Am. J. Dis. Child. 127:645, 1974.
5. Hanson, J. W., and Smith, D. W.: The fetal hydantoin syndrome, J. Pediatr. In press.
6. Senior, B.: Impaired growth and onychodysplasia, Am. J. Dis. Child. 122:7, 1971.

Dwarfism Syndromes with Premature Aging

PROGERIA
The Hutchinson-Gilford Syndrome

CARDINAL CLINICAL FEATURES

Premature aging, shortness of stature, craniofacial disproportion, micrognathia, prominent scalp veins, alopecia, acroosteolysis of the distal phalanges, resorption of the distal ends of the clavicles. Early there may be a stiff, infiltrated skin and a sculptured nasal tip.

CLINICAL PICTURE

In 1886 Hutchinson[1] first described a 3½-year-old boy who had a "withered, old-mannish look" and an absence of hair. He had no nipples, but there were little scars at their sites. In 1897 Gilford[2] reported a similar patient and provided a further description of the patient whom Hutchinson had discussed earlier. Gilford derived the name progeria from the Greek προγηρος to indicate premature aging.

Among the most characteristic features[3] of this syndrome is craniofacial disproportion. The cranium and vessels over it are prominent and there is frontal bossing. The facies is small and delicate, and the orbits are small, so that the eyes appear to protrude. The features are thin and pinched. There is severe micrognathia and a small peaked nose with a sculptured nasal tip, in which the outlines of the nasal cartilages can be seen.[2] The lips are thin. Hair loss is characteristic. As infants, these patients have fine, sparse, white or blond hair even if the original hair was dark. Eyebrows and eyelashes are sparse. This condition proceeds to total alopecia. Even lanugo disappears. The skin appears aged; it is thin and dry and transparent, and the subcutaneous veins are visible. There is paucity of subcutaneous fat and marked wasting of the musculature, with wrinkling and looseness of the skin. Loss of tissue in the ear lobe can be striking. With time, brown mottled areas of pigmentation develop. The nails often are hypoplastic. Dentition is delayed and there is crowding of the teeth. Absent breasts and nipples were reported in Hutchinson's patient.[1] Other patients have had hypoplastic nipples. There is an absence of sexual hair, and sexual development usually does not occur. Decreased sweating has been noted.

At birth or in early infancy a thickened skin is seen in some patients with progeria, which has been de-

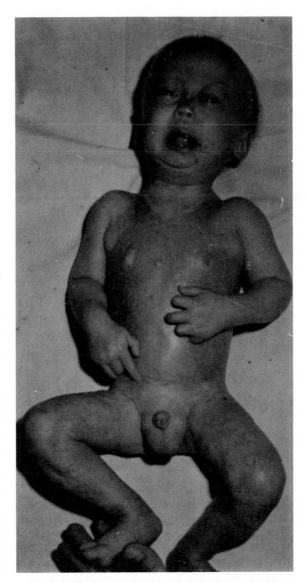

Fig. 1.—R. D. (Case Report). A 7-month-old boy with progeria. Even at this age a number of characteristic features were evident. There were striking alopecia and prominent veins. The eyes appeared to bulge. The mandible was hypoplastic.

scribed as sclerodermatous.[4] It really is not. Its thickened infiltrated appearance (Figs. 1 and 2) is virtually unique. Midfacial cyanosis also may be frequent early in life. These features and the glyphic nasal tip may permit an early diagnosis.[3]

Failure to thrive and shortness of stature are characteristic. The average birth weight of reported infants was 2,650 gm.[3] For the first few months they appear to be growing normally; then they develop a severe depression in the rate of growth. The deficit in weight may be more pronounced than that in height.[3] Bone age usually is equal to chronologic age. The ultimate height seldom exceeds that of a normal 5-year-old child.

Roentgenographic examination reveals a thin cranium. The sutures and anterior fontanel remain open longer than normal. Wormian bones and skull fractures are often seen. The distal ends of the clavicles become thin and are partially resorbed. There also is progressive loss of bone from the distal phalanges. Osteoporosis and thinness of the long bones is generalized. Coxa valga is regularly seen.

The joints appear large. They become stiff, and limitation of motion is progressive. The shoulders are narrow and the stance a wide, stooped, horseback-rider posture. The voice may be high pitched or piping. Intelligence is normal.

These patients age very rapidly and usually die in the first or second decade. Atherosclerotic changes become important in the first decade. Angina pectoris usually develops in association with atherosclerosis and may lead to death from coronary artery disease or a cerebrovascular accident. At autopsy it is common to see patchy and focal myocardial necrosis and fibrosis.

Some patients have been reported to have elevated levels of serum glycoproteins and cholesterol, whereas others have had normal concentrations. Goldstein[5] has reported that fibroblasts derived from a 9-year-old boy with progeria survived only to two subcultures, in contrast to the survival of fibroblasts derived from a normal control of the same age to 20–30 subcultures. Martin *et al.*[6] carried out a careful study of the numbers of doublings of fibroblasts in culture and found that a patient with progeria did not show such a strik-

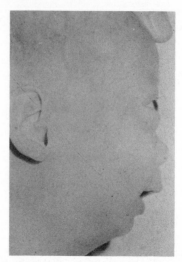

Fig. 3.—R. D. Six-week-old boy with progeria who presented because of stiffness of the skin. In retrospect, it was possible to see at this age some of the features illustrated at age 7 months in Figure 1.

ing impairment of growth potential. However, this patient ranked twenty-third of 26 in that age group, suggesting that the disorder diminished growth potential. Danes[7] found that fibroblasts from homozygous patients and their heterozygous parents showed decreased rates of cell growth in culture and reduced mitotic activity, DNA synthesis and cloning efficiency.

DIFFERENTIAL DIAGNOSES

The Cockayne syndrome[8] is similarly characterized by premature senility, which begins in infancy. Growth failure and loss of adipose tissue, as well as a light-sensitive skin, are seen. Mental retardation, deafness, retinosis pigmentosa and optic atrophy are characteristic, and cataracts develop later. Radiologic findings include intracranial calcifications, flattened vertebral bodies, thin, sclerotic ribs and clavicles and dense epiphyses. The skin is highly photosensitive. The Cockayne syndrome is transmitted via an autosomal recessive gene.

Fig. 2.—R. D. at 6 weeks of age. The marked thickening and stiffening of the skin is especially evident in the appearance of the nipple.

Fig. 4.—R. D. Roentgenogram of the chest at 12 months. The osteolysis of the lateral ends of the clavicles was evident in films taken at 6 weeks of age.

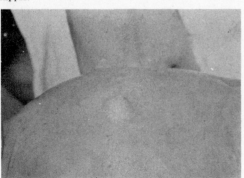

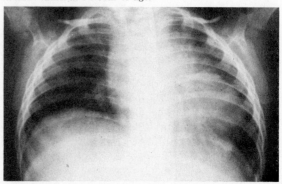

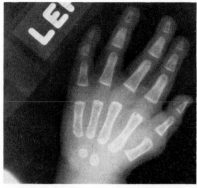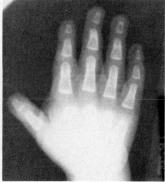

Fig. 5. – R. D. Roentgenograms of the hands. There was acroosteolysis of the phalanges, especially evident in the most distal.

The Werner syndrome[9] sometimes is thought of as an adult form of progeria. Features include short stature, loss of subcutaneous adipose tissue, premature graying, sparse hair and thin, dry, telangiectic skin with areas of hyper- and depigmentation on the extremities. Hyperkeratoses of the skin over bony prominences often become ulcerated. Juvenile cataracts and adult-type diabetes have been reported, as well as atherosclerosis with calcification. Ten per cent of these patients develop malignancies, especially sarcomas.[9]

The Rothmund-Thomson syndrome (see section on Dwarfism Syndromes with Premature Aging) also should be differentiated from progeria.

CASE REPORT

R. D., a 6-week-old white male, was the product of a normal pregnancy and delivery. The parents were normal, and he had three normal siblings. There was no history of consanguinity. The child was admitted to University Hospital because of progessive hardening of the skin, which had been noted after the first few days of life. He also had a history of poor feeding.

Physical examination revealed an irritable infant with sparse hair and prominent eyes (Fig. 3). There was craniofacial disproportion and micrognathia. The skin was stiff over the chest, abdomen, groin and upper thighs. The nipples were protuberant (see Fig. 2) but soft. The skin was thick, hard and transparent. Hip and knee joints appeared to be stiff because of stiffness of the skin.

Skin biopsy showed an increase in the amount of fibrous tissue. There were no significant changes in the amount of fat. Blood tests for glycoprotein and cholesterol were normal. Karyotype was normal.

A few months later the child was seen again. By this time he had complete alopecia, prominent veins over the skull and prominent eyes (see Fig. 1). The skin was still hard and thick but had in addition some areas of brown pigmentation.

Roentgenograms revealed resorption of the distal ends of the clavicles and the distal phalanges (Figs. 4 and 5).

Skin biopsy was done at 6 weeks of age and again at 9 months. The life span was approximately 9 subpassages, as compared to the normal adult range of 10–20 passages, or of newborn skin, which normally grows for 30 or more subpassages. Cloning efficiency was reduced.

GENETICS

The disorder is transmitted as an autosomal recessive. Two siblings have been reported with progeria who were the offspring of parents who were cousins.[10] Consanguinity has been present in three of 19 families in which this point was mentioned.[3] The incidence of the disorder in the United States appears to be one in 8,000,000 births.[3]

TREATMENT

None.

REFERENCES

1. Hutchinson, J.: Congenital absence of hair and mammary glands with atrophic condition of the skin and its appendages in a boy whose mother had been almost wholly bald from alopecia areata from the age of six, R. Med. Chir. Trans. 69:473, 1886.
2. Gilford, H.: Progeria: A form of senilism, Practitioner 73:188, 1904.
3. De Busk, F. L.: The Hutchinson-Gilford progeria syndrome, J. Pediatr. 80:697, 1972.
4. Feingold, M., and Kidd, R.: Progeria and scleroderma in infancy, Am. J. Dis. Child. 122:61, 1971.
5. Goldstein, S.: Lifespan of cultured cells in progeria, Lancet 1:424, 1969.
6. Martin, G. M., Sprague, C. A., and Epstein, C. J.: Replicative life-span of cultivated human cells. Effects of donor's age, tissue, and genotype, Lab. Invest. 23:86, 1970.
7. Danes, B. S.: Progeria: A cell culture study on aging, J. Clin. Invest. 50:2000, 1971.
8. Gamstorp, I.: Donohue's syndrome – leprechaunism – Cockayne's syndrome. A report of two patients and discussion of the relation between Donohue's syndrome and Cockayne's syndrome, Eur. Neurol. 7:26, 1972.
9. Epstein, C. J., Martin, G. M., Schultz, A. L., and Motulsky, A. G.: Werner's syndrome. A review of its symptomatology, natural history, pathologic features, genetics and relationship to the natural aging process, Medicine 45:177, 1966.
10. Mostafa, A. H., and Gabr, M.: Heredity in progeria with follow-up of two affected sisters, Arch. Pediatr. 71:163, 1954.

HALLERMANN-STREIFF SYNDROME

CARDINAL CLINICAL FEATURES

Congenital cataracts; microphthalmia; dwarfism; dental defects; a characteristic small face appearing prematurely aged with frontotemporal alopecia, hypotrichosis of eyelids and eyebrows, atrophy of the skin and a small, pinched nose.

CLINICAL PICTURE

The syndrome bears the names of Hallermann[1] and Streiff,[2] who reported patients with the syndrome and

Fig. 1.—D. A., a 4-year-old boy with the Hallermann-Streiff syndrome. The frontal alopecia, thin skin and prominent veins gave the face a prematurely aged appearance. The pinched nose is characteristic, as are blue sclerae and hypoplastic teeth. (Courtesy of Dr. Steven Dassel of the University of Washington, Seattle.)

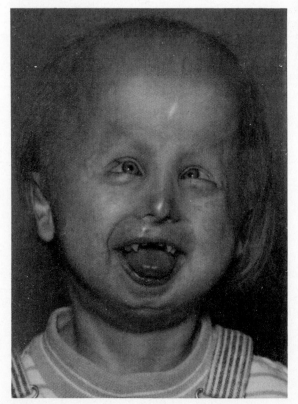

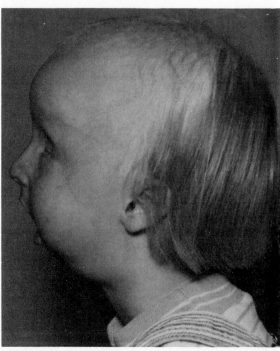

Fig. 2.—D. A. The lateral view illustrates the micrognathia and the characteristic distribution of hair posteriorly in the middle-aged male pattern with balding forehead and temples. (Courtesy of Dr. Steven Dassel of the University of Washington, Seattle.)

recognized it as a distinct disorder—distinct particularly from mandibulofacial dysostosis. The major features of the syndrome were set out by François[3] in 1958. He distinguished it clearly from progeria and anhidrotic ectodermal dysplasia. Patients with this disorder are quite tiny, but with body dimensions that are proportionate.

Patients also have a characteristic, rather aged facial appearance (Fig. 1). The skull is brachycephalic, and there is frontal and parietal bossing.[4] The hair is fine and sparse. Its growth appears to be restricted to the posterior part of the head, leaving a pronounced frontal and temporal alopecia (Figs. 2 and 3). There usually are no eyebrows and the eyelashes are short and sparse or absent.[5] The face is small and has been referred to as bird-like. In fact, Streiff[2] distinguished these patients from others who have been described as bird-headed, considering them to resemble more the

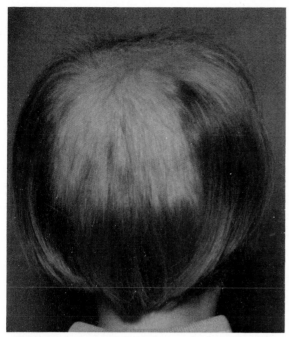

Fig. 3.—D. A. Occipital view illustrates the bald spot and the increased bitemporal diameter of the brachycephalic skull. (Courtesy of Dr. Steven Dassel of the University of Washington, Seattle.)

head of a parrot. We assume that he meant that the patient's nose suggested to him the beak of the bird. We tend to think of the face as pinched, especially the nose, which is small. This appearance results partly from malar hypoplasia and partly from a pronounced atrophy of the skin. In some patients it almost seems possible to see the cartilage of the tip of the nose through the translucent skin. Subcutaneous tissue is conspicuously absent. Mandibular hypoplasia also is striking and there may be microstomia and a narrow, high-arched palate. A variety of dental defects may be seen in these patients. Teeth may be present in the newborn infant. The deciduous and permanent teeth erupt irregularly and often late. They frequently are malformed or absent, especially the upper incisors. The teeth are hypoplastic. Malocclusion may be prominent.

Ocular anomalies are a regular feature of the syndrome.[6] Microphthalmia and bilateral congenital cataracts have been reported in most patients. Spontaneous rupture and resorption of the lens has been reported.[3, 4] The visual defect may be severe. Some of these patients are blind. Many are limited to light perception or counting of fingers.[7] Blue sclerae may be seen in this syndrome. In some patients there is an antimongoloid slant of the palpebral fissures, and there may be corneal bulging or buphthalmos. Glaucoma has been reported,[7] as has atrophy of the iris. Nystagmus may be a presenting symptom.[5]

Shortness of stature is characteristic of this syn-

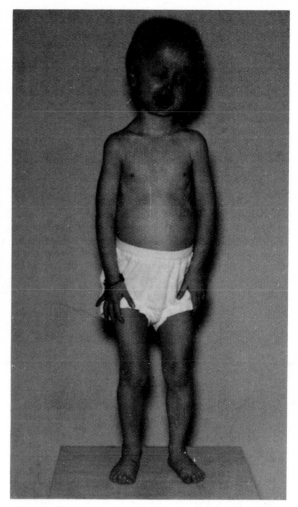

Fig. 4.—D. A. The patient had shortness of stature. (Courtesy of Dr. Steven Dassel of the University of Washington, Seattle.)

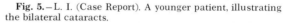

Fig. 5.—L. I. (Case Report). A younger patient, illustrating the bilateral cataracts.

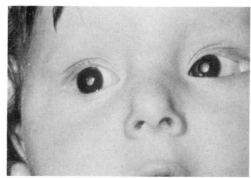

drome (Fig. 4). Most reported patients have been of normal intelligence, but a few have been severely retarded.

CASE REPORT

L. I. was a 14-month-old Mexican girl who was referred to University Hospital because of bilateral cataracts (Fig. 5). The cataracts had been present from birth. She had sat at 6 months and was walking.

Examination revealed shortness of stature and a body weight consistent with that of a 7-month-old infant. The height age was 7 months. The head circumference was 44 cm. She was virtually blind, but she did see light and moving objects. There was an intermittent searching nystagmus. The sclerae were blue. Bilateral cataracts were prominent. There was frontal bossing and a thin, pinched, or beaked nose. There was a high-arched palate, micrognathia and no teeth. The bone age was within normal limits.

GENETICS

The cause of this syndrome is not known. Most cases are sporadic. It has been suggested that the disease is determined by a dominant gene, which has appeared in most instances as a result of new mutation. The syndrome has been reported in a father and daughter.[9]

TREATMENT

Recognition of the syndrome serves to alert the physician to the possibility of congenital cataracts. It may be possible in some patients to avert blindness by prompt treatment of cataracts or glaucoma.

REFERENCES

1. Hallermann, W.: Vogelisicht und Cataracta congenita, Klin. Mbl. Augenheilkd. 113:315, 1948.
2. Streiff, E. G.: Dysmorphie mandibule-faciale (tête d'oiseau) et altérations oculaires, Ophthalmologica 120:79, 1950.
3. François, J.: A new syndrome. Dyscephalia with bird face and dental anomalies, nanism, hypotrichosis, cutaneous atrophy, microphthalmia and congenital cataract, Arch. Ophthalmol. 60:842, 1958.
4. Falls, H. F., and Schull, W. J.: Hallermann-Streiff syndrome. A dyscephaly with congenital cataract and hypotrichosis, Arch. Ophthalmol. 63:409, 1960.
5. Hoefnagel, D., and Benirschke, K.: Dyscephalia mandibulo-oculo-facialis (Hallermann-Streiff syndrome), Arch. Dis. Child. 40:57, 1965.
6. Ide, C. H., and Webb, R. W.: Hallermann-Streiff syndrome, Am. J. Ophthalmol. 67:151, 1969.
7. Hopkins, D. J., and Horam, E. C.: Glaucoma in the Hallermann-Streiff syndrome, Br. J. Ophthalmol. 54:416, 1970.
8. Hutchinson, D.: Oral manifestations of oculomandibulodyscephaly with hypotrichosis (Hallermann-Streiff syndrome), Oral Surg. 31:234, 1971.
9. Guyard, M., Perdriel, G., and Ceruti, F.: Sur deux cas de syndrome dyscéphalique à tête d'oiseau, Bull. Soc. Ophthalmol. Fr. 62:443, 1962.

ROTHMUND-THOMSON SYNDROME
Poikiloderma Congenita

Cardinal Clinical Features

Peculiar marmorization of the skin, irregular erythema, telangiectasia, dermal atrophy, sparse hair or alopecia, dystrophic nails, shortness of stature, hypogonadism, juvenile cataracts.

Clinical Picture

Poikiloderma congenita, or the Rothmund-Thomson syndrome, was first described by Rothmund in 1868.[1] He described eight children in three families in whom cataracts were associated with a peculiar degeneration of the skin. These families lived in an isolated area of the Bavarian Alps where intermarriage was common. In 1923 Thomson described one, and later two more, children with unusual skin changes and skeletal abnormalities.[2] It has since been apparent that these authors were describing the same syndrome, and it has come to be known by both of their names.[3]

This syndrome is characterized by peculiar marmorization of the skin.[1] The skin usually is normal at birth, but by 3–6 months of age the skin of the face becomes involved. The cheeks and ears become diffusely red or bright vermilion. Later the macular, reticulated appearance develops. The skin lesions spread to the extremities, buttocks and knees. The trunk may be spared, at least early. Involved areas may be swollen, shiny or tense, but soon develop punctate, linear or reticulated areas of atrophy of the skin. The fully developed picture is of alternating erythema and atrophy. Telangiectases appear, and there are irregular mottled areas of brownish pigmentation and depigmentation, along with striae. These colorations together are skin changes that are described as marble-like and are found in all cases. Photosensitivity is seen in some of these patients. The skin comes to look xerodermatous. In adults squamous cell carcinomas are seen.[4, 5] Histologically the skin shows hyperkeratosis, slight edema and acanthosis, along with areas of atrophy, loss of papillae and hyperpigmentation.

The hair is sparse, and in some patients there is total alopecia, with no eyelashes, eyebrows or scalp hair. Lanugo hair usually is also absent in the affected areas.[3] Nail dystrophy has been observed in about one fourth of the patients. Hyperkeratosis of the palms and soles is seen in some.

Juvenile cataracts have been reported in half of the patients with the Rothmund-Thomson syndrome. It is possible that some cataracts have been missed, especially in earlier years before the full definition of the syndrome was appreciated. Cataracts usually appear between 3 and 6 years of age, but they have been observed as early as 4 months of age.[6] Occasionally they have developed in the third or fourth decade of life. These cataracts may produce a complete loss of vision. Other ocular anomalies include corneal degeneration.

Shortness of stature[7] is seen in about half of the patients. Some are short enough to have been classified as dwarfs. The hands may be small, with short fingers and especially short terminal phalanges. Bony abnormalities of some sort are seen in about one third of patients. These include shortening of the long bones and even absence of some, as well as areas of sclerotic or cystic changes. The forearms frequently have been involved. Some patients have had microcephaly. Cleft hand, cleft foot and syndactyly have been observed, as has absence of the thumb.[8]

Dental abnormalities may include microdontia, delayed eruption or congenitally missing teeth.[9] The size and morphology of the teeth may be bizarre. The crowns of the teeth are abnormal and show an unusual wear pattern.

Hypogonadism has been reported in one fourth of the patients. It has been observed in both sexes.[10] Involved patients may have undescended testes, micropenis, amenorrhea or sterility.

Differential Diagnosis

The Rothmund-Thomson syndrome is unique. Once recognized, it should not be confused with other disorders. However, it may be useful to give a brief differential diagnosis of oculocutaneous disorders. The Werner syndrome is one of the syndromes of premature aging. It usually appears in the second decade of life and is characterized by shortness of stature; graying of hair; early adult onset of cataracts; retinal degeneration; calcification of the tendons; and waxy, hyperkeratotic, atrophic skin changes. These patients often have premature atherosclerosis and late-onset diabetes. They have a propensity toward malignancies, particularly sarcomas. Dyskeratosis congenita is characterized by cutaneous atrophy and pigmenta-

Fig. 1.—G. H. (Case Report). This boy with the Rothmund-Thomson syndrome had a generalized dermatosis (worse on the face), perineum and exposed surfaces of the extremities. He also had marked shortness of stature.

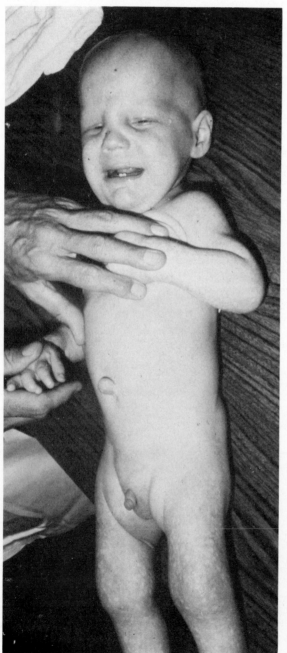

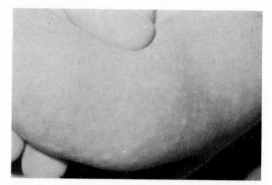

Fig. 2.—G. H. Close-up of the arm reveals the characteristic marble character of the skin lesions. This appearance seems to result from alternating patches of erythema and atrophy.

tion, dystrophy of the nails and teeth, shortness of stature, leukoplakia of the oral mucosa, atresia of the lacrimal duct, pancytopenia and testicular atrophy. The Hallermann-Streiff syndrome has a characteristic facies with a beaked, pinched nose, marked hypotrichosis, shortness of stature and congenital cataracts. Most patients are mentally retarded. The Cockayne syndrome is characterized by premature aging and dwarf-

Fig. 3.—G. H. There was alopecia of the eyebrows and lashes as well as of the scalp. Scabs over the forehead and lower face following scratching are typical.

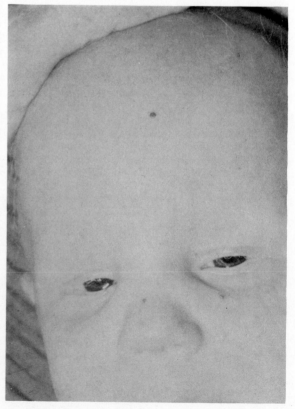

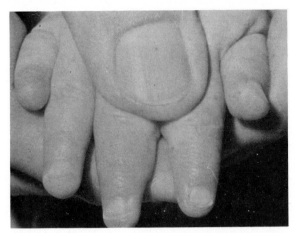

Fig. 4.–G. H. The nails were thin and flat and broke easily.

ism, leading to a wizened appearance. These patients have a lupus-like rash in the butterfly area, photosensitivity and dermal pigmentation and atrophy. They also have optic atrophy, pigmentary retinal changes and deafness. They often die from early atherosclerosis. Other oculocutaneous disorders that can be differentiated from the Rothmund-Thomson syndrome are keratosis follicularis spinulosa decalvans (see section on Neurocutaneous Disorders); the Shafer syndrome, in which there are cataracts; follicular hyperkeratosis; focal alopecia and thick nails; and ichthyosis with cataracts.

Case Report

G. H., a 13-month-old white male, was born 2 weeks early after a normal pregnancy and delivery. Birth weight was only 3 lb, 7 oz. His length was 16 inches. He was the first child of normal parents; there was no consanguinity. He appeared to be healthy and had a normal skin. There was no history of hypoglycemia during the neonatal period. He was seen at 2 months for evaluation of microphallus and undescended testes.

At the age of 6 months he developed skin lesions. This manifestation started as a mottled erythematous rash over the cheeks and then spread to involve the entire body. The buttocks and extremities were especially affected. The rash was vermilion. Telangiectases appeared, as well as linear reticulated atrophy.

The child had four teeth at 10 months. He had no eyebrows or eyelashes and had lost the scalp hair he had had at 6 months of age. He gained weight slowly. He had repeated ear infections and diarrhea.

On physical examination at 13 months of age he weighed 12 lb, 14 oz, and his length was 25 inches. Height age was 4 months. He had total alopecia. There was a mottled erythematous rash, telangiectasia and linear atrophy, as well as areas of

pigmentation and depigmentation (Figs. 1–3). The nails were thin and flat and broke easily (Fig. 4). He had small hands and short fingers. He still had only four teeth. Careful examination of the eye revealed no cataracts. His penis was small and there were small testes in the canals.

Genetics

The Rothmund-Thomson syndrome is transmitted as an autosomal recessive. Many of the patients reported were offspring of consanguineous marriages or had affected siblings. The genetics of the original three families has been carefully summarized.[1, 3, 7] There was extensive consanguinity. In reported patients there has been a female predominance of 70%.[3]

Treatment and Prognosis

There is no effective treatment for the skin lesions. Surgical treatment of cataracts may be important for the preservation of vision. Squamous cell carcinoma should of course be treated surgically. The patient with the Rothmund-Thomson syndrome should have a normal life span.

REFERENCES

1. Rothmund, A.: Ueber Cataractern in Verbindung mit einer eigenthümlichen Hautdegeneration, Graefes Arch. Ophthalmol. 14:159, 1868.
2. Thomson, M. S.: Poikiloderma congenitale, Br. J. Dermatol. Syph. 47:221, 1936.
3. Taylor, W. B.: Rothmund's syndrome–Thomson's syndrome, congenital poikiloderma with or without juvenile cataracts, a review of the literature, report of a case, and discussion of the relationship of the two syndromes, Arch. Dermatol. 75:236, 1957.
4. Rook, A., and Whimster, I.: Congenital cutaneous dystrophy (Thomson's type), Br. J. Dermatol. Syph. 61:197, 1949.
5. Sexton, G. B.: Thomson's syndrome (Poikiloderma congenitale), Can. Med. Assoc. J. 70:662, 1954.
6. Silver, H. K.: Rothmund-Thomson syndrome: An oculocutaneous disorder, Am. J. Dis. Child. 111:182, 1966.
7. Seefleder, R.: Über familiäres Auftreten von Katarakt und Poikilodermie (Poikilodermia vascularis atrophicans Jakobi), Z. Augenheilkd. 86:81, 1935.
8. Blinstrub, R. S., Lehman, R., and Sternberg, T. H.: Poikiloderma congenitale, report of two cases, Arch. Dermatol. 89:659, 1964.
9. Kraus, B. S., Gottlieb, M. A., and Meliton, H. R.: The dentition in Rothmund's syndrome, J. Am. Dent. Assoc. 81:894, 1970.
10. Bloch, B., and Stauffer, H.: Skin diseases of endocrine origin (dyshormonal dermatoses), Arch. Dermatol. 19:22, 1929.

Neurocutaneous Disorders

INCONTINENTIA PIGMENTI

CARDINAL CLINICAL FEATURES

Skin lesions that occur in different stages, first vesicular, later verrucous, and ultimately hyperpigmented and arranged in streaks and whorls; a variety of other abnormalities, including delayed or abnormal dentition, alopecia, skeletal abnormalities, mental retardation; occurrence almost exclusively in females.

CLINICAL PICTURE

Incontinentia pigmenti is a rare disease, characterized by its skin lesions. It was first described by Garrod[1] in 1906. It is sometimes known as the Bloch-Sulzberger syndrome because of classic descriptions of the same patient by Bloch and by Sulzberger in 1926 and 1927. Typical papules, vesicles or bullae are found in most patients at birth or develop soon thereafter (Figs. 1 and 2). There is a distinct tendency for the lesions to present in a linear pattern. Eosinophils may be seen on smears of the lesions. Examination of the blood usually reveals an eosinophilia that may be as high as 50% early in life.

Verrucous lesions follow the vesicular eruption. The stages are not clear-cut or mutually exclusive, but in general the vesicular pattern characterizes the first 4 months, and the verrucous lesions may be seen under 1 year. Hyperpigmentation may be seen as early as 1 month. It usually persists into adulthood. The bizarre whorled pattern of hyperpigmentation resembles veined marble (Figs. 3 and 4).

Ocular abnormalities are seen in about a third of patients. About a third of these have a mass in the posterior chamber resembling retrolental fibroplasia. Recognition of the syndrome can spare the patient an enucleation for suspected retinoblastoma. Other ocular lesions include congenital corneal opacities, cataracts, iris malformations, optic atrophy, microphthalmia, nystagmus and strabismus. In one patient reported, a 13-year-old girl with right hemiatrophy, there were clusters of dark brown pigmented lesions on the retina and conjunctiva.[2] Figure 3 shows a patient with facial asymmetry.

A variety of cerebral manifestations also have been observed in this syndrome.[3, 4] These problems have been reported in as high as 30% of patients with incontinentia pigmenti. Convulsive seizures, microcephaly, spastic diplegias and mental retardation are relatively common. Pneumoencephalography has been carried out in a few patients and has revealed cerebral atrophy. Progressive cerebellar ataxia and macular degeneration have been observed in three daughters of a consanguineous marriage who had incontinentia pigmenti.

The histopathology of the skin is that of blisters and hyperkeratosis, with infiltration of well-formed eosinophilic cells.[5] Biopsy of the hyperpigmented lesions shows keratotic whorls. There are large amounts of

Fig. 1.—D. K. (Case Report). Typical appearance of the vesicular lesions of incontinentia pigmenti in a young infant. The entire skin may be involved. A tendency to a linear pattern is evident.

Fig. 2.—Close-up of an extremity. The vesicular lesions are linear and confluent. Crusting occurs with the older lesions.

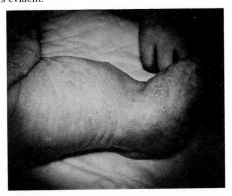

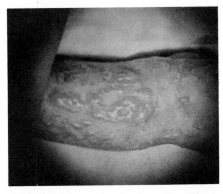

Fig. 3. – Older girl with incontinentia pigmenti. Facial asymmetry is evident. She had hemiatrophy involving all of one side of her body.

melanin synthesized in the basal melanocytes. It has been thought that these melanocytes were incontinent of their abundant product.

CASE REPORT

D. K. was a 17-day-old white female who was brought to University Hospital for an eruption that had been present

Fig. 4. – Feet of the patient seen in Figure 3. The pigmented lesions represent the typical veined marble appearance of the older patient with incontinentia pigmenti.

since birth. She was the product of normal pregnancy and full-term delivery and weighed 5 lb, 8 oz at birth. The mother had had no previous pregnancies.

On physical examination the infant was found to have papules and vesicles or blisters distributed over the upper and lower extremities (see Figs. 1 and 2). There was a distinct linear character to the clusters of lesions. There were no ocular abnormalities. Neurologic examination was negative.

She had a pronounced eosinophilia of 48%. The peripheral leukocyte count was 27,000/cu mm. A skin biopsy showed mild keratinization with infiltration of eosinophils.

GENETICS

The genetics of this condition are not established. It has been thought to be an X-linked disorder.[6] Virtually all surviving patients with the disease have been female: among 216 cases reported, 210 were female.[6] More than one patient has been reported in at least 22 families, and the disorder has been observed in single ovum twins. Transmission has been observed from involved mother to daughter. These findings have suggested that the gene is a sex-linked dominant in which there is lethality for the hemizygous male. It is consistent with this hypothesis that spontaneous abortion has been common in families in which there has been a child with incontinentia pigmenti, and the number of healthy male and female siblings has been about equal.

TREATMENT

None.

REFERENCES

1. Garrod, A. E.: Peculiar pigmentation of the skin in an infant, Trans. Clin. Soc. Lond. 39:216, 1906.
2. McCrany, J. A., and Smith, L.: Conjunctival and retinal incontinentia pigmenti, Arch. Ophthalmol. 79:417, 1968.
3. Reed, W. G., Carter, C., and Cohen, T. M.: Incontinentia pigmenti, Dermatologica 134:243, 1967.
4. O'Doherty, N. J., and Norman, R. M.: Incontinentia pigmenti (Bloch-Sulzberger syndrome) with cerebral malformation, Dev. Med. Child Neurol. 10:168, 1968.
5. Jackson, R., and Nigam, S.: Incontinentia pigmenti: A report of three cases in one family, Pediatrics 30:433, 1962.
6. Lenz, V. W.: Zur Genetik der Incontinentia pigmenti, Ann. Paediatr. 196:149, 1961.

KERATOSIS FOLLICULARIS SPINULOSA DECALVANS
Oculocerebral Syndrome with Aminoaciduria and Keratosis Follicularis

Cardinal Clinical Features

Keratosis follicularis, absence of lateral portions of the eyebrows, sparseness of eyelashes, localized alopecia of the scalp, corneal clouding; congenital glaucoma and lenticular cataract. A reported patient also had unusual hands and dermatoglyphs and aminoaciduria.

Clinical Picture

Keratosis follicularis spinulosa decalvans is an unusual ectodermal disorder involving the skin and eye. It may also involve the central nervous system.

The skin lesions are those of keratosis follicularis in which the cutaneous hair follicles have characteristic spiny elevations. The lesions are especially distributed over the extensor surfaces, and they show follicular widening on section. They are plugged with keratin. These follicular plugs are composed of keratin, which, though markedly increased in amount, is normal in appearance, even on electron microscopy.[1] The lesions are associated with alopecia of lanugo hair in the involved areas. In addition, there is particularly striking alopecia of the eyebrows in which the lateral portions usually are absent. The eyelashes are always sparse. Various forms of alopecia, usually localized, may develop on the head, especially under the occipital protuberances in older patients. There may be alopecia of the beard, axillary or pubic hair.

There are a variety of ocular abnormalities. Most patients have photophobia, lacrimation and corneal lesions. The latter include clouding, asymmetry, vascularization and an appearance of keratosis or degeneration. Associated abnormalities include blepharitis and ectropion. Many patients develop ruddy telangiectatic changes of the cheeks. Hypoplasia of the mandible has been seen in a few patients.[2, 3]

Our patient (Case Report) had in addition a lenticular cataract and nonprogressive congenital glaucoma.

He also had mental retardation. His fingers were delicate and tapered, giving an appearance of arachnodactyly without unusual length. He also had a distinctly increased space between the second and third fingers, and there were extra skin folds on the palmar surfaces of the digits. He had unilateral cryptorchidism. There was a generalized aminoaciduria of modest degree without glycosuria or acidosis.

Case Report

A. J. was a black boy who weighed 3 lb, 6 oz at birth. A cataract was observed in the left eye. In addition, he had congenital glaucoma; two goniotomies were performed prior to 1 month of age.

He was hospitalized at 7 months of age for pneumonia. He had two brief seizures associated with varicella at 18 months. He rolled over at 9 months, sat without support at 10 months, stood with support at 2 years and walked at 3 years of age. He had not been toilet trained by age 4 years. He had never spoken an intelligible word. Drooling was a continuous problem.

At 3½ years of age the left cornea measured 11–12 mm horizontally and 11 mm vertically. The right cornea was clear and measured 10 mm horizontally and 9 mm vertically. Tension was normal. A left lens aspiration was carried out because of the cataract.

Family history was noncontributory.

At 4 years of age, physical examination revealed a small, retarded boy with a striking facial appearance. The eyelashes and eyebrows were extremely sparse, and there was virtually no lateral half of each eyebrow (Fig. 1). His height of 39 inches was below the tenth percentile for that age, and his weight of 27½ lb was below the third percentile for that age. Over-all proportions were normal. Head circumference of 18¾ inches was in proportion to the height. Bilateral blepharitis was present. The left globe appeared larger than the right, and there was a persistent lateral nystagmus. The left cornea was larger than the right and cloudy; the cataract was clearly visible in this eye (see Fig. 1).

The skin was unusual in that hard, coarse, well-defined hyperkeratotic spicules were present in the region of the hair follicles (Fig. 2). These lesions were hyperpigmented and occurred over the arms, legs, buttocks and dorsal thorax. The

211

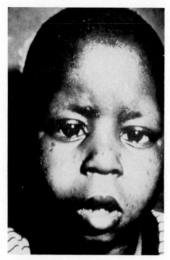

Fig. 1.—A. J. (Case Report). View of the face illustrating alopecia of the lateral portions of the eyebrows. There was a cataract in the left lens.

teeth were small and carious. The palatal arch was accentuated and the mandible somewhat small. A grade II systolic murmur was present over the left sternal border. The left testicle was not palpable.

The hands presented a rather peculiar appearance (Fig. 3). The fingers appeared elongated because they were both laterally and anteroposteriorly thin and markedly tapered. The space between the second and third fingers was accentuated. Extra skin folds could be seen on the digits and are illustrated in the dermatoglyphic pattern (Fig. 4). The nails were thin and narrowed. Deep tendon reflexes were 2+ and 3+ throughout. The patient had a widely spaced toddler gait, but cerebellar function appeared to be normal.

Repeated attempts at eliciting speech-related movements of the tongue and labial musculature failed. Estimation of intellectual function with the Denver Development Scale was between 18 and 20 months, so that the developmental quotient approximated 40.

Skin biopsy provided evidence of marked hyperkeratosis, with follicular widening and plugging. Electron microscopic

Fig. 2.—A. J. Extensor surface of the forearm showing keratotic follicular lesions.

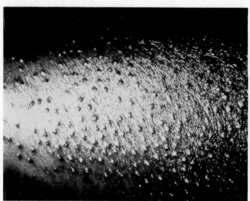

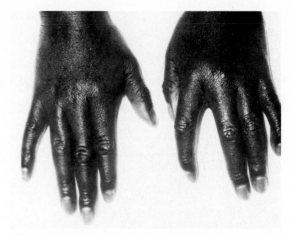

Fig. 3.—A. J. Dorsal view of the hands. The digits were elongated and tapered.

examination of sections of skin showed similar findings.

Roentgenographic examination of the skull showed asymmetry of the orbits. The mandible appeared infantile. There was a prominent aorta. The intravenous pyelogram and bone age were normal. All of the long bones appeared rather long, thin and delicate, especially in the hands.

Testing of the hair for fluorescence with acridine orange was negative. The EEG was normal. The ECG showed a sinus tachycardia and deep Q waves in leads V_5 and V_6, suggesting minimal left ventricular hypertrophy. Screening of the urine for metabolic disease suggested a generalized aminoaciduria. This was confirmed by quantitative chromatography on the automatic amino acid analyzer.

GENETICS

Keratosis follicularis spinulosa decalvans is transmitted as an X-linked trait. Two large kindreds have been reported from the Netherlands [2-4] and from Germany,[5] and a smaller one from Switzerland.[6] There is

Fig. 4.—A. J. Dermatoglyphs. Accessory transverse lines on the fingers are illustrated by the *dotted lines*.

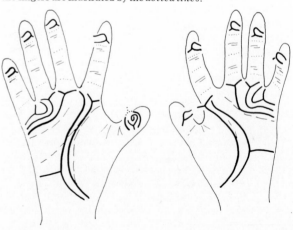

some information on a large family in the United States.[7] Milder abnormalities have been seen in some of the heterozygous females. The absence of male to male transmission has been well documented in two large kindreds.[5]

TREATMENT

Ocular abnormalities should be managed in conventional fashion. Otherwise no treatment is known.

REFERENCES

1. Adler, R. C., and Nyhan, W. L.: An oculocerebral syndrome with aminoaciduria and keratosis follicularis, J. Pediatr. 75:436, 1969.
2. Lameris, H. J.: Ichthyosis follicularis, Ned. Tijdschr. Geneeskd. 41:1524, 1905.
3. Rochat, G. F.: Familiare cornea degeneratie, Ned. Tijdschr. Geneeskd. 42:515, 1906.
4. Holthuis, P.: Keratosis follicularis spinulosa decalvans (Siemens), Ned. Tijdschr. Geneeskd. 87:1825, 1943.
5. Siemens, H. W.: Ueber einen, in der menschlichen Pathologie noch nicht beobachteten Vererbungsmodus: Dominant-geschlechtsgebundene Vererbung, Arch. Rass. Gesell. Biol. 17:47, 1925.
6. Franceschetti, P. A., Jaccottet, M., and Jadassohn, W.: Manifestations cornéennes dans la keratosis follicularis spinulosa decalvans (Siemens), Ophthalmologica 133:259, 1957.
7. Falls, H. F.: Clinical detection of the genetic carrier state in ophthalmic pathology, Am. Tr. Ophthalmol. 37:841, 1954.

McCUNE-ALBRIGHT SYNDROME
Polyostotic Fibrous Dysplasia

CARDINAL CLINICAL FEATURES

Polyostotic fibrous dysplasia, pathologic fractures, café-au-lait spots, sexual precocity in the female.

CLINICAL PICTURE

The syndrome was first described by McCune[1] in 1936 and by Albright[2] in 1937. A number of names were used in the early descriptions of the disease.[2, 3] The term polyostotic fibrous dysplasia was proposed by Lichtenstein in 1938[4] and appears to evoke a picture of the pathology. The disease typically involves multiple bony sites. There may be some tendency to laterality, but bony involvement is virtually random. The histologic appearance of the lesions is of small, poorly calcified bony trabeculae dispersed irregularly within a matrix of loosely arranged fibrous tissue of uniform appearance, which replaces the bony cortex and the medullary space. In the relatively vascular stroma, fibroblasts are abundant, arranged in whorls or bundles.

Girls with this syndrome may be first brought to the attention of a physician because of sexual precocity or vaginal bleeding. Patients of either sex may present with bone pain, pathologic fractures or bony deformity.

The large hyperpigmented areas of the skin usually are present by the time the patient is seen. The pigmentation often is apparent at birth, but it may develop in early childhood. In McCune's original patient it appeared in the second year of life.[1] These are café-au-lait spots. Over the years the pigment may become a darker brown.[5] They are patchy, irregularly distributed and usually quite large. Their characteristic jagged margins have led to their designation as "coast of Maine" spots (Fig. 1). This is in contradistinction to the smoother-bordered, so-called "coast of California" spots characteristic of neurofibromatosis. Actually, spots of either configuration may be found in either condition, and the two conditions may be indistinguishable on the basis of the pigmented spots.[5] However, it usually is possible to make the distinction clinically. Axillary freckles appear to be unique for neurofibromatosis. In this syndrome the macular skin lesions commonly occur over the base of the neck or in the lower spinal area, but they can occur anywhere. They are often on one side or the other of the midline. An entire side of the body may be involved, and there is a tendency for pigmentation to be more prominent on the side of major osseous involvement.[5] Areas of pigmentation have been noted on the palate[6, 7] and in the mucosa just inside the lip. Electron microscopy of the spots reveals melanin hyperpigmentation of all the layers of the epidermis without change in the numbers of melanocytes.[5, 8]

Any of the bones of the body may be affected with the fibrous dysplasia that is the most serious problem

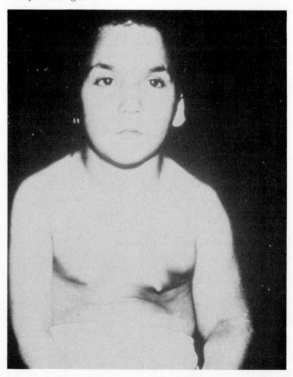

Fig. 1.—An 8-year-old girl with the McCune-Albright syndrome. She had two café-au-lait spots, one on the neck and one on the lower left rib cage. The latter is an excellent example of the "coast of Maine" spot, with its irregular, jagged outlines and many islands. The patient also had enlargement of the breast on one side. She had begun to have menstrual periods at 3 years of age.

in this syndrome. Bony symptoms may develop in childhood or not until adolescence.[4] The proximal femur and pelvis are characteristically involved. This leads to bowing and deformity of the legs and abnormalities of gait. Patients may limp or complain of pain in the hip. Shortening of an extremity may be the chief complaint. Pathologic fractures may result from minor trauma. When the skull and facial bones are involved, there may be asymmetry and disfiguring changes. Thickening and sclerosis of the base of the skull may lead to cranial nerve compression. Fibrous dysplasia localized in the temporal bone has been reported, with occlusion of the external auditory canal and consequent conductive deafness.[9] Similarly, involvement of the foramina of the optic nerve may lead to impairment of vision.[10] Unilateral exophthalmos has been reported.[11]

Roentgenographic characteristics are those of multiple cystic lesions of the bones.[12] They may be located in the shafts of the long bones, the flat bones or the tubular bones of hands and wrist. The lower extremities are particularly commonly involved. These lytic lesions are associated with expansion of the shafts of the bones as well as elongation; the overlying bony cortex is very thin.[12] Deformity of the femora assumes a rounded shepherd's crook appearance. The skull particularly may show thickening and sclerosis reminiscent of the Paget disease, as well as cystic lesions. Productive changes may be seen elsewhere in the skeleton. In females there is accelerated bony maturation.

Isosexual precocity is characteristic of the female with this disorder. Menstruation may occur in infancy or in early childhood. It precedes the development of breast tissue, axillary and pubic hair by an interval of months or years. Patients proceed to complete development of secondary sex characteristics, with rapid growth and bony maturation. They start growing ear-

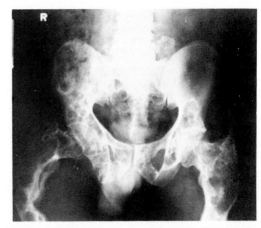

Fig. 3.—C. M. Roentgenogram of the pelvis. This is a classic example of polyostotic fibrous dysplasia of the bone. The multiple, irregular radiolucencies were present in both femora and throughout the pelvis. The involved bone was extremely thin, misshapen and easily fractured.

ly, but short stature is not a constant finding. The bleeding usually is anovulatory, as indicated by the lack of mature follicles or corpora lutea in ovarian biopsies.[13] Levels of follicle-stimulating hormone in the urine were normal in two patients,[13] but elevated in a 3-year-old girl.[14] Urinary estriol has been reported to be markedly elevated,[15] and in this patient there were peaks of luteinizing hormone excretion like that seen in adults at the time of ovulation. Ovulation occurs when patients reach normal pubertal age. Fertility appears to be normal.

Fig. 4.—C. M. The femora were involved throughout. The left femur illustrates the characteristic shepherd's crook appearance.

Fig. 2.—C. M. (Case Report). Irregular café-au-lait spots in 22-year-old man with the McCune-Albright syndrome.

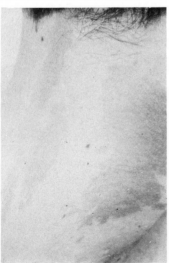

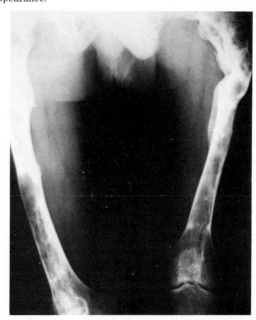

The occurrence of vaginal bleeding and other signs of precocity may antedate the presence of bone lesions for a number of years. Female patients with precocious menstruation should be followed for the possible occurrence of bony lesions. Some of these patients may have no skin pigmentation.[13]

In males puberty usually occurs at the normal age. One boy with polyostotic fibrous dysplasia has been described[16] who developed sexual precocity at 6 years of age. Biopsy of the testis revealed seminiferous tubules with numerous spermatozoa, as in the adult. Advanced bone age also has been reported in a boy with no signs of precocious puberty.[2] Hyperthyroidism has been reported in a number of patients with this disease.[1, 13, 17, 18] Multiple intramuscular myxomas have been reported.[19] These tumors are benign. Malignant transformation of an area of fibrous dysplasia into an osteosarcoma has been reported.[20-22] Most patients have been of normal intelligence, but mental retardation also has been reported.[1]

DIFFERENTIAL DIAGNOSIS

Polyostotic fibrous dysplasia can readily be differentiated from hyperparathyroidism. The concentrations of calcium and phosphorus in the serum are normal, and uninvolved areas of bone are of normal appearance. The level of alkaline phosphatase activity in the blood often is elevated. This condition has been confused with neurofibromatosis. These patients do not develop tumors of the skin or of peripheral nerves, and the café-au-lait spots in the two conditions are often distinctive.

CASE REPORT

C. M., a 24-year-old man admitted because of a fracture of the right femur, had been known to have polyostotic fibrous dysplasia since the age of 5 years. At that time he presented with a limp, and a skeletal survey showed multiple cystic lesions. Since then he had had numerous fractures. Both legs had previously been fractured. The patient also had had seizures from the age of 7. Treatment with phenobarbital and Dilatin had been discontinued a few years prior to admission. The patient underwent puberty normally at 14 years of age. There was no history of a similar condition in the family.

Physical examination showed the characteristic café-au-lait spots over the base of the spine and the back of the neck (Fig. 2). These brown pigmented areas were patchy and irregularly shaped and more prominent on the right. There was bowing of the lateral aspects of the femora.

Roentgenographic survey of the skeleton (Figs. 3–6) re-

Fig. 5.—C. M. Roentgenogram of the lower legs. Each tibia and fibula was involved. There was considerable asymmetry. There was an enormous fibrous lesion in the fibula on the right. Angulations and curvatures were present in all four bones.

Fig. 6.—C. M. Roentgenogram of the humerus. It is characteristic of this disease to involve the shafts of tubular bones, and in this patient all of the tubular bones were involved. The humerus was lucent with numerous lesions and was poorly modeled.

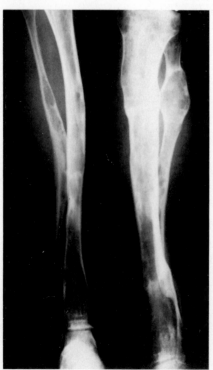

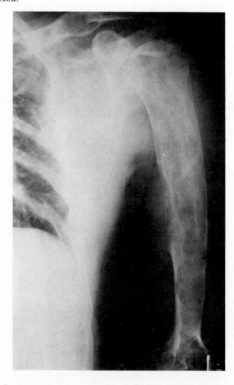

vealed extensive, expansive lytic lesions involving the right clavicle, nearly all of the right ribs and several of the left ribs. A process both lytic and sclerotic involved the right humeral head and the right scapula, and there was a pathologic fracture of the right humerus (see Fig. 6). There was a midshaft fracture of the right femur, with posterior and lateral displacement and angulation of the distal fragment. There were lesions in the proximal right radius and ulna, and a fracture through the radial head. X-rays of the skull showed sclerosis of the petrous bone and involvement of the sphenoid bone.

The concentration of calcium in the serum was 9.4 mg/100 ml, and of phosphorus 3.3. Alkaline phosphatase activity was 13.8 Bodansky units. The femur was treated by open reduction with insertion of a 10-mm rod and a bony graft. During this admission the patient had a generalized seizure, which resulted in the fracture of the right humerus.

GENETICS

The etiology of this syndrome is unknown. Most cases have been sporadic. There is no report of cases in siblings.

TREATMENT

The treatment program should involve early the care of an orthopedic surgeon. Treatment of fractures and prevention of deformities are major problems in these patients. Plastic surgery may be indicated in lesions on or around the face.

REFERENCES

1. McCune, D. J.: Osteitis fibrosa cystica; the case of a nine year old girl who also exhibits precocious puberty, multiple pigmentation of the skin and hyperthyroidism, Am. J. Dis. Child. 52:743, 1936.
2. Albright, F., Butler, A. M., Hampton, A. O., and Smith, P.: Syndrome characterized by osteitis fibrosa disseminata, areas of pigmentation and endocrine dysfunction, with precocious puberty in females. Report of five cases, N. Engl. J. Med. 216:727, 1937.
3. McCune, D. J., and Bruch, H.: Osteodystrophia fibrosa. Report of a case in which the condition was combined with precocious puberty, pathologic pigmentation of the skin and hyperthyroidism, with a review of the literature, Am. J. Dis. Child. 54:806, 1937.
4. Lichtenstein, L.: Polyostotic fibrous dysplasia, Arch. Surg. 36:874, 1938.
5. Benedict, P. H., Szabo, G., Fitzpatrick, T. B., and Sinesi, S. J.: Melanotic macules in Albright's syndrome and in neurofibromatosis, J.A.M.A. 205:618, 1968.
6. Bowerman, J. E.: Polyostotic fibrous dysplasia with oral melanotic pigmentation, Br. J. Oral Surg. 6:188, 1969.
7. Gorlin, R. J., and Pindborg, J. T.: Polyostotic Fibrous Dysplasia, Cutaneous Pigmentation, and Endocrine Disorders, in *Syndromes of the Head and Neck* (New York: McGraw-Hill Book Co., 1964), pp. 461–467.
8. Frenk, E.: Etude ultrastructurale des taches pigmentaires du syndrome d'Albright, Dermatologica 143:12, 1971.
9. Tembe, D.: Fibro-osseous dysplasia of temporal bone, J. Laryngol. Otol. 84:107, 1970.
10. Calderon, M., and Brady, H. R.: Fibrous dysplasia of bone. With bilateral optic foramina involvement, Am. J. Ophthalmol. 68:513, 1969.
11. Duplay, J., Kermarec, J.-F., Becle, J., and Grellier, P.: Une cause rare d'exophthalmie unilaterale chez l'enfant, Neurochirurgie 14:947, 1968.
12. Caffey, J.: Extremities, in *Pediatric X-ray Diagnosis* (5th ed.; Chicago: Year Book Medical Publishers, 1967), pp. 868–869.
13. Benedict, P. H.: Endocrine features in Albright's syndrome (Fibrous dysplasia of bone), Metabolism 11:30, 1962.
14. Harris, F.: Polyostotic fibrous dysplasia with a high FSH, Proc. R. Soc. Med. 60:11, 1967.
15. Husband, P., and Snodgrass, G. J. A. I.: McCune-Albright syndrome with endocrinological investigations, Am. J. Dis. Child. 119:164, 1970.
16. Benedict, P. H.: Sex precocity and polyostotic fibrous dysplasia, Am. J. Dis. Child. 111:426, 1966.
17. Moldawer, M., and Rabin, E. R.: Polyostotic fibrous dysplasia with thyrotoxicosis, Arch. Intern. Med. 118:379, 1966.
18. Samuel, S., Gilman, S., Maurer, H. S., and Rosenthal, I. M.: Hyperthyroidism in an infant with McCune-Albright syndrome: Report of a case with myeloid metaplasia, J. Pediatr. 80:275, 1972.
19. Wirth, W. A., Leavitt, D., and Enzinger, F. M.: Multiple intramuscular myxomas. Another extraskeletal manifestation of fibrous dysplasia, Cancer 27:1167, 1971.
20. Mazabraud, A., Toty, L., Roze, R., and Semat, P.: Un nouveau cas d'association de dysplasie fibreuse des os à un myxome des tissus mous, Sem. Hop. Paris 45:862, 1969.
21. Mazabraud, A., Semat, P., and Roze, R.: A propos de l'association de fibromyxomes des tissus mous à la dysplasie fibreuse des os, Presse Med. 75:2223, 1967.
22. Gimes, B., Thaisz, E., and Feher, L.: Beitrag zur malignen Entartung der fibrösen Knochendysplasia, Fortschr. Geb. Roentgenstr. Nuklearmed. 113:211, 1970.

XERODERMA PIGMENTOSUM

CARDINAL CLINICAL FEATURES

Hypersensitivity to sunlight; numerous freckles and areas of cutaneous atrophy, increased pigmentation and decreased pigmentation; multiple cutaneous malignancies. The de Sanctis-Cacchione form has in addition mental retardation, ataxia, shortness of stature, hypogonadism. All forms of xeroderma pigmentosum have a basic defect in DNA repair replication.

CLINICAL PICTURE

Hebra and Kaposi[1] first described the disorder in 1874 as a syndrome of skin manifestations including sunlight-hypersensitive skin, freckles and skin cancers. De Sanctis and Cacchione[2] reported three brothers in 1932 in whom an identical dermatologic disorder was associated with microcephaly, short stature, mental retardation and hypogonadism (Figs. 1 and 2). Cleaver[3] found in 1968 that, whereas normal skin fibroblasts can repair damage to DNA induced by ultraviolet irradiation, fibroblasts of patients with xeroderma pigmentosum have a defect in this repair mechanism.

The first manifestation is the hypersensitivity of the skin to sunlight. This usually occurs during the first year of life.[4] The first lesions appear in an area of exposure to the sun during infancy or early childhood and may be an unusually severe sunburn. Following the acute reaction there are pigmentary changes. The exposed skin is characterized by areas of increased and decreased pigmentation. There are small lesions that appear to be freckles. Atrophy of the skin develops in these areas (Fig. 3), as well as telangiectasia. Areas of hyperkeratosis may be seen.

Ocular defects are common. Most patients have photophobia and excessive tearing. Atrophy of the eyelid may lead to ectropion or entropion, drying of the conjunctiva and corneal ulceration.

These patients develop numerous skin cancers. Squamous cell carcinomas and keratoacanthomas are common. Patients also develop basal cell carcinomas, malignant melanomas, angiosarcomas and fibrosarcomas. Acute leukemia has been reported.[5] All of the cutaneous changes worsen with continued exposure even to limited sunlight. Repeated surgery on the face for removal of malignancy may be mutilative.

DE SANCTIS-CACCHIONE FORM. — In the de Sanctis-Cacchione syndrome[2, 5-7] the characteristic skin manifestations of xeroderma pigmentosum are associated with neurologic abnormalities. These include mental retardation and microcephaly. The patients may have spasticity, choreoathetosis or ataxia. There may be premature closure of the cranial sutures. Sensorineural deafness is common. The brain is small and the ventricles dilated. Cerebellar atrophy may be particularly prominent.[7] These patients also have retarded growth and delayed bone age. Sexual development also may be delayed,[2] although at least in childhood the genitals may appear normal.[4]

When fibroblasts are incubated with tritiated thymidine and then observed by radioautography, label is incorporated into the nucleus of only a few cells, which are in the S phase of DNA synthesis. When normal cells are subjected to ultraviolet irradiation, thymidine is incorporated in small amounts into a large number of cells, indicating the presence of a DNA repair mechanism. This DNA repair has been found to be defective in xeroderma pigmentosum.[3, 7, 8]

Fibroblasts of patients with the de Sanctis-Cacchione syndrome have a complete defect in this DNA repair enzyme. Patients with ordinary xeroderma pigmentosum have a lesser defect in this enzyme. This defect also has been demonstrated in the lymphocytes.[9]

Patients with xeroderma pigmentosum have been described[9, 10] in whom the skin characteristics were typical, including the presence of skin cancers, whose fibroblasts were indistinguishable from normal with respect to the ability to repair DNA damaged by ultraviolet. These cases appear to represent a new variant.

Cells of patients with xeroderma pigmentosum, like those of patients with the Down syndrome and Fanconi anemia, have a greatly increased sensitivity to transformation in vitro by the oncogenic virus SV40.[11] Heterozygous parents of patients with xeroderma pigmentosum also have higher than normal susceptibility to transformation.

CASE REPORT

B. H. was admitted at the age of 6½ years to the Jackson Memorial Hospital, Miami, Florida, because of abnormalities of the skin and eye. She had been born after an uncomplicated 7-month gestation weighing 4 lb, 12 oz. There was no family history of xeroderma pigmentosum, excessive freckling or consanguinity. One other pregnancy had terminated with a

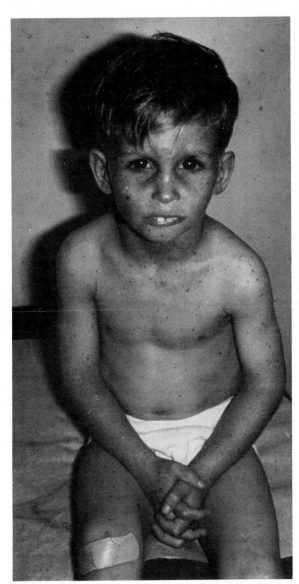

Fig. 1.—A 4-year-old boy with photosensitive skin and some extra freckling. He had pronounced shortness of stature. (Courtesy of Dr. Alan Schumacher of San Diego, California.)

miscarriage. The baby had poor neonatal weight gain and was kept in the hospital for 2 months. She continued to thrive poorly and weighed only 8 lb at the end of her first year of life.

A generalized skin rash and photophobia with lacrimation were first noted at 6 months of age during her first summer. Hyperpigmentation developed rapidly. It was recognized early to be accentuated by exposure to sunlight. Topical therapy of various sorts over the next 6 years failed to modify the skin condition.

Fig. 2.—Patient shown in Figure 1 at 8 years of age. Hypogonadism was evident. He remained very small; the height age was 6. (We are indebted to Drs. William Reed and Walter Nickel for the opportunity to observe and photograph this patient.)

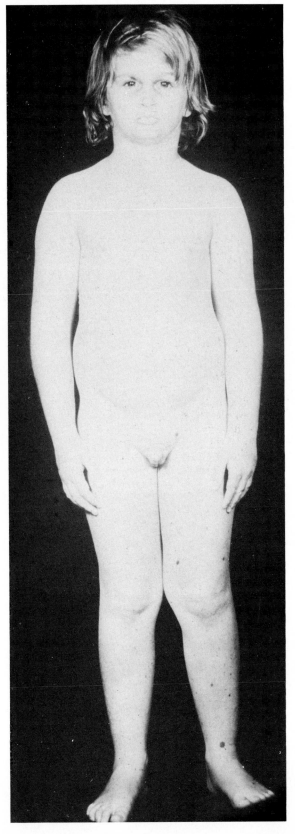

Physical examination revealed a small, black female whose height (43 inches), weight (32 lb) and head circumference (45 cm) fell below the third percentile for her age. Blood pressure was 105/70 mm Hg, pulse 88/minute and respiratory rate 20/minute.

The skin was dry, parchment-like and virtually hairless. Marked hyperpigmentation was observed (Figs. 4 and 5). Hypopigmented areas gave a salt and pepper appearance over the face (Fig. 6), neck, hands and lower legs. Scalp hair was sparse, and eyebrows and eyelashes were almost absent. The skin appeared aged and inelastic. Telangiectases were observed on the lateral aspect of the tongue.

The patient had photophobia and lacrimated freely on stimulation with light. There was mild microphthalmus. The corneas each measured 9 mm in diameter, compared with a norm of 11 mm for a child of her age. There was scleralization of the cornea presenting as a diffuse haziness. In the left eye there was a small, whitish nodule with increased vascularization.

Neurologic examination revealed bilaterally hyperactive deep tendon reflexes. The child appeared to be mentally retarded. She had a short attention span and a marked impairment in language development. Assessment of intellectual function with the Stanford Binet (Form L-M) revealed a mental age of 3 years to 3 years, 7 months, at a chronologic age of 6 years, 11 months. This gave an IQ between 39 and 48 and indicated a moderately retarded child.

Roentgenographic studies documented the patient's microcephaly. Skull measurements were 17×13 cm, average for a 9-month-old child. A bone age of 6 years approximated her chronologic age.

Laboratory studies showed a hemoglobin of 12.1 gm/100 ml and a hematocrit of 33.5. The WBC was 4,000/cu mm with 45% polymorphonuclear leukocytes, 27% lymphocytes, 2% monocytes and 26% eosinophils. A repeat differential revealed 10% eosinophils. The urinalysis was negative.

Fig. 3.—Patient shown in Figures 1 and 2. He clearly had xeroderma pigmentosum. The skin of the face showed atrophy as well as increased and decreased pigmentation. Some of the pigmented areas were brown; a few were blue. The atrophy over the nose gave him a pinched appearance.

A skin biopsy showed hyperkeratotic changes consistent with xeroderma pigmentosum.

GENETICS

Xeroderma pigmentosum and the de Sanctis-Cacchione syndrome appear to be transmitted by distinct autosomal recessive genes. Xeroderma pigmentosum is commonly seen in Arab countries, possibly a consequence of the high incidence of intermarriage and consanguinity. It is possible to detect the heterozygous carrier[10]; cells from parents of patients have reduced ability to repair ultraviolet-damaged DNA. Intrauterine diagnosis is feasible in this syndrome.[12] The test is relatively rapid and may be completed within 12 hours.

Exposure of DNA to ultraviolet light produces abnormal cyclobutyl pyrimidine dimers. These defects are repaired by excision of the dimers, repair replication and joining of the DNA strands.[13] The test for intrauterine diagnosis is based on the incorporation of bromodeoxyuridine into the repair regions. Subsequent ultraviolet irradiation will break the DNA at the sites containing this analogue, altering downward the pattern of DNA sedimentation in the ultracentrifuge. These changes occur in normal cells but not in cells containing the xeroderma pigmentosum defect. Studies using cell hybridization have indicated that ordinary xeroderma pigmentosum cells and those of patients with the de Sanctis-Cacchione syndrome are complementary.[8] It is now apparent that there are three different complementation groups, indicating

Fig. 4.—B. H. (Case Report). A 6-year-old retarded Negro girl with xeroderma pigmentosum of the de Sanctis-Cacchione type. She had multiple areas of increased black pigmentation and atrophy and many freckles. She had virtually no eyebrows. Height was 43 inches.

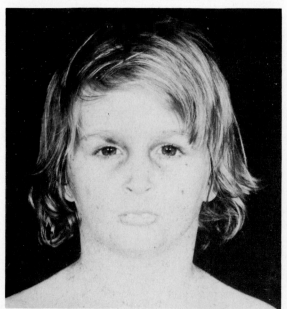

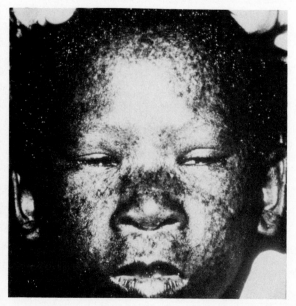

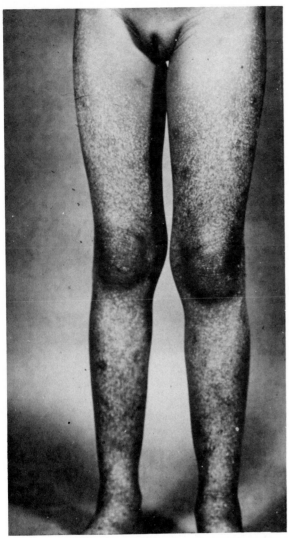

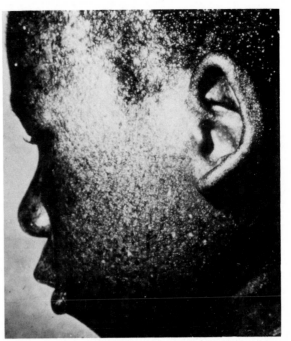

Fig. 6.– B. H. Profile. Some areas of depigmentation are evident as well as greatly increased pigmentation.

Fig. 5.– B. H. The lower extremities showed increased pigmentation and atrophy of the skin prominent in areas of exposure.

the presence of three different autosomal recessive genes.[14]

TREATMENT

It is advantageous for these patients to take extreme measures to avoid sunlight and to use sunscreen lotions whenever they must go outdoors. Early surgical removal of premalignant lesions and skin cancers is advisable. Skin grafts may be useful.

REFERENCES

1. Hebra, F., and Kaposi, M.: *On Diseases of the Skin* (London: New Sydenham Society, 1874), Vol. 3, p. 352.
2. de Sanctis, C., and Cacchione, A.: L'idiozia xerodermica, Riv. Sper. Freniat. 56:269, 1932.
3. Cleaver, J. E.: Defective repair replication of DNA in xeroderma pigmentosum, Nature 218:652, 1968.
4. Silberstein, A. G.: Xeroderma pigmentosum with mental deficiency. A report of two cases, Am. J. Dis. Child. 55:784, 1938.
5. Reed, W. B., May, S. B., and Nickel, W. R.: Xeroderma pigmentosum with neurological complications, Arch. Dermatol. 91:224, 1965.
6. Reed, W. B.: The genetics of the photodermatoses, in *The Clinical Delineation of Birth Defects, Part XII: Skin, Hair and Nails* (Baltimore: The Williams & Wilkins Company, 1971), pp. 129– 139.
7. Reed, W. B., Landing, B., Sugarman, G., Cleaver, J. E., and Melnyk, J.: Xeroderma pigmentosum. Clinical and laboratory investigation of its basic defect, J.A.M.A. 207:2073, 1969.
8. Der Kaloustian, V. M., and Kurban, A. K.: Xeroderma pigmentosum, Br. J. Dermatol. 88:513, 1973.
9. Burk, P. G., Yuspa, S. H., Lutzner, M. A., and Robbins, J. H.: Xeroderma pigmentosum and D.N.A. repair, Lancet 1:601, 1971.
10. Cleaver, J. E.: Xeroderma pigmentosum: Variants with normal DNA repair and normal sensitivity to ultraviolet light, J. Invest. Dermatol. 58:124, 1972.
11. Veldhuisen, G., and Pouwels, P. H.: Transformation of xerodermapigmentosum cells by SV40, Lancet 2:529, 1970.
12. Regan, J. D., Setlow, R. B., Kaback, M. M., Howell, R. R., Klein, E., and Burgess, G.: Xeroderma pigmentosum: A rapid sensitive method for prenatal diagnosis, Science 174:147, 1971.
13. Cleaver, J. E.: DNA damage and repair in light-sensitive human skin disease, J. Invest. Dermatol. 54:181, 1970.
14. Kraemer, K. H., Robbins, J. H., and Coon, H. G.: Three genetic forms of xeroderma pigmentosum, Clin. Res. 21:480, 1973.

Ectodermal Dysplasias

PACHYONYCHIA CONGENITA

Gerson Carakushansky

CARDINAL CLINICAL FEATURES

Piled-up, horn-like thickening of the nails and other abnormalities of keratinization, including thickened, hyperkeratotic skin and callosities of the feet.

CLINICAL PICTURE

In 1906 Jadassohn and Lewandowsky[1] described the abnormality of the nails to which they gave the name pachyonychia congenita. It is characterized by distinctive, excessively thickened nails. These patients also have palmar and plantar hyperkeratoses, follicular keratosis of the skin and leukokeratosis of the oral mucous membranes. Of these manifestations, only the nail changes are constantly present; they usually represent the earliest manifestation of the disorder.

At birth the nail may appear normal, but shortly thereafter the nail bed develops a yellow-brown discoloration. There is uplifting of the nail. The nails may be smooth and of normal appearance at the base, but at the free end they are wedge shaped and raised by a horny mass of material in the hyponychium. This leads to an angular projection up from the nail bed that is particularly prominent in the fingernails. The nail plates are extremely hard and are firmly attached to the nail beds. Ordinary measures were not sufficient to trim the nails of the patient originally reported by Jadassohn and Lewandowsky,[1] and a hammer and chisel were used for this purpose. The mother of the child described in the Case Report used a saw to reduce the size of the heaped-up, horny projections of his nails. Sohrweide[2] reported a patient in whom the nails were 0.7–1 cm in thickness at their free margins. Because of the projections, these nails are subject to injury. Paronychia may be a problem. Some babies with pachyonychia congenita have attracted attention because they were born with erupted teeth. Murray[3] first noted the association of erupted teeth at birth and pachyonychia congenita. Jackson and Lawler[4] and Soderquist and Reed[5] also observed this association, which may be familial and may be transmitted as a dominant through three or more consecutive generations.

Skin involvement may be noticeable in the presence of keratotic patches on the palms and soles. The entire skin may appear thickened. Follicular hyperkeratosis with well-developed keratin plugs may occur on the extensor surfaces around the large joints. There may be bullae, which may be associated with hyperhidrosis of the palms and soles. Areas such as the heels may readily form blisters. Bullae may develop over callosities and may be painful. In rare instances breakdown of the skin on weight-bearing surfaces may be so severe that admission to hospital and skin grafting are required. Occasionally an unusually dry kinky hair is seen on the scalp of white patients.[6] Nodular epithelial cysts have been described.[5]

Oral leukokeratosis is not found in all patients or kindreds but, when present, may be an early manifestation in infants. In some instances the oral lesions, which may be seen as early as at birth, have been mistaken for thrush. These lesions take the form of diffuse, white or gray-white plaques, most commonly involving the dorsal surface of the tongue. The lateral borders of the tongue may appear especially thickened. Other mucous membranes have on occasion shown leukokeratosis, including the nasal, tympanic and perianal mucosa.

Other reported defects, such as mental deficiency,

Fig. 1.–J. A. S. (Case Report). The nails were typical of pachyonychia congenita. Note the claw-like incomplete pointed cylinder or cone.

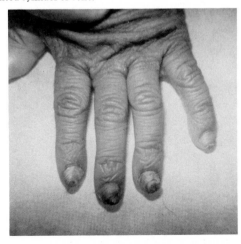

corneal changes and diffuse gastrointestinal polyposis, do not appear to be consistent features of the
syndrome.

DIFFERENTIAL DIAGNOSIS

The fully developed syndrome is seldom mistaken
for anything else. The differential diagnosis includes
hereditary onychogryposis, which is also inherited as a
dominant trait, but there are no abnormalities other
than the nail changes. In epidermolysis bullosa, blister
formation may be associated with nail thickening, but
the picture is really very different.

CASE REPORT

J. A. S., a 3-year-old black male, was admitted to the Pediatric Service of the Federal University of Rio de Janeiro because
of a chronic condition involving the skin and nails. He was the
second child of healthy parents. The first child was free of the
condition. In the second week of life all of the patient's fingernails and toenails had turned yellow. By 1 year of age the nails
were grossly thickened and had begun to curl. At the time of
admission the nails were remarkable. The piled-up nail
growth had led to curved, claw-like or horn-like, incomplete
pointed cylinders or cones. Our first pictures were lost; Figure
1 represents the appearance after the mother trimmed the
nails with a saw. Follicular keratotic lesions were present over
the knees (Fig. 2) and at the base of the toes (Fig. 3). This patient's entire skin was thickened and dry and had an ichthyotic
appearance. The ear was unusual (Fig. 4).

GENETICS

Cockayne[7] observed that the syndrome was in many
cases inherited directly from a parent who suffered
from the condition and that parental consanguinity
was absent. A single autosomal dominant gene with

Fig. 2. – J. A. S. Close-up of the knee showing follicular
keratosis.

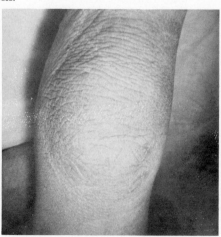

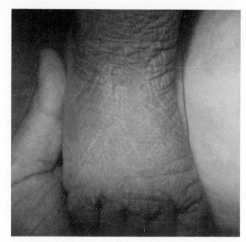

Fig. 3. – J. A. S. The foot illustrates the presence of keratotic
lesions at the base of the toes.

wide expressivity would appear to satisfy the evidence
of the pedigrees described.[4] Kumer and Loos[8] found 24
affected members in a pedigree comprising five generations. The varied expression in different kindreds
suggests that there may be a number of different dominant genes responsible for similar but distinct syndromes.

TREATMENT

Pachyonychia of the nails is of only cosmetic importance but is difficult to treat. Some rather vigorous attempts have been made because of the possible psychologic problems. Children are cruel, and patients
with this disorder have been labeled as "witches."
Even avulsion of the nail brings only temporary relief.
Amputation of the distal phalanges has been used, but
this does not appear justified. There is evidence that

Fig. 4. – J. A. S. This patient also had an unusual ear.

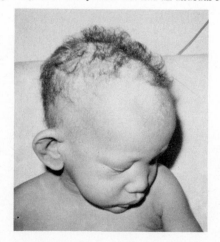

the distal matrix produces excessive amounts of keratin and therefore produces a material similar to that of a horse's hoof. Even after deep surgical excision of the nails,[9] a small portion of the distal matrix usually is left and nail again forms at the site.

REFERENCES

1. Jadassohn, J., and Lewandowsky, F.: Pachyonychia congenita, keratosis disseminata circumscripta (folliculosis): Tylomata, leukokeratosis lingual, Ikonogr. Dermatol. Tab. 6:29, 1906.
2. Sohrweide, A. W.: Pachyonychia congenita: Case report, Arch. Dermatol. Syph. 32:370, 1935.
3. Murray, F. A.: Congenital anomalies of nails associated with erupted teeth at birth, Br. J. Dermatol. 33:409, 1921.
4. Jackson, A. D. M., and Lawler, S. D.: Pachyonychia congenita: A report of 6 cases in one family with a note on linkage data, Ann. Eugenics 16:142, 1951–1952.
5. Soderquist, N. A., and Reed, W. B.: Pachyonychia congenita with epidermal cysts and other congenital dyskeratoses, Arch. Dermatol. 97:31, 1968.
6. Shrank, A. B.: Pachyonychia congenita, Proc. R. Soc. Med. 59:975, 1966.
7. Cockayne, E. A.: *Inherited Abnormalities of the Skin and Its Appendages* (London: Oxford University Press, 1933).
8. Kumer, L., and Loos, H. O.: Uber pachyonychia congenita (typus Riehl), Wien Klin. Wochenschr. 48:174, 1935.
9. Cosman, B., Symonds, F. C., Jr., and Crikelair, G. F.: Plastic surgery in pachyonychia congenita and other dyskeratoses, Plast. Reconstr. Surg. 33:226, 1964.

Classic Short-limbed Dwarfism

ACHONDROPLASIA

CARDINAL CLINICAL FEATURES

Short-limbed dwarfism, rhizomelic in type; a large head, with prominent forehead, and a saddle or scooped-out nose and midface; caudal narrowing of the spinal canal; trident hands.

CLINICAL PICTURE

Achondroplasia is the prototype of the short-limbed dwarfisms (Fig. 1). It is the most common type of chondrodystrophy.[1] The disease has been recognized for centuries and was known to the ancient Egyptians. The Egyptian god Bes was achondroplastic.

Achondroplasia is a congenital disease affecting the formation of endochondral bone.[2-4] Periosteal osteogenesis is unaffected. The histologic pathology of achondroplasia is not clearly abnormal.[4, 5] Biopsies of the iliac crest have been described as histologically normal.[4, 5] The chondro-osseous junctions of the ribs[5] and fibula[4] showed abnormal clustering of cells and partially calcified cartilage. The fibrous matrix was thought to be very slowly and irregularly resorbed by invading vasculature.[4] The changes are, however, subtle and the general picture in achondroplasia is one of regular well-organized endochondral ossification.[5] This is in marked contrast to the histopathology of other chondrodystrophies such as asphyxiating thoracic dystrophy or thanatophoric dwarfism.[5]

Achondroplasia is readily recognized and may be diagnosed at birth. The tubular bones are short and thick and as a consequence the extremities are short and bowed. Hypotonia may be very prominent in the infant.[3] The shortness of the limbs is the most striking feature of the disease, which is rhizomelic in type (Fig. 2). This term, which refers to the "root of the limb," indicates that the shortening is more marked in the humerus and the femur than in the more distal elements of the arm and leg. The finger tips of these patients barely reach their hips. In the normal individual the tips of the fingers reach the middle of the thigh. Some people have remembered the achondroplastic phenotype by considering the problems for such an individual in wiping the anus. The trunk is normal in size and the sitting height is normal. The mean adult height ranges from 48–52 inches.[1, 3]

Because of the short, bowed legs and pelvic tilt, pa-

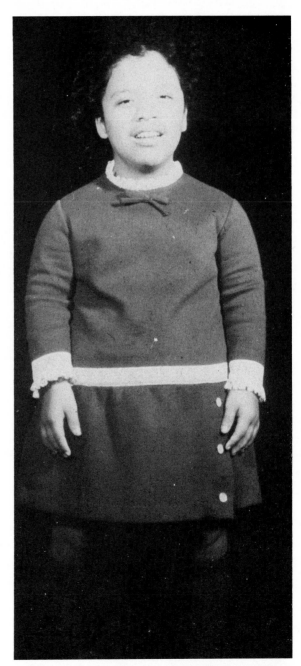

Fig. 1.–G. V. (Case Report). A 16-year-old girl with achondroplasia. She was short. The short-limbed character of the disorder is evident from the length of the arms.

231

tients may have an unusual duck-like gait. Ligamentous laxity and muscular hypotonia may contribute. Limitation of extension of the elbow is the rule. Pronation and supination also are limited. The hands are short and stubby. There is a wedge-shaped gap between the third and fourth fingers, which gives rise to the characteristic trident hand (Fig. 3). The fingers cannot be placed in a parallel position. When they are apposed, they will not touch throughout their length. The three prongs of the trident are made up of the thumb, the second and third fingers and the fourth and fifth fingers. These patients may find their legs tire easily or develop leg pains at night.[4] There is an exaggerated lumbar lordosis (Fig. 4) and the buttocks appear prominent. Thoracolumbar kyphosis (Fig. 5) or gibbus is common in infancy. It may persist into adult life or, with weight-bearing and ambulation, it may give way to lordosis.

The head and face of these patients are characteristic. The cranium is large (Fig. 6). The head circumference usually is well above the 97th percentile. The forehead is prominent, as are the brows (Fig. 7). The nasal bridge is depressed so that the nose has a scooped-out appearance. The maxilla is hypoplastic. In the presence of this midfacial hypoplasia the mandible may appear relatively prognathic. The membranous bone of the skull grows normally, whereas the cartilaginous bones of the base of the skull grow slowly. This disproportionate growth leads to a short cranial base with a small, misshapen or "crumpled" foramen magnum.

The roentgenographic findings in achondroplasia are highly characteristic.[2] The long bones are short and thick and lack normal modeling. There may be a ball-and-socket appearance of the epiphysis and metaphysis. This appearance of the epiphysis set into the metaphysis also has been referred to as a circumflex or chevron sign. In the hands the proximal and middle phalanges are short, broad and conically shaped. The metaphyses of the bones may flare. Genu varum is a frequent abnormality. In the vertebrae the interpedicular distances diminish progressively from the thoracic vertebrae to the sacrum. This is the reverse of the normal situation,[6] in which there is caudad widening of the spine. Excessive amounts of cartilage separate the ossification centers in the vertebrae. There may be anterior beaking of upper lumbar vertebrae. The ribs are short and have cupped ends. The sacrum is horizontal. The pelvis is short and broad and the acetabular roofs are flat. The femoral neck is short. The greater sciatic notches are characteristically narrow and deep.

Patients with this disorder may have a variety of neurologic complications. Hydrocephalus may develop as a result of the small foramen magnum.[7] The small, short base of the skull may lead to constriction of the subarachnoid space in the posterior fossa. Repeated episodes of increased intracranial pressure have been described.[8] The narrowed spinal cord in the lumbar region may result in compression of the cord or nerve roots by a ruptured, prolapsed intravertebral disk or deformity of the vertebral bodies themselves as osteoarthritis develops.[9] The involved patient may develop slowly progressive paresis, or weakness and numbness in the legs, or there may be sudden paraplegia.[9] We have seen one patient with altered school performance who was found to have decreased visual acuity and optic atrophy. Two patients with diminished visual acuity and optic atrophy also were seen by Walsh.[10] Intelligence is in the normal range.

DIFFERENTIAL DIAGNOSIS

Achondroplasia, as a common, rather widely known form of dwarfism, tends to be overdiagnosed. Many rarer disorders are originally diagnosed as achondroplasia.[3, 11, 12] In thanatophoric dwarfism (see chapter, this section),[11, 13] the findings are similar to those in achondroplasia but much more severe. These patients often are stillborn, or death occurs in the first 24 hours. The head is large, and hydrocephalus may be present at birth. The vertebral bodies are flat and appear in the posteroanterior view like an inverted U or an H. The ribs are very short and the rib cage small. Expansion is severely restricted, and this leads to respiratory distress and early death. Histologic examination reveals disorganized endochondral ossification, in contrast to regular, organized endochondral ossification in achondroplasia.

Diastrophic dwarfism (see chapter, this section) also is a short-limbed dwarfism. It should be readily distinguished by clubfoot, hitchhiker thumb and deformed ears.

Hypochondroplasia[14, 15] resembles achondroplasia but is much milder. It is a distinct entity with a dominant transmission and, as one would expect, more patients are found in affected families than in achondroplasia, in which a majority of cases are sporadic and appear to result from new mutation. Achondroplasia and hypochondroplasia never occur in the same family. In this disorder the head and face appear normal, although there may be minimal shortening of the base of the skull. The disease is not usually recognized at birth and the trident hand is not present. The gene appears to be allelic with that for achondroplasia.[1]

The pseudoachondroplastic form of spondyloepiphyseal dysplasia or pseudoachondroplastic dysplasia[11, 16] is a short-limbed dwarfism, with a trunk of normal size, which has been confused with achondroplasia. Head and face are normal. Roentgenographic findings include irregularity and fragmentation of the epiphyses and irregular mushroomlike metaphyses. Hall and Dorst[16] have pointed out that there is considerable heterogeneity in this disorder in which there seem to be four types, two autosomal dominant and two recessive.

This is obviously only a partial list of some of the conditions frequently mistaken for achondroplasia. The full list would be extensive indeed.[11, 12, 17]

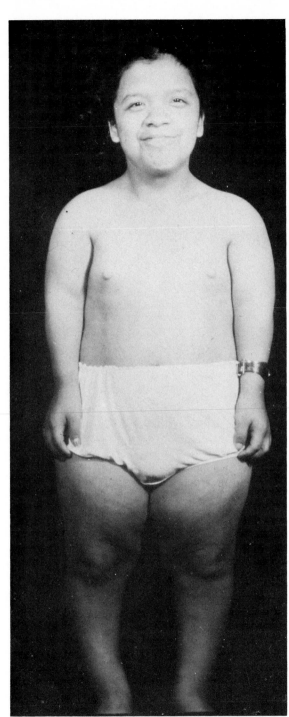

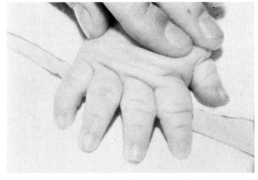

Fig. 3. – The trident hand in a 3-month-old boy with achondroplasia.

Fig. 4. – An adult patient with achondroplasia. The characteristic lordosis is partially hidden by the elbow.

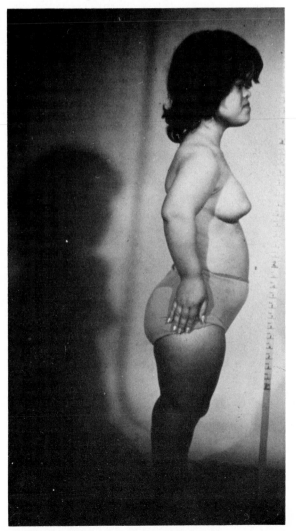

Fig. 2. – G. V. The rhizomelic shortening of the extremities is evident. The humeral and femoral shortening was more pronounced than that seen distally. The lower extremities were bowed.

Fig. 5. – A 1-year-old girl with achondroplasia. At this age patients have thoracolumbar kyphosis. They develop lordosis later.

Fig. 6. – Patient shown in Figure 3 had a large cranium and a markedly depressed nasal bridge, so that the base of the nose and nostrils are seen head on.

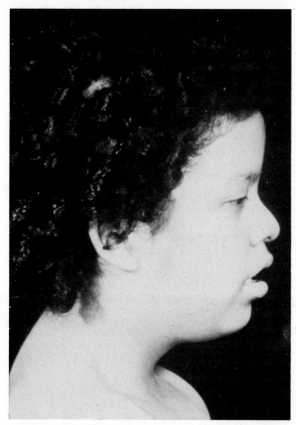

Fig. 8. – G. V. Profile illustrates the large cranium, prominent forehead and depressed nasal bridge.

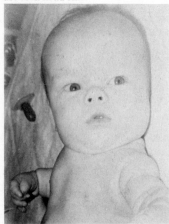

Fig. 7. – Profile of young patient illustrates the prominent occiput, large cranium and depressed nasal bridge.

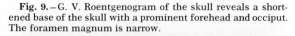

Fig. 9. – G. V. Roentgenogram of the skull reveals a shortened base of the skull with a prominent forehead and occiput. The foramen magnum is narrow.

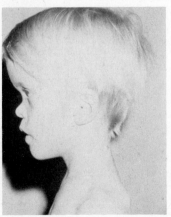

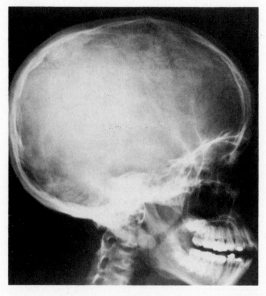

CASE REPORT

G. V. was a 14-year-old white girl known to have achondroplasia. When she was born, her mother was 28 years old and her father 36. There was no family history of a similar condition. An older brother was 6 feet tall. Over the years the patient's nasal passages were increasingly obstructed, and she had frequent otitis media recurrences and some hearing loss. She was an intelligent child and in the eighth grade at school. She appeared well adjusted to being short; yet she was not ready to meet other people with the same problem by contacting Little People of America.

Physical examination revealed a pleasant, giggly, somewhat obese girl. Height was 43¾ inches. The ratio of the upper to lower segment was 1.4, characteristic for short-limbed dwarfism (see Fig. 1). The head was large and the forehead prominent. The nasal bridge was depressed (Fig. 8). She had a pronounced lumbar lordosis and prominent buttocks. The limbs were quite short (see Fig. 2). The fingertips barely reached the hips, and the fingers were short and had the typical trident appearance. Neurologic examination was normal. Roentgenologic examination revealed a shortened base of the skull, with a prominent forehead and occiput (Fig. 9). A mild shortening of the pedicles of the cervical spine and fibrous occipitalization of the atlas were present. The thoracic and lumbar spine showed a gradually decreasing interpedicular distance on downward progression (Fig. 10). In the lateral view the pedicles of the entire lumbar spine were very shortened, and there was a general contraction of the dural space (Fig. 11). In the pelvis there was outward rotation of both iliac bones and flat acetabular roofs. The sacrum was horizontal; the femoral necks were short and the capital femoral epiphyses broad (Fig. 12). There were pronounced rhizomelic shortening of the extremities (Fig. 13), a lack of normal modeling of the long bones and brachydactyly with trident hands.

GENETICS

Achondroplasia is determined by an autosomal dominant gene. Some 80% of cases are sporadic, and it is thought that these are due to new mutation, as the gene is fully penetrant. Increased age of the father has been documented in sporadic cases of patients with achondroplasia, and this supports the conceptualization of new mutation.[18] This particular mutation has been estimated to occur at a rate of one in 20,000 births,[19] but this may be high.[3, 11] Patients with dwarfism frequently marry others with similar disorders, and this is particularly true of achondroplastic dwarfs. This of course creates the possibility of producing children homozygous for the gene. The homozygote for this gene has skeletal dysplasia so severe that it leads to death in utero or in early infancy.[20]

TREATMENT

These patients need understanding and support. In the event of spinal cord compression, laminectomy

Fig. 10.—G. V. Roentgenogram of the lumbar spine illustrates the classic change in interpedicular distance in the achondroplastic spine. Normally this distance increases from L1 to L5 so that the spine flares. In the achondroplast the spine actually tapers as the interpedicular distance decreases. In this film the distance was 23 mm at L1 and 20 mm at L5.

Fig. 11.—G. V. Lateral roentgenogram of the spine shows that the pedicles were extremely short—less than half the normal pedicular length. This feature of the disease compromises the spinal canal and may lead to cord compression.

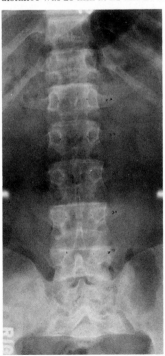

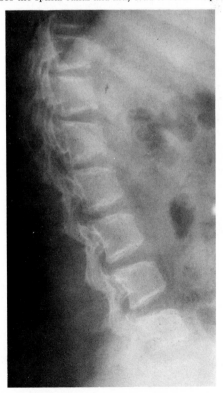

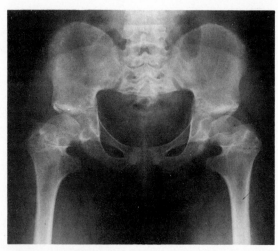

Fig. 12.—G. V. Roentgenogram of the pelvis is very characteristic. The shape has been compared to a valentine. Both iliac bones were outwardly rotated. The acetabular roof was flat and the sacrum horizontal. The femoral necks were very short and the capital femoral epiphyses broad. The bases of the ilia were stippled. The sacrosciatic notch was narrow.

Fig. 13.—G. V. Roentgenograms of the arm, illustrating rhizomelic shortening. The bones were shortened and flared terminally. The shafts were bowed. The upper extremities were more markedly involved than the lower.

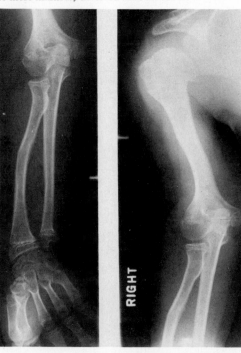

and bone fusion are recommended.[21] In following these patients, repeated careful examination of the neurologic status of the lower extremities is important.

REFERENCES

1. McKusick, V. A.: *Heritable Disorders of Connective Tissue* (4th ed.; St. Louis: C. V. Mosby Co., 1972), p. 750.
2. Caffey, J.: *Pediatric X-Ray Diagnosis* (5th ed.; Chicago: Year Book Medical Publishers, 1967), p. 819.
3. Maroteaux, P., and Lamy, M.: Achondroplasia in man and animals, Clin. Orthop. 33:91, 1964.
4. Ponseti, I. V.: Skeletal growth in achondroplasia, J. Bone Joint Surg. [Am.] 52:701, 1970.
5. Rimoin, D. L., Hughes, G. N., Kaufman, R. L., Rosenthal, R. E., McAlister, W. H., and Silberberg, R.: Endochondral ossification in achondroplastic dwarfism, N. Engl. J. Med. 283:728, 1970.
6. Simril, W. A., and Thurston, D.: The normal interpediculate space in the spines of infants and children, Radiology 64:340, 1955.
7. Wise, B. L., Sondheimer, F., and Kaufman, S.: Achondroplasia and hydrocephalus, Neuropediatrie 3:106, 1971.
8. Bergström, K., Laurent, U., and Lundberg, P. O.: Neurological symptoms in achondroplasia, Acta Neurol. Scand. 47:59, 1971.
9. Schreiber, F., and Rosenthal, H.: Paraplegia from ruptured lumbar discs in achondroplastic dwarfs, J. Neurosurg. 9:648, 1952.
10. Walsh, F. B.: *Clinical Neuro-Ophthalmology* (Baltimore: Williams & Wilkins Co., 1957), pp. 396–397.
11. Scott, C. I., Jr.: The genetics of short stature, Prog. Med. Genet. 8:243, 1972.
12. Silverman, F. N., and Brünner, S.: Errors in the diagnosis of achondroplasia, Acta Radiol. [Diagn.] (Stockh.) 6:305, 1967.
13. Harris, R., and Patton, J. T.: Achondroplasia and thanatophoric dwarfism in the newborn, Clin. Genet. 2:61, 1971.
14. Beals, R. K.: Hypochondroplasia. A report of five kindreds, J. Bone Joint Surg. [Am.] 51:728, 1961.
15. Walker, B. A., Murdoch, J. L., McKusick, V. A., and Beals, R. K.: Hypochondroplasia, Am. J. Dis. Child. 122:95, 1971.
16. Hall, J. G., and Dorst, J. P.: Pseudoachondroplastic Dysplasia (SED), Recessive Maroteaux-Lamy Type, in *The Clinical Delineation of Birth Defects. Part IV, Skeletal Dysplasias*, Birth Defects: Original Article Series (New York: The National Foundation-March of Dimes, 1969), p. 254.
17. Silverman, F. N.: A differential diagnosis of achondroplasia. Radiol. Clin. North Am. 6:223, 1968.
18. Murdoch, J. L., Walker, B. A., Hall, J. G., Abbey, H., Smith, K. K., and McKusick, V. A.: Achondroplasia—a genetic and statistical survey, Ann. Hum. Genet. 33:227, 1970.
19. Mörch, E. T.: Achondroplasia is always hereditary and is inherited dominantly, J. Hered. 31:439, 1940.
20. Hall, J. G., Dorst, J. P., Taybi, H., Scott, C. I., Langer, L. O., and McKusick, V. A.: Two Probable Cases of Homozygosity for the Achondroplasia Gene, in *The Clinical Delineation of Birth Defects. Part IV, Skeletal Dysplasias*, Birth Defects: Original Article Series (New York: The National Foundation-March of Dimes, 1969), p. 24.
21. Bailey, J. A.: Orthopaedic aspects of achondroplasia, J. Bone Joint Surg. [Am.] 52:1285, 1970.

THANATOPHORIC DWARFISM

CARDINAL CLINICAL FEATURES

Micromelic dwarfism due to chondrodystrophy, narrow thorax, large calvaria, uniformly fatal outcome very early in life.

CLINICAL PICTURE

The condition now known as thanatophoric dwarfism was thought to be part of the syndrome of achondroplasia until Maroteaux, Lamy and Robert[1] in 1967 recognized this group of patients as representing a distinct entity. The name was derived from the Greek θανατοφορος and implies that the disorder brings with it an early death.

Thanatophoric dwarfism is a rare congenital chondrodystrophy, which produces a very striking short-limbed dwarfism.[2] The cranium is disproportionately large in comparison to the face (Figs. 1 and 2) because the endochondral bone of the base of the skull is involved, whereas the membranous bone of the cranium grows normally. There is prominent frontal bossing. The anterior fontanel is large and the sutures widely separated. A saddle nose and scooped-out midface are characteristic. There may be hypertelorism and the eyes may seem to bulge. The very narrow thorax is an integral feature of the syndrome.

The extremities are exceedingly short and the long bones bowed. The shortening, although generalized, is rhizomelic.[3] There are numerous skin folds. The extremities tend to extend outward from the trunk. The thighs are abducted and externally rotated. The average length at birth is between 32 cm and 47 cm.[1, 3] The upper to lower ratio is increased, and the fingertips barely reach the area of the hips. The length of the trunk is relatively normal. The fingers and toes are very short and stubby (Fig. 3). The fingers have been described as sausage-like.[4]

The narrow thorax is a consequence of the very short ribs. Many of these babies are stillborn. Those born alive usually die soon after birth, generally of respiratory failure. They have hypotonia and respira-

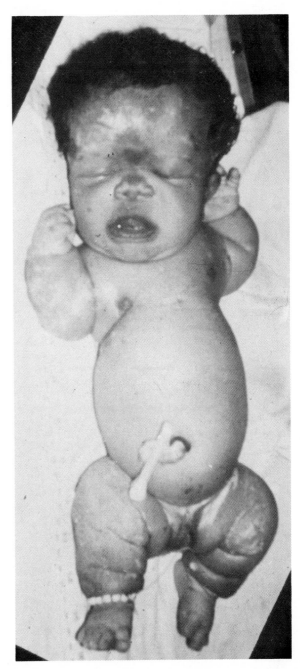

Fig. 1.—Infant with thanatophoric dwarfism. The micromelic dwarfism is illustrated; the femoral bowing was striking. The nasal bridge was depressed and the forehead prominent. The abdomen was protuberant and the chest markedly narrow. (Courtesy of Dr. Kenneth Lee Jones, of the University of California San Diego.)

tory distress. The deformity of the chest leads to tachypnea, retractions, cyanosis and severe respiratory acidosis, or congestive failure. There also may be atelectasis. The abdomen usually is protuberant. At autopsy the lungs are small and airless.

Roentgenographic findings[5, 6] are similar to those of achondroplasia but exaggerated. The discrepancy between the long thin trunk and the very short extremities is striking. The cranium is large and the base of the skull very small. The cranium is well mineralized; the bones of the face are small. The foramen magnum is narrow, and compression of the cord may occur at this level. The spinal canal is narrow. The cervical vertebrae are well developed, as is the odontoid process. However, the vertebral bodies are all very flat, poorly developed and diminished in height (Fig. 4). This is in contrast to the rectangular vertebral bodies seen in achondroplasia. In the frontal projection they have an H or inverted U appearance. The intervertebral space is increased. The interpedicular distances do not increase or widen distally from L1–L5. The thorax is very narrow both in anteroposterior and lateral directions, and the ribs are very short. There is flaring and cupping of the anterior ends of the ribs (Fig. 5). The clavicles project above the ribs; they may look like the handlebars of a bicycle. The scapula is square. The pelvis is small and shows a marked decrease in the vertical height of the iliac and ischial bones. The sacrosciatic notch is small. There is generalized micromelia, with shortening and bowing of the long bones. The metaphyses are irregular and there are thorn-like projections at the ends of the long bones and in the pelvis. The femur has been said to have the appearance of a telephone receiver[7] (see Fig. 5). Ossification is absent in the epiphyseal centers of the distal femur and proximal tibia. The metatarsals and metacarpals are very short, as are the phalanges of the feet and hands. The middle phalanges are especially short. Radioulnar synostosis has been observed.[4]

It is possible to make an antenatal diagnosis of this condition by x-ray. By the seventh month of gestation it is possible to see the large head and short, bowed limbs. Several mothers of infants with this disease have commented on a lack of sensation of fetal movement.[5] Polyhydramnios is common.

Histopathologic analysis of endochondral bone from an infant with thanatophoric dwarfism reveals disorganization of endochondral ossification.[8, 9]

DIFFERENTIAL DIAGNOSIS

Thanatophoric dwarfism must be differentiated from achondroplasia (see chapter this section).[10] The two conditions are quite similar, but this one is much more severe.

The homozygous achondroplastic dwarf who is the offspring of parents who are both achondroplastic has a similar very severe shortness of the limbs, a narrow thorax and a large cranium.[11] These patients also die

an early respiratory death. The roentgenographic findings are more like those of heterozygous achondroplasia than of thanatophoric dwarfism.[5] They are readily differentiated from infants with thanatophoric dwarfism by the family history; the parents of a thanatophoric dwarf always are normal.

Achondrogenesis[5, 9] is a lethal form of short-limbed dwarfism, with severe shortening of the limbs and a large cranium. Roentgenograms are diagnostic in this condition. They show very poor ossification of the bones. Ossification of the vertebral bodies is very retarded and may be virtually absent in the lumbar region and in the sacral, pubic and ischial bones. There is marked flaring of the proximal and distal metaphyses of the humeri. Histologic examination reveals disorganization and lack of any columnar arrangement of cartilage cells. A lack of cell matrix in the resting cartilage among the cartilage cells appears to be specific for achondrogenesis.[9] There also is a lack of normal development of the intervertebral disks. These infants are stillborn or die within a few days after birth. This is a rare disorder. It is thought to be inherited as an autosomal recessive.

Asphyxiating thoracic dystrophy has many features in common with thanatophoric dwarfism.[4] The absence of platyspondyly and normal interpedicular distance in the lumbar vertebrae permit its distinction, and the micromelia is nowhere near as severe as in thanatophoric dwarfism.

CASE REPORT*

Baby B. was born at the Johns Hopkins Hospital to a 34-year-old mother. The father was 36 years old. There was no consanguinity. Four previous pregnancies had been normal. Polyhydramnios was severe enough to warrant admission to hospital, and amniocentesis was performed. The infant was born by cesarean section, weighing 7 lb 1 oz. The Apgar score at 1 minute was 2, and at 5 minutes, 5.

Examination revealed a dramatically short-limbed Caucasian male dwarf with a large head (Fig. 6). The head circumference was 41.5 cm, that of the chest 28 cm. The length was 15¾ inches. The pulse was 200, and respirations 60/minute.

The short limbs contrasted with a relatively long trunk. The skin was plethoric. There were many extra folds on the arms and legs. There were bulging of the upper cranium and apparent facial hypoplasia. The anterior fontanel was 3 × 5 cm. The bridge of the nose was depressed and the nares anteverted and small. The ears were low set, and there was a high palatal arch. The chest was short and narrow. The lung fields were clear but respirations were very shallow. The abdomen was protuberant. The liver was palpable 8 cm below the costal margin, the spleen 6 cm. The shortening of all extremities was striking. The legs were bowed and the arms appeared to stick straight out from the body. A mild low thoracic gibbus was present. There was no Moro, grasp or sucking reflex. The cry

*We are indebted to Dr. Kenneth Lee Jones of the University of California San Diego, for permission to report the details of this patient.

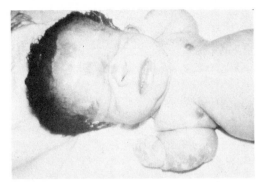

Fig. 2.—Close-up of patient in Figure 1 shows the dramatic narrowing of the chest and the characteristic facial appearance. The upper arms were short and bowed and the hands short and stubby. (Courtesy of Dr. Kenneth Lee Jones of the University of California San Diego.)

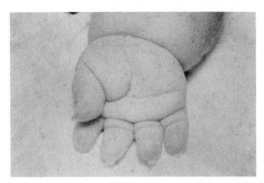

Fig. 3.—Short, broad hand with short, stubby fingers. (Courtesy of Dr. Kenneth Lee Jones of the University of California San Diego.)

Fig. 4.—Lateral roentgenogram of a patient with thanatophoric dwarfism, illustrating the complete vertebra plana. The little narrow tongues that represent the vertebral bodies are unique to this syndrome. (Courtesy of Dr. Michael Weller of the University of California San Diego.)

Fig. 5.—Roentgenogram showing extreme rhizomelic shortening of limbs, with tiny curved femora. The left femur particularly had the characteristic telephone receiver appearance. The metacarpals were very short. The vertebrae were flat. The ribs were very short and had distal flaring and cupping. The cranium was large. (Courtesy of Dr. Michael Weller of the University of California San Diego.)

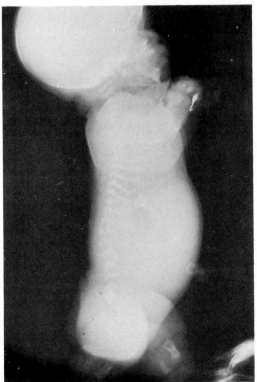

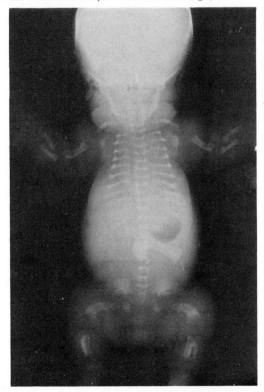

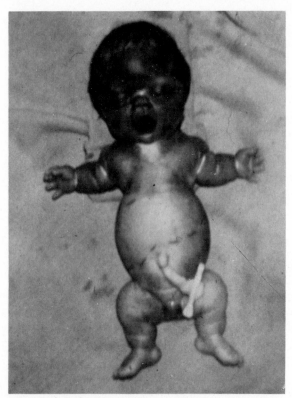

Fig. 6.–Baby B. (Case Report). Thanatophoric dwarfism. Postmortem picture illustrates the extreme nature of the short-limbed dwarfism in this syndrome. The skin folds of the extremities were increased. The narrow chest and proportionately large cranium are evident. There were hypertelorism and very prominent eyes. (Courtesy of Dr. Douglas Cunningham of the University of California San Diego.)

was poor. The infant moved all extremities but muscle tone was poor.

Roentgenograms showed marked flattening of the vertebral bodies (see Fig. 5). The ribs were very short and cupped anteriorly. The chondrocranium was short and the facial bones hypoplastic. The short, bowed long bones were classic.

The infant became increasingly tachypneic. Grunting respirations and retractions were prominent. He died at 27 hours of age. Autopsy showed a dilated heart and visceral congestion of the lungs as well as the findings of thanatophoric dwarfism.

GENETICS

Thanatophoric dwarfism usually has appeared sporadically in children of normal parents. Increased paternal age has never been reported. Two male siblings with thanatophoric dwarfism have been reported[12, 13] who were born to a Jewish woman from

Morocco married to her first cousin. Harris and Patton[10] also reported two other siblings with this syndrome born to normal parents. These multiple sibs and the consanguinity suggest that the disorder is transmitted on an autosomal recessive gene.

A predominance of males in reported series[1, 10] is interesting. There could be sex limitation or early loss of involved female embryos. However, the data may not differ significantly from a 1:1 ratio of males to females.[14]

TREATMENT

None is effective.

REFERENCES

1. Maroteaux, P., Lamy, M., and Robert, J.-M.: Le nanisme thanatophore, Presse Med. 75:2519, 1967.
2. Scott, Jr., C. I.: The genetics of short stature, Prog. Med. Genet. 8:243, 1972.
3. Kaufman, R. L., Rimoin, D. L., McAlister, W. H., and Kissane, J. M.: Thanatophoric dwarfism, Am. J. Dis. Child. 120:53, 1970.
4. Giedion, A.: Thanatophoric dwarfism, Helv. Paediatr. Acta 23:175, 1968.
5. Langer, L. O., Spranger, J. W., Greinacher, I., and Herdman, R. C.: Thanatophoric dwarfism. A condition confused with achondroplasia in the neonate, with brief comments on achondrogenesis and homozygous achondroplasia, Radiology 92:285, 1969.
6. Gwinn, J. L., and Lee, F. A.: Radiological case of the month–thanatophoric dwarfism, Am. J. Dis. Child. 120: 141, 1970.
7. Bailey, J. A.: Forms of dwarfism recognizable at birth, Clin. Orthop. 76:150, 1971.
8. Rimoin, D. L., Hughes, G. N. F., Kaufman, R. L., and McAlister, W. H.: The chondrodystrophies–clinical and histopathological correlations, J. Lab. Clin. Med. 74:1002, 1969.
9. Saldino, R. M.: Lethal short-limbed dwarfism: Achondrogenesis and thanatophoric dwarfism, Am. J. Roentgenol. Radium Ther. Nucl. Med. 112:185, 1971.
10. Harris, R., and Patton, J. T.: Achondroplasia and thanatophoric dwarfism in the newborn, Clin. Genet. 2:61, 1971.
11. Hall, J. G., Dorst, J. P., Taybi, H., Scott, C. I., Langer, L. O., and McKusick, V. A.: Two Probable Cases of Homozygosity for the Achondroplasia Gene, in *The Clinical Delineation of Birth Defects. Part IV, Skeletal Dysplasias*, Birth Defects: Original Article Series (New York: The National Foundation-March of Dimes, 1969), p. 24.
12. Chemke, J., Graff, G., and Lancet, M.: Familial thanatophoric dwarfism, Lancet 1:1358, 1971.
13. Graff, G., Chemke, J., and Lancet, M.: Familial recurring thanatophoric dwarfism. A case report, Obstet. Gynecol. 39:515, 1972.
14. Pena, S. D. J., and Goodman, H. O.: The genetics of thanatophoric dwarfism, Pediatrics 51:104, 1973.

ELLIS-VAN CREVELD SYNDROME
Chondroectodermal Dysplasia

Cardinal Clinical Features

Short-limbed dwarfism; polydactyly; congenital heart disease; ectodermal dysplasia involving nails, hair and teeth.

Clinical Picture

This syndrome was first described by Ellis and van Creveld in 1940.[1] An important characteristic of the syndrome is a pronounced short-limbed dwarfism, which affects mainly the distal portions of the extremities, making this predominantly a mesomelic shortening. Adult height in these patients ranges from 42–60 inches.[2] They also have genu valgum. The fingers are short and broad, and postaxial polydactyly, in which the extra digit is located on the ulnar side, is a consistent finding. This extra finger is lateral to the normal fifth finger and is well formed. It may be less than completely mobile. Adult patients usually cannot make a tight fist. More than six fingers on a hand is rare. Polydactyly of the feet also may occur in this syndrome, always on the fibular side. The nails are dysplastic. They are often small and may look like scales.

Patients with the Ellis-van Creveld syndrome are normal in intelligence. The skull is normal. They often have a partial harelip appearance due to a shortness of the upper lip and fusion of the lip to the gingivae, with multiple frenula. The gingival-labial sulcus may be entirely missing. Hypoplasia of the anterior maxilla may make the upper lip appear sunken. In most patients teeth have already erupted at birth, or they appear during the first month of life. The deciduous teeth are then lost prematurely. The secondary teeth are dysplastic and widely spaced or fail to erupt. The teeth often are conical. The occlusal surface may appear crenated. Fusion of teeth also has been seen. The hair often is very fine and may be fair and sparse, especially in infancy. The hair may also be normal.

Congenital heart disease is a frequent finding. It may be present in 50–60% of patients[2] and is the main determinant of prognosis in this syndrome.[3] The most prominent types are a single atrium or a large atrial septal defect.[3] Ventricular septal defect, patent ductus

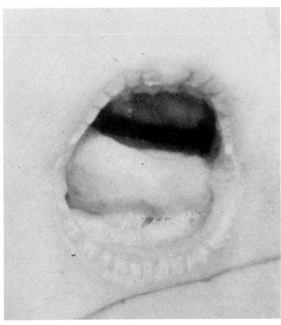

Fig. 1.—M. G. (Case 1.). Newborn infant with the Ellis-van Creveld syndrome. The upper lip illustrates broad (or multiple) tight frenulum, which obliterated the usual sulcus.

Fig. 2.—M. G. The lower gingivae showed prominences in the areas of the developing teeth.

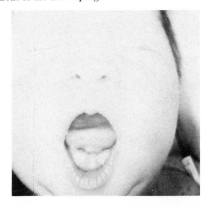

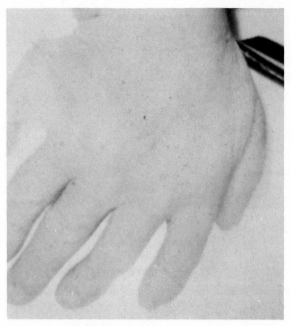

Fig. 3.—M. G. The polydactyly was as usual on the ulnar side. The extra finger was well formed. Its implantation appeared to be angular, and it was well separated from the fifth finger. The nails were dysplastic.

arteriosus, coarctation of the aorta, hypoplastic aorta, hypoplastic left ventricle or total anomalous venous return also are seen.[4] Patients with severe congenital heart disease may succumb soon after birth. Stillbirth also may be common in this syndrome. The other major cause of death in the neonatal period is a mechanical disorder of respiration due to hypoplasia of the bronchial cartilage or pulmonary hypoplasia. The thorax often is narrow in these patients. Epispadias has been observed, as has hypospadias. Testes may be undescended.

Roentgenographic findings are helpful in diagnosis.

Fig. 4.—M. G. The toes show markedly dysplastic nails.

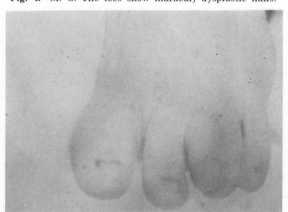

The bones of the extremities are short and thick. The pattern of shortening is one of progressive increase in shortening, from proximal to distal. Thus, the shortening is mesomelic and prominent in the ulna, radius, fibula and tibia, which may be unusual in shape. This pattern is the opposite of that found in achondroplasia. The proximal phalanges are normal in length in contrast to the middle phalanges, which are short and wide, and the distal phalanges, which are hypoplastic. Fusion of the hamate and capitate bones is very characteristic. There may develop other synostoses, as, for instance, between the distal fibula and tibia, and there may be an ankylosis that prevents flexion of the index finger. The polydactyly is seen as a well-formed sixth finger with a small, hypoplastic sixth metacarpal partially fused to the fifth metacarpal. There is markedly accelerated maturation of the epiphyseal ossification centers. Changes in the proximal portion of the tibiae lead to the genu valgum. The proximal tibia characteristically has a long lateral slope and a short medial one, and it is capped by a thin, medially ectopic ossification center. Exostoses from the medial aspects of the proximal tibial metaphyses have been reported.[5] The proximal ends of the tibia and ulna and the distal end of the radius are enlarged. Shortening of the ulna relative to the radius may lead to dislocation of the radial head.

CASE REPORTS

CASE 1.— M. G., a 2-day-old Mexican-American female, was referred to University Hospital for evaluation of congestive heart failure and polydactyly. She was the product of a normal pregnancy and delivery, the first child of a healthy 18-year-old mother and 22-year-old father. There was no consanguinity and no family history of malformation.

Physical examination revealed a mildly cyanotic infant with mild respiratory distress. Her heart rate was 160/minute and respiratory rate 60. She was 20 inches long and had an upper to lower ratio of 1.7. There was distal shortening of the extremities. She weighed 7 lb. The skull was normal. The up-

Fig. 5.— J. R. (Case 2). The hands of an adult male with the Ellis-van Creveld syndrome. The dysplastic nails were very striking. He also had polydactyly on the ulnar side. (Courtesy of Dr. Dantae Ayala of Parana, Argentina.)

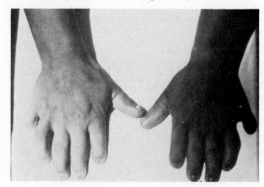

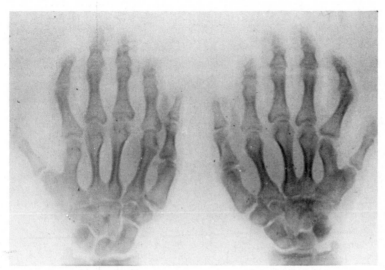

Fig. 6. – J. R. Roentgenogram of the hands, illustrating the characteristic fusion of the capitate and hamate bones. There were six metacarpals, but the fifth and sixth were fused proximally and separated distally. It is this that creates the angulation of the sixth digit. (Courtesy of Dr. Dantae Ayala of Parana, Argentina.)

per lip was short, and fusion to the gum was striking (Fig. 1). A few peg-shaped, lower teeth were present (Fig. 2). She had a sixth finger on each hand, on the ulnar side (Fig. 3). The nails of both hands and feet were dysplastic (Fig. 4).

The chest was symmetrical. There was a 2+ right ventricular lift. The second heart sound was split, and there was a third sound at the apex. There was a grade III/VI holosystolic murmur best heard at the left lower sternal border. The liver was palpable 6 cm below the right costal margin. The ECG showed right ventricular hypertrophy. The chest x-ray revealed cardiomegaly and increased pulmonary vascularity. Cardiac catheterization indicated a total anomalous pulmonary venous return, as well as a ventricular septal defect and a patent ductus arteriosus. The infant died during cardiac surgery.

CASE 2*. – J. R. was a 21-year-old student of agronomy. His height was 49 inches. He had fine hair. Teeth were dysplastic and had abnormal implantation. The thorax was narrow and appeared dystrophic.

The patient had short legs with thick thighs. There was bilateral genu valgum. Polydactyly was present in the fingers (Fig. 5) and toes. The nails were dysplastic. Roentgenograms revealed synostosis of the hamate and capitate (Fig. 6).

GENETICS

The disorder is transmitted as an autosomal recessive. Multiple involved siblings born to normal parents

*We are indebted to Dr. Dantae Ayala of Parana, Argentina, for providing us with the details of this patient.

have been reported. There is a high incidence of consanguinity among the parents of these patients. The disorder has been reported to be common in several ethnic groups, including the Old Amish Order of Lancaster County, Pennsylvania.[2]

TREATMENT

The major problem is the management of congenital heart disease.

REFERENCES

1. Ellis, R. W. B., and van Creveld, S.: A syndrome characterized by ectodermal dysplasia, polydactyly, chondroplasia and congenital morbus cordis, Arch. Dis. Child. 15:65, 1940.
2. McKusick, V. A., Egeland, J. A., Eldridge, R., and Krusen, D. E.: Dwarfism in the Amish. I. The Ellis-van Creveld syndrome, Johns Hopkins Med. J. 115:306, 1964.
3. Giknis, F. L.: Single atrium and the Ellis-van Creveld syndrome, J. Pediatr. 62:558, 1963.
4. Lynch, J. I., Perry, L. W., Takakuwa, T., and Scott, L. P.: Congenital heart disease and chondroectodermal dysplasia, Am. J. Dis. Child. 115:80, 1968.
5. Caffey, J.: Chondroectodermal dysplasia (Ellis-van Creveld disease), Am. J. Roentgenol. Rad. Ther. Nucl. Med. 68:875, 1952.

DIASTROPHIC DWARFISM

CARDINAL CLINICAL FEATURES

Short-limbed dwarfism; clubfoot; limited joint mobility; distinctive thickening of the external ears; short, broad fingers and hands; hitchhiker thumb.

CLINICAL PICTURE

In 1960, Lamy and Maroteaux[1] reported a group of children with micromelic dwarfism who had clubfoot and scoliosis. Three of the patients were their own, and 11 were drawn from the literature. At least 121 patients have now been reported.[2] Lamy and Maroteaux called this form of dwarfism "diastrophique," a word they derived from the Greek word διαστροφος, in order to imply tortuous or twisted. The term diastrophic is used in geology to describe the process of the bending of the crust of the earth by which mountains, continents and ocean basins are formed.

The dwarfism is notably severe. The ultimate adult height is usually about 44 inches, although, rarely, some individuals may reach 56 inches.[2] The shortening of the limbs is of the rhizomelic variety, that is, the most pronounced shortening occurs in the most proximal bones (femur and humerus). The extremities are not only short, but deformed, and the degree of deformity increases with age. In its most extreme form the fingertips may barely reach beyond the costal margin. There is an ulnar deviation at the wrists. The fingers and the hands are short and broad. The thumbs are hypermobile and proximally placed, and there is often a subluxation at the metacarpal phalangeal joints, with abduction, giving an appearance reminiscent of the thumb of a hitchhiker[2] (Figs. 1–3). The feet and toes are also short and broad. Most patients have an equinovarus deformity of the foot. There is a shortening of the Achilles tendon, and patients usually stand on their toes. There is a wide gap between the great toe and the second (Fig. 4). In some patients there is a hitchhiker malformation of the great toe. The joints of the shoulder, hip, elbow and knee are limited in motion. The fingers are stiff in extension, especially at the proximal interphalangeal joints, whereas the metacarpal phalangeal joints are hypermobile.

The limitations of joint motion are a major feature of the disease. They often are accompanied by shortening and contracture of the tendons. The hips appear normal at birth but, as the child begins to walk, he develops a dysplasia of the hip, often associated with dislocation and a coxa vara deformity. This deformity is progressive, and the changes become more pronounced with age. Genu valgum and contractures of the knee also are common, as are contractures of the elbow and dislocation of the radial head. Subluxation of the patella is a frequent problem.

Scoliosis is not present at birth but usually becomes evident in early childhood and is progressive. Kyphosis in the cervical area may be prominent and progressive with age; serious neurologic complications may result from cord compression.[3] The skull is normal.

Similarity in the facies of these patients has been pointed out.[2] This is particularly true of the area around the mouth. The distance between the nose and upper lip is broad and the circumoral area prominent. The arch of the palate may be wide, and many teeth become visible when the patient smiles. Because of these features, these patients have been called "cherub dwarfs." The nose may have a narrow bridge, a broad midarea and flaring nostrils.

The deformity of the ear is unique. It is seen in over 80% of patients. It appears in the neonatal period with a swollen, inflamed-appearing, often cystic pinna, Aspiration at this stage yields a serofibrinoid sterile fluid.[3] The swelling usually diminishes spontaneously in a few weeks but may have a deformity of the pinna. With time there is a progressive, thickened, crumpled appearance, and ultimately calcifications appear in the thickened areas.

Cleft palate occurs in about half of the patients. Variants of this may be a cleft of the soft palate, a submucous cleft or a high-arched palate. The newborn infant may have a narrow trachea and a hoarse cry because of abnormalities in cartilage similar to those in the pinna.[2] Death in infancy from respiratory failure and aspiration pneumonia is relatively common.[4]

Roentgenographic examination reveals short, but massive tubular bones with broad, flared metaphyses. The rhizomelic shortening is readily apparent. The bones are impressive for the degree of irregularity and deformity. The ends of the long bones are lucent as well as enlarged. Epiphyseal development is delayed, but the carpal bones show accelerated maturation. The first metacarpal is very short and is ovoid or triangular. The other metacarpal bones are short and wide,

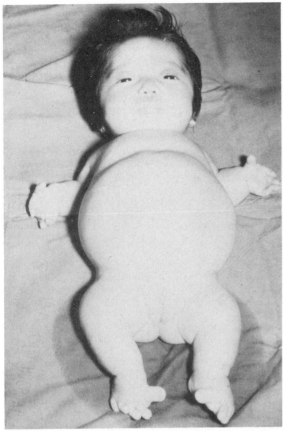

Fig. 1.—An infant with diastrophic dwarfism. The short-limbed dwarfism is striking. She had the hitchhiker deformity of thumbs and great toes.

Fig. 3.—The hitchhiker thumb and short, wide fingers and hands.

Fig. 4.—Hitchhiker deformity in the feet. The great toes were angulated and widely separated from the rest of the toes.

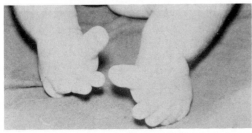

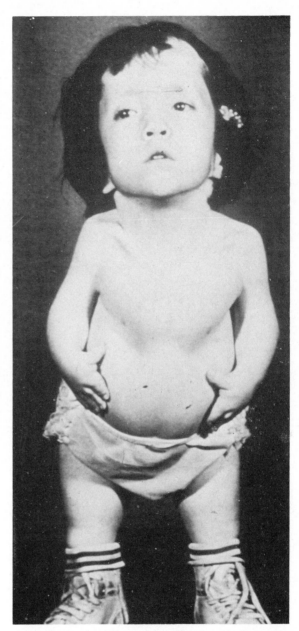

Fig. 2.—Six-year-old, 26-inch tall girl, with diastrophic dwarfism. Height age was 7 months. Legs were very short; arms were short and curved. The bilateral hitchhiker thumbs are well demonstrated. There also was deformity of the chest. (Courtesy of Dr. John Mace of Loma Linda University, and reproduced with permission from Am. J. Dis. Child. 116:525, 1968.[8])

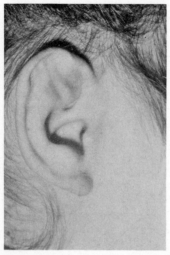

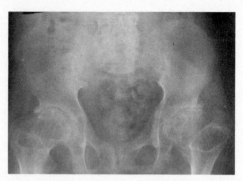

Fig. 7.—S. M. Roentgenogram of the pelvis. The ends of the femora appeared irregularly swollen. The acetabular cavities were also irregularly and enormously enlarged. Articular cartilage was thinned.

Fig. 5.—S. M. (Case 1). The ear of the proband of the family whose pedigree is shown in Figure 9. The cartilage of the ear is thickened throughout the superior portion of the pinna.

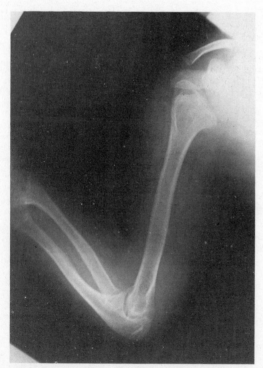

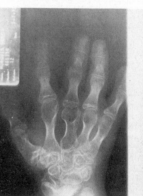 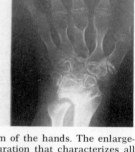

Fig. 8.—S. M. Roentgenogram of the hands. The enlargement and distortion of configuration that characterizes all of the long bones in this disease was particularly evident in the metacarpals. They were knurled, rounded and lucent. The first metacarpal was particularly short and somewhat triangular.

Fig. 9.—Case Reports. Pedigree of the family illustrating autosomal recessive inheritance. *Arrow* indicates S. M., the proband.

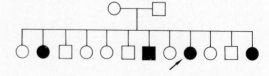

Fig. 6.—S. M. Roentgenogram of the arms illustrates the mesomelic shortening that is characteristic of this syndrome. The forearms were short. The radius was bowed. The ends of all of the bones of the arm were enlarged, misshapen, flared and relatively radiolucent.

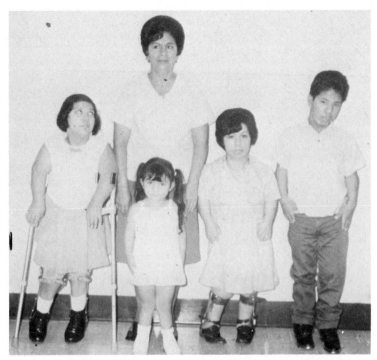

Fig. 10.—Case Reports. Family in which the four siblings shown all had diastrophic dwarfism. The shortness of stature was marked. The mother was normal.

Fig. 11.—M. M. (Case 2). The 24-year-old sister of S. M. These patients seem to grow shorter as they get older. She had the bilateral club feet characteristic of the disorder. The hands were short and broad. She had limitation of joint motion, particularly striking in the hips, elbows and knees.

Fig. 12.—W. W. Roentgenogram in an adult with the diastrophic variant. The calcification of the costal cartilages is striking.

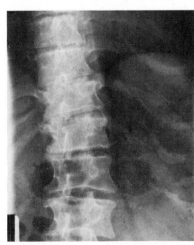

and the distal ends are wider than the proximal ends. The height of the vertebrae is normal. The interpedicular distances between L1 and L5 may narrow somewhat[5] but never to the degree seen in achondroplasia. The deep sacrosciatic notches characteristic of achondroplastic dwarfism are not present. The iliac bones may be distorted and the pelvis abnormally developed.[5] The acetabular cavities are wide and deep, but the femoral head is hypoplastic, wide and flat, and the neck short and wide. Subluxations of a number of joints may be seen. Precocious calcification of the costal cartilages is a striking characteristic of this dwarfism. It may reflect a general disorder of cartilage, seen most clearly in the ear. In the newborn period the epiphyses of the tubular bones may be stippled.[6]

A type of diastrophic variant has been described (Fig. 12). Patients with this disorder usually are taller than the usual diastrophic dwarf. Club feet may be absent or not as resistant to treatment. Patients have similar deformities of the external ears.

Differential Diagnosis

The diastrophic dwarf should be distinguished primarily from the achondroplastic dwarf. This is an important distinction, because there is a greater morbidity in diastrophic dwarfism and more difficulties are encountered in the management of their bony abnormalities. The distinction also is important for genetic counseling. Achondroplasia is an autosomal dominant trait and, when it occurs with normal parents, represents a new mutation with essentially no risk of recurrence. Diastrophic dwarfism is inherited as an autosomal recessive from normal parents, and there is a 25% chance of recurrence.

The differentiation from achondroplasia may not be easy to make in early infancy, and a majority of patients reported have been so diagnosed at first, but the progression of deformity in diastrophic dwarfism and its associated abnormalities are unique. Its recognition should not be difficult. The combination of a gibbus and an anterior protrusion of the sternum that is sometimes seen may suggest the Morquio disease. Diastrophic dwarfism also is confused with arthrogryposis multiplex congenita. Patients with arthrogryposis have joint contractures, but they have a striking decrease in muscle mass and none of the distinctive physical or roentgenographic abnormalities of diastrophic dwarfism.

Case Reports

Case 1.—S. M. was 12 years old when first seen at University Hospital for shortness of stature. She had carried a diagnosis of the Morquio disease for many years. Her height was 47

inches. Thus, her height age of 6¾ years was half her chronologic age. The upper to lower body ratio was 1.46, which indicated, as did her appearance, that she had short-limbed dwarfism. These observations excluded a diagnosis of the Morquio disease, which is characteristically a short-trunked dwarfism. The length of her trunk was not abnormal.

She was born, after a normal pregnancy and delivery, with bilateral clubfoot. At 12 she had limitation of joint motion, especially in the knees, hips and elbows. The cartilage of the right pinna was thickened (Fig. 5); the left was normal. She did not have hitchhiker thumb and there was no cleft palate. The skull was normal, as was her intelligence. There was no scoliosis or kyphosis.

Urinalysis for mucopolysaccharides was normal. A skeletal survey revealed that there was generalized shortening of all of the tubular bones, prominently rhizomelic, and there was flaring of the metaphyses. Endochondral growth of bone was abnormal throughout. The ends of the bones were enlarged, misshapen and relatively radiolucent (Fig. 6). The femoral heads were flattened. The pelvis was narrowed (Fig. 7). The metacarpals were conspicuously short. The rounded, knurled, lucent appearance at the ends of the bones was striking (Fig. 8). The proximal interphalangeal joints were very narrow or appeared to be absent. There were calcifications in the cartilage of the ear.

This patient was one of 12 children (Fig. 9). Three siblings were affected (Fig. 10; S. M. on left). There was no consanguinity of the parents. There was a certain amount of variability of expression in this family.

Case 2.—M. M., the 24-year-old sister, was 47 inches tall. She had very short, broad hands, but no hitchhiker thumb. The ears were normal. There was limitation of supination at the elbows and the hip deformity was so great that she had undergone surgery at age 13. She still had limited motion at this joint. Both feet were clubbed (Fig. 11). She drove a car and was gainfully employed sorting dials. There was very little motion in the right ankle. Both feet were short and broad and had very broad toes. The left knee was fully mobile, but the right knee was limited in flexion and extension.

Case 3.—F. M., the 15-year-old brother, was 54¾ inches tall and was the least deformed of the family. He had short, broad hands and somewhat short limbs, but both less so than his siblings; he also had very limited motion of the elbows, which appeared relatively very large. He had no scoliosis or club feet.

Case 4.—A. M., the 4-year-old sister, was 38¾ inches tall. She looked grossly normal, but on closer inspection had short arms, somewhat limited elbow motion and short, broad hands. She had genu valgum and pes valgus but no clubfoot. Her ears were very impressive. Both pinnae were thick and crumpled.

All four siblings in this family were diagnosed as having diastrophic dwarfism: all had short-limbed dwarfism, with limited motion of the joints. Two had clubfoot, and two the ear anomaly. It is of interest to note such intrafamilial variability, as well as the apparent progression from the mild presentation seen in the youngest to the more severe in the oldest. Presumably, further progression will be the rule.

Genetics

This disorder is transmitted as an autosomal recessive. Consanguinity has been observed.[5]

TREATMENT

The clubfoot is said to be particularly difficult to treat.[7] Orthopedic surgery for these and other areas may have to be repeated because the disease is progressive. Kyphoscoliosis is progressive, and treatment should be carried out to prevent neurologic complications. Early death may occur in infancy because of respiratory failure and aspiration pneumonia. Early diagnosis could lead to a preparedness to treat these problems vigorously, as otherwise the life span should be normal. On the other hand, the degree of deformity is sufficiently great that walking becomes difficult and, in some patients, impossible.

REFERENCES

1. Lamy, M., and Maroteaux, P.: Le nanisme diastrophique, Presse Med. 68:1977, 1960.
2. Walker, B. A., Scott, C. I., Hall, J. G., Murdoch, J. L., and McKusick, V. A.: Diastrophic dwarfism, Medicine 51:41, 1972.
3. Langer, L. O., Jr.: Diastrophic dwarfism in early infancy, Am. J. Roentgenol. Radium Ther. Nucl. Med. 93:399, 1965.
4. Wilson, D. W., Chrispin, A. R., and Carter, C. O.: Diastrophic dwarfism, Arch. Dis. Child. 44:48, 1969.
5. Taybi, H.: Diastrophic dwarfism, Radiology 80:1, 1963.
6. Silverman, F. N.: A differential diagnosis of achondroplasia, Radiol. Clin. North Am. 6:223, 1968.
7. Kite, J. H.: Achondroplasia: The clubfoot problem in achondroplasia, South. Med. J. 54:577, 1961.
8. Gellis, S. S., and Feingold, M.: Diastrophic dwarfism, Am. J. Dis. Child. 116:525, 1968.

MULTIPLE EPIPHYSEAL DYSPLASIA

Cardinal Clinical Features

Maldevelopment of the epiphyseal ossification centers, bilateral and symmetrical, leading to shortness of stature; genu valgum; abnormal gait; pain and stiffness of the hips; short, stubby digits.

Clinical Picture

Multiple epiphyseal dysplasia was first described by Jansen[1] in 1934. Fairbank[2] in 1935 characterized it as a distinct clinical entity involving a number of different patients and gave it the name dysplasia epiphysealis multiplex. It is characterized by shortness of stature, or dwarfism, and short, stubby digits with blunt ends. An involved patient may appear completely normal except for the presence of genu valgum and may have completely normal activity. Others may have difficulty in walking, climbing stairs or even rising from a lying position. They may have pains in the hips, knees or ankles. A broad-based, duck-like gait is common. Some patients have severe deformities, including coxa vara or flexion contractures of the knees, elbows or hips. Tibial bowing also may occur. Pes planus is common. There may be limited abduction of the shoulders. Subluxation of the radial and tibial heads has been seen rarely.

Most descriptions have been of children aged 5–14 years. These patients appear normal as infants. The gait usually is abnormal from the beginning, but this often is subtle. The patients are brought to the physician because of the onset of pain or deformity. The condition affects both sexes. Intelligence usually is normal. The prognosis is good. Manifestations expected in the adult are shortness of stature and early degenerative arthritis in the weight-bearing joints.

The diagnosis is made on the basis of the roentgenographic appearance.[3] Development of the centers of ossification of the tubular bones is delayed and irregular in ossification. The epiphyses may appear mottled. With weight-bearing, the epiphysis may become compressed, moulded or fragmented. Involvement is bilateral and symmetrical.

The knees and ankles are always more severely involved than the wrists and elbows. The patella frequently is mottled. Changes in the distal tibia are valuable in establishing the diagnosis of epiphyseal dysplasia, especially in an adult with advanced bilateral osteoarthritis of the hips. As the patient grows older, the ossification centers mineralize normally. After ossification has occurred, there may be dense central nuclei of bone surrounded by less dense areas, or there may be vertical striations at the ends of the bones. As changes develop in the epiphyseal ossification centers of the tibia, there is a deficiency of the lateral portion of the distal center. This produces a triangular appearance in the frontal view. Tibial obliquity or slanting persists.

Metaphyseal changes are observed rarely. The carpal and tarsal bones may be hypoplastic, and they may have irregular outlines. The metacarpals often are short and irregular. The vertebrae are important in the differential diagnosis. In most reported cases the vertebrae were normal. Maudsley[4] reported vertebral changes in three of 14 cases. These consisted of irregularity of the surfaces, Schmorl's nodules, and anterior wedging of isolated vertebrae. Patients also have been seen with universal vertebra plana or platyspondyly.

Differential Diagnosis

The Legg-Perthes disease, when it is bilateral, is limited to the proximal femoral epiphyseal ossification centers. The femoral head is flattened, and there may be destructive changes in the neighboring metaphysis. The major finding in congenital hypothyroidism (cretinism) is retardation of skeletal maturation. There may be spotty mottling of the epiphyseal centers and irregular ossification in this cretinoid epiphyseal dysgenesis.

Patients with multiple epiphyseal dysplasia congenita (chondrodystrophia calcificans congenita; the Conradi syndrome), are recognizably abnormal in the immediate postnatal period. The infant has short limbs. He may have contractures with flexion deformities at the hips. Stippling of the epiphyses is prominent in the shoulders, hips and knees. There also is stippling of vertebrae, sternum and hyoid bone. The outstanding feature in spondyloepiphyseal dysplasia congenita (the Morquio disease) is the platyspondyly. Involvement of the thoracic and lumbar vertebrae make for a dwarfism with short trunk. Irregularity of the epiphyses is present. Patients often have genu val-

250

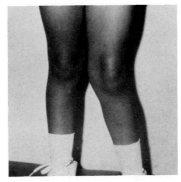

Fig. 2.—R. B. Closer view of the knees, illustrating the marked angulation of the tibia.

gum, but they also have severe kyphosis. Keratosulfate is found in the urine. In the pseudoachondroplastic form of spondyloepiphyseal dysplasia there are small, irregular epiphyses, irregular metaphyses and flattening and/or anterior beaking of the vertebrae.

CASE REPORT

R. B., a 3-year-old girl, was referred to University Hospital for evaluation of joint pain. The child was the product of a

Fig. 3.—R. B. Roentgenographic appearance of the knees. The distal femoral and proximal and distal tibial ossification centers all show poor development and flattening. Changes are most marked in the distal femoral epiphyses.

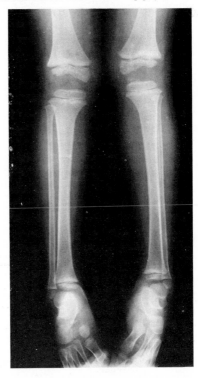

Fig. 1.—R. B. (Case Report). A 3-year-old girl with multiple epiphyseal dysplasia. Genu valgum was pronounced. Otherwise she appeared normal.

normal pregnancy and delivery. The father was 29 years old at the time of her birth. She sat at 6 months and walked at 11 months. She developed pain and stiffness of the knees. She had some difficulty in walking and climbing stairs. The mother was concerned about her knock-kneed appearance. There was no family history of short stature, joint pain or arthritis.

Examination revealed a well-developed black female. Her intelligence was normal, as was her general appearance. She weighed 33 lb. Her height was 37½ inches (10th percentile). Her only abnormality was a rather pronounced bilateral genu valgum (Figs. 1 and 2). The spine was normal. The digits were normal.

Roentgenograms revealed multiple irregularities and hypoplasia of the ossification centers of the long bones that were bilateral and symmetrical. Changes were most prominent in the knees (Fig. 3). They also were prominent in the hips (Fig. 4) and ankles. Bone age was normal. The spine was normal. Calcium, phosphate and alkaline phosphatase in the serum were normal. Roentgenographic examinations of the epiphyseal centers in the mother, father and 11-year-old sibling were normal.

GENETICS

This disorder is transmitted as an autosomal dominant. Most reported cases have been sporadic, representing new mutations. Involved patients were reported in three generations of two families. In other families reported, there was involvement in a mother and four children, in a mother and three children, in a mother and a son, in a mother and twin boys and in a father and a son.

TREATMENT

The disease is to a considerable extent self-limited. Normal trabeculation is ultimately achieved after fusion of the epiphysis has occurred. Deformity, once present, tends to persist. Early degenerative osteoarthritis is common. No therapy of proved value is available. Angulations can be corrected surgically after ossification and fusion have occurred.

Fig. 4. – R. B. Roentgenogram of the hips. The hips show moderate epiphyseal irregularity, with deformity of the articular surfaces of the capital femoral epiphysis and little or no change in the acetabular roof.

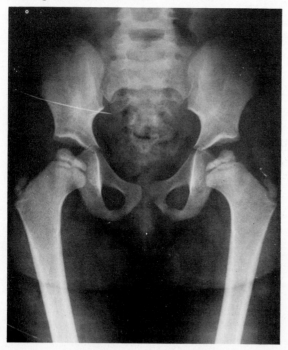

REFERENCES

1. Jansen, M.: Uber atypische Chondrodystrophie (Achondroplasie) und über eine noch nicht beschriebene angeborene Wachstums störung des Knochensystems: Metaphysäre Dysostosis, Z. Orthop. Chir. 61:253, 1934.
2. Fairbank, H. A. T.: Dysplasia epiphysealis multiplex, Proc. R. Soc. Med. 39:315, 1946.
3. Christensen, W. R.; Lin, R. K., and Berghout, J.: Dysplasia epiphysealis multiplex, Am. J. Roentgenol. Radium Ther. Nucl. Med. 74:1059, 1955.
4. Maudsley, R. H.: Dysplasia epiphysealis multiplex, J. Bone Joint Surg. [Br.] 37:228, 1955.
5. Freiberger, R. H.: Multiple epiphyseal dysplasia, a report of three cases, Radiology 70:379, 1958.
6. Shephard, E.: Multiple epiphyseal dysplasia, J. Bone Joint Surg. [Br.] 38:458, 1956.
7. Leeds, N. E.: Epiphyseal dysplasia multiplex, Am. J. Roentgenol. Radium Ther. Nucl. Med. 84:506, 1960.
8. Caffey, J.: *Pediatric X-ray Diagnosis* (5th ed.; Chicago: Year Book Medical Publishers, Inc., 1967).
9. Rubin, P.: *Dynamic Classification of Bone Dysplasias* (Chicago: Year Book Medical Publishers, Inc., 1964).

Dwarfism with Short Trunk

COSTOVERTEBRAL DYSPLASIA
Spondylocostal Dysplasia,
Spondylothoracic Dysplasia

CARDINAL CLINICAL FEATURES

Short-trunked dwarfism; developmental segmental anomalies of spine and ribs; short, webbed neck; increased anteroposterior diameter of the chest; protuberant abdomen.

CLINICAL PICTURE

The fundamental defects in this syndrome are anomalies of the ribs and vertebrae. This produces a short-trunked dwarfism (Fig. 1). Caffey[1] published pictures and roentgenograms of a brother and sister with multiple hemivertebrae and rib anomalies who were typical examples of this syndrome. These patients often have short, webbed necks with a limitation of motion and a low posterior hairline. The disproportion in this type of dwarfism is similar to that of the Morquio disease in which the short trunk is due to vertebra plana. In this syndrome the spine often is irregularly curved. When the patient stands up, his fingertips may each his knees. Patients have protuberant abdomens (Fig. 2). The thorax also may appear protuberant because of the shortness of the thoracic spine, but most often it actually is small. Pulmonary function usually is diminished, and frequent respiratory infections may be a problem and even a cause of death. Death from respiratory infection prior to 1 year of age is common in this syndrome. Congenital hypoplasia of a lung or a lobe may be seen.

Roentgenographic examination reveals hemivertebrae, cleft and butterfly vertebrae, as well as striking segmentation defects of the entire vertebral column from the cervical to the coccygeal spine.[1-5] There also may be partial absence of some and a reduction in the total number of these bones. Rib anomalies include errors in segmentation, costal fusion, bifurcations and a decreased total number. The articulations of ribs and vertebrae are abnormally close together, so that the ribs may radiate from the spinal column in a fan-like fashion.[2] The spacing of the ribs is irregular. Hemivertebrae have been recognized in this syndrome in roent-

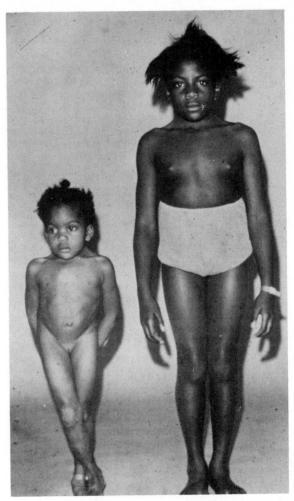

Fig. 1.—Two sisters with costovertebral dysplasia. Both had short necks and short trunks. Their faces were normal. The normal extremities and very short trunk give an appearance that the contents of the chest and abdomen are excessive for the size of the child. This is similar to the over-all appearance of a patient with the Morquio syndrome.

255

genograms obtained before birth. The remainder of the skeletal survey is normal.

The spinal and costal deformities occur very early in fetal life. In the fourth to seventh week of development,[2] sclerotomes give rise to intervertebral disks and connective tissue vertebrae. Centers of chondrification appear around the notocord, originating in two centers for each vertebral body and one for each side of the incomplete arch. All of the centers normally grow and fuse to form the vertebra. Dysplasia in these centers, and failures of fusion, would lead to the picture observed.

DIFFERENTIAL DIAGNOSIS

The differential diagnosis includes the Morquio disease, in which the appearance of the trunk is similar but which is readily distinguished by x-ray or examination of the urine for keratosulfate. The Treacher Collins and Goldenhar syndromes have similar vertebral anomalies. An oculovertebral syndrome with similar vertebral malformation has been described by Weyers and Thier.[6] Spondyloepiphyseal dysplasia may give a similar initial appraisal.

CASE REPORTS

CASE 1.—Baby K. was the product of a normal, full-term pregnancy. She was a breech presentation, and delivery was by cesarean section. Her 17-year-old mother was normal and healthy, as was the father. This was their first child. There was no consanguinity or family history of a similar disorder. A week prior to delivery, a roentgenogram of the fetus was obtained because of the breech presentation. At that time malformations of the ribs and vertebrae were observed. Following delivery the baby had a heart beat of 20 for a few minutes and then died.

The trunk was disproportionately short and the abdomen

Fig. 2.—M. S. (Case 2). The prominent chest and abdomen are evident.

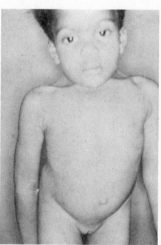

protuberant (Fig. 3). Length was 14½ inches: the lower segment, 7½ inches, and the upper segment, 7 inches, giving an upper to lower ratio of 0.9. The normal ratio at birth is 1.7. This is the classic pattern for short-trunked dwarfism. The birth weight was 6 lb. The head circumference was 14.25 inches, that of the chest 12 inches. The neck was very short and webbed, and the head appeared to arise directly from the shoulders. There were no facial anomalies. Severe lordosis was present. The chest was narrow. The liver and spleen were palpable, but the size of both organs was normal at autopsy. The lower extremities showed no bony defects on x-ray, although the left leg had a genu recurvatum and the right leg had a flexion contracture at the knee joint. There were bilateral club feet.

Postmortem x-rays revealed classic costovertebral dysplasia. The spine was unusually short. The number of vertebrae was decreased, and these were poorly segmented. Multiple

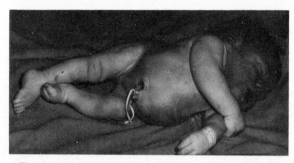

Fig. 3.—Baby K. (Case 1). A stillbirth with short-trunked dwarfism. The neck was very short: the head appeared to be sitting on the shoulders. The abdominal and thoracic contents were normal in size but so out of proportion to the spaces they occupied that it was originally thought that the patient had hepatomegaly.

Fig. 4.—Baby K. There were poorly segmented vertebrae and many hemivertebrae. The ribs were markedly malformed, demonstrating many errors of segmentation or fusion.

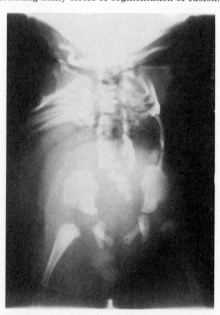

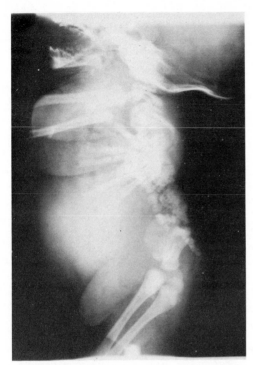

Fig. 5.— Baby K. Lateral view of the same patient. The ribs were distributed in a fan shape. The vertebral anomalies are also seen in this projection.

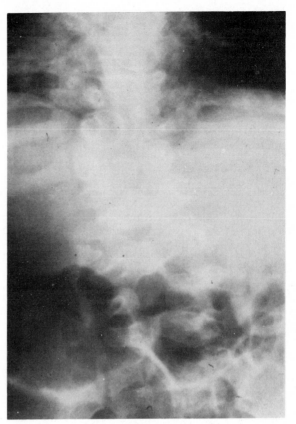

Fig. 6.— M. S. Roentgenogram shows the many hemivertebrae, which have produced a spine that is short, curved and misshapen.

hemivertebrae were present, some of which were fused (Fig. 4). The posterior neural arches were deficient, and there were multiple spinae bifidae. The ribs also had segmentation defects and were distributed in a fan-like array (Fig. 5). The rest of the roentgenographic examination was normal.

CASE 2.— M. S. was a 4-year-old black girl admitted to Mercy Hospital, San Diego, for pneumonia. She was found to have short-trunked dwarfism. Family history revealed that she had a 9-year-old sister with an identical syndrome; the sister was 53 inches tall and had a sitting length of 37 inches. The parents were normal. There were no other siblings. There was no consanguinity.

The patient had a short, webbed neck with limited movement and kyphoscoliosis. The hairline seemed low. Her length was 37.5 inches, sitting length 29 inches. In a standing position her finger tips reached her knees. She appeared to have a normal supply of thoracic and abdominal organs squeezed in a space that was abnormally short (see Fig. 2), and thus the anteroposterior diameter of the trunk was increased.

Roentgenographic examination revealed multiple segmentation anomalies of the cervical, thoracic and lumbar spine. There were many hemivertebrae (Fig. 6). The patient also had multiple rib anomalies, including bifurcation, fusion and reduction in number. X-rays of the sister revealed the same problem.

Chromosomal studies revealed normal female karyotypes.

GENETICS

This syndrome may represent genetic heterogeneity in which the same phenotype may result from autoso-mal recessive or dominant transmission. The patients described in the Case Reports seemed likely to be examples of autosomal recessive inheritance, as did Caffey's.[1] The picture seemed even clearer in the patients reported by Cantú and colleagues,[2] in which there were two involved siblings and three such cousins in a highly consanguineous family. The patients of Norum[3] were from a population isolate, and those of Lavy, Palmer and Merritt[4] were the offspring of a consanguineous marriage. Four affected persons were reported in each of these families. In contrast, Van de Sar[7] reported multiple hemivertebrae in a mother and daughter. These patients and that of Rimoin et al.[8] appear to represent autosomal dominant inheritance of the same syndrome. In general, patients with autosomal recessive cases are more severely affected and are prone to early death. It is possible that some patients with this phenotype may be nongenetic, as a similar phenotype has been produced in experimental animals by maternal hypoxia[9] or hypoglycemia[10] during pregnancy.

TREATMENT

None.

REFERENCES

1. Caffey, J. (ed.): *Pediatric X-Ray Diagnosis* (5th ed.; Chicago: Year Book Medical Publishers, Inc., 1967), p. 1113.
2. Cantú, J. M., Urrusti, J., Rosales, G., and Rojas, A.: Evidence for autosomal recessive inheritance of costovertebral dysplasia, Clin. Genet. 2:149, 1971.
3. Norum, R. A.: Costovertebral Anomalies with Apparent Recessive Inheritance, in *The First Conference on the Clinical Delineation of Birth Defects. Part IV, Skeletal Dysplasia*, Birth Defects: Original Article Series (New York: The National Foundation-March of Dimes, 1969), p. 326.
4. Lavy, N. W., Palmer, C. G., and Merritt, A. D.: A syndrome of bizarre vertebral anomalies, J. Pediatr. 69:1121, 1966.
5. Moseley, J. E., and Bonforte, R. J.: Spondylothoracic dysplasia—a syndrome of congenital anomalies, Am. J. Roentgenol. Radium Ther. Nucl. Med. 106:166, 1969.
6. Weyers, H., and Thier, C. V.: Malformations mandibulofaciales et délimitation d'un "syndrome oculo-vertebral," J. Genet. Hum. 7:143, 1958.
7. Van de Sar, A.: Hereditary multiple hemivertebra, Doc. Med. Geograph. Trop. 4:23, 1952.
8. Rimoin, D. L., Fletcher, B. D., and McKusick, V. A.: Spondylocostal dysplasia, a dominantly inherited form of short-trunked dwarfism, Am. J. Med. 45:948, 1968.
9. Ingalls, T. H., and Curley, F. J.: Principles governing the genesis of congenital malformations induced in mice by hypoxia, N. Engl. J. Med. 257:1121, 1957.
10. Theiler, K.: Redaktionelle Ubersicht, das Wirbel-Rippen-Syndrom, Schweiz. Med. Wochenschr. 98:907, 1968.

SPONDYLOEPIPHYSEAL DYSPLASIA

CARDINAL CLINICAL FEATURES

Short-trunked dwarfism; dysplasia of vertebral bodies and pelvis; odontoid hypoplasia and thoracic protrusion such that the chin appears to rest on the chest; hypertelorism, prominent eyes, myopia, retinal detachment; cleft palate.

CLINICAL PICTURE

This condition was first described in 1966 by Spranger and Wiedemann.[1, 2] Spranger and Langer[3] made a study of the x-rays of 29 of these patients and characterized the radiologic features of the syndrome.

The disorder is recognizable at birth. The newborn infant has a short-trunked dwarfism and, in addition, a flat face, hypertelorism and a short neck. There may be an upward slant of the palpebral tissues and a cleft palate. It is especially important to look repeatedly for the ocular anomalies associated with the syndrome, as these patients are extremely myopic and may have retinal detachments that can lead to blindness.[4, 5] Other ocular abnormalities have been observed in these patients,[4] including corneal opacities, cataracts and glaucoma. Deafness has been observed in some patients. Psychomotor retardation may be seen.

The spine is very short. The thorax combines a barrel-shape with a pronounced pectus carinatum. There is marked thoracic kyphosis and lumbar lordosis. The hands and feet appear normal. Genu valgum occurs and sometimes bowlegs are observed. The patient may have a waddling gait. Problems with ambulation may be aggravated by hypotonia.

The most striking roentgenographic findings are in the spine, pelvis and proximal femur. Roentgenograms in infancy reveal a generalized delay in ossification.[3] The vertebral bodies appear flat and pear shaped. The pelvis is underossified, particularly in the pubis. The distal femoral and proximal tibial epiphyses usually are not ossified in the newborn infant.

Later in childhood the ossification of the pubic bone is still poor. The iliac wings are low and rounded and are reminiscent of the ears of Mickey Mouse. The acetabula have flattened roofs and appear empty because of the marked underossification of the femoral heads and necks. The femora develop a marked varus deformity. The vertebral bodies retain their flat dysplastic appearance. Odontoid hypoplasia is a characteristic feature of the syndrome. Epiphyseal dysplasia is more marked proximally, and the long tubular bones are short. The carpal bones and phalanges of the hands and feet may be only slightly involved and usually appear normal. In the adult the odontoid remains only partly ossified, and the flattened, irregular vertebral bodies determine the characteristic shortness of stature. Coxa vara may be extreme.

DIFFERENTIAL DIAGNOSIS

This condition should be differentiated from the Morquio syndrome (see section on Mucopolysaccharidoses), which also produces short-trunked dwarfism, but this is not evident at birth. The appearance of the vertebral bodies in the two conditions differs, as does that of the pelvis. In the Morquio disease there is a distinct keratosulfaturia not seen in spondyloepiphyseal dysplasia. The other syndrome from which spondyloepiphyseal dysplasia congenita must be distinguished is spondyloepiphyseal dysplasia tarda,[6] a genetic disorder transmitted as an X-linked recessive, in which the short trunk does not become evident until the age of about 10 years. Patients with this disorder also have flat vertebrae, kyphoscoliosis, mild epiphyseal irregularities and short femoral necks.

CASE REPORT*

R. G. was a 21-month-old Mexican female who had been admitted to Mercy Hospital, San Diego, for pneumonia. She had had several previous admissions for similar problems. She was the product of a normal pregnancy, but was delivered by cesarean section because of a facial presentation. She appeared at birth to be dwarfed. She also had a cleft palate. The birth weight was 6 lb, 1 oz. She was first admitted to hospital at a few weeks of age because of difficulty in feeding and pneumonia and later readmitted at 3 months for the same problems.

She was the seventh child of a 40-year-old mother and a 42-year-old father. Both parents were of normal height, as were all six siblings. The paternal uncle was a short-limbed dwarf, as was his son. A skeletal survey revealed that the uncle had classic achondroplasia.

---

*We are indebted to Dr. David Allan of Mercy Hospital, San Diego, for permission to report on this patient.

On physical examination R. G. was very short (Fig. 1). Her height at 20 months was 24 inches, giving a height age of 4 months. She had a flat face (Figs. 1 and 2), a flat nasal bridge and a cleft palate. The neck was particularly short. The chest was barrel-shaped and keeled (Fig. 3). She was hypotonic. At 22 months she sat but did not stand or speak. Ophthalmologic examination was normal.

Roentgenographic survey of the skeleton at a few weeks of age showed complete absence of ossification of the pubis and the femoral head (Fig. 4). The odontoid was hypoplastic. At 22 months (Fig. 5) there was continued lack of ossification of the pubis and underossification of the femoral head. There was marked flaring of the metaphyseal end of the proximal femur and a varus deformity. Metaphyseal flaring was seen as well in the other long bones of the extremities, but the hands and feet were normal. The vertebrae were flat and pear shaped (Fig. 6). These x-rays were classic for a diagnosis of spondylo-epiphyseal dysplasia congenita. The films clearly ruled out a diagnosis of achondroplasia and indicated that in this family spondyloepiphyseal dysplasia was due to a new mutation, related perhaps to the advanced paternal age.

GENETICS

Spondyloepiphyseal dysplasia is transmitted as an autosomal dominant. Mother to child transmission has been reported. A significant number of cases results from new mutation. The effect of advanced paternal age on the incidence of new mutation suggested in our patient (Case Report) has been observed in a series of sporadic cases in which the average age of the fathers was 6 years older than that of a control group.[4]

TREATMENT

None. With counseling these patients can learn to live with the deformity. Careful ophthalmologic monitoring and appropriate management of retinal detachment are important.

Fig. 2. – R. G. At age 6 weeks, she appeared very short. The neck was short and the chin rested on the sternum, but this appearance is not as unusual at this age as later. The nasal bridge was depressed. (Courtesy of Dr. David Allan of Mercy Hospital, San Diego.)

Fig. 1. – R. G. (Case Report). Spondyloepiphyseal dysplasia in a 22-month-old infant. The over-all appearance of this child was classic. The dwarfism was of the short-trunk variety. The chest was narrow and flared below, and the chin rested squarely on the sternum. (Courtesy of Dr. David Allan of Mercy Hospital, San Diego.)

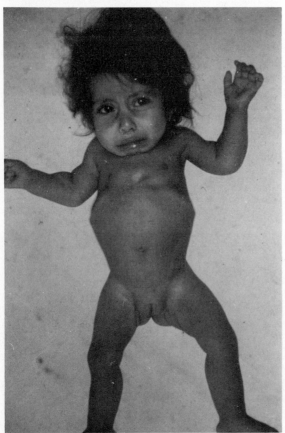

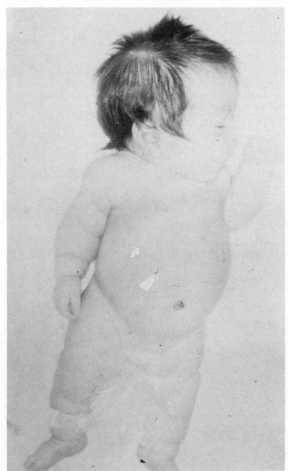

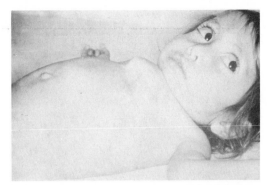

Fig. 3.—R. G. Pectus carinatum. The eyes appeared prominent. The patient had hypertelorism and a cleft palate. (Courtesy of Dr. David Allan of Mercy Hospital, San Diego.)

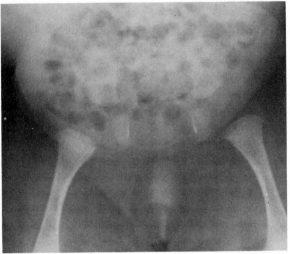

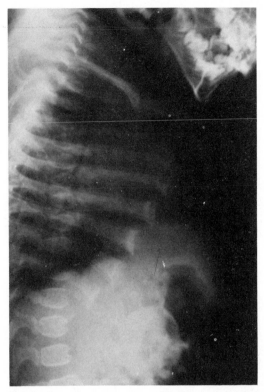

Fig. 6.—Lateral roentgenogram illustrates the vertebral malformation. The thoracic vertebrae were flattened and pearshaped. The ribs were cupped and flared anteriorly. The sternum protruded anteriorly so that it was almost horizontal at the point of the chin. (Courtesy of Dr. David Allan of Mercy Hospital, San Diego.)

Fig. 4.—R. G. Roentgenogram of the pelvis at 6 weeks. The lack of centers of ossification is striking. (Courtesy of Dr. David Allan of Mercy Hospital, San Diego.)

Fig. 5.—R. G. Roentgenogram of the pelvis at 22 months. The lack of ossification of the pubis is a characteristic of the syndrome. The femoral head was not yet ossified. The femoral necks were wide and flat. The normal flaring of the ilium was lacking. The acetabulum appeared wide and flattened. (Courtesy of Dr. David Allan of Mercy Hospital, San Diego.)

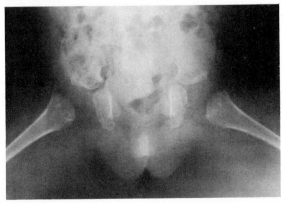

REFERENCES

1. Spranger, J., and Wiedemann, H. R.: Dysplasia spondyloepiphysaria congenita, Helvet. Paediatr. Acta 21:598, 1966.
2. Spranger, J., and Wiedemann, H. R.: Dysplasia spondyloepiphysaria congenita, Lancet 2:642, 1966.
3. Spranger, J. W., and Langer, L. O.: Spondyloepiphyseal dysplasia congenita, Radiology 94:313, 1970.
4. Fraser, G. R., Friedmann, A. I., Maroteaux, P., Glen-Bott, A. M., and Mittwoch, U.: Dysplasia spondyloepiphysaria congenita and related generalized skeletal dysplasias among children with severe visual handicaps, Arch. Dis. Child. 44:490, 1969.
5. Roaf, R., Longmore, J. B., and Forrester, R. M.: A childhood syndrome of bone dysplasia, retinal detachment and deafness, Dev. Med. Child Neurol. 9:464, 1967.
6. Bach, C., Maroteaux, P., Schaeffer, P., Bitan, A., and Crumiere, C.: Dysplasie spondylo-epiphysaire congénitale avec anomalies multiples, Arch. Fr. Pediatr. 24:23, 1967.
7. Langer, L. O.: Spondyloepiphysial dysplasia tarda, Radiology 82:833, 1964.

DYGGVE-MELCHIOR-CLAUSEN SYNDROME

CARDINAL CLINICAL FEATURES

Short-trunked dwarfism, platyspondyly, small ilia with lacy margins on the iliac crests, mental retardation.

CLINICAL FEATURES

Dyggve, Melchior and Clausen[1] described three sibs in 1962, two boys and a girl, from Greenland whom they thought to have the Morquio disease but who turned out to have a distinct syndrome. The father and mother were related: the father was the mother's paternal uncle. They had eight children, of whom three were affected with a disorder in which a short-limbed dwarfism, remarkably like the Morquio disease, was unassociated with keratosulfaturia or corneal opacities.

It was reported that these patients excreted large amounts of an unknown substance, thought possibly to be a desulfated mucopolysaccharide.[2, 3] The syndrome has been listed as mucopolysaccharidosis type VII. However, more recent studies indicate clearly that these patients do not have mucopolysacchariduria, and cultured fibroblasts have a normal pattern of isotopic sulfate incorporation into cellular acid mucopolysaccharide. These data indicate that the syndrome is not a mucopolysaccharidosis, and it should be removed from the list of mucopolysaccharidoses.[4, 5]

The most striking feature of the syndrome is short-trunked dwarfism. The neck is short. There is an exaggerated lumbar lordosis and thoracic kyphosis. The chest is barrel shaped and the sternum protrudes. The patients most often appear normal at birth, although available neonatal measurements indicate usually that they are small. The disorder generally is first recognized at the age of 1–1½ years. The initial complaint may be feeding difficulty or thoracic deformity. Shortness of stature is prominent. There is no corneal clouding, there are no dental defects and hearing is normal. A small cranium is the rule. Most patients have retarded speech development. Mental retardation may be severe. IQ values of 32–61 have been recorded. Patients have been described as very kind, good-natured and easy to deal with. Moreover, some patients have been described as being without mental retardation. A waddling gait is characteristic. Limitation of joint mobility is striking, particularly in the elbows and the hips. The extremities also are usually short. The hands may be broad and short, and there may be limitation of motion at the finger joints.

Roentgenographic findings are diagnostic. The most pronounced finding is platyspondyly, particularly of the lumbar spine. The vertebrae have a unique shape in that they are flat, pointed anteriorly and pinched in the center.[4] There is hypoplasia of the odontoid process. The intravertebral spaces are wide. The pelvis is small, particularly the ilia. The iliac crests have an irregular lacy margin.[4, 6] These changes are unique to this syndrome.

The ribs are broad anteriorly. The posterior portions are normal. The sternum is wide and may have an irregular margin. The long bones tend to be short and to have wide shafts. The epiphyses often appear late and are flat and irregular, occasionally fragmented. The metaphyses also may be irregular and flared. The femoral necks are short and the femoral heads hypoplastic. The femoral head may be fragmented. The acetabular roof often is irregular, and there may be lateral displacement of the capital femoral epiphysis. Bilateral coxa valga may be present. Late ossification of the epiphyses may be especially prominent in roentgenograms of the carpal bones. The metacarpals and phalanges are short, but the proximal metacarpals are not pointed, and the distal ends of the ulna and radius do not slant toward each other as in dysostosis multiplex.

DIFFERENTIAL DIAGNOSIS

Patients with this syndrome bear some resemblance to those with the Morquio syndrome (see section on Mucopolysaccharidoses). Both syndromes have dwarfism of the short-trunk variety and platyspondyly. Patients with the Morquio syndrome differ in the occurrence of cloudy corneas, enamel hypoplasia and late cardiac defects. Furthermore, intelligence is normal. Patients excrete excessive amounts of keratan sulfate in the urine. Those with the Dyggve-Melchior-Clausen syndrome clearly do not have keratan sulfate in the urine.[4, 7] An original patient of Melchior's has recently been found to excrete normal amounts of the usual mucopolysaccharides.[4] These patients have been

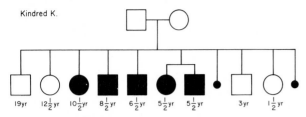

Kindred K.

19yr 12½yr 10½yr 8½yr 6½yr 5½yr 5½yr 3yr 1½yr

Fig. 1.—(Case Report). Pedigree of the K. family with the Dyggve-Melchior-Clausen syndrome. Each member had a very short trunk and a broad, barrel chest and protruding sternum reminiscent of the Morquio syndrome. There was a pronounced limitation of mobility of the elbows. They were patients of Dr. Vazken M. Der Kaloustian of the American University of Beirut.

shown further to have no intracellular storage vacuoles in a variety of cell types.[4,8] The normal metabolism of radioactive sulfate establishes the fact that this is not a mucopolysaccharidosis. The association of a skeletal disorder with mental retardation suggests that there is an underlying metabolic abnormality, but its nature remains to be determined.

CASE REPORT*

The K. family. In this family (Fig. 1) five of nine sibs were affected with the Dyggve-Melchior-Clausen syndrome. The

Fig. 2.—Six-and-one-half- and 8½-year-old brothers of K. K. (Case Report). (Figures 2 and 3 courtesy of Dr. Vazken M. Der Kaloustian of the American University of Beirut and reprinted with permission from J. Neurol. Sci.[13] and Radiology.[4])

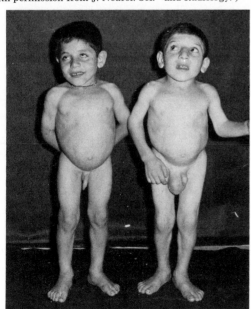

*We are indebted to Dr. Vazken M. Der Kaloustian of the American University of Beirut for providing us with the clinical details of this family.

parents were normal. Each of the affected children appeared normal at birth, but each manifested skeletal abnormalities by the age of 1½ years.

K. K. was a 5½-year-old male, one of dizygotic twins. His twin sister was similarly affected. They were the product of a normal pregnancy and delivery and appeared normal at birth. The patient was very short as a consequence of his short trunk. This shortness of stature is illustrated in his 6½- and 8½-year-old brothers (Fig. 2.) The patient's height was 34 inches, and the height age was 2 years at 5½ years of age. The upper to lower ratio was 46/40, or 1.15, which is characteristic of short-trunk dwarfism. The span was 35 inches. The neck was short; the mandible often appeared to rest on the chest. The sternum was protruding and the anteroposterior diameter of the chest increased. There was an exaggerated lumbar lordosis and thoracic kyphosis, and the abdomen also protruded. There was no hepatosplenomegaly. The patient was microcephalic: head circumference was 47 cm. There were limitations of joint mobility, especially at the elbow joints. The patient had a waddling gait. He was severely retarded and had no speech. Hearing was normal, as was vision. The teeth were normal. Slit-lamp examination of the eyes revealed no corneal

Fig. 3.—Roentgenogram of the spine of K. K.'s 6½-year-old brother. There was general platyspondyly. The anterior bodies of the vertebrae were pointed. There were broad notches on the superior and inferior surfaces.

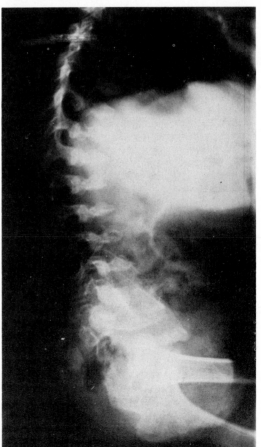

clouding. Examination of the urine for mucopolysaccharide was normal. Pneumoencephalogram was normal.

Roentgenographic survey of the skeleton revealed platyspondyly, with changes typical of the Dyggve-Melchior-Clausen syndrome. Films of the patient's 6½-year-old brother are shown in Figure. 3. The odontoid process of the C2 vertebrae was hypoplastic.

GENETICS

The disorder appears to be carried on an autosomal recessive gene. The initial report of Dyggve and colleagues[1] was of three sibs of normal but consanguineous parents. Several other families have since been reported in which there were multiple involved sibs.

TREATMENT

None.

REFERENCES

1. Dyggve, H. V., Melchior, J. C., and Clausen, J.: Morquio-Ullrich's disease. An inborn error of metabolism?, Arch. Dis. Child. 37:525, 1962.
2. Clausen, J., Dyggve, H. V., and Melchior, J. C.: Mucopolysaccharidosis, Arch. Dis. Child. 37:525, 1962.
3. Clausen, J., Dyggve, H. V., and Melchior, J.C.: Chemical and enzymatic studies of a family with skeletal abnormalities associated with mental retardation, Clin. Chim. Acta 29:197, 1970.
4. Spranger, J., Maroteaux, P., and Der Kaloustian, V. M.: Dyggve-Melchior-Clausen syndrome, Radiology 114:415, 1975.
5. McKusick, V. A.: *Heritable Disorders of Connective Tissues* (4th ed.; St. Louis: C. V. Mosby Co., 1972), p. 610.
6. Dyggve, H. V., Melchior, J. C., and Clausen, J.: Dyggve-Melchior-Clausen's syndrome (DMC's syndrome), Ugeskr. Laeger 135:127, 1973.
7. Linker, A., Evans, L. R., and Langer, L. O.: Morquio's disease and mucopolysaccharide excretion, J. Pediatr. 77: 1039, 1970.
8. Kaufman, R. L., Rimoin, D. L., and McAlister, W. H.: The Dyggve-Melchior-Clausen Syndrome, in Bergsma, D. (ed.): *Second Conference on the Clinical Delineation of Birth Defects. Part VI: Nervous System* (New York: The National Foundation-March of Dimes, 1971), pp. 144–149.
9. Barylak, A., Kozlowski, K.: Dyggve disease, Aust. Pediatr. J. 8:338, 1972.
10. Gwinn, J. L., Barnes, G. P.: Radiological case of the month. Morquio-Brailsford's disease, Am. J. Dis. Child. 115:347, 1968.
11. Zellweger, H., Ponseti, I. V., Pedrini, V., Stanler, F. S., and Von Moorden, G. K.: Morquio-Ullrich's disease, J. Pediatr. 59:549, 1961.
12. Spranger, J., Langer, L. O., and Wiedemann, H. R.: Bone Dysplasias (Stuttgart/Philadelphia: Fischer/Saunders, 1974).
13. Afifi, A. K., Der Kaloustian, V. M., Bahuth, N. B., and Mire-Salman, J.: Concentrically laminated membranous inclusions in myofibres of Dyggve-Melchior-Clausen syndrome, J. Neurol. Sci. 21:335, 1974.

Distinct Syndromes with Very Short Stature

NOONAN SYNDROME

CARDINAL CLINICAL FEATURES

A Turner phenotype with normal karyotype, including short stature, webbed neck, low hairline and hypertelorism; cardiac abnormality, most commonly pulmonary valvular stenosis.

CLINICAL PICTURE

Noonan and Ehmke,[1] in the course of a study reported in 1963 of 835 children with congenital heart disease, recognized a syndrome in nine patients who had pulmonary valvular stenosis. These patients had short stature, hypertelorism and mental retardation. Some had ptosis, undescended testicles and skeletal anomalies. Six of these patients were male, and three female. Summitt, Opitz and Smith[2] assembled information on patients thought to have a "male Turner's syndrome" and commented on the similarity of phenotype as well as some distinctions from the classic Turner female. They first coined the term Noonan syndrome. The syndrome is relatively common.

Among the distinctive features were the pattern of growth, the types of congenital heart disease and a greater frequency of mental retardation. Of course, it is now easier to distinguish patients with the Noonan syndrome from those with the syndromes of Turner and Bonnevie-Ullrich, as those with the Noonan syndrome have an XX or XY chromosomal complement.

Patients with the Noonan syndrome have an impressive similarity of facies (Fig. 1). They usually have pulmonic valvular stenosis. They have a variable shortness of stature[1-3]: some are very short; others may be of normal height. The neck is short and the hairline is characteristically very low, especially posteriorly (Fig. 2). Varying degrees of webbed neck are seen. The ears are prominent and low set (Figs. 3 and 4). There may be nerve deafness. Patients have hypertelorism, a slight antimongoloid slant, micrognathia and usually a downward curve to the angles of the mouth. They may have epicanthal folds. Ptosis of one or both upper lids is common. There may be strabismus and a prominence of the eyes or appearance of exophthalmos. The bridge of the nose may be depressed and broad. The palate is high arched, and dental malocclusion is common. Deformities of the sternum are frequent, particularly pectus excavatum of the scooped-out shoe type

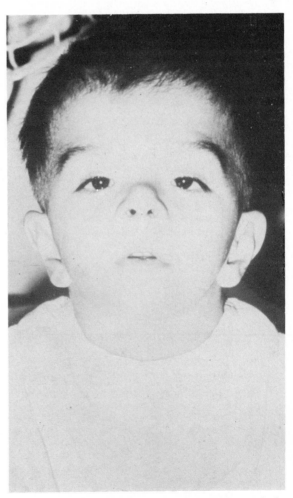

Fig. 1.—D. M. (Case Report). A 2-year-old boy with the Noonan syndrome. Hypertelorism, ptosis and strabismus were prominent. The ears were large and low set. The bridge of the nose was broad and depressed, the tip anteverted. The mouth was down-turned.

267

or pectus carinatum; the sternum also may be abnormally short.

Patients with the Noonan syndrome, in common with patients with the Turner syndrome, have in addition to the webbed neck a characteristic shield chest with widely spaced (see Fig. 3) and hypoplastic, inverted or accessory nipples. They have cubitus valgus, clinodactyly of the fifth fingers and hypoplastic nails. Lymphedema of the dorsa of the hands and feet may be present in the newborn period. This is the presentation described by Bonnevie and Ullrich. It may persist until somewhat later.

Congenital heart disease is a characteristic of the syndrome. It usually occurs on the right side and, in fact, is most often an isolated stenosis of the pulmonary valve. Alternatively, there may be a supravalvular stenosis of the pulmonary artery or of a branch, a patent ductus arteriosus or an atrial septal defect. The presentation of pulmonic stenosis may be unusual.[3] The systolic murmur may be best heard at the lower left sternal border. The typical systolic ejection click may be absent. Post-stenotic dilatation of the pulmonary artery rarely is seen in the roentgenogram. The ECG may show a pattern of right ventricular hypertrophy more like that of the tetralogy of Fallot. Cardiac catheterization or angiography is of course diagnostic. It is important that cardiologists recognize the phenotype in order to suspect the presence of severe pulmonic stenosis in spite of atypical roentgenographic, electrocardiographic and auscultatory findings.

Males with the Noonan syndrome characteristically have some problem in testicular development.[4] The most common abnormality is that of small undescended testes, which on section reveal hypoplasia or aplasia of germinal tissue. The scrotum often is hypoplastic. Anorchism may also be seen, whereas in some patients the testes appear normal. Anorchism in a phenotypic male indicates that testes were present early in fetal life in order to have induced the masculine development of internal and external genitalia. These patients usually have both a vas deferens and epididymis. Hypoplastic testes may be found in the inguinal canal or they may be completely cryptorchid. These patients are infertile. Libido has been reported to be absent or reduced.

On the other hand, females with the Noonan syndrome have normal ovaries and may be fertile. Puberty usually is delayed in these girls. They have normal secondary sex characteristics but their development also is often delayed. Renal anomalies occur in this syndrome. Rotational errors or duplication of a kidney, as well as hydronephrosis, have been reported. It is often a good rule in patients with congenital heart disease to obtain an abdominal film following cardiac angiography in order to visualize the kidneys. It is particularly advisable in the Noonan syndrome.

Dermatoglyphs in the Noonan syndrome[5] may reveal low ridge counts. The axial triradius may be distally placed, and there may be bilateral hypothenar patterns.[6] Patients may have simian creases. Some patients have vertebral anomalies. Osteoporosis has been reported.[7] Mental retardation, a regular but not constant finding in this syndrome, varies from mild to severe.

DIFFERENTIAL DIAGNOSIS

The significant distinction is the differentiation of this syndrome from the Turner syndrome (see section on Cytogenetic Syndromes). Patients with the Noonan syndrome have been referred to as having the male Turner syndrome, XX and XY Turner phenotype and even the female male Turner syndrome. Actually there are considerable differences in phenotype. Patients with the classic Turner syndrome have a complete or partial monosomy for the X chromosome and usually an XO karyotype. They are always short and usually are not mentally retarded. They have high dermal ridge counts. The characteristic shortening of the fourth metacarpal is not seen in the Noonan syndrome.[7] Heart disease involving the left side is characteristic of the Turner syndrome, particularly coarctation of the aorta or aortic stenosis. No patients with an XO karyotype have been reported with pulmonary stenosis.

CASE REPORT*

D. M., a 2½ year old white male, was seen in genetic counseling clinic for evaluation of multiple congenital anomalies and a heart murmur. The child was the product of a normal pregnancy and a premature delivery. Birth weight was 3 lb, 6 oz. There was an early history of poor feeding, and poor weight gain. At 6 months of age he weighed 8 lb, 13 oz and his length was 22½ inches. The heart murmur was noted at that time. At 9 months he weighed 9 lb, 3 oz and was 25½ inches in length. Developmental milestones were delayed.

Physical examination revealed a short, shy, retarded boy. His height of 31¾ inches at 2½ years gave him a height age of 17 months. His weight of 20 lb, 3 oz was that of an infant of 8 months. The child had a distinctive facies. This was a consequence of the presence of hypertelorism (see Fig. 1), prominent epicanthal folds (Fig. 5), bilateral ptosis, low-set prominent ears and a depressed nasal bridge. He had a high-arched palate. The neck was short and webbed (Fig. 6) and had a low posterior hairline. There was a shield chest with pectus excavatum and widely spaced nipples. Examination of the heart showed a diminished S2 at the second left interspace. The pulmonic component could not be heard, and an ejection sound preceded a grade III/VI harsh, systolic murmur, heard best at the second left interspace. He had bilateral cubitus valgus, simian creases and clinodactyly. The scrotum was hypoplastic, and both testes were in the inguinal canal. The ECG showed right ventricular hypertrophy. Cardiac catheterization revealed mild stenosis of the pulmonic valve and of the peripheral right pulmonary artery. An intravenous pyelogram was normal.

The mother of this child had mild hypertelorism and a uni-

*We are grateful to Drs. O. W. Jones and William F. Friedman of the University of California San Diego, for permission to report on this patient.

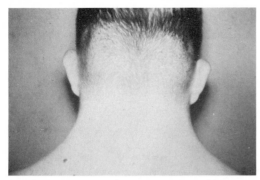

Fig. 2.—Patient with the Noonan syndrome illustrates the low hairline despite a recent haircut. The low-set ears also are apparent. (Courtesy of Dr. Doralys Arias of Orlando, Florida.)

Fig. 3.—This patient had a shield chest, and the nipples were widely spaced. Cubitus valgus was striking.

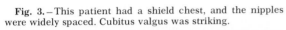

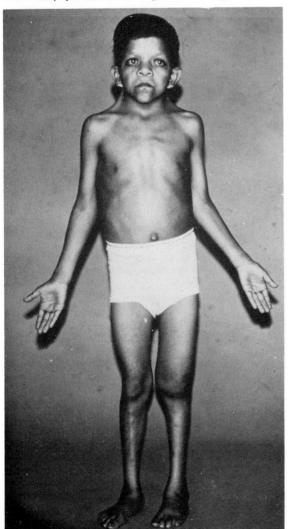

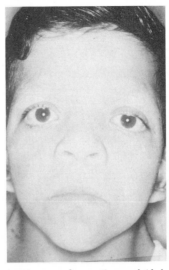

Fig. 4.—Low-set ears and an antimongoloid slant and characteristic facies.

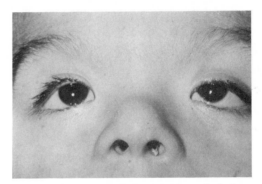

Fig. 5.—D. M. The eyes had an antimongoloid slant and prominent epicanthal folds.

Fig. 6.—D. M. Short, webbed neck.

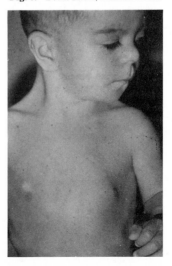

lateral simian line. An older sister was normal, as was a younger half-sibling.

GENETICS

The evidence is that the Noonan syndrome is transmitted by an autosomal dominant gene with varying expressivity. Patients have normal karyotypes, 46XX and 46XY. Partial expression, particularly of minor stigmata of the syndrome, is frequently seen in first degree relatives of clear-cut patients. It is more common to find that the mother has evidence of the disorder, but transmission from the father to children of both sexes also has been observed.[6-10]

TREATMENT

Congenital cardiac abnormalities, particularly pulmonic stenosis, are often surgically correctable.

REFERENCES

1. Noonan, J. A., and Ehmke, D. A.: Associated noncardiac malformations in children with congenital heart disease, J. Pediatr. 63:468, 1963.
2. Summitt, R. L., Opitz, J. M., and Smith, D. W.: Noonan's syndrome in the male, J. Pediatr. 67:936, 1965.
3. Noonan, J. A.: Hypertelorism with Turner phenotype, Am. J. Dis. Child. 116:373, 1968.
4. Heller, R. H.: The Turner phenotype in the male, J. Pediatr. 66:48, 1965.
5. Nora, J. J., Torres, F. G., Sinha, A. K., and McNamara, D. G.: Characteristic cardiovascular anomalies of XO Turner syndrome, XX and XY phenotype and XO/XX Turner mosaic, Am. J. Cardiol. 25:639, 1970.
6. Levy, E. P., Pashayan, H., Fraser, F. C., and Pinsky, L.: XX and XY Turner phenotypes in a family, Am. J. Dis. Child. 120:36, 1970.
7. Riggs, W., Jr.: Roentgen findings in Noonan's syndrome, Radiology 96:393, 1970.
8. Nora, J. J., and Sinha, A. K.: Direct male-to-male transmission of the XY Turner Phenotype?, Lancet 1:25, 1970.
9. Nora, J. J., and Sinha, A. K.: Direct familial transmission of the Turner phenotype, Am. J. Dis. Child. 116:343, 1968.
10. Summitt, R. L.: Turner syndrome and Noonan's syndrome, J. Pediatr. 74:155, 1969.
11. Collins, E., and Turner, G.: The Noonan syndrome—a review of the clinical and genetic features of 27 cases, J. Pediatr. 83:941, 1973.

DE LANGE SYNDROME

CARDINAL CLINICAL FEATURES

Distinctive appearance created by a tiny face, brachycephaly, hirsutism, synophris, a small nose with anteverted nostrils, elongated area from nose to upper lip, thin lips, micrognathia; micromelia or phocomelia; hypoplasia of dermal ridges; retardation of growth and mental development.

CLINICAL PICTURE

The syndrome, which bears the name of Cornelia de Lange,[1] is one of the most common of the malformation syndromes. It is best recognized in the gestalt. Patients look enough like one another that, once a case has been seen, recognition of most patients should present little difficulty. The appearance that makes these patients so readily identifiable is made up of a complex of minor malformations.[2-7, 13] There are at the same time major malformations, among the most serious of which is the degree of mental retardation.

The facial appearance is striking (Figs. 1 and 2). The face is diminutive (Fig. 3), in a small, brachycephalic

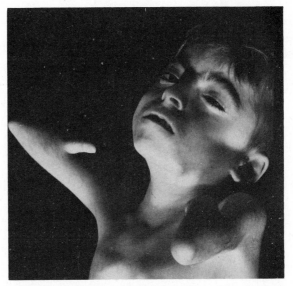

Fig. 2.—A boy with the de Lange syndrome illustrating the similarity in facies to that of patient in Figure 1. The forehead was hirsute. The malformations of the upper extremities were striking.

Fig. 1.—W. R. An 18-month-old infant with the de Lange syndrome, illustrating the classic appearance of this syndrome. The hair, eyebrows and lashes, nose, area between the nose and lip, lip and chin are very well demonstrated.

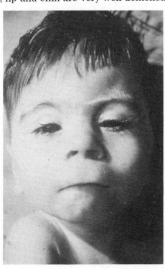

skull. Patients generally are hirsute. The hairline is low both in front and in back. The eyebrows are thick and meet in the midline (synophris) (see Fig. 3). The hair also is prominent over the back and the extremities (Fig. 4). The eyelashes are long and beautiful. The nose is small and the nostrils often anteverted. The area between the nose and the vermilion border of the upper lip is quite long. The lips themselves are thin, and the angles of the mouth turn down. The mandible is small. The face is singularly devoid of expression.

These children are primordially very small. Birth weights are low for gestational ages. Patients grow poorly, and height age tends to approximate 50% of the gestational age. Bone age also is usually retarded. An IQ over 50 is rare. Speech development is particularly poor. As infants, these patients do not have a cry but rather a low-pitched, growling sound that is raucous but feeble. They do not develop speech.

The trunk tends to be cylindrical. The rib cage is barrel shaped. The ribs are pitched horizontally, and the upper rib cage is wider than the lower. The sternum is short. The nipples and umbilicus often appear hypoplastic.

271

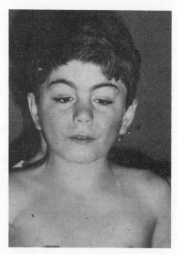

Fig. 3.—An older patient (16 years) with the de Lange syndrome. She illustrates the diminutive face, low hairline, synophris and long, black eyelashes. The nose was anteverted, the area from nose to lips increased and the lips thin.

The extremities also are characteristic. They are hypoplastic and they usually tend to taper conically. The most common abnormality is micromelia (Fig. 5), but these patients may have severe phocomelia (Figs. 2 and 6) and varying degrees of hypoplasia in between. A lobster claw anomaly occasionally is seen. Patients often have restricted motion at the elbow. Roentgenograms of the upper extremity reveal malformed or dislocated radial heads, the pattern being similar to that of radioulnar synostosis. Roentgenograms also are characteristic in showing small or rudimentary second phalanges of the fifth fingers and short, thick first metacarpals. In infancy the acetabular angles are low.

Simian creases often are seen. Clinodactyly of the fifth finger is the rule. The thumbs are proximally placed. Most patients have cutaneous syndactyly of the second and third toes.

The most subtle of the abnormalities in the extremities is the hypoplasia of the dermal ridges. This is regularly encountered and is particularly striking in the hypothenar area. Other common dermatoglyphic abnormalities include the related reduction in total ridge

Fig. 4.—Hirsutism of the back and a low hairline in an infant with the syndrome.

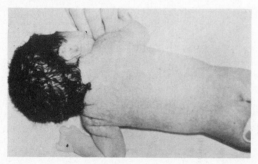

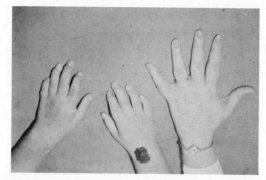

Fig. 5.—Micromelia. The hand on the left is that of a patient age 18. The hand on the right is that of a normal adult female. The patient's hands were generally short. The thumbs were proximally placed, and there was clinodactyly of the fifth fingers.

count, ulnar loops and radial loops. There often is a distal axial triradius. An open field hallucal pattern frequently is noted on the foot.

Patients usually have small teeth, and their eruption is delayed. The palate is high arched and occasionally cleft. The ears may be low set. Cutis marmorata tends to persist. Capillary hemangiomas are common. A few patients have had webbing of the neck, and some have been thought early in infancy to have the Turner syndrome. Congenital heart disease occurs in about 10% of patients. Genitalia most often are hypoplastic. Males are usually cryptorchid.

Ocular abnormalities are irregularly seen. They include optic atrophy, myopia, coloboma, eccentric pupils, blue sclerae, nystagmus and microphthalmia.

Infants with this syndrome may fail to initiate respiration. They often feed poorly and must be tube fed for early survival. They aspirate frequently. They have frequent infections, especially pulmonary infections. They commonly die in infancy, and most of those to be seen are less than 10 years of age. Those who survive infancy seem to have a behavioral phenotype that is almost as stereotyped as their physical characteristics. They are expressionless and speechless. They are autistic in the sense that they do not relate to parents or to others caring for them. They indulge in behavioral stereotypies, especially those of turning behavior and hand posturing.[8] Many of them develop self-mutilative behavior.[9] This usually is less aggressive than that seen in the Lesch-Nyhan syndrome, but an occasional bitten lip is indistinguishable.

Some variability has been described.[10] It is possible that the syndrome may occur without severe mental retardation. Certainly the arm manifestations can vary from phocomelia to a proximally placed thumb. Possibly any of the manifestations may have a spectrum of expression from minimal to severe. It also is possible that there is some development of the manifestations. At least one case has been reported in which the syndrome appeared not to be recognizable at 9 months but was typical by 21 months.[11]

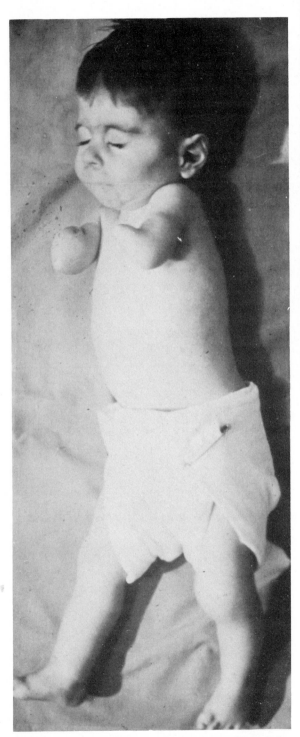

Fig. 6.—W. R. The infant shown in Figure 1. He had phocomelia. He could also dislocate his arms at will.

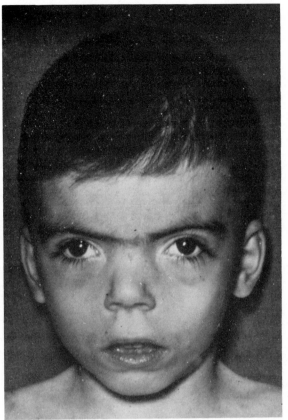

Fig. 7.—J. E. (Case Report). Front view of the face at 6 years of age, illustrating the synophris and characteristic craniofacial morphology of the de Lange syndrome.

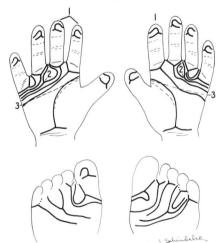

Fig. 8.—J. E. Diagrammatic representation of the dermatoglyphics, illustrating (1) radial loops frequently found on the second or third digits in patients with the de Lange syndrome; (2) b' (prime) triradii and (3) simian creases. Dermatoglyphic patterns are designated by continuous *solid lines*, flexion creases by *broken lines*. (Dermatoglyphic analysis and drawings courtesy of Mrs. Joan Davies.)

CASE REPORT

J. E. was 18 months old when first seen at the Child Development Center of the University of Miami School of Medicine in Florida. He had the physical appearance characteristic of the de Lange syndrome (Fig. 7). He was small. His weight was 16 lb (50th percentile for age 6 months), and his height was 30 inches (75th percentile for 12 months). Head circumference was 44.2 cm (50th to 75th percentile for 6 months). He had heavy black eyebrows, which met in the midline, and very long eyelashes on both upper and lower lids. The hairline was low, extending downward posteriorly. The right ear lobe was abnormally lobulated and had a cleft dividing it into two malformed parts. He had a broad, flat chest with an increased anteroposterior diameter. The hands and feet were noticeably small. Both hands appeared short and broad, with clinodactyly of the fifth finger. The patient had simian transverse palmar creases. The thumbs were placed proximally. There was cutaneous webbing of the second and third toes bilaterally. He had a deep coccygeal dimple with an overlying soft tissue tag. The skin was diffusely carotenemic. Developmental retardation was profound; at 5 years, 11 months of age, psychometric testing placed his intellectual function between 8 and 12 months with a Psyche-Cattell IQ score of 13.

Occasional staring spells or akinetic seizures were observed from 18 months to 3 years of age. He was treated with Tridione for 2 years with satisfactory control, and no further seizure activity was reported.

Roentgenographic examination at 15 months revealed microbrachycephaly, micrognathia and small facial bones. The metacarpals of both hands were broad and short, with marked hypoplasia and proximal displacement of the first metacarpals. The middle phalanx of each fifth finger was triangular and hypoplastic, resulting in radial deviation of the distal phalanges (bilateral clinodactyly). The roentgenograms revealed premature fusion of the sternal ossification centers, a slight C-shaped deformity of the thoracolumbar spine, and a defect of the posterior neural arch of S_2 corresponding to the external skin dimple. The acetabular angles were markedly reduced, and at 6 years there was bilateral shortening of the femoral necks and a slight valgus deformity. The bone age was delayed at 15 months of age and subsequently.

Dermatoglyphic analysis (Fig. 8) showed radial loops on the second and third digits of the left hand and on the third digit of the right hand. An interdigital third loop was associated with a b' triradius on both palms. The bilateral simian creases were evident. The total ridge count was 122, only slightly less than control (145), but there was ridge dysplasia.

By 4 years of age he was noted usually to be sucking his middle three fingers except when he had a toy in his mouth. Furthermore, he pulled or scratched at his chin and chest, often leading to excoriation and bleeding, particularly over the sternum. He had learned to dislocate his hips while standing, which usually produced a strikingly audible click and was followed by crying as if in pain. At about 5 years he began biting his thumbs. He sometimes persisted until the skin was broken. His lip soon became a part of the biting pattern. At 6 years of age he produced a large multilative laceration of the left side of his lower lip, which healed with loss of the vermilion border. A program of intensive operant behavioral training, including aversive stimulation, was found to be quite useful in management.

GENETICS

The etiology of this syndrome is not known. Most cases are sporadic. It had been thought to be de-

termined by a single autosomal recessive gene, but is is much too common to represent the mutation rate at a single locus, as in the Apert syndrome, which occurs predominantly as a sporadic case in patients who do not reproduce. When a patient with the Apert syndrome reproduces, it can be seen that the condition is a dominant. In the de Lange syndrome the parents are normal. On the other hand, a number of instances have been observed in which there were more than one case in a sibship.[12] It has been observed in monozygous twins. Minor abnormalities are occasionally seen in a progenitor. Recurrence risk in a sibship following the birth of an involved infant has been calculated to be between 2% and 5%.

TREATMENT

No effective treatment has been developed. Tube feeding may be required for alimentation early. Operant methods may be useful in handling behavioral problems.

REFERENCES

1. de Lange, C.: Sur un type nouveau de dégénération (typus Amstelodamensis), Arch. Med. Enf. 36:713, 1933.
2. Ptacek, L. J., Opitz, J. M., Smith, D. W., Gerritsen, T., and Waisman, H. A.: The Cornelia de Lange syndrome, J. Pediatr. 63:1000, 1963.
3. Jervis, G. A., and Stimson, C. V.: De Lange syndrome: The "Amsterdam type" of mental defect with congenital malformation, J. Pediatr. 63:634, 1963.
4. Silver, H. K.: The de Lange syndrome. Typus Amstelodamensis, Am. J. Dis. Child. 108:523, 1964.
5. Lee, F. A., and Kenny, F. M.: Skeletal changes in the Cornelia de Lange syndrome, Am. J. Roentgenol. Radium Ther. Nucl. Med. 100:27, 1967.
6. Abraham, J. M., and Russell, A.: De Lange syndrome: A study of nine examples, Acta Paediatr. Scand. 57:339, 1968.
7. Opitz, J. M., Segal, A. T., Lehrke, R. L., and Nadler, H. L.: The etiology of the Brachmann-de Lange syndrome, Lancet 2:1019, 1964.
8. Nyhan, W. L.: Behavioral phenotypes in organic disease, Pediatr. Res. 6:1, 1972.
9. Shear, C. S., Nyhan, W. L., Kirman, B. H., and Stern, J.: Selfmutilative behavior as a feature of the de Lange syndrome, J. Pediatr. 78:506, 1971.
10. Pashayan, H., Whelan, D., Guttman, S., and Fraser, F. C.: Variability of the de Lange syndrome: Report of three cases and genetic analysis of 54 families, J. Pediatr. 75:853, 1969.
11. Pashayan, H., Levy, E. P., Fraser, F. C.: Can the de Lange syndrome always be diagnosed at birth?, Pediatrics 46:940, 1970.
12. Beratis, N. G., Hsu, L. Y. F., and Hirschhorn, K.: Familial de Lange syndrome. Report of three cases in a sibship. Clin. Genet. 2:170, 1971.
13. Berg, J. M., McCreary, B. D., Ridler, M. A., and Smith, G. F.: The de Lange Syndrome, Monograph 2 (Oxford/New York Toronto/Sydney/Braunschweig: Pergamon Press, 1970.).

INFANTILE HYPERCALCEMIA SYNDROME
Elfin Facies and Supravalvular Aortic Stenosis,
Supravalvular Aortic Stenosis Syndrome,
Williams Elfin Facies Syndrome

CARDINAL CLINICAL FEATURES

Elfin facies, supravalvular aortic stenosis, peripheral systemic and pulmonary artery stenosis, transient hypercalcemia in infancy, anomalies of dental development, low birth weight, growth retardation, mental retardation.

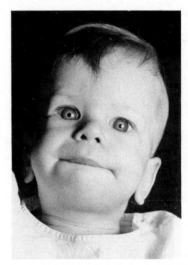

Fig. 2.—A typical infant with the hypercalcemia syndrome. (Courtesy of Dr. William Friedman of the University of California San Diego.)

Fig. 1.—An 11-month-old boy with the idiopathic hypercalcemia syndrome. He had supravalvular aortic stenosis. Growth and development were delayed. He illustrates the characteristic elfin facies. The ears were prominent. (Courtesy of Dr. William Friedman of the University of California San Diego.)

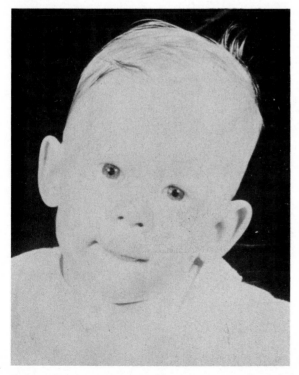

CLINICAL PICTURE

Fanconi et al.[1] described a severe form of chronic infantile hypercalcemia in 1952 in which involved infants had failure to thrive, a characteristic facies, osteosclerosis, a systolic murmur and mental retardation. In 1961 Williams and colleagues[2] described several patients with supravalvular aortic stenosis whose facial appearances resembled each other sufficiently to suggest the diagnosis; they proposed that these features, along with the mental retardation that these patients manifested, constituted a distinct syndrome. A connection between the two syndromes of infantile hypercalcemia and supravalvular aortic stenosis was first suggested by Black and Bonham Carter[3] and documented by Garcia, Friedman, Kaback and Rowe.[4]

Patients with this syndrome have an elfin-like facies (Figs. 1 and 2). The face is memorable and, although

275

unmistakable, is not easy to describe. It also has been said that this facial appearance is less strikingly shown in photographs than it is in real life.[2] The forehead is prominent. Hypertelorism and prominent epicanthal folds are regularly seen. Strabismus is less common. The nose is short and upturned and has a depressed bridge. The ears often are large and low set (see Fig. 1), and the mandible small. The chin may appear pointed.

Among the most characteristic features are large, prominent, thick lips, a wide, slack, usually open mouth and small maxilla and mandible. The upper lip may form a bow, especially in infancy. These are unusually happy, readily smiling children. The smile, which may be somewhat horizontal, is worth emphasis. We have had the experience of looking at a child with this syndrome in post infancy, realizing that this was an unusual-appearing child but not yet appreciating what syndrome was before us, when all of a sudden the child smiled, and it hit us that this was the hypercalcemia syndrome.

The head may be small or asymmetrical. Fanconi's patient had microcephaly and premature synostosis of the cranial sutures. Dental anomalies include small, hypoplastic teeth, enamel hypoplasia, dysgnathia and anodontia. Patients often have malocclusion. Their eyes frequently are blue. They may appear to have a "baby," or developing, eye color. A stellate pattern in the iris also is seen (Fig. 3). The retinal vessels may be unusually tortuous, as they are in coarctation of the aorta.

Males with this syndrome often have undescended testes and inguinal hernias. Females sometimes have precocious development of secondary sexual characteristics. In early infancy these patients present with low birth weight and subsequent failure to thrive. Adults may be quite short, or they may catch up in growth. They often have anorexia, feeding difficulties, persistent vomiting, constipation and hypotonia. They also may have polydypsia and polyuria. These symptoms appear to be related to the hypercalcemia; they may be associated with nephrocalcinosis, azotemia and renal failure. Since the hypercalcemia generally

disappears spontaneously, patients usually are spared permanent renal dysfunction. However, careful study often reveals some abnormality of creatinine clearance or renal biopsy.[5] Calcium deposits have been seen in the cornea.

A somewhat hoarse, deep, metallic or raucous voice is characteristic. Deep tendon reflexes may be increased. The degree of retardation generally is mild. The usual IQ is in the range of 60–75. IQs have ranged as low as 30 and one patient had an IQ of 118.[5] These children have been described as affectionate and charming characters.

The most important laboratory finding in infancy is hypercalcemia. The value may be as high as 14 mg/100 ml. Patients also may have hyperphosphatemia and increased excretion of calcium and phosphate. Some patients in addition have had hypercholesterolemia and hyperglyceridemia. All of the abnormal chemical findings usually disappear by the second or third year of life.

The cardiac findings are those of left ventricular outflow obstruction. There usually is a systolic ejection murmur at the left or right upper sternal border, which may be heard over the entire precordium. There may be a precordial lift and a thrill palpable over the suprasternal notch. The murmur often is loud and transmitted clearly to the neck. The diffusely localized

Fig. 4.—Lateral view of the aorta following selective injection of contrast material into the left ventricle in a patient with supravalvular aortic stenosis. There was marked dilatation of the sinuses of Valsalva and constriction of the aorta at the superior margin of the sinuses. The over-all caliber of the entire aorta was small beyond the point of the supravalvular stenosis. (Courtesy of Dr. William Friedman of the University of California San Diego.)

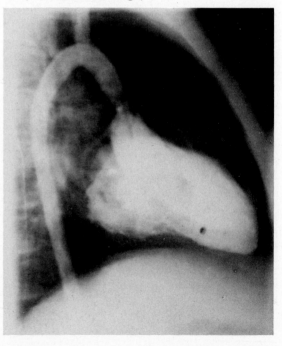

Fig. 3.—Stellate pattern in the iris of a patient with the hypercalcemia syndrome.

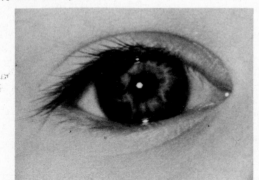

murmur of peripheral pulmonary stenosis also may be audible. Blood pressures in the upper extremities may be unequal, the right arm usually higher than the left. The blood pressure often is elevated. Cardiac catheterization with selective angiography can localize the aortic stenosis to a supravalvular narrowing[4] (Fig. 4). There also may be a mild stenosis of the pulmonary artery or one of its branches. Hypoplasia of the aorta is common. Coarctation of the aorta has been reported.[6] Mitral insufficiency has been observed, as has histologic evidence of myocardial calcifications.

Roentgenographic examination in infancy may reveal sclerosis of the base of the skull or of the long bones and vertebrae. Synostosis of the cranial sutures may occur. Beyond infancy, bands of increased density may be seen at the metaphyses and at the epiphyseal margin of the vertebrae. Patients may have kyphoscoliosis. Occasionally, a flat film of the abdomen will reveal nephrocalcinosis.

The cause of the syndrome is not known. However, all of the craniofacial and the cardiac characteristics of the syndrome have been induced in fetal rabbits by the administration of large parenteral doses of vitamin D during pregnancy.[7, 8] It is thought that the syndrome may result from maternal hypervitaminosis D or an abnormality in vitamin D metabolism.

CASE REPORTS

CASE 1.—T. H., a 10-year-old white male, had a history of delayed motor developmental milestones. However, he was

Fig. 5.—T. H. (Case 1). A 10-year-old boy with the syndrome. He also had the classic facies. The mouth was broad and the smile flattened horizontally.

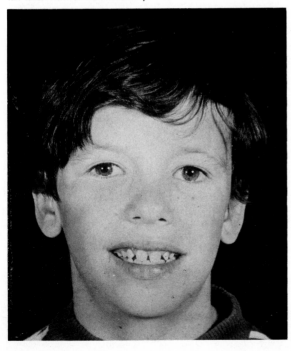

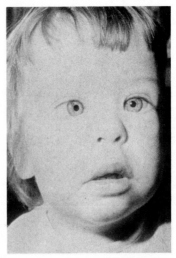

Fig. 6.—A. A. (Case 2). The characteristic facial features include epicanthal folds, strabismus and a depressed nasal bridge with acute upturn of the tip of the nose. The patient had a prominent philtrum and upper lip. (Courtesy of Dr. William Friedman of the University of California San Diego.)

attending the third grade in a regular school. Feeding difficulties in infancy had been associated with poor weight gain and constipation. He had undergone a laparotomy for undescended testes and none were found. He had always been asymptomatic with regard to the cardiovascular system.

Physical examination revealed a short, pleasant boy with the characteristic elfin facies of the infantile hypercalcemia syndrome (Fig. 5). He had prominent ears, hypertelorism, epicanthi, an anteverted nose, a full, prominent upper lip and an asymmetrical expression. The teeth were widely spaced and dysplastic. Blood pressures were equal in both upper extremities, and he was normotensive. A systolic thrill was palpable over the right carotid artery. There was a grade II/VI systolic ejection murmur maximal at the upper right sternal border and radiating to the suprasternal notch. The abdomen was soft and there was no organomegaly. The scrotum was flat, and testicles could not be palpated.

Important laboratory data included normal serum calcium and cholesterol concentrations. The ECG and chest roentgenogram were unremarkable. Skeletal survey and flat film of the abdomen were normal, as was the chromosomal karyotype. The degree of supravalvular aortic stenosis was considered mild, and routine cardiac followup without treatment was arranged.

CASE 2.—A. A. was a 22-month-old white female who was admitted to University Hospital for evaluation of delayed developmental milestones, unusual facies and a heart murmur. She was the product of a normal, full-term pregnancy and delivery but weighed only 1 lb, 6 oz at birth. Feeding difficulties began in infancy, and a heart murmur was detected shortly after birth.

Physical examination revealed the height to be 31 inches and weight 20 lb. She had normal vital signs. The blood pressure in the right arm was 105/60, in the left arm 90/55 and in the right leg 115/65 mm Hg. She had a rather characteristic elfin facial appearance (Fig. 6). Examination of the heart revealed right and left ventricular precordial hyperactivity. A loud systolic ejection murmur was maximal in the suprasternal notch and transmitted widely to the precordium. A separate late systolic murmur was detected posteriorly. A Denver development test put her at the 12-month level.

Important laboratory data included a serum calcium of 9.4 mg/100 ml. Chest roentgenogram demonstrated slight cardiomegaly with a left ventricular cardiac contour. ECG showed combined ventricular hypertrophy. Cardiac catheterization revealed moderately severe supravalvular aortic stenosis and moderate bilateral branch stenosis of the pulmonary arteries. It was decided to consider cardiac surgery if the severity of the left ventricular outflow tract obstruction increased.

GENETICS

There is no evidence that this syndrome is genetic. All cases in patients with the complete syndrome have been sporadic. However, there certainly have been familial instances of supravalvular aortic stenosis, and the pattern of inheritance has been autosomal dominant.[9]

TREATMENT

Supravalvular aortic stenosis may be treated directly by surgical intervention. In patients with milder disease the condition may improve spontaneously with age.[10]

REFERENCES

1. Fanconi, G., Giradet, P., Schlesinger, B., Butler, N., and Black, J.: Chronische Hypercalcämie, kombiniert mit Osteosklerose, Hyperazotämie, Minderwuchs und kongenitalen Missbildungen, Helv. Paediatr. Acta 4:314, 1952.

2. Williams, J. C. P., Barratt-Boyes, B. G., and Lowe, J. B.: Supravalvular aortic stenosis, Circulation 24:1311, 1961.

3. Black, J. A., and Bonham Carter, R. E.: Association between aortic stenosis and facies of severe infantile hypercalcemia, Lancet 2:745, 1963.

4. Garcia, R. E., Friedman, W. F., Kaback, M. M., and Rowe, R. D.: Idiopathic hypercalcemia and supravalvular aortic stenosis, N. Engl. J. Med. 271:117, 1964.

5. Antia, A. I., Wiltse, H. E., Rowe, R. D., Pitt, E. L., Levin, S., Ottesen, O. E., and Cooke, R. E.: Pathogenesis of the supravalvular aortic stenosis syndrome, J. Pediatr. 71:431, 1967.

6. Eie, H., Semb, G., Müller, O., and Bramness, G.: Localized supravalvular aortic stenosis combined with mental retardation and peculiar facial appearance, Acta Med. Scand. 191:517, 1972.

7. Friedman, W. F., and Roberts, W. C.: Vitamin D and the supravalvular aortic stenosis syndrome: The transplacental effects of vitamin D on the aorta of the rabbit, Circulation 34:77, 1966.

8. Friedman, W. F., and Mills, L. F.: The relationship between vitamin D and the craniofacial and dental anomalies of the supravalvular aortic stenosis syndrome, Pediatrics 43:12, 1969.

9. McCue, C. M., Spicuzza, T. J., Robertson, L. W., and Mauck, H. P.: Familial supravalvular aortic stenosis, J. Pediatr. 73:889, 1968.

10. Fraser, D., Kidd, B. S. L., Kooh, S. W., and Paunier, L.: A new look at infantile hypercalcemia, Pediatr. Clin. North Am. 13:503, 1966.

11. Beuren, A. J., Schulze, C., Eberle, P., Harmjanz, D., and Apitz, J.: The syndrome of supravalvular aortic stenosis, peripheral pulmonary stenosis, mental retardation, and similar facial appearance, Am. J. Cardiol. 13:471, 1964.

12. Jones, K. L., and Smith, D. W.: The Williams elfin facies syndrome, J. Pediatr. 86:718, 1975.

RUBINSTEIN-TAYBI SYNDROME

CARDINAL CLINICAL FEATURES

Broad thumbs, broad great toes, characteristic facies with beaked nose and downward extension of the nasal septum, ocular defects, mental retardation.

CLINICAL PICTURE

Rubinstein and Taybi[1] in 1963 first called attention to a syndrome that is very striking and relatively common. Seven children originally reported had retardation of mental, motor, language and social development. The hallmark of this syndrome is the occurrence of broad thumbs and great toes. The terminal phalanges are broad, and on x-ray they may show abnormal segmentation and partial duplication. There may be angulation or radial displacement of the thumb or similar deformity of the first toe as well as abnormal shape of the proximal phalanx or first metatarsal.

Patients also have a readily recognizable facial abnormality (Figs. 1–3). The forehead is usually prominent and the nasal bridge broad. The nasal septum extends well below the alae. The septum often is deviated. Commonly the ears are abnormally shaped or low set. The palate is high arched and the mouth very small.

A considerable variety of ocular abnormalities is seen in this syndrome.[3] An antimongoloid slant of the palpebral fissures is common, as are strabismus and epicanthal folds. Patients usually have long eyelashes and highly arched brows. Refractive errors also are common and may be either myopic or hyperopic. These patients also tend to develop cataracts. Colobomas, blepharoptosis, obstruction of the nasal lacrimal duct, enophthalmos and optic atrophy are less commonly encountered problems. These patients often are microcephalic. Nevertheless, the anterior fontanel is commonly enlarged or late in closing. The foramen magnum also may be enlarged. Capillary hemangiomas often are seen on the forehead or the nape of the neck or back. Hirsutism is prominent.

Vertebral anomalies include spina bifida, kyphosis, lordosis or scoliosis. Pectus excavatum or carinatum may be seen, as well as fusion of the ribs. Other anomalies associated with the syndrome include supernumerary nipples, congenital heart disease, undescended testes and renal anomalies. Shortness of stature is the rule. Patients commonly have convulsive seizures. Feeding difficulties are frequent in infancy, as are repeated episodes of respiratory infection in early childhood. Paronychia is common. The IQ usually is less than 50, although a few in the 80s have been reported. The patients usually are pleasant and cooperative but may be shy with strangers.

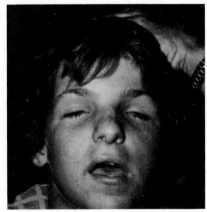

Fig. 1 (above).—C. H. (Case 2.) Facial appearance includes the prominent nose with lower extension of the septum, antimongoloid slant of the eyes and low-set ears.

Fig. 2 (left).—C. N. The configuration of the face was characteristic of the Rubinstein-Taybi syndrome, particularly the nose.

Fig. 3 (right).—C. N. Lateral view of the face.

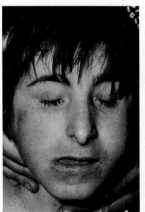 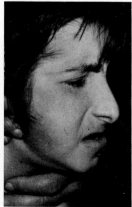

279

Fig. 4 (above).—C. N. (Case 1.) Shortness of stature. The facial characteristics, large thumbs and great toes also are evident.

Dermatoglyphic analysis in the Rubinstein-Taybi syndrome is of some interest. There is an increased frequency of the arch pattern, leading to a low total ridge count. There also are distally placed axial triradii and simian lines. Extremely unusual dermal patterns are found in the thenar/first interdigital area of the palm and also in the hypothenar area.[4, 5] The prints of individuals with the syndrome have a true pattern, with loops or whorls in the thenar/interdigital 1 area. This pattern is usually on both but is at least on one hand in approximately 90% of patients. In control individuals only 11% have a true pattern in this area. Occasionally the rare double pattern of a loop and a whorl or a double loop are seen. Dermatoglyphic analysis may be particularly useful in the very young patient in whom the diagnosis may not be so obvious as in older individuals.

Differential Diagnosis

The differential diagnostic possibilities are not numerous. The appearance of the face may suggest a first arch syndrome, but the colobomas, middle ear abnormalities, micrognathism and macrostomia described by Treacher and Collins and by Franceschetti and Klein are absent. Many of the acrocephalopolysyndactylies have broad great toes or thumbs, often with duplication of these phalanges (see section on Craniosynostoses). It is of interest in this regard that one of the patients, reported by Salmon[7] as having characteristics of the Rubinstein-Taybi syndrome, even on dermatoglyphic analysis had craniosynostosis with scaphocephaly and a facial appearance similar to that of the Crouzon disease. Patients with D–1 trisomy may resemble those having the Rubinstein-Taybi syndrome, especially in early infancy. Three patients with broad thumbs and great toes and a large beaked nose were described by Wilson.[8] Antimongoloid palpebral fissures, transverse palmar creases and double distal phalanges on x-ray of the great toe were other fea-

Fig. 5 (left).—C. N. The terminal phalanges of the thumbs were broad.
Fig. 6 (center).—C. N. Broad terminal phalanges of the great toes and evidence of paronychia, which is common.
Fig. 7 (right).—C. H. (Case 2.) The terminal phalanges of the thumb were broad.

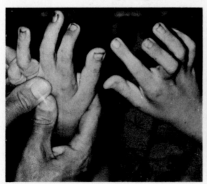
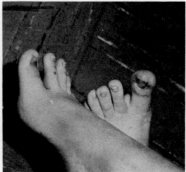
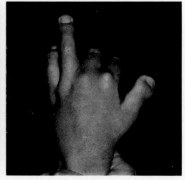

tures. In one of these infants the rest of the phenotype was characteristic of D–1 trisomy, but in two the diagnosis appeared to be the Rubinstein-Taybi syndrome until the karyotype was examined.

CASE REPORTS

CASE 1.–C. N., a 24-year-old white female, was seen at Fairview State Hospital, Costa Mesa, California. She had profound mental retardation. She was the product of a normal pregnancy and delivery. There was no history of exposure to drugs, illness or radiation during pregnancy and no consanguinity. She had four normal siblings. She had a history of repeated respiratory infections and recurrent blepharitis. There was no history of seizures.

She was a small, ambulatory girl (Fig. 4). She did not speak, although she did make sounds. Height was 49.75 inches, and height age was 7.5 years. She was microcephalic: head circumference was 47 cm. She had heavy eyebrows, which met in the midline, a prominent forehead and a broad nasal bridge. The nose was beaked and the nasal septum extended below the alae (see Figs. 2 and 3). She had low-set ears. The mouth was small and the palate highly arched. Examination of the eyes revealed an antimongoloid slant and a cataract of the right eye, as well as blepharitis and esotropia. She had spinal kyphosis.

Broad thumbs (Fig. 5) and great toes (Fig. 6) were striking, and there was paronychia on both sides of the right great toe. Hirsutism was prominent.

X-rays of the skull showed an enlarged foramen magnum. The terminal phalanges of the thumbs were short and broad, and a similar malformation was observed in the terminal phalanges of the great toes. She had renal calculi and a nonfunctioning right kidney.

CASE 2.–C. H. was a 9-year-old patient at the Fairview State Hospital, Costa Mesa. Her height was 49 inches. She had characteristic broad thumbs (Fig. 7) and great toes (Fig. 8). The facies also was typical (see Fig. 1) of the Rubinstein-Taybi syndrome. Mental retardation was severe. Family history was negative. The karyotype was normal. She had a double renal pelvis.

GENETICS

Cases in most patients reported have been sporadic. There appears to be a small preponderance of females in this syndrome. Johnson[9] has described affected sibs. In one patient[10] the parents were second cousins. Another was the product of father-daughter incest.[11] These findings suggest that the condition may be genetic and that it may be transmitted as an autosomal recessive. Minor clinical findings have been reported by Rubinstein[12] in relatives of patients with the syndrome. Broad thumbs were found in the father of two patients, the paternal grandfather of two patients and in another patient in the father, a paternal uncle, three paternal aunts and a paternal great grand-uncle.

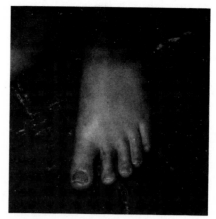

Fig. 8.–C. H. The broad great toe.

TREATMENT

No specific treatment is available.

REFERENCES

1. Rubinstein, J. H., and Taybi, H.: Broad thumbs and toes and facial abnormalities, Am. J. Dis. Child. 105:588, 1963.
2. Coffin, G. S.: Brachydactyly, peculiar facies and mental retardation, Am. J. Dis. Child. 108:351, 1964.
3. Roy, F. H., Summitt, R. L., Hiatt, R. L., and Hughes, J. G.: Ocular manifestations of the Rubinstein-Taybi syndrome, Arch. Ophthalmol. 79:272, 1968.
4. Robinson, G. C., Miller, J. R., Cook, E. G., and Tischler, B.: The syndrome of broad thumbs and toes and mental retardation, Am. J. Dis. Child. 111:287, 1966.
5. Giroux, J., and Miller, J. R.: Dermatoglyphics of the broad thumb and great toe syndrome, Am. J. Dis. Child. 113:207, 1967.
6. Taybi, H., and Rubinstein, J. H.: Broad thumbs and toes and unusual facial features, a probable mental retardation syndrome, Am. J. Roentgenol. Radium Ther. Nucl. Med. 93:362, 1965.
7. Salmon, M. A.: The Rubinstein-Taybi syndrome: A report of two cases, Arch. Dis. Child. 43:102, 1968.
8. Wilson, M. G.: Rubinstein-Taybi and D₁ Trisomy syndromes, J. Pediatr. 73:404, 1968.
9. Johnson, C. F.: Broad thumbs and broad great toes with facial abnormalities and mental retardation, J. Pediatr. 6:942, 1966.
10. Jeliu, G., and Saint-Rome, G.: Le syndrome de Rubinstein-Taybi. A propos d'une observation, Union Med. Can. 96:22, 1967.
11. Padfield, C. J., Partington, M. W., and Simpson, N. E.: The Rubinstein-Taybi syndrome, Arch. Dis. Child. 43:94, 1968.
12. Rubinstein, J. H., The broad thumbs syndrome–progress report 1968, in *The First Conference on the Clinical Delineation of Birth Defects. Part II, Malformation Syndromes* (New York: The National Foundation-March of Dimes, 1969), p. 25.

Phakomatoses

STURGE-WEBER SYNDROME
Encephalotrigeminal Angiomatosis

CARDINAL CLINICAL FEATURES

Facial vascular nevus (port-wine stain), ipsilateral intracranial hemangioma and calcifications, ipsilateral angioma of the choroid and glaucoma, convulsive disorder, hemiplegia, mental retardation.

CLINICAL PICTURE

Encephalotrigeminal angiomatosis (the Sturge-Weber syndrome) was first described by Schirmer in 1860[1] in a patient who had a port-wine facial nevus and ipsilateral glaucoma. In 1879 Sturge[2] described a similar patient with a right facial hemangioma, a large right eye with angiomatosis of the retina and choroid and focal seizures on the opposite side. He reasoned that there were vascular lesions on the right side of the brain. Weber[3] pointed out the syndromic association of facial and cerebral angiomatosis and described the characteristic wavy, double-contoured calcifications seen in roentgenograms of the skull.

The syndrome is characterized at birth by a cutaneous hemangiomatous lesion on the face. This nevus may be pink or red, but ultimately it assumes a purple or port-wine appearance (Fig. 1). The color fades on pressure. It may be unilateral or bilateral; it usually conforms to the sensory distribution of the trigeminal nerve. The lesions are sharply demarcated and not elevated (Fig. 2). On the other hand, there may be an overgrowth of tissues in the hemangiomatous area, leading to localized gigantism, such as hemihypertrophy of the lip or tongue. The buccal mucosa, pharynx and uvula sometimes are involved in the angiomatous process.

Ocular manifestations may include buphthalmos or congenital glaucoma due to angiomatosis of the choroid. Some patients develop glaucoma at a later age. Other ocular manifestations include coloboma of the iris, heterochromia iridis and hemianopsia due to involvement of the optic radiations or visual cortex by a cerebral angioma.

The most common cerebral symptoms are convulsions. These may be focal or generalized and are seen in most patients. They usually begin in infancy. Focal seizures usually occur on the side of the body opposite the facial lesion. Hemiplegia, hemiparesis and hemiatrophy often are present, again on the side opposite the cutaneous lesion. Mental retardation may be present. There is wide variation from normal intelligence to severe retardation. An occasional patient may be violent or have other behavioral abnormalities. Pathologic examination reveals that the calcium deposits are in the outer layers of the cerebral cortex, not in the vessels.[4] There are vascular malformations and

Fig. 1.—A patient with the Sturge-Weber syndrome.

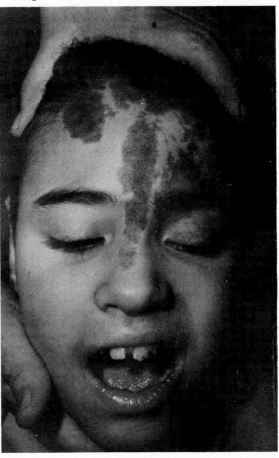

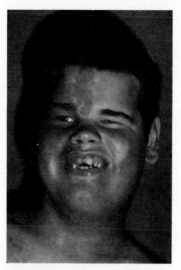

Fig. 2.—A mentally retarded boy with the Sturge-Weber syndrome. He had facial asymmetry and an extensive port-wine stain covering most of one side of his face. The hemangioma was not strictly unilateral and extended down over the neck and chest.

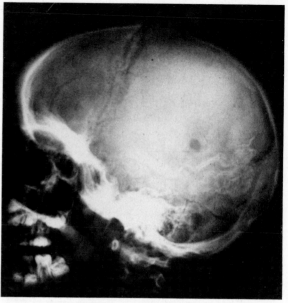

Fig. 4.—F. C. Roentgenogram of the skull. The characteristic, wavy, double-lined pattern was most prominent in the occipital area.

angiomatosis. Cerebral cortical atrophy may develop. Cerebral manifestations are due to degeneration of the cerebral cortex as a consequence of the vascular anomalies. Hemangiomas also may be found in the kidney or other viscera.

Roentgenographic examination of the skull in these patients is characteristic.[2, 5] Intracranial calcifications assume railroad track configuration, with sinuous parallel lines. This appearance is due to the fact that the calcifications correspond to the shape of the cerebral gyri and sulci. This picture is pathognomonic of this syndrome, but absence of the characteristic roentgenographic features does not rule out the disease.

Fig. 3.—F. C. (Case Report). A severely retarded boy with the Sturge-Weber syndrome. The unilateral hemangioma included the forehead, eye, malar area and gingivae.

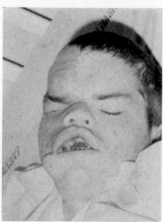

These findings are not seen in infancy; the calcification increases with age; it has been seen as early as 13 months of age.

CASE REPORT

F. C. was a 19-year-old white male institutionalized at Fairview State Hospital, Costa Mesa, California, for severe mental retardation and repeated seizures. He was the product of a normal, full-term gestation and had a birth weight of 7 lb, 3 oz. At birth a port-wine nevus was noted on the right side of his face. He also had congenital glaucoma of the right eye, which was removed in infancy. He developed his first seizure at 20 months. Since that time he had had two or three seizures a day, lasting 3–8 minutes each.

He sat alone at 8 months and walked at 1 year. He spoke a few words at 2 years but none thereafter. At 19 years he was severely retarded. He had never been successfully toilet trained, even though he had been at Fairview since the age of 7 years. His seizures were well controlled with Mysoline and phenobarbital. There was no family history of this disorder.

On physical examination he had a right-sided port-wine stain (Fig. 3) that was trigeminal in distribution. The buccal mucosa of the mouth was involved. The remainder of the physical examination was unremarkable.

The patient could walk short distances around the ward, holding onto the wall. He fed himself with his hands. X-ray of the skull showed the characteristic railroad track calcifications (Fig. 4).

GENETICS

The Sturge-Weber syndrome is not known to be genetically determined. On the other hand, the occasional occurrence of significant hemangiomas in other members of the family suggests the possibility of a dominant, with problems of penetrance. It is sometimes associated with another hemangiomatous condition, known as the Klippel-Trenaunay-Weber syndrome[6] (see chapter, this section), which is characterized by cutaneous hemangiomas, varicosities, localized hypertrophy of an extremity or side and osteohypertrophy. Patients with the Sturge-Weber syndrome also have been reported to have tuberous sclerosis and neurofibromatosis.[7]

PROGNOSIS AND TREATMENT

The prognosis depends on the extent of the cerebral lesion. Control of seizures by the use of anticonvulsant medication may not be successful. However, about half of the patients lead comfortable lives with the aid of anticonvulsant therapy. Surgical removal of a small lesion may be feasible.[8]

REFERENCES

1. Schirmer, R.: Ein Fall von Telangiektasie, Graefes Arch. Ophthalmol. 7:119, 1860.
2. Sturge, W. A.: A case of partial epilepsy apparently due to a lesion of one of the vaso-motor centers of the brain, Trans. Clin. Soc. London 12:162, 1879.
3. Weber, P. F.: A note of the association of extensive haemangiomatosus naevus of the skin with cerebral (meningeal) hemangioma, Proc. R. Soc. Med. 22:431, 1929.
4. Krabbe, K. H.: Facial and meningeal angiomatosis associated with calcifications of the brain cortex, Arch. Neurol. Psychiatr. 32:737, 1934.
5. Bielawski, J. G., and Tatelman, M.: Intracranial calcification in encephalotrigeminal angiomatosis, Am. J. Roentgenol. Radium Ther. Nucl. Med. 62:247, 1949.
6. Furukawa, T., Igata, A., Toyokura, Y., and Ikeda, S.: Sturge-Weber and Klippel-Trenaunay syndrome with nevus of Ota and Ito, Arch. Dermatol. 102:640, 1970.
7. Chao, D. H-C.: Congenital neurocutaneous syndromes of childhood, III. Sturge-Weber disease, J. Pediatr. 55:635, 1959.
8. Peterman, A. F., Hayles, A. B., Dockerty, M. B., and Love, J. G.: Encephalotrigeminal angiomatosis (Sturge-Weber disease), clinical study of thirty-five cases, J.A.M.A. 167: 2169, 1958.

KLIPPEL-TRENAUNAY-WEBER SYNDROME
Giant Hemangioma with Localized Hypertrophy or Gigantism

CARDINAL CLINICAL FEATURES

Port-wine hemangioma involving an extremity, varicose veins, hypertrophy of soft tissues and bony structures.

CLINICAL PICTURE

Klippel and Trenaunay[1] described the syndrome in 1900 and called attention to the classic triad of varicose veins, cutaneous hemangioma and bony and soft tissue hypertrophy. In 1907 Weber[2] provided the first description in the English literature.

A variety of manifestations has been observed in patients with this syndrome. Vascular abnormalities usually include a nevus flammeus, capillary hemangioma and cavernous hemangioma, or all three. Varicosities are the rule and may be congenital. A lymphangioma also may be present. Some patients have had a recognizable arteriovenous fistula or aneurysm, but in most this is not demonstrable. Among the trophic changes observed, atrophy of soft tissues or bone may be seen, as well as hypertrophy.

Other anomalies have been reported in association with this syndrome. These include syndactyly and polydactyly as well as spina bifida. The combination of the Klippel-Trenaunay-Weber syndrome and the Sturge-Weber syndrome has been reported. The syndrome also has been observed in association with other phakomatoses and also with ichthyosis. Other patients have been reported with coloboma of the iris and with scleral melanosis.

Complications of the process itself include edema, phlebitis, thrombosis, stasis dermatitis and ulcers. Repeated infections may occur in the involved areas even without a break in the skin. There also may be localized paresthesias or hyperhidrosis. Compensatory scoliosis may occur, and some patients have had damage to the hip joint.

CASE REPORT

L. A., an 8-year-old white male, was first admitted to University Hospital for rectal bleeding. Hemoglobin was 5.6 gm/100

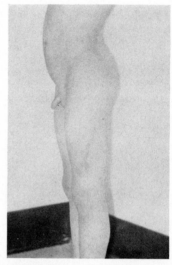

Fig. 1.—L. A. (Case Report). The left leg was larger than the right. The hemangioma, readily visible on the lateral thigh, was port wine in color but very soft, and it infiltrated subcutaneous tissues and muscles deeply.

ml and the hematocrit 20. Reticulocyte count was 3.4%. Stool specimens were repeatedly guaiac positive. The patient was found to have a congenital angioma involving the sigmoid. Surgical resection of the sigmoid and descending colon were carried out, and there was no further rectal bleeding.

Fig. 2.—L. A. Close-up of the lesion on the thigh illustrates the varicose veins as well.

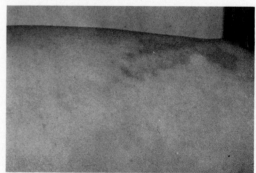

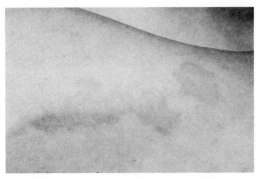

Fig. 3. – L. A. Another view of the same lesion.

Examination revealed a hemangioma of the lateral aspect of the left thigh (Figs. 1–3), as well as varicose veins in the left lower extremity that had been present since birth. The circumferences measured 10 cm above and below the patella were as follows: right thigh 30 cm, left thigh 30.5 cm, right leg 26 cm and left leg 27 cm. There was no edema, and no bruit could be heard.

X-rays of the left lower extremity revealed increased soft tissue density in the subcutaneous tissues on the lateral aspect of the left distal thigh and proximally below the knee. The left tibia was approximately 1 cm longer than the right.

COMMENT. – This patient had bony and soft tissue hypertrophy in association with multiple hemangiomas and varicosities. To our knowledge he is the first patient with this syndrome to present with rectal bleeding.

GENETICS

This syndrome is not known to be genetically determined. In some instances the disease has appeared to be inherited in an irregularly dominant way, as well as in a recessive pattern.[3, 4] An unusual incidence of twins in the families of patients with this syndrome has been commented upon.[4]

TREATMENT

Some hemangiomas are accessible to surgical removal. Most cannot be completely removed. Demonstrated arteriovenous connections should certainly be treated surgically. Varicosities may be treated with elastic hose or bandages. Infections and ulcers should be treated appropriately. Some patients may need orthopedic appliances.

REFERENCES

1. Klippel, M., and Trenaunay, P.: Du noevus variqueux ostéo-hypertrophique, Arch. Genet. Med. (Paris) 3:641, 1900.
2. Weber, F. P.: Angioma-formation in connection with hypertrophy of limbs and hemihypertrophy, Br. J. Dermatol. 19: 231, 1907.
3. Waardenburg, P. J., Franceschetti, A., and Klein, D.: *Genetics and Ophthalmology* (Assen, The Netherlands: Royal Van Gorcum, Publisher, 1961), pp. 1381–1390.
4. Koch, G.: Zur Klinik, Symptomatologie, Pathogenese und Erbpathologie des Klippel-Trenaunay-Weber-Schen Syndroms, Acta Genet. Med. Gemellol. (Roma) 5:326, 1956.
5. Mullins, J. F.: The Klippel-Trenaunay-Weber syndrome (Naevus vasculosis osteohypertrophicus), Arch. Dermatol. 86:202, 1962.

Unilateral Diseases

HEMIHYPERTROPHY

Cardinal Clinical Features

Asymmetry in which there is enlargement of an entire side of the body; associated abnormalities including Wilms' tumor, hepatoma, adrenocortical carcinoma, mental retardation.

Clinical Picture

Hemihypertrophy is a congenital disorder in which there is distinct asymmetry because of enlargement of one entire side or half of the body. The hypertrophy involves the skeleton and the soft tissues.

The human form is never completely symmetrical from side to side. This observation is most easily made by careful assessment of the face. This variation in nature has long been recognized, and asymmetry of the human body can be seen in early Greek and Roman sculpture. Congenital hemihypertrophy has been generally recognized and has been reported under many names, including Curtius syndrome, Steiner syndrome, congenital hypertrophy, hemigigantism, hemimacrosomia and partial gigantism. It is of interest that histologic study of one case in order to determine whether the large size was due to an increase in cell size or cell number revealed that it was the number of cells that increased. If this is generally true, the condition should more properly be known as hemihyperplasia.

Congenital, localized hypertrophy may involve as little as a single digit or as much as the entire half of the body. There may be segmental, unilateral or crossed hypertrophies, but the term hemihypertrophy generally implies a syndrome in which one side of the body is symmetrically larger than the other side. The right side is more often affected than the left, and males seem to be affected more frequently. The asymmetry is present at birth and continues throughout life. It usually becomes less prominent with growth and development, although it may increase at puberty.

Facial asymmetry often is the most striking manifestation and often the most upsetting to the patient and parents. The tongue is enlarged and thickened on the involved side. The fungiform papillae on that side may be enlarged. The lips, uvula, maxilla and mandible, as well as some teeth, are enlarged. The transition at the midline is abrupt. The nostril usually is larger on the involved side, and the nasal septum may deviate to the other side.

The skeleton on the involved side is uniformly larger than that of the other side. The condition can be documented roentgenographically by measuring the length of the bones, as long as standard conditions have been used. There may be unilateral enlargement of the chest, which may promote the development of scoliosis. The hair may be thicker or coarser on the involved side. The hair on the two sides may be of different colors, for instance, red and black.[1] Nail growth may be abnormal. The skin may appear thicker and have increased sweat gland or sebaceous gland activity. There may be increased vascularity, large or multiple telangiectases, multiple hemangiomas or venous varicosities. Areas of increased pigmentation or increased numbers of nevi or café-au-lait spots may be seen. A supernumerary nipple or an enlarged breast may be seen. Internal organs such as the kidney or adrenal are commonly enlarged on the involved side, as are the testis and labia majora. The cerebral hemisphere on the involved side may be considerably larger than on the other side.[2]

Mental retardation occurs in 15–20% of patients with hemihypertrophy. Some patients have had seizures. Dilatation of the pupil on the involved side has been observed. Heterochromia may be present in which the iris is of a different color in each eye. The patient may have some difficulty coordinating the two sides, one of which is much stronger than the other, particularly when participating in athletics.

Dermatoglyphic patterns may be asymmetrical, with abnormal patterns on the involved side. The presence of abnormal dermal patterns may suggest that in such cases the disordered growth may have its onset before the eighth week of gestation.[3] Increased bone age[4] and delayed bone age[1] have been observed.

Associated abnormalities. — The most important of the problems that occur in children with hemihypertrophy is the development of malignant neoplasms. In fact, one of the real benefits of early diagnosis is prompt treatment of malignancies that would otherwise be lethal. Three types of tumors observed in association with congenital hemihypertrophy involve the kidney, liver and adrenal cortex.

The most significant association is that between hemihypertrophy and Wilms' tumor of the kidney.[4-8] At least 26 patients have been reported with this combination of disorders.[4] It has been recommended that infants and children with hemihypertrophy be examined by intravenous pyelography at least once every 6 months in order to detect the presence of Wilms' tumor. The tumor may be on the hypertrophied or on the other side.[4, 8] There is a marked female preponderance among patients with both hemihypertrophy and Wilms' tumor.

Hemihypertrophy also has been reported with adrenal carcinoma and adenoma.[5, 7, 9] The adrenal tumor does not necessarily affect the same side of the body as the hemihypertrophy. Riedel[5] reported a patient in whom an adrenogenital syndrome was due to an adrenocortical tumor in a male child with left-sided hemihypertrophy. The child subsequently developed Wilms' tumor. Hemihypertrophy may occur with either mixed or epithelial hepatoblastoma.[10] One case has been described in which hemihypertrophy was associated with a neuroblastoma.[11] Another patient has been reported in whom there was an embryonal sarcoma of the lungs.[12]

A variety of skeletal abnormalities, most of them minor, has been seen in patients with hemihypertrophy. These include macrodactyly, polydactyly, syndactyly, clubfoot and lobster claw deformity. Scoliosis as well as pelvic tilt are common. Minor genitourinary anomalies such as hypospadias and cryptorchidism have been reported.

Differential Diagnosis

In the Silver syndrome[13] (see chapter, this section), congenital hemihypertrophy is associated with shortness of stature, low birth weight for gestational age, a short or incurved fifth finger, triangular facies, turned down corners of the mouth, and café-au-lait spots. Elevated urinary gonadotropins have been observed in some cases, as has a tendency to early sexual maturation.

In the Klippel-Trenaunay-Weber syndrome (see section on Phakomatoses), localized areas of gigantism are associated with large cutaneous and deep hemangiomas. There is hypertrophy of bone as well as of soft tissues. Many varicose veins are prominent. The disorder often involves an entire extremity and may involve two extremities on one side; however, the enlargement is seldom symmetrical, even within a single extremity.

Congenital arteriovenous fistula may result in overgrowth of one extremity. In neurofibromatosis, hypertrophy of an entire limb often is seen. The enlargement usually is not symmetrical but may be so. Tuberous sclerosis also may be associated with large areas of localized gigantism. It is sometimes difficult to distinguish hemihypertrophy from hemiatrophy. The latter usually is associated with some neurologic lesion. The Milroy disease, or idiopathic lymphedema, usually involves both lower extremities.

Case Reports

Case 1.—C. T., a white female infant, was the product of a normal pregnancy and delivery. There was no history of viral disease, exposure to drugs or x-irradiation during pregnancy. There was no family history of congenital anomaly or asymmetry. Birth weight was 8 lb, 4½ oz, height 20¼ inches and head circumference 36.8 cm. She was thought to be normal until the age of 7 weeks when her mother noticed that the right side was larger than the left.

On physical examination at 7 weeks of age, weight was 11 lb, 3 oz and height 22 inches. There was marked asymmetry, with the right side larger than the left. There was mild asymmetry of the face. The right side of the chest was larger than the left. The lungs were clear and the heart appeared normal.

Measurements of the extremities (in inches) were as follows:

	Right	Left
Forearm circumference	6	5⅜
Leg length	9½	9
Leg circumference	6	5
Thigh circumference	8	7½

A diagnosis of hemihypertrophy was made. Because of the association of hemihypertrophy and Wilms' tumor, hepatoblastoma and adrenal carcinoma, physical examination and intravenous pyelogram were repeatedly carried out and found to be normal over a 10-month period.

Case 2.—J. G. was an 8-month-old white female infant who was first seen for evaluation of hemihypertrophy. She was the product of a normal pregnancy and delivery. Her birth weight was 7 lb, 5 oz. She was the first child of a 22-year-old mother. There was no family history of congenital asymmetry.

The infant had been noted to have a large tongue since birth (Fig. 1). By 5 weeks it appeared to interfere with feeding. The mother then observed that the right side of the child was larger than the left (Figs. 2 and 3). The child also was noted to have a midline pigmented nevus on the forehead. At 3 months of age a partial glossectomy was performed and the nevus removed. An intravenous pyelogram showed a large right kidney. There was no evidence of Wilms' tumor.

Physical examination revealed marked asymmetry of the face. The right cheek and lip were prominent, and both the maxilla and mandible were larger on the right. The pupils were equal and reactive. The right ear measured 2 inches vertically, the left ear 1¾ inches. The tongue remained markedly enlarged.

Fig. 1.—J. G. (Case 2). A 3-year-old girl with an asymmetrical tongue produced by hemihypertrophy. (Courtesy of Dr. Lewis Turner of Kaiser-Permanente Hospital, San Diego.)

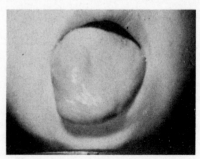

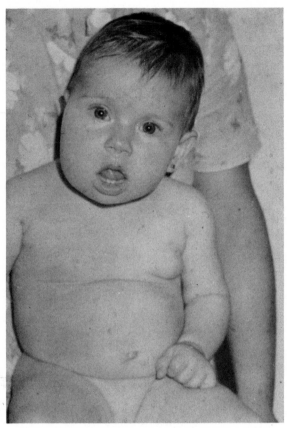

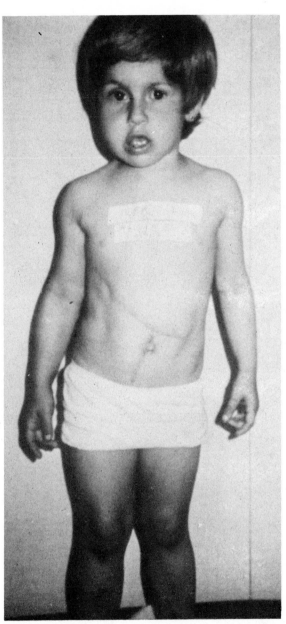

Fig. 2. – J. G. The entire right side of the infant was larger than the left.

Fig. 3. – J. G. The lower right extremity was considerably larger than the left.

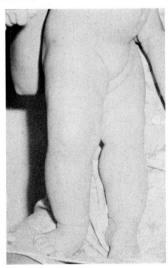

Fig. 4. – J. G. At 3 years of age. The right arm in this view is very clearly symmetrically larger than the left. The right leg and the right side of the face also were larger. (Courtesy of Dr. Lewis Turner of Kaiser-Permanente Hospital, San Diego.)

Measurement of the extremities (in inches) revealed the following values:

	Right	Left
Upper arm length	6⅝	5½
Forearm length	6	5⅛
Calf circumference	9	7
Thigh length	9½	8⅞
Leg length	11	9⅝

A diagnosis of hemihypertrophy was made. An intravenous pyelogram revealed the large right kidney and no evidence of Wilms' tumor. Chromosome studies of skin biopsies taken from each leg showed no cytogenetic difference between the affected and unaffected limbs.

When the child was age 2½ years, her mother noted that the right flank was becoming larger (Fig. 4). There was a large, hard, right flank mass that extended to the iliac crest. An intravenous pyelogram revealed a large right upper pole renal mass. At surgery a large Wilms' tumor was removed, and postoperatively the child received irradiation, actinomycin D and vincristine.

CASE 3.—J. M., a 14-year-old girl, was admitted to Fairview State Hospital, Costa Mesa, California, at 4 years of age because of mental retardation. She had been delivered normally after an uneventful pregnancy to a 25-year-old unmarried mother. Subsequent development was delayed.

She was a very pleasant child with an IQ of 42. She had hemihypertrophy in which the left side was consistently larger than the right (Fig. 5). The face, extremities and trunk were involved. Measurements of the extremities (in inches) were as follows:

	Right	Left
Upper arm circumference	8½	10⅛
Forearm circumference	7½	8¼
Thigh circumference	17⅝	18⅝
Calf circumference	10½	12⅛

The skin of the left side was unusually pigmented (Figs. 6 and 7), the brown pigment being distributed in irregular whorls and flecks. Its appearance was reminiscent of marble and appeared characteristic of incontinentia pigmenti. These areas of unusual skin were separated from the normal skin

Fig. 5.—J. M. (Case 3). A 14-year-old girl with hemihypertrophy of the entire left side. She also was mentally retarded.

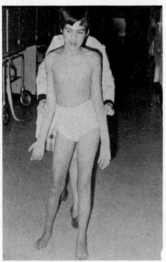

pigmentation of the right side by a sharp demarcation at the midline.

The patient had a history of seizures. X-rays of the skull were normal. The EEG was compatible with seizure activity. She had never had an abdominal tumor.

GENETICS

The etiology of hemihypertrophy is not known. It is thought that faulty cell division in the zygote, resulting in two daughter cells of unequal size, could represent a form of incomplete twinning. The frequent occurrence of neoplasms in this syndrome suggests that the hemihypertrophy may be a disorder in the regulation of growth similar to that involved in the genesis of embryonal neoplasms.

The observation of chromosomal mosaicism[3] in some patients has suggested that the problem may be related to the arrangement of genetic material during early zygotic cleavage. In diploid-triploid mosaicism, asymmetry could result from an asymmetrical chro-

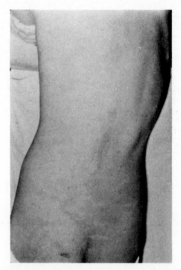

Fig. 6.—J. M. Brown pigmentation in the pattern reminiscent of late incontinentia pigmenti.

Fig. 7.—J. M. Close-up of the brown pigmentation.

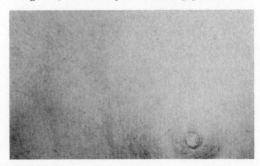

mosome pattern present during early embryogenesis. Trisomy 18 mosaicism also has been reported in hemihypertrophy. This also has been reported in a patient with Wilms' tumor and other anomalies but no hemihypertrophy. However, this cytogenetic abnormality is only a rare concomitant of asymmetry. In most patients studied the chromosomes have appeared normal.

Hemihypertrophy has been thought not to be genetically determined. On the other hand, documentation of cytogenetic abnormalities in even a few patients suggests that the others may be determined by a genetic event. Seven families have been observed in which there were two or more members with hemihypertrophy.[4] Fraumeni et al.[4] reported a brother and sister with right-sided hemihypertrophy; furthermore, the right leg of their maternal uncle had been longer than the left since childhood.

Treatment

Congenital hemihypertrophy serves to alert the physician to the possible presence of a malignant neoplasm. Frequent examination with this possibility in mind permits early aggressive treatment of any tumors encountered. Glossectomy is rarely required for the management of the enlargement of the tongue. Orthopedic treatment may be required for scoliosis or other consequences of hemihypertrophy.

REFERENCES

1. Ringrose, R. E., Jabbour, J. T., and Keele, D. K.: Hemihypertrophy, Pediatrics 36:434, 1965.

2. Ward, J., and Lerner, H. H.: A review of the subject of congenital hemihypertrophy and a complete case report, J. Pediatr. 31:403, 1947.

3. Johnston, A. W., and Penrose, L. S.: Congenital asymmetry, J. Med. Genet. 3:77, 1966.

4. Fraumeni, J. F., Jr., Geiser, C. F., and Manning, M. D.: Wilms' tumor and congenital hemihypertrophy: Report of five new cases and review of the literature, Pediatrics 40:886, 1967.

5. Riedel, H. A.: Adrenogenital syndrome in a male child due to adrenocortical tumor, Pediatrics 10:19, 1952.

6. Björklund, S.: Hemihypertrophy and Wilms' tumor, Acta Paediatr. 44:287, 1955.

7. Benson, P. F., Vulliamy, D. G., and Taubman, J. O.: Congenital hemihypertrophy and malignancy, Lancet 1:468, 1963.

8. Miller, R. W., Fraumeni, J. F., Jr., and Manning, M. D.: Association of Wilms's tumor with aniridia, hemihypertrophy and other congenital malformations, N. Engl. J. Med. 270:922, 1964.

9. Fraumeni, J. F., Jr., and Miller, R. W.: Adrenocortical neoplasms with hemihypertrophy, brain tumors, and other disorders, J. Pediatr. 70:129, 1967.

10. Geiser, C. F., Baez, A., Schindler, A. M., and Shih, V. E.: Hepatoblastoma, hemihypertrophy and cystathioninuria, Pediatrics 46:66, 1970.

11. Lenstrup, E.: Eight cases of hemihypertrophy, Acta Paediatr. Scand. 6:205, 1926.

12. Gorlin, R. J., and Meskin, L. H.: Congenital hemihypertrophy, J. Pediatr. 61:870, 1962.

13. Silver, H. K., Kiyasu, W., George, J., and Deamer, W. C.: Syndrome of congenital hemihypertrophy, shortness of stature, and elevated urinary gonadotropins, Pediatrics 12:368, 1953.

SILVER SYNDROME
Silver-Russell Syndrome, Russell-Silver Syndrome

CARDINAL CLINICAL FEATURES

Intrauterine growth retardation, short stature, relatively large head and smaller triangular face, clinodactyly, asymmetry.

CLINICAL PICTURE

The syndrome was first described in 1953 by Silver and colleagues,[1] who reported two unrelated children with shortness of stature, low birth weight for gestational age, congenital "hemihypertrophy" and elevated urinary gonadotropins. In 1954 Russell,[2] described five unrelated children with intrauterine growth retardation, craniofacial dysostosis, a characteristic facial appearance and short arms. Asymmetry was a feature in two of the patients. Several authors[3-5] have suggested that the patients reported by Silver and Russell represent variation within the same clinical entity.

The syndrome is characterized by a small size and low weight at birth following a full-term gestation. The height throughout infancy and childhood tends to parallel the normal curve but to remain below the third percentile. Therefore, shortness of stature is a continuing characteristic. There is little information on ultimate height. Rimoin[3] reported two identical twins with the syndrome; at 21 years of age their heights were 58 and 60 inches, respectively. Szalay's patient was 58 inches at 18 years.[5]

The cranium appears large in relation to the face, although the head circumference usually is within normal limits for the size of the patient. This is what has been referred to as craniofacial dysostosis.[2] Closure of the fontanels is delayed. The appearance of a large head, a small face and wide-open fontanels may lead the physician to suspect hydrocephalus; unnecessary pneumoencephalograms have been performed. Patients also have a characteristic facial appearance. The face appears triangular, with the broad forehead and wide biparietal diameter forming the base, and the narrow receding chin the apex of the triangle. The corners of the mouth turn down and may extend into short lateral folds. The lips are thin. A short incurved fifth finger usually is present. There may be limited supination of the arm.[2, 3] Patients occasionally may have café-au-lait spots,[6] but usually only a few, and syndactyly of the second and third toes.[7]

Asymmetry is a variable feature of this syndrome. Some patients have significant asymmetry that involves an entire side of the body. Silver's original patients[1, 6, 7] were thought to have hemihypertrophy, because one side of the body was considerably larger than the other. In patients with a considerable inhibition of growth beginning long before birth, it appears more appropriate to consider the phenomenon an asymmetry or laterality in which growth is inhibited more on one side than on the other. Several reported patients have not had any asymmetry.[2, 8, 9] Others have had localized asymmetries restricted to the skull, the spine or a portion of the extremities. Bilaterally symmetrical, distally placed dermatoglyphic axial triradii suggest that asymmetry may have its onset relatively late in intrauterine life.[3]

Abnormalities of endocrine function occasionally have been reported, but these are not the rule in the syndrome. Some patients have had elevated levels of urinary gonadotropins. Precocious sexual development and premature estrogenization of the vaginal mucosa also have been seen. Hypospadias has been reported.[3] The mechanism for the short stature is not clear; assays for growth hormone have been normal.[10] Some reported patients have had fasting hypoglycemia and increased sweating not correlated with hypoglycemia.[8] Patients with this syndrome may have normal or delayed bone age, and some have had delayed adolescence. Mental retardation has been reported, but intelligence usually is normal.

DIFFERENTIAL DIAGNOSIS

The syndrome should be differentiated from other syndromes in which there is intrauterine growth retardation. In the Bloom syndrome[11] intrauterine growth retardation and short stature are associated with fa-

298

cial telangiectatic, sun-sensitive erythema. These patients have chromosomal breaks and a high incidence of malignancy. The Bloom syndrome is transmitted as an autosomal recessive. The so-called bird-headed dwarfism associated with the name of Seckel[12] also is characterized by intrauterine growth retardation and short stature. These patients are microcephalic, and the face appears large in relation to the small head and body. Mental retardation is a regular part of the syndrome.

The asymmetry of the hemihypertrophy syndrome (see chapter, this section) is clearly very different; these patients have normal birth weights and normal facial appearance. Significant asymmetry also may be seen in the Klippel-Trenaunay-Weber syndrome (see section on Phakomatoses) and neurofibromatosis. It is a characteristic of the unilateral psoriasis and ectromelia syndrome (see chapter, this section).

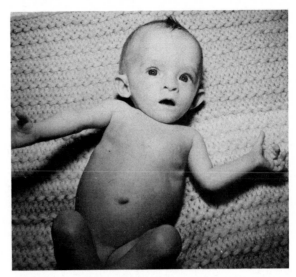

Fig. 2.—Twin B. In this view the triangular face is illustrated, as well as the down-turned corners of the mouth and the large, rather simple, low-set ears.

Fig. 1.—(Case Report). Identical twins, one of whom had the Silver syndrome. At 6 months Twin B, on the left, weighed 6 lb, 7 oz, and Twin A weighed 12 lb.

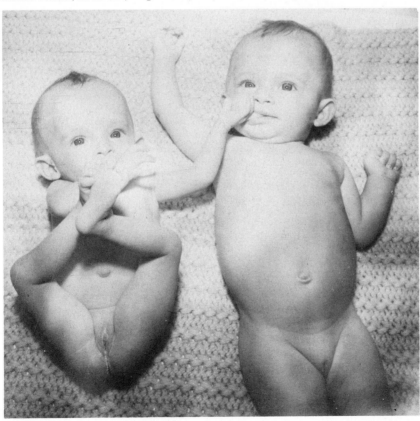

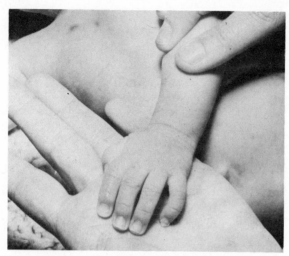

Fig. 3.—Twin B. Clinodactyly of the fifth finger.

CASE REPORT*

A set of twins (Fig. 1) was born at the U. S. Naval Hospital, San Diego, the products of a normal pregnancy and delivery. Twin A weighed 4 lb, 14 oz, Twin B 3 lb. Twin A appeared normal in every way and had a normal growth curve. At 5 months she weighed 11 lb and her height was normal for her age; Twin B weighed only 6 lb, 6 oz at 5 months of age. Her length was 52 cm, a height age of 1–2 months. Her developmental progress was normal.

Twin B had the characteristic triangular facies of the Silver syndrome (Fig. 2). The cranium appeared large, although the head circumference of 39 cm was below the third percentile for her age. The anterior fontanel was 5 × 4.5 cm. There was down-turning of the corners of the mouth and very thin lips. There were a preauricular skin tag on the right, a café-au-lait spot on the left flank and clinodactyly of the fifth fingers (Fig. 3). Also present was a slight syndactyly of the second and third toes. There was a very mild but noticeable asymmetry. Circumferences at the wrist were 6.5 cm on the right and 7 cm on the left, and 7 cm and 7.5 cm, respectively, at the elbow. The length from the acromion to the elbow on the right was 7.5 cm, and on the left, 8.5 cm. From shoulder to fingertip, the lengths were 19.5 cm and 20.5 cm, respectively. Similarly, a 1-cm difference in length was found in the measurements from the anterior superior spine to the patella and from the patella to the lateral malleolus. Bone age was considerably retarded. Extensive testing of the blood groups was carried out and the twins appeared to be identical.

GENETICS

The cause of this syndrome is not known. It may be nongenetic, and it seems to us that the patients reported have been heterogeneous. Examination for mosaicism of karyotypes in fibroblasts derived from the two sides of the body of typical patients has been negative

*We are indebted to Drs. William McCord, Earl Harley and James Campbell of the U. S. Naval Hospital, San Diego, for providing the details of this report and for permission to see and photograph these children.

in our hands and those of others. The syndrome has been reported to be concordant in a pair of monozygotic identical twins[3] and discordant in a pair of dizygotic twins.[7] It also has been observed in a mother and a daughter and in three brothers[9] whose mother was short. Most reported cases have been sporadic. These observations are consistent with the proposition[8] that the syndrome is due to a single dominant gene, the sporadic cases resulting from a new mutation. The discordant twins reported in this chapter who appeared to be monozygotic suggest a nongenetic etiology.

TREATMENT

There is no treatment. The disorder is compatible with a normal life. In counseling it is important to distinguish this disorder from hemihypertrophy, in which there is a substantial risk of neoplasia. Determination of the blood glucose concentration may be helpful in understanding episodes that may result from hypoglycemia. A small amount of catch-up growth has been reported to follow the administration of growth hormone.[13]

REFERENCES

1. Silver, H. K., Kiyasu, W., George, J., and Deamer, W. G.: Syndrome of congenital hemihypertrophy, shortness of stature and elevated urinary gonadotropins, Pediatrics 12: 368, 1953.
2. Russell, A.: A syndrome of "intra-uterine" dwarfism recognizable at birth with craniofacial dysostosis, disproportionately short arms, and other anomalies (5 examples), Proc. R. Soc. Med. 47:1040, 1954.
3. Rimoin, D. L.: The Silver syndrome in twins, Birth Defects 5:183, 1969.
4. Black, J.: Low birth weight dwarfism, Arch. Dis. Child. 36: 633, 1961.
5. Szalay, G. C.: Pseudohydrocephalus in dwarfs: The Russell dwarf, J. Pediatr. 63:622, 1963.
6. Silver, H. K.: Congenital asymmetry, short stature and elevated urinary gonadotropin, Am. J. Dis. Child. 97:768, 1959.
7. Silver, H. K.: Asymmetry, short stature, and variations in sexual development, Am. J. Dis. Child. 107:495, 1964.
8. Gareis, F. J., Smith, D. W., and Summitt, R. L.: The Russell-Silver syndrome without asymmetry, J. Pediatr. 79: 775, 1971.
9. Fuleihan, D. S., Der Kaloustian, V. M., and Najjar, S. S.: The Russell-Silver syndrome: Report of three siblings, J. Pediatr. 78:654, 1971.
10. Eeckels, R., van der Schuren-Lodeweyckx, M., and Wolter, R.: Plasma growth hormone determination in Silver-Russell syndrome, Helv. Paediatr. Acta 25:363, 1970.
11. Bloom, D.: Congenital telangiectatic erythema resembling lupus erythematosus in dwarfs, Am. J. Dis. Child. 88:754, 1954.
12. McKusick, V. A., Mahloudji, M., Abbott, M. H., Lindenberg, R., and Kepas, D.: Seckel's bird headed dwarfism, N. Engl. J. Med. 277:279, 1967.
13. Tanner, J. M., and Ham, T. J.: Low birth weight dwarfism with asymmetry (Silver's Syndrome): Treatment with human growth hormone, Arch. Dis. Child. 44:231, 1969.

UNILATERAL PSORIASIS AND ECTROMELIA

CARDINAL CLINICAL FEATURES

Strictly unilateral psoriasiform erythroderma with generalized musculoskeletal dysplasia on the same side. The central nervous system may or may not be involved.

CLINICAL PICTURE

Among the disorders that affect only one side of the body, one of the most striking is this combination of psoriasis, ectromelia and generalized musculoskeletal dysplasia. All lesions are sharply delimited and do not cross the midline.

Five patients have been reported, two of them siblings.[1-5] In all of them there has been unilateral ectromelia. The severity of the lesions resembles the thalidomide embryopathy. In one patient the lower extremity was virtually absent.[3] Dysplasia or hypoplasia of the extremities may be more marked proximally, or distally. The bones involved may have thin cortices as well

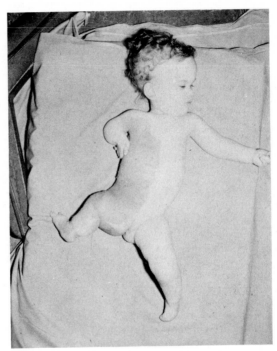

Fig. 2.—H. D. At 20 months the psoriatic dermatosis was still sharply delimited. Ectromelia of the right arm and leg was striking. The chest also was smaller on the right, as was the face.

Fig. 1.—H. D. (Case Report). At age 5 months. The skin lesion was strictly unilateral. Deformities of the arm and leg are apparent on the same side. (Reprinted with permission from Arch. Dermatol.[2])

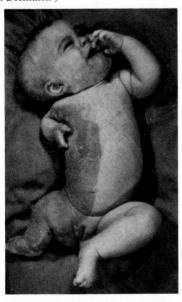

as shortening or deformity. Some bones, of course, may be absent. Muscular tissue in the involved extremities usually is also markedly reduced, whereas there may be an actual increase in amorphous fat.

Most patients have axial as well as skeletal abnormalities on the involved side. The rib cage, pelvis and vertebrae may be hypoplastic. Bony growth on the involved side usually is slow. One patient had hydronephrosis and hydroureter on the involved side, as well as unilateral bronchopulmonary abnormality.[3] The two siblings had fatal congenital heart disease.

The skin disorder is strictly delimited to the involved side. The border between involved and uninvolved skin can be detected by palpation. The disorder has been variously described as psoriasiform or ichthyosiform erythroderma. It appears bright red (Figs. 1 and 2) and edematous and bleeds very easily. The patient has a distinct odor of keratin. Large quantities of thin

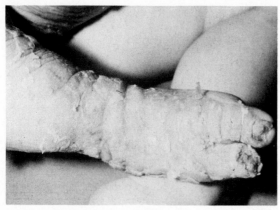

Fig. 3.—H. D. Close-up of the lower limb illustrates the large, thick sheets of desquamating keratin that characterized this patient. The appearance of the skin in the other illustrations reflected vigorous local therapy. There were two irregularly shaped toes separated by a deep cleft.

flakes of keratinized scale form and are desquamated, looking like potato chips or corn flakes (Fig. 3). The skin may be sensitive.

The histologic appearance is that of acanthotic skin with hyperkeratosis and parakeratosis. The stratum corneum is increased and the epidermis is markedly cellular. Kinetic studies of the skin have characterized

TABLE 1.—CELLULAR KINETICS OF THE SKIN

	LABELED CELLS PER CM
Patient: left side	45
right side	907
Normal Control	86
Ichthyosis	231
Psoriasis	959

the disorder as psoriatic (Table 1).[1] The rates of formation of skin cells can be assessed by timed intradermal injection of tritiated thymidine followed by biopsy and autoradiography (Fig. 4). In this way the proliferative disorders of the skin can be characterized because the numbers of cells in which DNA is being synthesized (s-phase) can be quantitated by counting those that have incorporated radioactive thymidine into their nuclei and thus generated black autoradiographic grains. Ichthyotic skin proliferates much more rapidly than normal skin. Psoriatic skin proliferates even more rapidly. In the patient studied the kinetics were those of psoriatic skin. The association of disordered embryonic development with increased cutaneous cellular DNA synthesis is interesting. There is, of course, no information as to whether or not the two mechanisms are similar.

Fig. 4.—H. D. Histologic sections of the epidermis of the normal side (**left**) and psoriatic side (**right**), following injection of tritiated thymidine. The involved epidermis and stratum corneum were thickened. The basal cell layer on the affected side was densely labeled, indicating rapid formation of new cells. (Reprinted with permission from Shear, *et al.*[1])

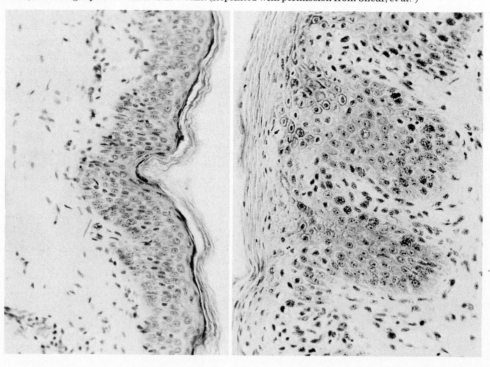

The uniformity of cerebral defect is not clear in this syndrome. Two patients[1, 4] have had similar abnormalities, including ventricular dilatation on pneumoencephalography. It is of considerable interest that the cerebral cortical dysplasia is on the same side as all of the other manifestations. It is important not to assume a severity of cerebral disease from the obvious severity of cutaneous and musculoskeletal abnormalities. These patients may function quite well in spite of their deformities.

CASE REPORT

H. D., a female, was born by breech delivery weighing 5 lb, 4 oz. A previous sibling had died at 6 years of age of acute lymphoblastic leukemia. The maternal grandmother had had psoriasis from childhood, and the mother had mild psoriatic patches on both elbows.

At birth the infant appeared to have right-sided hemiatrophy: the right side of the body was covered by a yellowish discoloration. Her appearance at 5 months of age is illustrated in Figure 1. Her early psychomotor development was slow, and she was admitted to the Sunland State Hospital for the retarded in Orlando, Florida, at about 1 year of age. However, she continued to develop and, following evaluation at the University of Miami Child Development Center, she was returned to her home.

Examination at 20 months (see Figs. 2 and 3) revealed the characteristic skin lesion, as described above. The involved finger- and toenails were hyperplastic and cornified. She had facial and cranial asymmetry and a tendency to tilt the head to the right. The rib cage on the right was hypoplastic and the

Fig. 5. – H. D. **Left,** roentgenogram of the right lower extremity. Muscle mass was strikingly reduced and fat density increased. The femur was markedly dysplastic. **Right,** the normal side is shown for comparison. (Reprinted with permission from Shear, et al.[1])

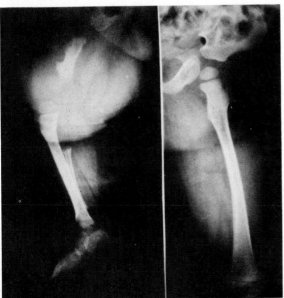

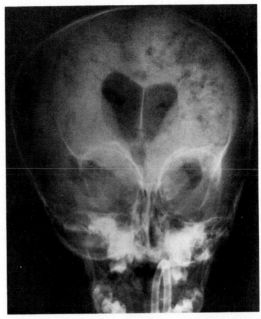

Fig. 6. – H. D. Pneumoencephalogram. Both lateral ventricles actually were large, but the major enlargement, as well as major atrophy of the brain, was on the right, the same side as the peripheral deformities. (Reprinted with permission from Shear, et al.[1])

pectoralis major muscle was absent. The right upper limb was small, with marked shortening of both proximal and distal portions. The elbow was partially webbed and restricted in motion. The small, deformed hand had four digits and pedunculated nubbins of tissue on the ulnar side. Mobility at the shoulder was sufficient for bimanual tasks, and the thumb was functional.

Roentgenograms revealed the bones of the right arm to be hypoplastic, with thin cortices. Phalanges were absent. The right leg was profoundly shortened, especially proximally. Its circumference was much greater than that of the normal side. The limb was maintained in dorsiflexion, and motion at the hip was limited. The foot had a lobster-claw appearance, with two large toes separated by a deep cleft. X-rays revealed a very hypoplastic femur (Fig. 5). Facial bones, ribs, pelvic bones and vertebrae were hypoplastic on the right.

Pneumoencephalography revealed cortical atrophy and ventricular enlargement, predominantly on the right (Fig. 6). Electroencephalography revealed 4/second θ activity on the right. At 2 years and 9 months, she was functioning at a mental age of 2 years and 4 months, giving her an IQ of 84. She was fitted with a prosthesis and learned to walk unassisted.

A 2-week trial of therapy with methotrexate brought demonstrable improvement in the skin. Erythema and edema were generally reduced, and some previously involved areas cleared completely.

GENETICS

The numbers of cases are too few to provide real information, but it seems likely that this condition is genetically determined. It has been seen in both sexes.

TREATMENT

The skin may generally be managed using topical treatment designed to soften and remove the scale. Systemic treatment with methotrexate may be useful, but its potential hazards must be weighed. Orthopedic treatment and prostheses may be useful in the management of the skeletal abnormalities. From our experience it is important for those caring for the child not to be put off by immediate appearances and to provide a chance for development.

REFERENCES

1. Shear, C. S., Nyhan, W. L., Frost, P., and Weinstein, G. D.: Syndrome of Unilateral Ectromelia, Psoriasis and Central Nervous System Anomalies, in *Skin, Hair, and Nails,* Birth Defects Original Article Series (New York: The National Foundation-March of Dimes, 1971), p. 197.
2. Cullen, S. I., Harris, D. E., Carter, C. H., and Reed, W. B.: Congenital unilateral ichthyosiform erythroderma, Arch. Dermatol. 99:724, 1969.
3. Rossman, R. E., Shapiro, E. M., and Freeman, R. G.: Unilateral ichthyosiform erythroderma, Arch. Dermatol. 88:567, 1963.
4. Zellweger, V. H., and Uehlinger, E.: Ein Fall von halbseitiger Knochenchondromatose (Ollier) mit Naevus ichthyosiformis, Helv. Paediatr. Acta 2:153, 1948.
5. Falek, A., Heath, C. W., Ebbin, A. J., and McLean, W. R.: Unilateral limb and skin deformities with congenital heart disease in two siblings: A lethal syndrome, J. Pediatr. 73: 910, 1968.

Gigantism Syndromes

BECKWITH SYNDROME
Beckwith-Wiedemann Syndrome

CARDINAL CLINICAL FEATURES

Macroglossia, omphalocele, gigantism, hypoglycemia, mild microcephaly, visceromegaly.

CLINICAL PICTURE

Beckwith[1] reported in 1963 three unrelated infants who had been found at autopsy to have omphalocele, muscular macroglossia, bilateral cytomegaly of the fetal adrenal cortex, hyperplasia of the gonadal interstitial cells and renal medullary dysplasia. In 1964[2] he described two living children with this syndrome who had microcephaly, hypoglycemia and gigantism.

At about the same time, Wiedemann et al.[3] reported three siblings with this syndrome. The syndrome has come to be known as the Beckwith syndrome, as well as the Beckwith-Wiedemann syndrome. Wiedemann has called it the E. M. G. syndrome, which might serve as a reminder of the cardinal features exomphalos, macroglossia and gigantism. However, people do not seem to remember syndromes, or many other things, well from initials. The disorder will probably continue to be called the Beckwith syndrome.

Macrosomia is a prominent characteristic of this syndrome. These infants are large at birth, and postnatal growth takes place at or above the 90th percentile. Skeletal maturation also is advanced. Hemihypertrophy has been seen in at least four patients.[1]

Macroglossia is a hallmark of the syndrome (Figs. 1 and 2). It usually is present but may be variable in expression. The enlargement of the tongue has been found to be due to muscular hypertrophy. The tongue is uniformly enlarged and usually protrudes from the mouth. Rarely it may cause respiratory embarrassment in the supine position, but it usually is manageable in a side or prone position. Feeding rarely is a problem. In such a case a soft nipple may be helpful or, very rarely, a brief period of tube feeding. Macroglossia appears to regress over the years with growth of the oral cavity and gradual accommodation of the large tongue. In some patients this has not occurred, and surgical correction has been done.

Umbilical anomalies may occur in varying degrees of severity. The classic lesion is an omphalocele (Fig.

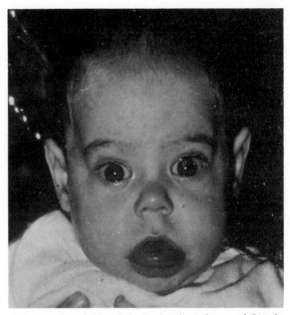

Fig. 1. – This child with the Beckwith syndrome exhibits the large, protruding tongue.

3). Some patients may have simply an umbilical hernia or a diastasis recti.

Mild microcephaly may be noticeable in the presence of somatic gigantism. Microcephaly also may be absent. The occiput often is prominent. Many patients have a nevus flammeus of the face, which involves the center of the forehead and the upper lids, occasionally extending down to the nose and upper lip. Ear lobe anomalies are relatively common. Irving[4] found a linear indentation of the lobe (Fig. 4) in seven of 11 patients. This minor malformation may take the form of an inverted Y.

Hypoglycemia probably is the most important manifestation of the syndrome. It is not always present but is found in one third to one half of the infants with this syndrome. The structural malformations in these patients should serve to alert the physician to the possibility of hypoglycemia. Reduction in blood sugar usually is severe. It most often occurs on the second or third day of life and usually persists beyond the neonatal period. Hyperinsulinism has been documented in some but not all patients studied. Autopsied patients have uniformly had enlargement of the pancreas and hyperplasia of the islet cells, which would provide a structural basis for hyperinsulinism. A lack of immunoreactive glucagon also has been reported in the serum of a child with this syndrome.[5] Inadequate secretion of glucagon would of course also promote hypoglycemia. Patients with this syndrome generally have normal intelligence. Mental retardation, seen in some, may be attributed to episodes of neonatal hypoglycemia.

Organomegaly or cellular hyperplasia may be widespread. Hepatomegaly is common. Enlarged kidneys also may be palpable. Neonatal polycythemia may be present. Enlargement of the adrenals or hyperplasia of the pancreas, ovary or testis is not generally recognized clinically but may be observed grossly as well as histologically. Cytomegaly of the fetal adrenal contrasts with hyperplasia in other organs. Renal and ureteral malformations have been reported.[4] These include bilateral hydroureter with left-sided hydronephrosis and double right kidney. Renal medullary dysplasia and microscopic cortical cysts have been seen regularly in autopsied patients dying early in life. Clitoral enlargement has been reported, as have cryptorchidism and supernumerary toes. Diaphragmatic eventration with its dome-like elevation has been observed on x-ray. Intestinal malrotation is common, as it is in other patients with omphalocele. Study of the chromosomes has been normal.

The development of malignant neoplasms has been recognized as a characteristic of the syndrome. Children with the syndrome are at high risk for the development of intra-abdominal malignancy. Adrenal carcinoma and Wilms' tumor have been reported.

Patients surviving the neonatal period have generally done well. Mild craniofacial defects become apparent later in childhood so that these children tend to resemble each other. They have malocclusion, prognathism and a bite that is open anteriorly. There is a slight exophthalmic appearance, a short overbite and underdevelopment of the middle third of the face. Many patients have a midline frontal ridge.

DIFFERENTIAL DIAGNOSIS

The complete syndrome of omphalocele macroglossia, gigantism and hypoglycemia is unique. Generalized muscular hypertrophy occurs also in the de Lange syndrome, type II. Muscular development also appears striking in generalized lipodystrophy (the Berardinelli or Seip syndrome) and in a syndrome in which acanthosis nigricans is associated with diabetes mellitus, with high insulin levels, poor growth, dental anomalies and reduced subcutaneous fat. The peripheral muscles do not appear striking in the Beckwith syndrome, as subcutaneous tissues are well developed and many patients are somewhat obese. Omphalocele may of course occur in the absence of this syndrome. Of 94 patients with exomphalos seen in Liverpool, 11 had macroglossia.

Macroglossia may occur as a primary disorder in the absence of the other manifestations seen in this syndrome. Macroglossia also is seen in congenital hypothyroidism, the Debré-Semelaigne syndrome. It also is seen in type II glycogen storage disease, amyloidosis, mucopolysaccharidosis and neurofibromatosis, as well as in the Thomsen myotonia congenita. Patients with the Down syndrome appear to have a large tongue, but this probably reflects a problem of accommodation to a very small mouth.

CASE REPORTS

CASE 1.—T. P. was the product of a premature delivery following an uncomplicated pregnancy in an 18-year-old Mexican gravida I para I mother. There was no evidence of consanguinity. She was born by cesarean section, apparently because of difficulty with delivery. The birth weight was 9 lb, 6 oz. A large tongue was seen protruding from her mouth, especially when she cried. The head circumference was 33 cm, the body length 21¼ inches.

There was an approximately 5-cm omphalocele, (see Fig. 3), which was intact. The external genitalia were those of a normal female. Enlarged kidneys were palpable bilaterally. There was no hepatomegaly or other abdominal mass.

She had a poor cry, but the Moro reflex was active and the deep tendon reflexes normal.

Blood sugar was 66 mg/100 ml. Hemoglobin was 22 gm/100 ml and hematocrit 58. The infant underwent surgery 4 hours after delivery, and the omphalocele was repaired. She did well. She had no apnea, lethargy or seizures. Subsequent blood sugar concentrations were 96, 74, 78 and 90 mg/100 ml. Postnatal gigantism has been observed with followup. Height and weight have followed the 90th percentile. The intravenous pyelogram was normal.

CASE 2.—B. F. was born at term of a group A, Rh negative, primiparous mother. Pregnancy and delivery were uncomplicated. There was no consanguinity. Birth weight was 11 lb, 4 oz. Length was 23 inches, head circumference 34 cm and chest 35.5 cm.

This boy was an enormous baby with prominent macroglossia. He had a broad nasal bridge and an anomalous ear with an angulated canal.

There was no omphalocele. Instead, a large diastasis recti and a small umbilical hernia were present. Cryptorchidism was observed on the left, and the right testis was abnormally large. The liver was large; it was felt three fingerbreadths below the costal margin. Both kidneys were palpably enlarged.

Blood sugar was 30 mg/100 ml and was corrected with oral sugar. Subsequent blood sugar concentration was 75 mg/100 ml. Roentgenograms revealed a fracture of the midshaft of the left clavicle.

The patient was observed on examination to have a poor urinary stream. BUN was 17.8 mg/100 ml. Intravenous pyelogram revealed delayed opacification with the first contrast at 30 minutes. There was marked dilatation of both renal collection systems. The ureters were distended down to the bladder, which also appeared to be distended.

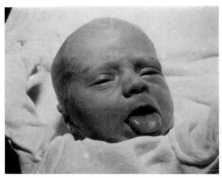

Fig. 2.—A neonatal infant with the classic facial appearance and large tongue of the Beckwith syndrome. (Courtesy of Dr. Douglas Dechairo.)

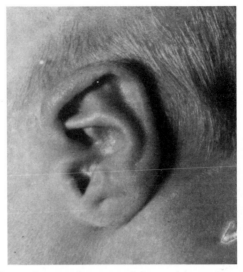

Fig. 3.—T. P. (Case 1). The overall size of this newborn baby was impressive, as was the omphalocele. The tongue was large.

Fig. 4.—Linear indentation of the ear lobe in the infant shown in Figure 2.

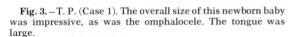

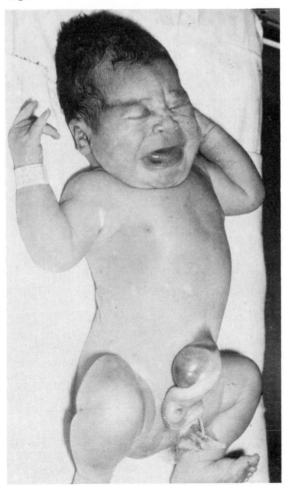

GENETICS

Autosomal recessive inheritance is suspected. The occurrence of the syndrome in siblings has been recognized repeatedly. It also has been observed in second cousins.[4] The parents of the three siblings reported by Wiedemann[3] were apparently not consanguineous, although they were at first thought to be.

TREATMENT

A major problem in this syndrome is the management of hypoglycemia. The large tongue should alert the physician to this syndrome, and any newborn infant with an omphalocele should have his blood sugar determined. If hypoglycemia is detected, the patient should be treated with oral and/or intravenous glucose. Cortisol, diazoxide or glucagon also may be useful in the management of severe hypoglycemia.

Surgical repair of the omphalocele should be undertaken promptly. Partial glossectomy has been performed in some patients, but surgery on the tongue should be delayed, as with growth it is often found to be unnecessary.

Children with this syndrome should be closely observed for intra-abdominal malignancy. Intravenous pyelography should be routine.

REFERENCES

1. Beckwith, J. B.: Macroglossia, Omphalocele, Adrenal Cytomegaly, Gigantism, and Hyperplastic Visceromegaly, in *The First Conference on the Clinical Delineation of Birth Defects. Part II, Malformation Syndromes*, Birth Defects: Original Article Series (New York: The National Foundation-March of Dimes, 1969), p. 188.

 2. Beckwith, J. B., Wang, C., Donnell, G. N., and Gwinn, J. L.:
    Hyperplastic fetal visceromegaly with macroglossia,
    omphalocele, cytomegaly of the adrenal fetal cortex, post-
    natal somatic gigantism and other abnormalities; a newly
    recognized syndrome. Abstract read at Annual Meeting of
    the American Pediatric Society, Seattle, June 16–18,
    1964; p. 56.
 3. Wiedemann, H. R., Spranger, J., Moghavei, M., Kubler, W.,
    Tolksdorf, M., Bontemps, M., Drescher, J., and Gun-
    schera, H.: Uber das Syndrom Exomphalos-Makroglosie-
    Gigantismus, uber generalizierte Muskelhypertrophie,
    progressive Lipodystrophie und Miescher Syndrom in
    Sinne diencephaler Syndrom, Z. Kinderheilkd. 102:
    1, 1968.
 4. Irving, I.: Exomphalos with macroglossia: A study of 11
    cases, J. Pediatr. Surg. 2:499, 1967.
 5. Lazarus, L., Young, J. D., and Friend, J. C. M.: E. M. G.
    syndrome and carbohydrate metabolism, Lancet 2:1347,
    1968.
 6. Combs, J. T., Grunt, J. A., and Brandt, I. K.: New syndrome
    of neonatal hypoglycemia. Association with visceromega-
    ly, macroglossia, microcephaly and abnormal umbilicus,
    New Engl. J. Med. 275:236, 1966.
 7. Filippi, G., and McKusick, V. A.: The Beckwith-Wiede-
    mann syndrome, Medicine 49:279, 1970.
 8. Wiedemann, H. R.: E. M. G. syndrome and carbohydrate
    metabolism, Lancet 2:104, 1968.
 9. Sotelo-Avila, C., and Singer, D. B.: Syndrome of hyperplas-
    tic fetal visceromegaly and neonatal hypoglycemia
    (Beckwith's syndrome), Pediatrics 46:240, 1970.
10. Sheman, F. E., Bass, L. W., and Fetterman, G. H.: Congeni-
    tal metastasizing adrenal cortical carcinoma associated
    with cytomegaly of the fetal adrenal cortex, Am. J. Clin.
    Pathol. 30:439, 1958.

Genetic Syndromes with Neoplasia

NEVOID BASAL CELL CARCINOMA SYNDROME

Cardinal Clinical Features

Multiple nevoid cutaneous basal cell carcinomas, cysts of the jaw, rib anomalies, pits on hands and soles, intracranial calcifications.

Clinical Picture

The syndrome was described by Binkley and Johnson[1] in 1951 in a woman with basal cell nevi, dental cysts, agenesis of the corpus callosum, an ovarian fibroma and a bifid rib. Gorlin and Goltz[2] first characterized the disorder as a syndrome and suggested its autosomal dominant inheritance. Penetrance of the syndrome is high. As many as 95% of individuals with the gene will have some manifestation of the syndrome.[3] On the other hand, expression is variable. Involved individuals may have involvement ranging from a single feature to the complete syndrome.

The cutaneous features of the disease usually are those that come first to the attention of the physician.[3] Tumors are seldom present at birth. They usually appear in childhood, especially around puberty, and certainly by the second or third decade. Flesh-colored or brownish pigmented papules, soft nodules or flat plaques may appear in the first years of life. They tend to be numerous, are somewhat variable in size (from 1 mm – 1 cm in diameter) and color and tend to appear in crops periodically throughout childhood and later. These tumors grow very slowly and may be quiescent for months or years. Some patients have had as many as 1,000 of these lesions. There is a predilection of multiple nevoid basal cell carcinomas for the face (Fig. 1). They may be found on the nose, eyelids, periorbital areas, malar regions and upper lip as well as on the neck and trunk (Fig. 2). Exposure to sunlight appears to play no role in this skin cancer. The tumors may appear on exposed and unexposed areas of skin. Even in a single patient the lesions may take many forms: some superficial, some pigmented, some solid, some cystic.[4] They appear multicentric in origin. It is impressive how benign these tumors appear clinically. This and their slow growth lead patients to come tardily for treatment. In one series the average age at onset was 14 years, whereas the average age at which the patient sought treatment was 21 years. The histologic appearance is in startling contrast to the clinical impression. These tumors are basal cell cancers from the start and can readily be recognized as such microscopically. Bone or osteoid may be seen in some.[5] Malignant clinical behavior is characterized by increase in size, ulceration, crusting and bleeding. These cancers are potentially destructive to the nose and face. Patients have been described who committed suicide[4] in reaction to the disfigurement caused by numerous skin cancers.

There also may be a variety of benign tumors of the skin. These include epithelium-lined cysts, sebaceous cysts and milia.[6, 7] Fibromas and lipomas also may be seen. Tumors may be seen in other areas as well. There have been a number of ovarian fibromas,[8] and medulloblastoma of the central nervous system is a recognized complication of the syndrome.[5] Astrocytomas and hydrocephalus have been reported. Lymphatic mesenteric cysts also have been described in a number of patients.[8]

Mandibular cysts are a feature of this disease and may be scattered throughout the jaw. These usually appear in the first decade of life and often develop before the skin lesions. They occasionally involve the maxilla. The cysts are odontogenic keratocysts or primordial cysts and have an epithelial lining. They vary from microscopic to several centimeters in diameter[4] or they may be large enough to cause facial deformity. Pain or tenderness, or inability to open the mouth develops in some patients, as well as infection, with drainage into the mouth. The cysts may contain an unerupted tooth but usually do not. Ultimately they occur in about 65% of patients.[3] It is the rule for cysts to recur after curetting. Rarely an ameloblastoma has been seen in conjunction with a jaw cyst.

Pits of palms and soles are a unique feature of this syndrome. They appear between the second and third decades and vary in number from a few to a great many. They are a few millimeters in diameter and superficial. They may have an erythematous telangiectatic base[3] and are permanent. These changes have been interpreted as a dyskeratosis of the palms and soles.[9] The histopathology is that of a defect in keratinization. The facial features of patients with this syndrome also may be characteristic.[5] There may be pagetoid parietal bossing, and a cranial vault that may appear large. The supraorbital ridges may be prominent and the eyes may appear sunken. The bridge of the nose may be broad. Mandibular prog-

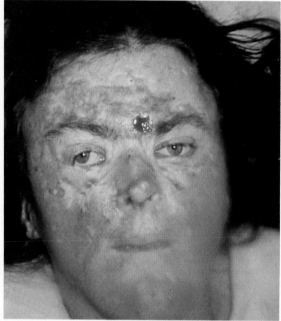

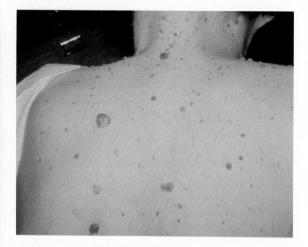

Fig. 1 (left).—A 30-year-old woman with the basal cell nevus syndrome. The face shows extensive scarring following the removal of many lesions. There was an active lesion of the lower midforehead.

Fig. 2 (right).—In the patient shown in Figure 1, the back, at an earlier age, illustrates some of the thousands of fleshy, fibroma- or mole-like forms. Histologically these were basal cell epitheliomas.

(Figures 1 and 2 courtesy of Dr. Dennis A. Weigand of the University of Oklahoma.)

nathism usually is present but mild. Some patients have had dystopia canthorum.[4]

In patients with this syndrome the skeleton may be involved elsewhere. Rib defects, usually a bifid rib, are characteristic. There also may be fusion, agenesis or deformity of ribs, or wide, flattened anterior ends.[10] Spina bifida occulta occurs as well as synostosis and partial agenesis of vertebrae, kyphosis and scoliosis. Bridging of the sella turcica has been reported in which the anterior and posterior clinoid have been fused.[5, 10]

Brachymetacarpalism in this syndrome takes the form of a short fourth metacarpal. Albright's sign, in which a dimple occurs instead of a knuckle, may be seen. This sign may be correlated with the presence of ovarian fibromas.[8] Patients with these tumors may develop calcification in a fibroma or in the uterus. Ovarian cysts and occasionally mesenteric cysts also have been reported. Ectopic calcifications may occur, most commonly in the falx cerebri. This ossification is lamellar and distinctive. It also may be seen in the tentorium cerebelli and dura. A short fourth metacarpal is generally considered a sign of pseudohypoparathyroidism. Some patients have been reported to have a poor phosphaturic response to administered parathormone, much as do patients with pseudohypoparathyroidism. On the other hand a recent careful study of six patients with this syndrome failed to reveal any with an abnormal response to parathyroid hormone.[11] Hypogonadism has been reported in the male. These patients have had cryptorchidism, scant facial hair or a female pubic hair distribution.

Some patients with this syndrome have had borderline or slightly lowered intelligence. Some have had electroencephalographic abnormalities, and a few have had seizures. Schizophrenia also has been reported.[5] Agenesis of the corpus callosum has been

Fig. 3.—C. G. (Case Report). A 13-year-old girl with the syndrome. Evident are hypertelorism, the repaired cleft of the lip and the large mandible. (Courtesy of Dr. Akram Tamer of the University of Miami, Florida.)

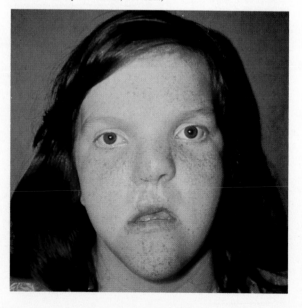

Fig. 4. – C. G. Roentgenogram of the jaw illustrates the large cystic lesions in the mandible. (Courtesy of Dr. Akram Tamer of the University of Miami, Florida.)

reported.[1] Ophthalmologic findings include congenital cataracts, colobomas and blindness as well as hypertelorism. The basal cell carcinoma may occur first on the eyelids.[12] Multiple chalazia and strabismus also are seen. It is probable that cleft lip and palate occur with increased frequency in this syndrome.

Case Report*

C. G., a 13-year-old white girl, was admitted to the Jackson Memorial Hospital, Miami, Florida, because of multiple mandibular and maxillary cysts. The cysts were found incidentally in dental x-rays taken 18 months prior to admission.

The patient was the product of a full-term, uneventful pregnancy. Birth weight was 10 lb. She was found to have a complete left cleft of the lip and palate, the lip was operated on at 1 month and the palate at 2 years of age. The patient had always been large for her age. At 4 years of age she developed "moles" over the face, the upper trunk and the proximal part of the extremities that did not appear to increase in number or to change. She had had repeated upper respiratory infections, ear infections and sinusitis from the age of 6 years. Her teeth were abnormally placed and she had to have constant dental care. She attended regular school and was an "average" student in the sixth grade.

Family history revealed that the patient's grandfather had had basal cell carcinomas of the skin for 15 years.

On physical examination the patient was a tall, heavy-set adolescent. She was 6 feet tall, weighed 213 lb and had a head circumference of 60 cm. There were marked frontal bossing, moderate hypertelorism and exotropia of the right eye. The bridge of the nose was depressed. Healed scars were noted on the lip (Fig. 3) and palate. The teeth were badly displaced, and there was marked prognathism.

*We are grateful to Dr. Akram Tamer of the University of Miami School of Medicine, Miami, Florida, for providing us with the details of the patient's clinical course.

On the skin there were multiple brown, raised lesions, about 2.5 mm in diameter over the face, upper trunk and proximal parts of the arms. Physical examination was otherwise normal.

Roentgenograms of the skull showed calcified dura and calcified falx cerebri. There was a cyst in the left maxilla and three cysts in the mandible. She had a bifid sixth rib on the left. Biopsy of a mandibular cyst (Fig. 4) showed it to be an odontogenic cyst. Biopsy of a skin lesion revealed a hollow nevoid structure.

Genetics

The disorder is determined by an autosomal dominant gene. It has a high penetrance and variable expression. Both sexes are equally affected.

Treatment

All nevoid basal cell carcinomas should be removed. The usual approach to small lesions is by electrodesiccation and curettage. Tumors may be expected to continue to appear throughout the life of the patient. Some lesions may be successfully treated with topical 5-fluorouracil.[13]

References

1. Binkley, G. W., and Johnson, H. H., Jr.: Epithelioma adenoides cysticum; basal cell nevi, agenesis of corpus callosum and dental cysts, Arch. Dermatol. Syph. 63:73, 1951.
2. Gorlin, R. J., and Goltz, R. W.: Multiple nevoid basal cell epithelioma, jaw cysts and bifid rib syndrome, N. Engl. J. Med. 262:908, 1960.
3. Howell, J. B., Anderson, D. E., and McClendon, J. L.: Multiple cutaneous cancers in children: The nevoid basal cell carcinoma syndrome, J. Pediatr. 69:97, 1966.
4. Gorlin, R. J., Yunis, J. J., and Tuna, N.: Multiple nevoid basal cell carcinoma, odontogenic keratocysts and skeletal anomalies, Acta Derm. Venereol. (Stockh.) 43:39, 1963.
5. Gorlin, R. J., and Sedano, H. O.: The Multiple Nevoid Basal Cell Carcinoma Syndrome Revisited, in *The Clinical Delineation of Birth Defects. Part XII, Skin, Hair and Nails,* Birth Defects: Original Article Series (New York: The National Foundation-March of Dimes, 1971), p. 140.
6. Anderson, D. E., McClendon, J. L., and Howell, J. B.: Genetics and Skin Tumors, with Special Reference to Basal Cell Nevi, in M. D. Anderson Hospital and Tumor Institute: *Tumors of the Skin* (Chicago: Year Book Medical Publishers, Inc., 1964), pp. 91–127.
7. Howell, J. B., and Caro, M. R.: The basal-cell nevus, its relationship to multiple cutaneous cancers and associated anomalies of development, Arch. Dermatol. 79:67, 1959.
8. Clendenning, E. E., Block, J. B., and Radde, I. C.: Basal cell nevus syndrome, Arch. Dermatol. 90:38, 1964.
9. Ward, W. H.: Naevoid basal cell carcinoma associated with a dyskeratosis of the palms and soles. A new entity, Aust. J. Dermatol. 5:204, 1950.
10. Pollard, J. J., and New, P. F. J.: Hereditary cutaneomandibular polyoncosis. A syndrome of myriad basal cell nevi of the skin, mandibular cysts, and inconstant skeletal anomalies, Radiology 82:840, 1964.

11. Kaufman, R. L., and Chase, L. R.: Basal Cell Nevus Syndrome: Normal Responsiveness to Parathyroid Hormone, in *The Clinical Delineation of Birth Defects. Part XII, Skin, Hair and Nails,* Birth Defects: Original Article Series (New York: The National Foundation-March of Dimes, 1971), p. 149.

12. Markovits, A. S., and Quickert, M. H.: Basal cell nevus, Arch. Ophthalmol. 88:397, 1972.

13. Moynahan, E. J.: Multiple basal cell naevus syndrome — successful treatment of basal cell tumours with 5-fluorouracil, Proc. R. Soc. Med. 66:627, 1973.

14. Hashimoto, K., Howell, J. B., Yamanishi, Y., Holubar, K., and Bernhard, R., Jr.: Electron microscopic studies of palmar and plantar pits of nevoid basal cell epithelioma, J. Invest. Dermatol. 59:380, 1972.

15. Neblett, C. R., Waltz, T. A., and Anderson, D. E.: Neurological involvement in the nevoid basal cell carcinoma syndrome, J. Neurosurg. 35:577, 1971.

16. Maddox, W. D., Winkelmann, R. K., Harrison, E. G., Jr., Devine, K. D., and Gibilisco, J. A.: Multiple nevoid basal cell epitheliomas, jaw cysts, and skeletal defects, J.A.M.A. 188:106, 1964.

17. Mason, J. K., Helwig, E. B., and Graham, J. H.: Pathology of the nevoid basal cell carcinoma syndrome, Arch. Pathol. 79:401, 1965.

18. Ferries, P. E., and Hinrich, W. L.: Basal cell carcinoma syndrome, Am. J. Dis. Child. 113:538, 1967.

19. Graham, J. K., McJimsey, B. A., and Hordin, J. C.: Nevoid basal cell carcinoma syndrome, Arch. Otolaryngol. 87:72, 1968.

20. Anderson, D. E., Taylor, W. B., Falls, H. F., and Davidson, R. T.: The nevoid basal cell carcinoma syndrome, Am. J. Hum. Genet. 19:12, 1967.

21. Rater, C. J., Selke, A. C., and Vanepps, E. F.: Basal cell nevus syndrome, Am. J. Roentgenol. Radium Ther. Nucl. Med. 103:589, 1968.

22. Mills, J., and Ponkles, J.: Gorlin's syndrome. Radiologic and cytogenic study of nine cases, Br. J. Radiol. 40:366, 1967.

PEUTZ-JEGHERS SYNDROME
Intestinal Polyposis and Mucocutaneous Melanin Pigmentation

CARDINAL CLINICAL FEATURES

Melanin spots on the lips, buccal mucosa and digits; multiple polyps anywhere in the gastrointestinal tract, particularly in the jejunum; intussusception; intestinal bleeding.

CLINICAL PICTURE

Melanin deposits of the lips and buccal mucosa were first recognized to be associated with polyps of the intestinal tract by Peutz[1] in 1921. Jeghers, McKusick and Katz[2] provided a definitive description of the condition and its transmission, which led to the eponymic designation. The pigmented spots vary from light tan to pale bluish to deep black. Characteristically they are found on the lips, especially the lower lips and the buccal mucosa (Fig. 1). They sometimes are seen on the hard palate and gingivae, as well as at other locations such as perioral, perinasal or periorbital. They also may occur on the fingers (Fig. 2) and toes. The spots resemble freckles, but they are found in areas in which freckles do not occur. They may be seen within

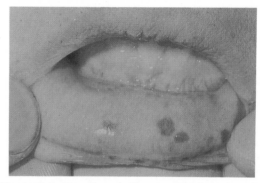

Fig. 1.—I. A. (Case Report). Close-up of the buccal mucosa of a 7-year-old girl with the Peutz-Jeghers syndrome, illustrating the deposits of melanin.

the first 6 months of life and do not vary with exposure to sunlight.

Polyps are most commonly found in the small intestine. Jejunal polyps are a particularly regular feature of the disease. However, polyps may be seen anywhere in the intestine, including the stomach and the colon.

Fig. 2 (left).—I. A. Melanin spots were also visible on the palmar surfaces of the fingers.
Fig. 3 (right).—I. A. At age 10 her pallor is evident, particularly around the lips. Closer inspection of the lips revealed spots of melanin pigmentation.

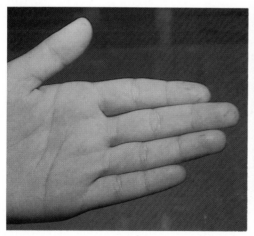

Bloody stools frequently occur and may be the first sign of disease. Recurrent colicky abdominal pain is common, and intussusception is a prominent complication. Anemia has been observed in about 25% of reported patients due to loss of blood into the intestinal tract. Nutritional problems, edema and inanition may occur.

In a review of the literature[3] multiple polyps were found in the gastrointestinal tract in 54 patients, and a single polyp in 12 patients. Polyps were found in the jejunum in 54 patients and in the ileum in 43. Colonic polyps were seen in 26 patients. Polyps were found in the rectum in 22 patients, in the stomach in 17, in the duodenum in eight and in the appendix in three.

Histologically the polyps have been described as benign adenomatous polyps. They are characterized by a multiplicity of cell types much like that of the surrounding gastric or intestinal tissue. This has suggested that these tumors are developmental anomalies or hamartomas rather than true neoplasms.[3] Doubt was cast on the diagnosis of malignancy reported in some cases, for it had been based on the histologic appearance of the lesion and none of the patients described had died of disseminated carcinoma.[3.] However, it is now clear that malignancy with death from disseminated gastrointestinal carcinoma has been reported.[4] Ovarian tumors, particularly granulosa cell tumors, have been observed in patients with this syndrome.

DIFFERENTIAL DIAGNOSIS

A syndrome has been described by Cronkhite and Canada[5] that is similar to the Peutz-Jeghers syndrome but appears to be distinct. In this disease gastrointestinal polyposis is associated with ectodermal changes. It appears to develop late in life and to be very rare; only nine cases have been reported since the first description. Involved patients have had no family history of gastrointestinal disturbance or ectodermal changes. They presented with diarrhea, alopecia and atrophy of the nails of the fingers and toes. Pigmentation of the skin was diffuse, brown and ill defined, with streaks and patches on the face, body folds and dorsa of the hands and fingers. Hypoproteinemia, edema and nutritional deficiency were prominent. Polyposis was found to involve the stomach, small intestine and colon. Biopsy revealed benign adenomatous polyps. The disease was fatal within 6–8 months of onset in most patients.

Familial polyposis of the colon or polypoid adenomatosis of the colon also is transmitted as an autosomal dominant. In marked contradistinction to the Peutz-Jeghers syndrome, this is a regularly premalignant condition. Carcinoma may occur in the second decade or as late as the seventh.

In the Gardner syndrome,[6] familial polyposis of the colon, and sometimes of the stomach and small intestine, is associated with osseous tumors. The colonic

polyps frequently undergo malignant degeneration; this syndrome is transmitted as an autosomal dominant trait.

Familial polyposis also has been recognized to be associated with tumors at other anatomic sites. Among the most important of these is the central nervous system,[7] where medulloblastoma and glioblastoma have been observed. Some patients also have been seen with tumors of the soft somatic tissues. These observations suggest that familial polyposis is a dominantly inherited disorder in which there is a potential for the development of neoplasia in a wide variety of tissues.

CASE REPORT

I. A., 7-year-old Mexican girl, was admitted to University Hospital for evaluation of a cardiac murmur.

She had a history of pallor and fatigue for 6 months' duration. More recently she had complained of anorexia and headache. There was no history of recurrent colic, melena or abdominal mass. A physician heard a grade II systolic murmur and referred her to Pediatric Cardiology.

On admission she was noted to have pigmented lesions over the lips (Fig. 3) and inside the mouth (see Fig. 1). Melanotic

Fig. 4. – I. A. Late small bowel series revealing the coiled-spring appearance of an intussusception.

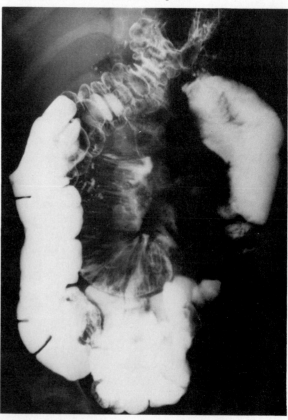

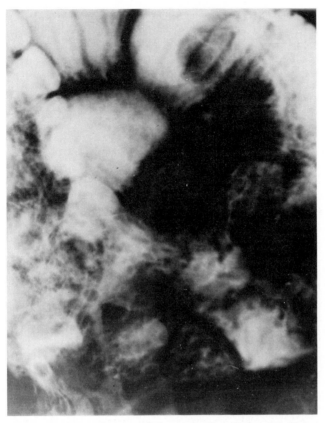

Fig. 5.—I. A. Barium in the intestine has outlined multiple polypoid masses.

spots also were seen on the fingers (see Fig. 2). She was found to have a hemoglobin of 6 gm/100 ml and a hematocrit of 22. The heart was not abnormal, and the murmur was interpreted to be hemic.

Because of these findings a gastrointestinal series of roentgenograms was undertaken. A transient jejunal intussusception was found (Fig. 4), which was reduced during barium enema. Multiple polyps were outlined in the barium study (Fig. 5). The mother was found to have similar pigmentation on the lips and oral mucosa.

GENETICS

The Peutz-Jeghers syndrome is transmitted as an autosomal dominant character. There appears to be a high degree of penetrance, but it is clear that some involved members of a family may have pigmentary lesions without polyps. Bartholomew and Dahlin[3] made a survey of reported patients in which 40% had relatives with both polyps and pigmentation and another 15% had relatives with pigmentation alone.

TREATMENT

The serious nature of the complications of this disorder highlight the importance of the diagnosis and of the pigmentary lesions in suggesting further study. Intussusception may be successfully treated by barium enema. When surgical intervention is required for intussusception, the leading polyp should certainly be removed. A small number of accessible polyps might well be removed at the same time. Bleeding also may require surgical removal of polyps. Routine removal of polyps or of large sections of intestine because of fear of malignant degeneration is not warranted. Anemia usually can be treated with iron. Transfusion is occasionally necessary.

REFERENCES

1. Peutz, J. L. A.: Very remarkable case of familial polyposis of mucous membrane of intestinal tract and nasopharynx accompanied by peculiar pigmentations of skin and mucous membrane, Med. Maandschr. Geneesk. 10:134, 1921.
2. Jeghers, H., McKusick, V. A., and Katz, K. H.: Generalized intestinal polyposis and melanin spots of the oral mucosa, lips and digits. A syndrome of diagnostic significance. N. Engl. J. Med. 241:993, 1031, 1949.
3. Bartholomew, L. G., Dahlin, D. C., and Waugh, J. M.: Intestinal polyposis associated with mucocutaneous melanin pigmentation (Peutz-Jeghers syndrome). Review of the literature and report of six cases with special reference to pathologic findings, Gastroenterology 32:434, 1957.

4. Horn, R. C., Payne, W. A., and Fine, G.: The Peutz-Jeghers syndrome, Arch. Pathol. 76:29, 1963.

5. Cronkhite, L. W., Jr., and Canada, W. J.: Generalized gastrointestinal polyposis: An unusual syndrome of polyposis, pigmentation, alopecia and onychotrophia, N. Engl. J. Med. 252:1101, 1955.

6. Gardner, E. J.: Follow-up study of a family group exhibiting dominant inheritance for a syndrome including intestinal polyps, osteoma, fibroma and epidermal cysts, Am. J. Hum. Genet. 14:376, 1962.

7. Turcot, J., Despres, J. P., St. Pierre, F.: Malignant tumors of the central nervous system associated with familial polyposis of the colon: Report of two cases, Dis. Colon Rectum 2:465, 1959.

MULTIPLE INTESTINAL POLYPOSIS AND PROTEIN-LOSING ENTEROPATHY

CARDINAL CLINICAL FEATURES

Benign intestinal polyps, melena, anemia, cachexia, protein-losing enteropathy.

CLINICAL PICTURE

Polyps of the intestine are common in children and adults and are commonly associated with other clinical manifestations, some of which can be directly related to the polyps and others of which cannot. Juvenile intestinal polyps occur the most frequently and are distinct pathologically from the more neoplastic adenomatous polyps. Juvenile polyps also are known as inflammatory polyps and occur in adults as well as in children. Patients with one or two such polyps may have no symptoms, a little melena or an occasional episode of rectal bleeding. Rectal prolapse or protrusion of a polyp through the anus, as well as abdominal cramps or pain, may occur. Intussusception is a common complication that may require therapeutic intervention. Prolonged blood loss may lead to iron deficiency anemia and cachexia. Large numbers of polyps may lead to severe inanition. We have observed hypoproteinemia and edema due to a protein-losing enteropathy (Case Report).[1]

Multiple polyps of this type are found in adults with the Cronkhite-Canada syndrome.[2] In these patients diffuse polyposis involves much of the gastrointestinal tract. They also have ectodermal changes including dystrophy of the nails, alopecia and diffuse patches of hyperpigmentation. Only nine cases have been reported since 1955. The syndrome usually is fatal.

DIFFERENTIAL DIAGNOSIS

This syndrome differs from the Peutz-Jeghers syndrome (see chapter, this section) in which melanin spots around the mouth or on the digits are associated with multiple adenomatous polyps. Protein-losing enteropathy has been reported in the Cronkhite-Canada syndrome.[3]

Ruymann[4] reported an infant with a fatal syndrome analogous to that of Cronkhite and Canada except that there were no changes in the nails or skin pigment. This infant had massive polyposis of the entire intestine, alopecia, clubbing and progressive cachexia, as well as hypoalbuminemia, anemia and ascites. Hypoproteinemia and anasarca also were observed in an infant described by Arbeter et al.[5] So far, the only consistent differences between infants and adults has been in the ectodermal changes, but these may simply be rare secondary results of the primary problem engendered by massive polyposis. Our patient (Case Report) differed from others with hypoproteinemia and edema in that protein enteropathy occurred with a

Fig. 1.– N. A. (Case Report). Examination of the small bowel, illustrating numerous polyps in the ileum.

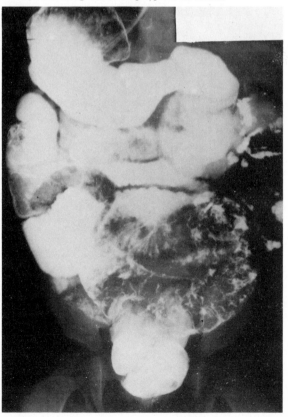

few well-localized intestinal polyps. However, massive losses can occur from even a single large polyp. Patients with intestinal polyposis may develop dehydration, hyponatremia and hypokalemia because of intestinal losses of water and electrolyte. Secondary hypokalemia nephropathy has been observed.

Juvenile intestinal polyposis differs from familial polyposis of the colon, which is transmitted by a mendelian dominant gene and is a premalignant condition. It is clear from a large experience that juvenile polyps do not undergo malignant transformation.[6]

CASE REPORT

N. A., an 18-month-old black female, was admitted to University Hospital because of edema of the face and extremities, abdominal pains and anemia. She was well until 4 months of age when she began to have problems with the bowel, including diarrhea, with straining and abdominal distention. She developed puffy eyelids. She was seen by several physicians and underwent numerous workups without a diagnosis. She was treated with iron for anemia. There was no history of a similar condition in the family.

On physical examination she appeared well developed but pale. She was not wasted. Her height was 34 inches, and her dry weight after loss of the edema was 28 lb — both normal for her age. Her face and extremities were edematous. There was no abnormal pigmentation of the skin or mucous membranes. The nails were normal, and there was no alopecia. There was a soft systolic ejection murmur over the precordium. The abdomen was distended but otherwise normal.

Laboratory data included hemoglobin 11 gm/100 ml., hematocrit 37, WBC 20,000/cu mm, BUN 9 and creatinine 0.5 mg/100 ml, serum iron 28 μg/100 ml (normal 60–140) and iron-binding capacity 330 mg/100 ml (normal 200–400), total protein 2.9 and albumin 1.2 gm/100 ml. Urinalysis revealed no proteinuria. Following intravenous injection of ^{51}Cr-labeled albumin, a 96-hour stool collection contained 2.9% of the radioactivity administered. The normal value is less than 1%. This test documented the presence of a protein-losing enteropathy and provided an explanation for the hypoproteinemia and edema. An upper gastrointestinal series revealed multiple polypoid lesions in the ileum (Figs. 1 and 2). Intermittent intussusception was demonstrated. A barium enema with air contrast showed no polyps in the colon.

While the child was in hospital, she passed a large quantity of bright red blood per rectum, and her hemoglobin fell to 8.5 gm/100 ml. A mass was felt in the right side of the abdomen. Laparotomy disclosed an intussusception of the proximal jejunum, with multiple polyps providing a leading point. A second group of polyps was found in the distal ileum. The intussusception was reduced, and both polyp-containing areas were resected. Her postoperative course was uneventful. A repeat test with ^{51}Cr-albumin was normal.

Fig. 2. — N. A. Higher magnification of the involved area.

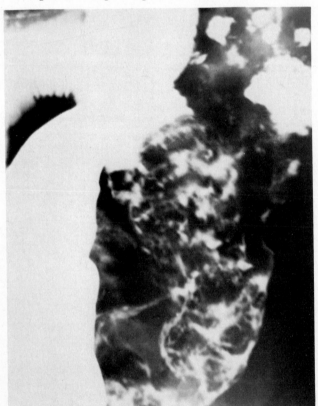

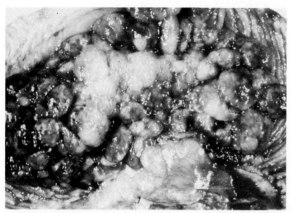

Fig. 3.—N. A. Specimen removed from the ileum. Multiple polyps are visible and appear to have coalesced.

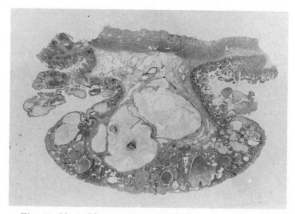

Fig. 4.—N. A. Microscopic section of a polyp. The inflammatory stroma is visible, as are numerous retention cysts.

Examination of the polypoid mass (Fig. 3) removed revealed firm, waxy, white polyps, with evidence of old hemorrhage in some. On section (Fig. 4) there were numerous small cystic areas in the larger polyps that were filled with pink serous or gelatinous fluid. Microscopically the surface of the polyps was columnar epithelium. The cysts were lined with mucous-secreting columnar cells. The stroma of the polyp was composed of vascular granulation tissue with lymphocytes, polymorphonuclear cells and plasma cells. These were the characteristic features of inflammatory, retention or juvenile polyps.

GENETICS

The etiology of this condition is unknown. It appears that it may be nongenetic.

TREATMENT

Therapy is essentially that of management of single or multiple polyps. Surgical removal of the polyps is the only definitive treatment.

REFERENCES

1. Berk, R. N., Rush, J. L., and Elson, E. C.: Multiple inflammatory polyps of the small intestine with cachexia and protein-losing enteropathy, Radiology 95:611, 1970.
2. Cronkhite, L. W., and Canada, W. J.: Generalized gastrointestinal polyposis: An unusual syndrome of polyposis, pigmentation, alopecia and onychotrophia, N. Engl. J. Med. 252:1101, 1955.
3. Orimo, H., Fujita, H., Yoshikawa, M., Takemoto, T., Matsuo, Y., and Nakao, K.: Gastrointestinal polyposis with protein-losing enteropathy, abnormal skin pigmentation and loss of hair and nail (Cronkhite-Canada syndrome), Am. J. Med. 47:445, 1969.
4. Ruymann, F. B.: Juvenile polyps with cachexia. Report of an infant and comparison with Cronkhite-Canada syndrome in adults, Gastroenterology 57:431, 1969.
5. Arbeter, A. M., Courtney, R. A., and Gaynor, M. F., Jr.: Diffuse gastrointestinal polyposis associated with chronic blood loss and anasarca in an infant, J. Pediatr. 76:609, 1970.
6. Roth, S. I., and Helwig, E. B.: Juvenile polyps of the colon and rectum, Cancer 16:468, 1963.

MULTIPLE MUCOSAL NEUROMA SYNDROME

CARDINAL CLINICAL FEATURES

Multiple mucosal neuromas, medullary carcinoma of the thyroid, pheochromocytoma, marfanoid habitus.

CLINICAL PICTURE

Williams and Pollock[1] in 1966 described two unrelated patients, both of whom had multiple neuromas in mucosal areas as well as carcinomas of the thyroid and pheochromocytomas. Gorlin[2,3] recently has summarized the findings in 34 published patients. The similarity of appearance among these patients was striking.

The hallmark of the syndrome is the neuroma, which always occurs in a number of mucosal sites around the lips, eyelids and tongue (Figs. 1–4). The buccal, palatal, nasal, gingival, conjunctival or laryngeal mucosa may be involved. The lips usually are large and may appear nodular (Fig. 3), blubbery or negroid. Both lips usually are involved. There may be thickening and eversion of the upper eyelids because of the presence of nodules on the eyelid margins. These may be pedunculated, up to 0.6 cm in diameter, and present in most patients with the syndrome. Lingual lesions usually are limited to the anterior dorsal surface of the tongue. They appear as pink pedunculated nodules and may give the tongue a serrated appearance.

Biopsy of lesions on the lips or tongue has shown the histologic appearance of a plexiform neuroma in which there is an unencapsulated mass of nerves, composed largely of convoluted axis cylinders. The lesion is a true neuroma or possibly an axonal hamartoma. It is quite distinct from a neurofibroma.

The oral lesions usually are the first component of the syndrome to appear. In most reported cases the neuromas were congenital or appeared in the first year of life, in some in the first decade. Other mucosal areas also may be involved early. Medullated nerve fibers may cross the cornea, anastomosing in the pupillary area. They appear as white fibers under the biomicroscope.

Patients with this syndrome are tall and thin and have long thin extremities (Fig. 5). A prognathic jaw is characteristic.[4] Pes cavus and dorsal kyphosis are common. Patients demonstrate the so-called Steinberg sign in which the infolded thumb under the clenched fingers extends beyond the ulnar side of the narrow hand. This sign is seen in about 1% of whites, 3% of blacks and most patients with the Marfan syndrome. Patients with the mucosal neuroma syndrome look quite a bit like patients with the Marfan syndrome (see Fig. 5). However, they do not have ectopia lentis or cardiovascular abnormalities. Muscle wasting has been described in patients with this syndrome,[3] especially in the limbs. Numerous pigmented, often pedunculated, soft papules have been described on the skin, as well as multiple freckle-like lesions.[4]

Some patients may have severe prolonged diarrhea, with frequent watery stools and diverticulosis or diverticulitis of the colon or megacolon.[4] In these patients there is intestinal ganglioneuromatosis, a hyperplasia of the neuroenteric ganglion cells that may involve the wall of the entire gastrointestinal tract. The bladder, the bronchi and the spinal dorsal nerve roots have been involved in some patients.

The pheochromocytomas that occur in association with the multiple mucosal neuroma syndrome usually appear at puberty or later in the second, third, fourth or fifth decade. The patient may present with episodic hypertension; with diaphoresis, choking sensations, weakness, flushing or blanching, headaches, dyspnea or palpitations; or with multiple episodes of nausea, vomiting and intractable diarrhea. The tumors may be bilateral. They may be very tiny or several centimeters in diameter. Rarely they metastasize. Histologically

Fig. 1.—Tongue of a patient with the multiple mucosal neuroma syndrome. The small neuromas are best seen at the edges of the tongue. (Courtesy of Dr. Harold O. Perry of the Mayo Clinic.)

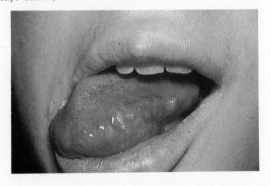

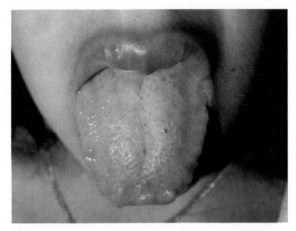

Fig. 2.—M. H. Patient with the multiple mucosal neuroma syndrome. The neuromas have formed nodules along the margins of the tongue. (Courtesy of Dr. Ronald Stein of Burbank, California.)

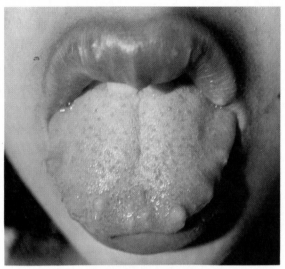

Fig. 4.—M. H. Close-up of the mucosal neuromas of the tongue. (Courtesy of Dr. Ronald Stein of Burbank, California.)

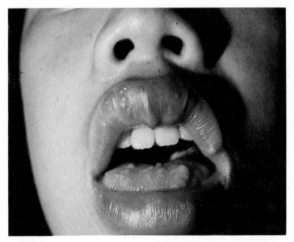

Fig. 3.—M. H. Her lips were quite full throughout. Nodules are evident in the upper lip. (Courtesy of Dr. Ronald Stein of Burbank, California.)

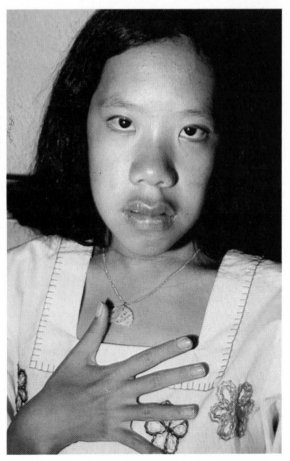

Fig. 5 (right).—M. H. She had a long, thin, marfanoid face and long slender fingers. Her lips were full, and she had previously undergone plastic surgery for blubbery lips. This patient had had two operations for medullary carcinoma of the thyroid. (Courtesy of Dr. Ronald Stein of Burbank, California.)

they are composed of small polyhedral cells. They assume a dark brown color on exposure to dichromate. The symptoms are due to the release of epinephrine or norepinephrine, or their products, from the tumor.

Medullary thyroid carcinoma appears in these patients in the second decade. The patient may notice a mass in the neck. The thyroid may be enlarged or nodular. Medullary thyroid carcinoma is an unusual tumor. It usually is multicentric and has an amyloid stroma. The tumor arises presumably from the parafollicular cells originating from the ultimobranchial body. The medullary cell of the thyroid may be derived from the neural crest. Thus, it is possible that the entire syndrome stems from a heritable defect in one cellular system. Virtually all familial thyroid neoplasms are medullary carcinomas. On the other hand, this type of tumor makes up less than 10% of all thyroid carcinomas. These tumors occasionally produce thyrocalcitonin, ACTH, prostaglandin, serotonin or histaminase.[6] They may metastasize. Determination of histaminase activity in the serum may have diagnostic utility. It also may be possible to diagnose the presence of a tumor prior to the presence of a palpable mass by determination of the serum content of calcitonin. Release of thyrocalcitonin may cause hypocalcemia and secondary hyperparathyroidism or parathyroid adenoma.

Every patient with the multiple mucosal neuroma syndrome does not necessarily manifest every component of the syndrome. On the other hand, of major importance in the recognition of the syndrome is the alerting of the physician to the possibility of a pheochromocytoma or a carcinoma of the thyroid.

GENETICS

The multiple mucosal neuroma syndrome is inherited as an autosomal dominant.[1] Medullary thyroid carcinoma may occur in a familial pattern, as may pheochromocytoma. In each instance the pattern of inheritance is that of an autosomal dominant trait. Medullary carcinoma and pheochromocytoma also may occur together with or without parathyroid adenoma. The association of pheochromocytoma and neurofibromatosis is well known. So is the association in the von Hippel-Landau syndrome of pheochromocytoma with angiomatosis of the retina and hemangioblastoma of the cerebellum. In each of the syndromes, inheritance is dominant. It is not clear whether the various syndromes represent the expression of different genes or whether some of them represent the variable expression of a single gene.[7] However, it seems likely that the former is the case. Certainly the multiple mucosal neuroma syndrome is a distinct entity.

TREATMENT

The treatment of pheochromocytoma and thyroid carcinoma is surgical.

REFERENCES

1. Williams, E. D., and Pollock, D. J.: Multiple mucosal neuroma with endocrine tumours: A syndrome allied to Von Reckinghausen's disease, J. Pathol. 91:71, 1966.
2. Gorlin, R. J., Sedano, H. C., Vickers, R. A., and Cervenka, J.: Multiple mucosal neuromas, pheochromocytoma and medullary carcinoma of the thyroid—a syndrome, Cancer 22: 293, 1968.
3. Gorlin, R. J., and Vickers, R. A.: Multiple Mucosal Neuromas, Pheochromocytoma Medullary Carcinoma of the Thyroid and Marfanoid Body Build with Muscle Wasting: Reexamination of a Syndrome of Neural Crest Malmigration, in *The Clinical Delineation of Birth Defects. Part X, The Endocrine System*, Birth Defects: Original Article Series (New York: The National Foundation-March of Dimes, 1971), p. 69.
4. Schimke, R. N., Hartmann, W. H., Prout, T. E., and Rimoin, D. L.: Syndrome of bilateral pheochromocytoma, medullary thyroid carcinoma and multiple mucosal neuromas, N. Eng. J. Med. 279:1, 1968.
5. Mielke, T. E., Becker, K. L., and Gross, J. B.: Diverticulitis of the colon in a young man with Marfan's syndrome associated with carcinoma of the thyroid gland and neurofibromas of the tongue and lips, Gastroenterology 48:379, 1965.
6. Baylin, S. B., Beaven, M. A., Engelman, K., and Sjöerdsma, A.: Elevated histaminase activity in medullary carcinoma of the thyroid gland, N. Engl. J. Med. 283:1239, 1970.
7. Schimke, R. N.: Familial Tumor Endocrinopathies, in *The Clinical Delineation of Birth Defects. Part X, The Endocrine System*, Birth Defects: Original Article Series (New York: The National Foundation-March of Dimes, 1971), p. 55.

LINEAR SEBACEOUS NEVUS SYNDROME
Nevus Linearis Sebaceus

CARDINAL CLINICAL FEATURES

Linear sebaceous nevi, convulsions, mental retardation.

CLINICAL FEATURES

Naevus sebaceus is often associated with the name Jadassohn,[1] because he first used the term in 1895 to describe benign nevoid tumors composed prominently of sebaceous glands. These lesions, which are hamartomatous or dysplastic, usually are found around the head. A syndrome in which unusual linear lesions of this type were associated with convulsions and mental retardation was first described by Feuerstein and Mims.[2] It may be classified among the neurocutaneous disorders in which lesions of the skin, usually of the face, are associated with developmental abnormality of the central nervous system. Sometimes known as phakomatoses, these disorders appear to reflect an abnormality of ectodermal development early in embryogenesis. These disorders point up the importance of cutaneous facial signs in indicating the presence of underlying cerebral defects.[3]

The characteristic feature of this syndrome, and one which permits its ready recognition, is the skin lesions. They are present at birth. A midline distribution is striking, and the lesions are well-circumscribed, slightly raised, yellowish orange, tan or dark brown plaques or nodules. They may develop a verrucous appearance. The lesions are commonly located on the forehead and may extend to the tip of the nose.[4, 5] There may be a tendency to laterality,[6] and the lesions may extend down one side of the nose.[6] They may occur on the scalp, face or neck (Fig. 2). There may be extension over the trunk and extremities (Figs. 1 and 3). Other types of skin lesions that may be associated with this syndrome include blue nevi, pigmented nevi, café-au-lait spots and capillary hemangiomas. Hirsutism may be a feature. A striking, large, depigmented spot over the face has been described.[3] This was over the left eye and left midportion of the face.

Biopsy of a linear nevoid lesion shows thickening and hyperkeratosis of the epidermis, with hyperplasia of the sebaceous glands.[2] The sweat glands may ap-

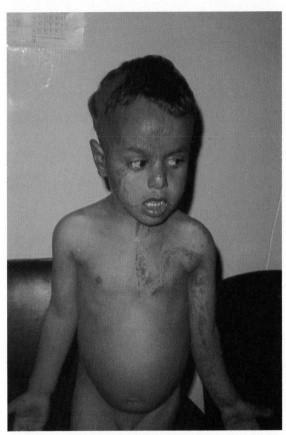

Fig. 1.—A 4½-year-old Saudi Arabian boy with the linear sebaceous nevus syndrome. In this view the linear pigmented lesions may be seen extending down the arm. (Courtesy of Dr. M. Mange of Dharan Health Center, Arameo, Saudi Arabia.)

pear dysplastic. The many sebaceous glands may be normal in structure. Many glands lie free in the cutis and some may have no connections with hair follicles or with the surface. Cystic areas and degenerative changes may occur. A benign apocrine tumor, or syringocystadenoma, has developed within a nevus.[6] The nevus sebaceus is a potentially premalignant lesion.[1]

Patients with this syndrome regularly develop sei-

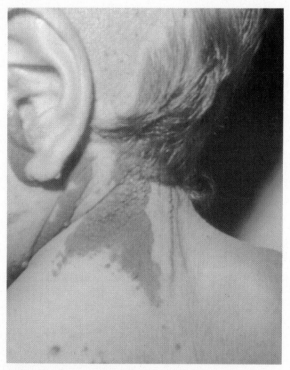

Fig. 2.—Close-up of the pigmented, raised, verrucous lesions and alopecia.

zures at an early age. The seizures may be grand mal, with generalized tonic or clonic seizures, or they may be focal. Focal abnormalities of the EEG often are found, and they may be ipsilateral to the skin lesion. Seizures also may be myoclonic. One patient has been reported in whom there was a hypsarrhythmic pattern.[7] Pneumoencephalography may reveal cortical atrophy,[3] which may be unilateral. Deterioration may be seen, especially in those with hypsarrhythmia.[7]

Mental retardation may be profound, the most severely retarded sometimes failing to thrive, or it may be moderate. Some patients have behavior problems or hyperactivity and are difficult to control. Cortical blindness has been seen. Spasticity and an increase in deep tendon reflexes may be present. Babinski responses and atonic neck reflex have been described.[5]

A variety of other anomalies may be associated with this syndrome. These include asymmetry of the cranium, hydrocephalus or premature closure of the sphenofrontal suture.[4] There may be an antimongoloid slant; nystagmus; lipodermoid of the conjunctiva; vascularization of the cornea and colobomas of the irides, choroid or retina. The bridge of the nose may be broad. The teeth may be hypoplastic. Extension of a nevus to the hard and soft palate may lead to local thickening of the mucosa. Coarctation of the aorta has been reported,[5] as has hamartoma or mesoblastic nephroma of the kidney.[8]

Differential Diagnosis

The so-called nevus unis lateris[9] may look similar to the linear sebaceous nevus and also may be associated with neurologic abnormality, including seizures and mental retardation. The former condition may be distinguished histologically by the presence of hyperkeratosis, papillomatosis and acanthosis of the epidermis without involvement of the sebaceous glands or other dermal structures. It could be that both conditions result from a similar or identical event.

Case Report*

C. T., a 17-year-old Chinese girl with the linear sebaceous nevus syndrome, was reported at 12 years of age by Sugarman and Reed.[3] She was severely mentally retarded and had been institutionalized since early childhood.

She was the product of a normal pregnancy and delivery. At birth she was found to have yellowish papules extending down the midline of the nose and over the right side of the face and neck. She also had a congenital defect of the scalp. A biopsy at 3 months of age was diagnostic. Seizures developed early and were frequent over the first 2 years. Control of seizures with Dilantin and phenobarbital was fair until about 10 years of age, after which major seizures were resistant to therapy. Mental retardation was profound and appeared to be progressive.

Physical examination at 17 years revealed her to be unresponsive to stimuli. The facial cutaneous nevus was dramatic (Figs. 4–8). There was an unpleasant sebaceous odor. She was spastic and quadriplegic. There were marked muscle wasting, joint contractures and hyperactive deep tendon reflexes. The EEG was abnormal. Pneumoencephalogram revealed cerebral atrophy and enlargement of all of the ventricles.

Fig. 3.—Genu valgum of patient shown in Figure 1. There were rachitic lesions on x-ray, resistant to vitamin D. (Courtesy of Dr. M. Mange of Dharan Health Center, Arameo, Saudi Arabia.)

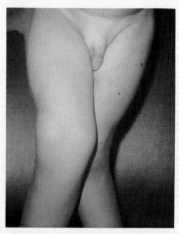

*Dr. William M. Morris of Pacific State Hospital, Pomona, California, kindly permitted us to study and photograph this patient and to report the clinical details.

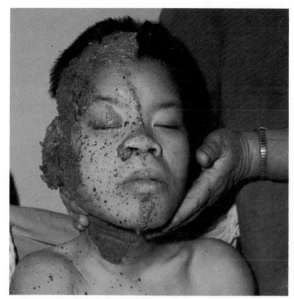

Fig. 4.—C. T. (Case Report) A 17-year-old retarded Chinese girl with the linear sebaceous nevus syndrome. Over the trunk and extremities she had a unilateral distribution of many pigmented nevi. The most striking lesions were on the face, where the characteristic sebaceous nevus ran parallel to, but just to one side of, the midline.

Fig. 5.—C. T. The face, illustrating the unilateral linear sebaceous nevus. The lesions over the forehead and nose appeared orange and verrucous. There were numerous pigmented nevi. The lesion at the nasolabial fold contained a basal cell carcinoma.

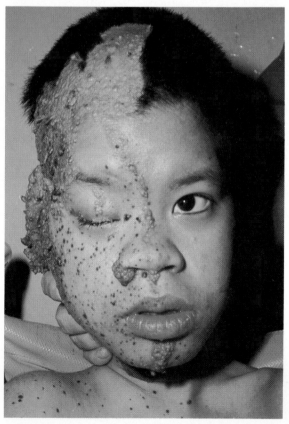

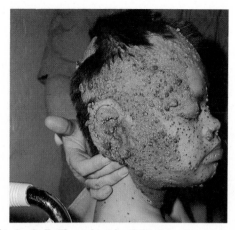

Fig. 6.—C. T. The right side of the face. There also were large areas of alopecia. The lesion under the mandible was darker in color.

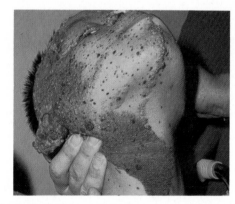

Fig. 7.—C. T. Close-up of the linear nevus of the neck.

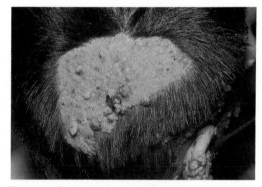

Fig. 8.—C. T. Posterior scalp with alopecia and nevus sebaceus tumor of the scalp.

GENETICS

All of the reported cases have been sporadic. There is no information that would permit a decision as to whether or not the syndrome is genetic. It appears to result from a developmental defect very early in fetal life. This could be as early as the third week of fetal life when the neural folds appear, the precursors of the skin lie just below and the anlagen of the nose and forebrain are close together.[7]

TREATMENT

Seizures should be treated with appropriate anticonvulsant medication; they may be very difficult to control. Sebaceous nevi are susceptible to malignant transformation and should be excised.

REFERENCES

1. Robinson, S.S.: Naevus sebaceus (Jadassohn). Report of four cases, Arch. Dermatol. Syph. 26:663, 1932.
2. Feuerstein, R. C., and Mims, L. C.: Linear nevus sebaceus with convulsions and mental retardation, Am. J. Dis. Child. 104:675, 1962.
3. Sugarman, G. I., and Reed, W. B.: Two unusual neurocutaneous disorders with facial cutaneous signs, Arch. Neurol. 21:242, 1969.
4. Gellis, S. S., and Feingold, M.: Linear nevus sebaceus syndrome, Am. J. Dis. Child. 120:139, 1970.
5. Marden, P. M., and Venters, H. D., Jr.: A new neurocutaneous syndrome, Am. J. Dis. Child. 112:79, 1966.
6. Bianchine, J. W.: The nevus sebaceus of Jadassohn, Am. J. Dis. Child. 120:223, 1970.
7. Herbst, B. A., and Cohen, M. E.: Linear nevus sebaceus, Arch. Neurol. 24:317, 1971.
8. Lansky, L. L., Funderburk, S., Cuppage, F. E., Schimke, N., and Diehl, A. M.: Linear sebaceous nevus syndrome, Am. J. Dis. Child. 123:587, 1972.
9. Holden, K. R., and Dekaban, A. S.: Neurological involvement in nevus unis lateris and nevus linearis sebaceus, Neurology 22:879, 1972.

Thumb and Radial Dysplasias

FANCONI ANEMIA

Cardinal Clinical Features

Pancytopenia, absent or rudimentary thumbs, hypoplastic radii, skin pigmentation, shortness of stature, renal anomalies, chromosomal breaks.

Clinical Picture

Fanconi[1] described this syndrome first in 1927 in a report of three brothers with progressive lethal anemia who also had brown pigmentation of the skin. Two years later Uehlinger[2] reported a patient who had pancytopenia and malformations of the thumbs and kidneys, and related this patient to those reported by Fanconi. The disorder has come to be known as the Fanconi anemia.[3] More than 160 patients have been described.[3, 4] A reason for an association between pancytopenia and the characteristic congenital anomalies is not immediately apparent, but the systems involved in the disease undergo embryonic differentiation in the same period between the 25th and 34th days of fetal life.[5]

Patients with this disease present with hematologic manifestations in childhood, usually between 4 and 10 years of age.[6] The first symptoms of marrow involvement usually are those of easy bruisability. There may be nosebleeds or internal bleeding, as well as evidence of hemorrhage into the skin.[5] These patients also may present with severe anemia and attendant pallor, fatigue or dyspnea. Examination reveals the presence of pancytopenia with a profound anemia as well as thrombocytopenia and leukopenia. In the white cell series it is the granulocytes that are reduced; there may be a relative lymphocytosis. The fully developed pancytopenia may present gradually so that the patient may at first appear to have an isolated anemia, leukopenia or thrombocytopenia. Red cells are abnormal. Patients may be macrocytic and may have a high color index. Erythrocyte life span may be shortened. The erythrocytes contain a large proportion of fetal hemoglobin; its content has been reported to range from 28–85%.[3] The bone marrow is atrophic. The marrow cavity may be filled largely with fat, and the number of all of the blood-forming elements is reduced to isolated nests of cells.

The skeletal anomalies are the alerting signs of this syndrome and provide for early diagnosis. The major finding is an abnormality of the thumb, and a majority of patients have this feature.[6] There may be absent thumbs (Fig. 1) or hypoplastic, rudimentary (Fig. 2) thumbs. The lesions may be, but often are not, bilaterally symmetrical. The first metacarpal may be hypoplastic or absent. The thenar area usually is flat. Rarely, a patient with this disorder has a duplication of the thumb. Radial abnormalities also are seen in this syndrome and include hypoplasia or even absence of the radius (see Fig. 1). Carpal ossification centers may be missing on the radial side. Skeletal anomalies seen rarely have included syndactyly, congenital dislocation of the hips, clubfoot, cervical rib and Sprengel's deformity of the scapula.[7]

Hyperpigmentation of the skin is a regular finding in this syndrome. The pigmentation is brownish (Figs. 2 and 3). It occurs most often on the trunk but may involve the groin, neck or axillae. It may occur in small mottlings, or there may be large patches with diffuse boundaries, often with smaller patches of vitiligo within. It may develop with time.[7]

Patients with the Fanconi anemia are characteristically small (Fig. 4). The birth weight often is low as a consequence of intrauterine growth retardation, and patients continue to grow slowly postnatally. Renal anomalies are seen in about one third of patients. There may be absences or fusions, as in the skeleton. Aplasia or absence of a kidney is common. Horseshoe

Fig. 1.—Absent thumbs in a boy with Fanconi anemia. There was absence of the radius on the left arm, leading to the radial deviation of that hand.

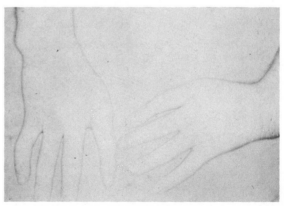

333

Fig. 2.—The left hand of another patient with Fanconi anemia illustrates the rudimentary thumb. The forearm also was markedly bowed. In this view the brown pigmentation on the chest and abdomen also is evident.

kidney also is frequent. Other anomalies include crossed renal ectopia and duplication of the collecting system. Hydronephrosis may occur. Any patient in whom the Fanconi anemia is diagnosed should have an intravenous pyelogram. Hypogenitalism may be prominent in almost half of male patients. The penis may be small, and the testes may be small or cryptorchid. Hypoplasia of the uterus and vagina has been reported in the female.[7]

Ocular abnormalities such as microphthalmia, ptosis, strabismus and nystagmus are found occasionally. Deafness is seen in some patients. Some patients have brisk deep tendon reflexes. Rarely there may be mental retardation, but intelligence usually is normal.

Cytogenetic analysis has become an integral part of the workup of a patient with the Fanconi anemia because chromosomal breaks are regularly found.[8-11] The number of chromosomes is normal, but they manifest a structural instability, leading to breaks, gaps, constrictions and rearrangements. Abnormalities may be seen in metaphase, including endoreduplication and a variety of exchanges of chromatid material or translo-

Fig. 3.—Close-up of pigmentation in patient shown in Figure 2.

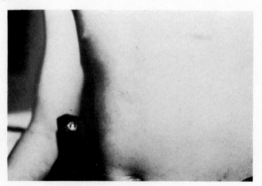

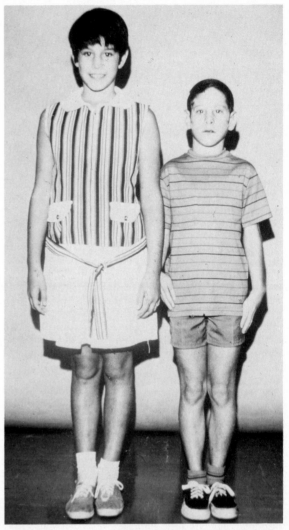

Fig. 4.—The shortness of stature shown in the boy (*right*) pictured in Figures 2 and 3 illustrates a characteristic of the syndrome. His sister (*left*) was normal. (Courtesy of Drs. Sandra Miner and David Allan of Mercy Hospital, San Diego.)

cations.[9] As many as 60% of mitoses are found to be abnormal in these patients.[11] Abnormalities have been seen directly in bone marrow smears, indicating that chromosomal malfunction is indeed occurring in vivo.[10] The autosomal recessive nature of the Fanconi anemia suggests that alteration in the responsible gene determines the chromosomal susceptibility to breakage.

Patients with the Fanconi anemia have a high incidence of leukemia and other malignant neoplasms.[12-14] The incidence of leukemia is high in relatives of patients with the Fanconi anemia.[15] A propensity for neoplasia may relate to the occurrence of chromosome breaks, as a similar cytogenetic picture also is seen in the Bloom syndrome, in which patients are unusually prone to leukemia and other malignant neoplasms. Fibroblasts derived from these patients and grown in cell culture show a much greater than normal susceptibility to transformation when treated with SV40 virus.[13] This is a propensity suggestive of malignant cells and it also is seen in the cells of the parents of these patients.

DIFFERENTIAL DIAGNOSIS

A variety of disorders present with skeletal and hematologic abnormalities, most of them involving some form of marrow hypoplasia. Thrombocytopenia with absent radius or radial aplasia (see chapter, this section) presents with thrombocytopenia in the newborn infant. There is an absence or hypoplasia of the megakaryocytes in the marrow. The radius is absent or hypoplastic, but these patients always have five fingers.

Another related syndrome is constitutional anemia with triphalangeal thumbs.[15]

Dyskeratosis congenita is an additional condition that may present with hyperpigmentation of the skin and pancytopenia.[16] These patients have ectodermal dystrophy in which there may be cutaneous hypopigmentation as well as hyperpigmentation, areas of atrophy and hyperkeratosis of the palms and soles. Occasionally, they have telangiectatic lesions and leukoplakia of the mucous membranes. Signs of marrow hypoplasia usually present in the second or third decade.

Carcinomas commonly develop in the third, fourth or fifth decade. There also is a high incidence of leukemia among individuals in the kindred. In this disorder there are no skeletal anomalies. It is transmitted as an autosomal recessive.

Other disorders in which there are chromosomal breaks and a propensity for malignant disease include the Bloom syndrome and ataxia telangiectasia. Patients with the Bloom syndrome have intrauterine growth retardation, short stature and butterfly-telangiectatic erythematous skin. There is no skeletal deformity and no anemia. It is transmitted as an autosomal recessive. Patients with ataxia telangiectasia (see section on Immunologic Problems) also have intrauterine growth retardation and short stature. They develop progressive cerebellar ataxia and telangiectasia of the bulbar conjunctiva and the skin.

The classic disorder in which there are absent or rudimentary thumbs is the Holt-Oram syndrome (see chapter, this section). These patients have congenital heart disease with either atrial or ventricular septal defect. There is no hematologic component. It is inherited as an autosomal dominant.

Several chromosomal abnormalities are associated with absent thumbs. These include the ring D chromosome syndrome and deletion of the long arm of chromosome 18, or 18q–trisomy (see section on Cytogenetic Syndromes).

GENETICS

The Fanconi anemia is a rare genetic disorder, transmitted as an autosomal recessive. Multiple siblings with normal parents have been reported. In some families in which a typical patient has had the complete syndrome, siblings have been observed who had only the hematologic manifestations.[6] It also is possible in an involved family to see a child with skeletal abnormalities and even chromosomal abnormalities and no hematologic problems, but it always remains possible that pancytopenia will develop subsequently. A high incidence of consanguinity has been observed. A high incidence of leukemia in other members of families in which Fanconi anemia occurs suggests that the heterozygous carrier of the mutant gene may be predisposed to malignancy.

TREATMENT

The prognosis in the Fanconi anemia is not good. Death usually is expected to occur within a few years after the appearance of the anemia. The leading causes of death are severe anemia, hemorrhage or infection. Intracranial, gastrointestinal and even endopericardial hemorrhages have been seen. Repeated blood transfusion is the mainstay of treatment. Treatment with testosterone and corticosteroids is considered beneficial and may give these patients a longer survival. Oxymetholone probably is preferable to testosterone because of smaller masculinizing side-effects.[14, 17] It also is thought to have a more favorable effect on the marrow,[14] but this is not always easy to document.[17]

REFERENCES

1. Fanconi, G.: Familiäre infantile perniziosaartige Anämie (perniziöses Blutbild und Konstitution), Z. Kinderheilkd. 117:257, 1927.
2. Uehlinger, E.: Konstitutionelle infantile (perniciosaartige) Anämie, Klin. Wocherschr. 8:1501, 1929.
3. Fanconi, G.: Familial constitutional panmyelocytopathy. Fanconi's anemia, Semin. Hematol. 4:233, 1967.
4. Fanconi, G.: Die familiare Panmyelopathie, J. Suisse Med. 94:1309, 1964.
5. Althoff, H.: Zur Panmyelopathie Fanconi als Zustandsbild multipler Abartungen, Z. Kinderheilkd. 72:267, 1953.
6. Nilsson, L. R.: Chronic pancytopenia with multiple congenital abnormalities (Fanconi's anaemia), Acta Paediatr. 49:518, 1960.
7. Dawson, J. P.: Congenital pancytopenia associated with

multiple congenital anomalies (Fanconi type), Pediatrics 15:325, 1955.

8. Schmid, Von W., Schärer, K., Baumann, T., and Fanconi, G.: Chromosomenbrüchigkeit bei der familiären Panmyelopathie (Typus Fanconi), Schweiz. Med. Wochenschr. 95: 1461, 1965.

9. Bloom, G. E., Warner, S., Gerald, P. S., and Diamond, L. K.: Chromosome abnormalities in constitutional aplastic anemia, N. Eng. J. Med. 274:8, 1966.

10. Schroeder, T. M.: Cytogenetische und cytologische Befunde bei enzymopenischen Panmyelopathien und Pancytopenien. Familiäre Panmyelopathie Typ Fanconi, Glutathionreduktasemangel-Anämie und megaloblastäre Vitamin B$_{12}$-Mangel-Anämie, Humangenetik 2:287, 1966.

11. Schmid, W.: Familial constitutional panmyelocytopathy, Fanconi's anemia (F. A.). II. A discussion of the cytogenetic findings in Fanconi's anemia, Semin. Hematol. 4:241, 1967.

12. Garriga, S., and Crosby, W. H.: The incidence of leukemia in families of patients with hypoplasia of the marrow, Blood 14:1008, 1959.

13. Dosik, H., Hsu, L. Y., Todaro, G. J., Lee, S. L., Hirschhorn, K., Selirio, E. S., and Alter, A. A.: Leukemia in Fanconi's anemia: Cytogenetic and tumor virus susceptibility studies, Blood 36:341, 1970.

14. Bernstein, M. S., Hunter, R. L., and Yachnin, S.: Hepatoma and peliosis hepatis developing in a patient with Fanconi's anemia, N. Eng. J. Med. 284:1135, 1971.

15. Aase, J. M., and Smith, D. W.: Congenital anemia and triphalangeal thumbs: A new syndrome, J. Pediatr. 74:471, 1969.

16. Steier, W., Van Voolen, G. A., and Selmanowitz, V. J.: Dyskeratosis congenita: Relationship to Fanconi's anemia, Blood 39:510, 1972.

17. Crossen, P. E., Mellor, J. E. L., Adams, A. C., and Gunz, F. W.: Chromosome studies in Fanconi's anaemia before and after treatment with oxymetholone, Pathology 4:27, 1972.

18. Poznanski, A. K., Garn, S. M., and Holt, J. F.: The thumb in the congenital malformation syndromes, Radiology 100: 115, 1971.

THROMBOCYTOPENIA WITH ABSENT RADIUS
Radial Aplasia, Thrombocytopenia,
Thrombocytopenia with Bilateral Absence of the Radius,
Radius Platelet Hypoplasia

CARDINAL CLINICAL FEATURES

Congenital thrombocytopenia with absent, reduced or immature megakaryocytes, bilateral absence of the radius.

CLINICAL PICTURE

HEMATOLOGIC ABNORMALITIES. – The onset of symptoms may be as early in life as the neonatal period or may begin in adulthood. These patients have petechiae and easy bruising, often at birth. Melena and hematuria may occur. Recurrent crops of petechiae, recurrent purpura or melena are the rule.

Hematologic studies reveal a marked thrombocytopenia. Bone marrow aspiration reveals a cellular marrow containing normal erythroid and myeloid precursors, but no megakaryocytes or a markedly reduced number of megakaryocytes. Those megakaryocytes present are immature in appearance and do not produce platelets. A moderate leukocytosis and an increased number of eosinophils usually are seen. There may be a leukemoid reaction, and differentiation from leukemia may require examination of the marrow. Granulocytosis may occur in response to bleeding, and anemia may result from blood loss. There may be an association between infection and the degree of thrombocytopenia.

The course of the thrombocytopenic symptoms may be mild or lethal. Menorrhagia may be prominent. Deaths from hemorrhage have been recorded. This usually is the result of intracranial bleeding, but exsanguination also may occur. Patients who maintain platelet counts over 30,000/cu mm generally have a good prognosis, whereas at least half of those who have counts less than 10,000/cu mm die. The critical period for survival usually is from birth to 1 year.

SKELETAL ABNORMALITIES. – Absence or hypoplasia of the radius is the hallmark of this syndrome. Because of this anomaly, the hands usually are radially deviated. The hand may appear to arise from the elbow or even from the shoulder. There may be associated hypoplasia or absence of the ulna. The humerus often is hypoplastic and may be absent. The shoulder girdle and clavicle are poorly formed in one third of patients reported. Hypoplasia of the shoulder girdle may lead to an appearance suggestive of webbing of the neck. Five fingers have been present on both hands in all cases. However, there often is limitation of motion of the fingers and flexion contractures. Mild syndactyly may be present. The thumb often is held in a transverse position across the palm, but it has a full range of motion. The muscle mass of the hands and fingers is reduced. There may be fusion of the metacarpals. Deformities of legs and hips are less common. They include dislocated hips, dislocated patellae and tibial torsion. Abnormality of the feet is fairly frequent. This includes calcaneal valgus and over-riding toes, especially the fifth toes. Absence of the toenails has been observed.[1]

Embryologically, it is of interest that radial aplasia or hypoplasia is associated with defects of the thumb and its metacarpal. Similar lesions of the ulna are associated with aplasia or hypoplasia of the other digits, the metacarpals and the carpals.[2]

OTHER ANOMALIES. – Patients with this syndrome also may have congenital heart disease, low-set ears and urogenital anomalies. Mental retardation occurs but is not common. Micrognathia, Menkel's diverticulum and uterine anomalies have been observed. These patients often seem to have unexplained diarrhea. Melena is common under these circumstances.

Chromosome studies have been normal. A specific search for breaks or fragmentations of chromosomes has been negative. Dermatoglyphic analysis has been abnormal in five cases studied. Abnormalities include absent palmar triradii, radially placed palmar triradii, an absence or decreased number of flexion creases on the fingers and simian creases.

DIFFERENTIAL DIAGNOSIS

Congenital defects of the upper extremities and aplastic anemia characterize the Fanconi anemia[3] (see

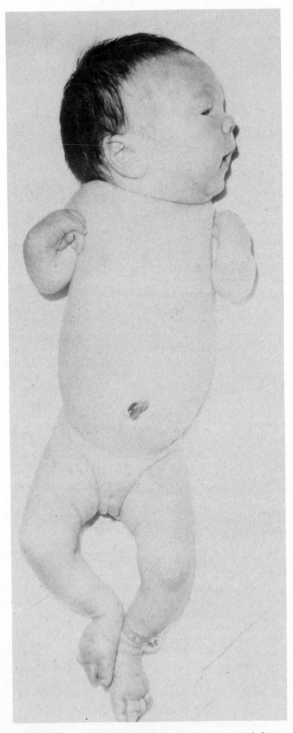

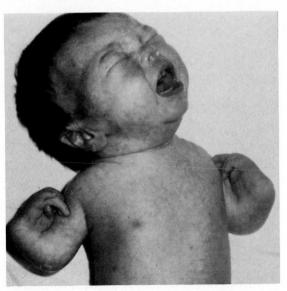

Fig. 2.–S. L. There were petechiae over the entire body. Forearms were very short. The hands were radially deviated on the forearms.

Fig. 1.–S. L. (Case Report). Infant with congenital thrombocytopenia and absent radii. The ecchymotic swollen feet are evident as well as the deformities of the upper extremities.

chapter, this section). It is transmitted via a recessive gene. Onset of hematologic symptoms usually is between the ages of 5 and 10 years. These patients have thrombocytopenia; they also have anemia, leukopenia and extreme hypoplasia of the bone marrow. Patients also may have brown pigmentation of the skin, microcephaly and urinary tract anomalies such as absence of a kidney or genital hypoplasia. Deformity of the thumb is characteristic. They may have low birth weight and, later, shortness of stature. Cytogenetic analysis may reveal chromosome breaks.

In the Holt-Oram syndrome (see chapter, this section), hypoplasia of the radius is associated with abnormal thumbs, which may be triphalangeal, hypoplastic or aplastic, and proximally placed. The patient may have phocomelia. Congenital cardiac defects are part of the syndrome and may be atrial septal defect or ventricular septal defect.

In the deletion syndrome, absent thumb with a ring D chromosome[4] (see section on Cytogenetic Syndromes), mental and physical retardation are associated with absence of the thumbs. Multiple congenital anomalies include fusion of the fourth and fifth toes, a small trigonocephalic head, with micrognathia, congenital dysplasia of the hip and congenital heart disease. Ptosis of the eyelids also may be present. Bilateral simian creases occur, and dermatoglyphs show absence of the palmar axial triradius.

Thalidomide embryopathy may produce identical upper limb anomalies but without hematologic abnormalities.

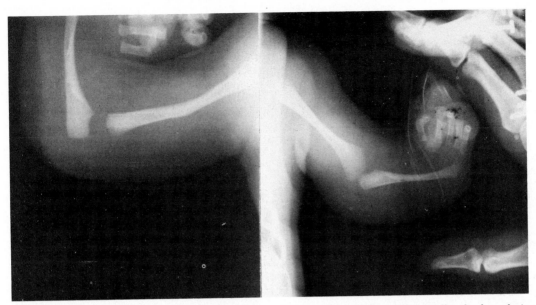

Fig. 3.—S. L. Roentgenograms of the upper extremities. **Left** and **right,** absence of the radii bilaterally; also, hypoplasia of the first metacarpal on the left.

Fig. 4.—S. L. By 3 years of age the patient had prominent epicanthal folds and somewhat broad nasal bridge.

Fig. 5.—S. L. at 3 years of age. The characteristic bowing of the lower extremities is evident.

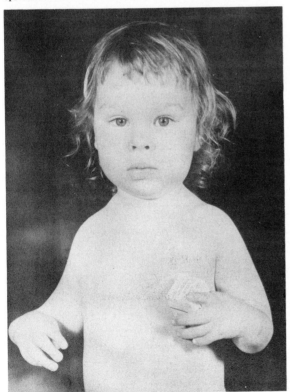

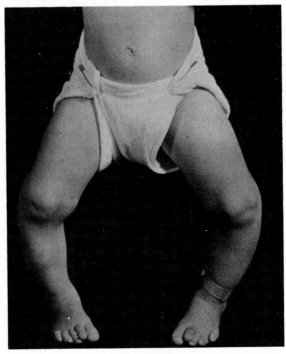

Case Report

S. L., a 2-day-old white female infant, was transferred to University Hospital for evaluation of ecchymosis, a low platelet count and deformities of both arms. She was the product of a full-term gestation and normal delivery and had a birth weight of 7 lb. The Apgar score was 9 at 1 minute. The mother denied any drugs or illness during gestation. Four siblings were completely normal and without skeletal abnormalities, anemia or bleeding tendencies. There was no consanguinity.

The anomalous upper extremities were immediately apparent. The forearms were short, and there was a clubhand appearance, with the hands radially deviated and approximating the forearm (Figs. 1 and 2).

The patient had very short upper extremities. There was a striking radial deviation of the hands, as well as bilateral low insertion of the thumbs. All of the fingers were present.

The baby was observed to have ecchymotic, swollen feet. Petechiae were present in generalized distribution, particularly over the trunk and buttocks. Hematologic investigation showed a hemoglobin of 16.5 gm/100 ml and a hematocrit of 49. The leukocyte count was 52,000/cu mm, with 3% eosinophils. The reticulocyte count was 4%, the platelet count 10,000/cu mm. Bone marrow aspiration revealed normal cellularity and 1.5% erythroblasts, 9% lymphocytes, 81.5% myeloid cells and 8% erythroid precursors. There were no megakaryocytes. The ECG was normal. Chromosome study was normal, indicating 46,XX. Roentgenographic skeletal survey revealed absence of the radii bilaterally (Fig. 3, left). There was radial deviation of the hands. The first metacarpal bone was present but somewhat hypoplastic (Fig. 3, right). The remainder of the bony structures were normal.

During 3 weeks in the hospital the patient was observed to have diarrhea and intermittent melena. She was treated with platelet and whole blood transfusions, and she stabilized. In follow up she continued to have thrombocytopenia with platelet counts of 30,000–39,000/cu mm. A pronounced tibial torsion was successfully treated by surgery. She continued over the first 2 years to have intermittent episodes of diarrhea. She developed large, almond eyes with a pronounced epicanthus (Fig. 4). Bowing of the lower extremities was striking (Fig. 5). Development was normal.

Genetics

Transmission is considered to be as an autosomal recessive. Parent to child transmission has not been observed. A family has been reported in which there were four involved children out of 15 in one generation.[1] There was no known consanguinity, but both parents had ancestors in Ireland named Kelley. The ratio of affected females to males is 2:1.

Trisomy 18 has been observed with esophageal atresia or other intestinal malformations, absence or hypoplasia of the radius and congenital thrombocytopenia.[5, 6]

Treatment

No therapeutic procedures have been of lasting hematologic benefit. Anemia due to blood loss may be treated with iron replacement. Blood transfusion or platelet transfusion may be helpful in emergency. Splenectomy has been employed with transient improvement. Steroids have been used during severe thrombocytopenic or symptomatic episodes with some transient improvement. Orthopedic treatment is necessary to correct the radial deviation of the hands. Splinting can be helpful, but surgical correction may be necessary.

References

1. Hall, J. G., Levin, J., Kuhn, J. P., Ottenheimer, E. J., Peter, K. A., Van Berkum, P., and McKusick, V. A.: Thrombocytopenia with absent radius, Medicine 48:411, 1969.
2. Kanavel, A. B.: Congenital malformations of the hands, Arch. Surg. 25:1, 1932.
3. Cowdell, R. H., Phizackerly, P. J. R., and Pyke, D. A.: Constitutional anemia (Fanconi's syndrome) and leukemia in two brothers, Blood 10:788, 1955.
4. Sparkes, R. S., Carrel, R. E., and Wright, S. W.: Absent thumbs with a ring D_2 chromosome. A new deletion syndrome, Am. J. Hum. Genet. 19:644, 1967.
5. Rabinowitz, J. G., Moseley, J. E., Mitty, H. A., and Hirschhorn, K.: Trisomy 18, esophageal atresia, anomalies of the radius, and congenital hypoplastic thrombocytopenia, Radiology 89:488, 1967.
6. Vorhees, M. L., Aspillaga, M. G., and Gardner, L. I.: Trisomy 18 syndrome with absent radius, varus deformity of hand, and rudimentary thumb, J. Pediatr. 65:130, 1964.
7. Shaw, S., and Oliver, R. A. M.: Congenital hypoplastic thrombocytopenia with skeletal deformities in siblings, Blood 14:374, 1959.
8. Bayrakci, G., and Walsh, J. R.: Amegakaryocytic thrombocytopenia and bilateral absence of the radii, Postgrad. Med. 33:401, 1963.
9. Hall, J. G., and Levin, J.: Congenital Amegakaryocytic Thrombocytopenia with Bilateral Absence of the Radius: Radius Platelet Hypoplasia or RPH, in The First Conference on the Clinical Delineation of Birth Defects. Part III, Limb Malformations, Birth Defects: Original Article Series (New York: The National Foundation-March of Dimes, 1969), p. 190.
10. Emery, J. L., Gordon, R. R., Rendle-Short, J., Varadi, S., and Warrack, A. J.: Congenital amegakaryocytic thrombocytopenia with congenital deformities and a leukemoid blood picture in the newborn, Blood 12:567, 1957.
11. Kato, K.: Congenital absence of the radius, J. Bone Joint Surg. 6:588, 1924.

HOLT-ORAM SYNDROME

Cardinal Clinical Features

Atrial septal defect or ventricular septal defect, cardiac arrhythmia, malformations of the upper extremities that are predominantly radial and include triphalangeal finger-like thumbs.

Clinical Picture

The syndrome is sometimes referred to under the headings of ventriculoradial dysplasia, atriodigital dysplasia or the upper limb cardiovascular syndrome, but it is unlikely that anyone will remember these terms. It usually is designated after Holt and Oram,[1] who observed a family in which atrial septal defect and congenital anomalies of the hand occurred in nine individuals in four generations.

The most important of the skeletal anomalies are those of the thumb. The thumbs may be finger-like, vestigial or absent (Fig. 1). The most common anomaly is one in which the thumbs lie in the same plane as the fingers. The thumb may have a rudimentary middle phalanx. Deformities of the carpal bones also are present: they usually are slender. The metacarpals, especially the first and second, may be elongated. The most minimal involvement of the upper extremity is a flattening and distal displacement of the thenar eminence, which produces a slender appearance of the hand and which usually occurs with diminished opposability of the thumb. In more extreme involvement, phocomelia or seal-limb deformity of the arm, with three digits, is seen. In this situation there is a rudimentary humerus, fused radius and ulna and radial hypoplasia (Fig. 2) or aplasia. Intermediate forms between this extreme and involvement only of the hand are seen, usually including hypoplasia of the radius. Such patients may have difficulty with supination of the forearm. Among other skeletal anomalies are cubitus valgus, a small scapula, shallowness of the glenoid fossa and a short humerus. Hypoplasia of the clavicle and narrow shoulders give an appearance of the arms hanging away from the body. In addition, there may be hemivertebrae, spina bifida, Sprengel's deformity or scoliosis. Narrow shoulders may be prominent, and there may be absence of the pectoralis major. These patients also may have anomalies such as a high-arched palate or pectus excavatum. Syndactyly occasionally is observed between thumb and first finger.

The dermatoglyphics in the Holt-Oram syndrome are of interest.[2-4] There are three prominent features: (1) an increased frequency of whorl patterns of the fingers, and thus a high total ridge count; (2) a radial loop on the involved thumb, as observed in two patients and seen in only 0.3% of normal individuals; and (3) distally located axial triradii and a hypothenar pattern on the hands with the finger-like thumbs. Simian lines or their equivalents also are seen.

The cardiovascular abnormalities in involved patients have varied. The cardiac lesion in the family reported by Holt and Oram[1] was an atrial septal defect. In the family reported by McKusick,[5] a mother and daughter had atrial septal defects. In addition, all of the patients had a tendency to develop cardiac arrhythmia, often of bizarre type. On the other hand, the cardiac lesion in one of the families reported by Holmes[2] was a ventricular septal defect. In the other family some patients had atrial septal defects and one had a ventricular septal defect. Atrial and ventricular septal defects have been seen in the same patient. Patients also have had pulmonary hypertension. Additional cardiac malformations, such as anomalous venous return, anomalous coronary artery or retroesophageal right subclavian artery, have been observed. Family members of the same sex tended to have cardiac and skeletal anomalies that were more alike than those of the opposite sex. However, considerable variation of expression is observed even in the same family. Harris and Osborne[6] believed that there was a particular relationship between hypoplasia or absence of the radius and ventricular septal defect, and it may be that this ventriculoradial dysplasia is a separate syndrome. However, at this point it appears that there is a single disorder with a spectrum of expression. Primary pulmonary hypertension has been described by Kuhn, Schaaf, and Wagner[9] in association with lesions of the hand and arm that are identical to those seen in the Holt-Oram syndrome, and this may represent a separate syndrome.

Patients with the Holt-Oram syndrome usually are acyanotic. They may have cardiomegaly. Systolic murmurs are present and usually best heard over the left intercostal spaces close to the sternum. Roentgen-

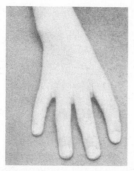

Fig. 1.—Hand of a 26-year-old female with the Holt-Oram syndrome. She had no thumb and an atrial septal defect of the ostium secundum type. (Courtesy of Dr. William Friedman of the University of California San Diego; reprinted with permission from Perkoff, J. K.[10])

ographically there may be enlargement of both ventricles, with increased vascular marking in the lung fields.

Cardiac arrhythmias include atrial fibrillation, multifocal ventricular ectopic beats and bradycardia. Bundle branch block or other forms of heart block may be seen, as well as a simple prolongation of the P–R interval. Rarely in an involved family skeletal abnormalities may be seen in a patient in whom there is no cardiac abnormality.

Intellectual development is normal.

ASSOCIATION OF CARDIAC AND SKELETAL ABNORMALITIES.—Abnormalities of the heart and musculoskeletal system are associated in a number of syndromes, including trisomies of chromosomes 13 and 18, as well probably as 21. They also are seen in the Turner syndrome and in patients with an XXXXX or XXXXY chromosomal constitution. They are essential concomitants of the Fanconi syndrome of aplastic anemia and the thrombocytopenia with absent radius syndrome. They also are seen in the Marchesani, Marfan, Ellis-van Creveld and Laurence-Moon-Bardet-Biedl syndromes, as well as in the syndrome described by Carpenter and similar syndromes.

Patients with the 13 trisomy have narrow distal phalanges, retroflexible, hyperconvex thumbs and polydactyly. Patients with 16–18 trisomy have an abnormality of fisting in which the fist clenches and the index finger appears to beckon. They also have syndactyly and hypoplastic nails. In 21 trisomy the hand and fingers are broad and short, with a proximally placed thumb and clinodactyly of the fifth finger. Clinodactyly of the fifth finger also has been observed in patients with multiple X chromosomes along with small delicate hands and simian lines, or radioulnar synostosis. In the Turner syndrome there is cubitus valgus and clinodactyly. A very short fourth metacarpal is characteristic. The nails may be small and hyperconvex. Hypoplastic nails and polydactyly occur in the Ellis-van Creveld syndrome, as do thickened, distally shortened bones. Brachydactyly is seen in the Marchesani syndrome and, of course, arachnodactyly in that of Marfan. Polydactyly, brachydactyly and syndactyly are seen in the Laurence-Moon-Bardet-Biedl syndrome as well as in that of Carpenter.

CASE REPORT

J. S. was a 7-month-old white female referred to the Genetic Counseling Clinic of University Hospital for evaluation of multiple congenital anomalies. She was the product of a normal pregnancy and delivery. She had a birth weight of 6 lb, 14½ oz and a length of 17½ inches. She was found at birth to have congenital heart disease and was diagnosed as having a ventricular septal defect. Mild congestive heart failure was treated with Digoxin. Rudimentary thumbs were observed at birth, and she had an imperforate hymen. She was hospitalized at the age of 7 weeks for pyloric stenosis. The patient developed normally. At 7 months she had two teeth, could hold objects and could transfer from hand to hand and to her mouth. She was able to sit alone and appeared alert and happy.

Family history was unremarkable. The parents were healthy and had no cardiac or skeletal abnormalities. The father was 17 years old, the mother 16. There was no consanguinity.

Fig. 3.—J. S. (Case Report). Hand of an infant with the Holt-Oram syndrome. She had a unilateral rudimentary thumb.

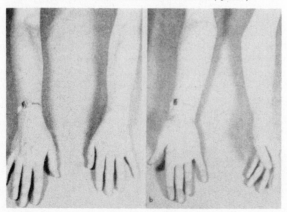

Fig. 2.—Hands of a 34-year-old woman with the Holt-Oram syndrome. Left, the hypoplastic, crooked left thumb and the left arm which is shorter than the right. Right, failure of supination on the left, which results from hypoplasia of the radius. She had an atrial septal defect of the ostium secundum type. (Reprinted with permission from Perkoff, J. K.[10])

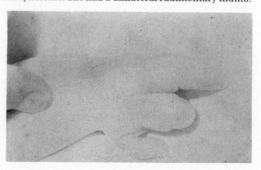

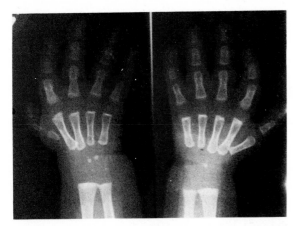

Fig. 4. – J. S. Roentgenograms of the hands. The left thumb showed generalized hypoplasia. It was especially marked in the metacarpal.

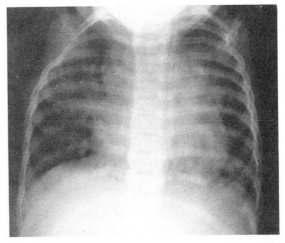

Fig. 5. – J. S. Roentgenogram of the chest indicating the cardiomegaly.

Physical examination was notable for the heart murmur, which was holosystolic and heard best at the lower left sternal border. The rhythm was normal. She had a rudimentary left thumb (Fig. 3), and the right thumb was hypoplastic and proximally placed. The rest of the digits and the arms were normal. The elbow joints were normal and she had no problems with pronation or supination. Lower extremities also were normal.

Roentgenographic examination revealed hypoplasia of the first metacarpals, more marked on the left side (Fig. 4). The thumbs had a digitalized appearance, especially on the left side. Chest x-ray revealed mild cardiomegaly with a moderate degree of left-to-right shunt at the level of the ventricular septum (Fig. 5). There was no evidence of congestive heart failure. The shoulder and clavicle were normal, as were the humerus, radius and ulna and the rest of the metacarpals and digits.

GENETICS

The syndrome is transmitted as an autosomal dominant. Male to male transmission has been observed, ruling out X-linkage. A greater severity of skeletal abnormalities has been observed in females, whereas the severity of cardiac malformations was the same in the two sexes.[4] Studies of the chromosomes are normal. In the large family reported by Gall and colleagues,[4] an excess of involved children of affected mothers over that expected suggested the possibility of abnormal segregation, that is, uneven distribution of chromosomes to the gametes during meiosis.

Embryologically both the upper limb and the primitive heart are undergoing important structural development during the fourth and fifth weeks of human fetal life. The action of the mutant gene presumably interferes with normal differentiation at that particular time, leading to upper limb and cardiovascular abnormalities. The lower extremities are never affected in this syndrome. This is consistent with the fact that the lower limb buds differentiate at a later stage of embryogenesis.[5]

TREATMENT

The cardiac malformations are subjects for surgical correction involving closure of the septal defects.

REFERENCES

1. Holt, M., and Oram, S.: Familial heart disease with skeletal malformations, Br. Heart J. 22:236, 1960.
2. Holmes, L. B.: Congenital heart disease and upper extremity deformity, N. Engl. J. Med. 272:436, 1965.
3. Rosner, F., and Aberfeld, D. C.: Dermatoglyphics in the Holt-Oram syndrome, Arch. Intern. Med. 126:1010, 1970.
4. Gall, J. C., Stern, A. M., Cohen, M. M., Adams, M. S., and Davidson, R. T.: Holt-Oram syndrome: Clinical and genetic study of a large family, Am. J. Hum. Genet. 18:187, 1966.
5. McKusick, V. A.: Medical genetics, 1960, J. Chronic Dis. 14:100, 1961.
6. Harris, L. C., and Osborne, W. P.: Congenital absence or hypoplasia of the radius with ventricular septal defect: Ventriculo-radial dysplasia, J. Pediatr. 68:265, 1966.
7. Lewis, K. B., Bruce, R. A., Baum, D., and Motulsky, A. G.: The upper limb-cardiovascular syndrome. An autosomal dominant genetic effect on embryogenesis, J.A.M.A. 193:1080, 1965.
8. McKusick, V. A.: Holt Oram Syndrome: Digital and Heart Anomalies in Three Generations, in *The First Conference on the Clinical Delineation of Birth Defects. Part III, Limb Malformations,* Birth Defects: Original Article Series (New York: The National Foundation-March of Dimes, 1969), pp. 187–189.
9. Kuhn, E., Schaaf, J., and Wagner, A.: Primary pulmonary hypertension, congenital heart disease and skeletal anomalies in three generations, Jap. Heart J. 4:205, 1963.
10. Perkoff, J. K. (ed.): *The Clinical Recognition of Congenital Heart Disease* (Philadelphia: W. B. Saunders Co., 1970).

THE VATER ASSOCIATION
Say Syndrome,
A Syndrome of Vertebral Anomalies, Imperforate Anus and Digital Abnormalities

CARDINAL CLINICAL FEATURES

Vertebral anomalies; imperforate anus; tracheo-esophageal fistula; skeletal anomalies that may consist of absent or rudimentary thumbs, polydactyly or absent radii; renal anomalies.

CLINICAL PICTURE

Say and Gerald[1] described a new syndrome in 1968 in which 10 unrelated patients had multiple malformations that included polydactyly, vertebral anomalies and imperforate anus. All of these cases were sporadic and the patients had no family histories of similar defects. Rib anomalies also were prominent in this series: two patients had absence or hypoplasia of the first metacarpal; one had a deformed femur; one had a tracheoesophageal fistula and renal anomalies; and a number had defects of the pelvis, especially hypoplasia of the ilium. Later in the same year Fuhrmann[2] described a 6-week-old boy with imperforate anus whose older brother had the same defect, as well as unilateral right radial polydactyly and congenital heart disease. Their mother had radial polydactyly of the right hand. Fuhrmann commented on the similarity of these patients to those described by Say and Gerald and raised the possibility of a familial basis for these deformities. Tünte,[3] in a study of 103 patients with anal atresia, found one patient with bilateral duplication of the thumb but several in whom the thumb was absent or rudimentary. Vertebral anomalies, absence of the first metacarpal and tracheoesophageal fistula also were observed. This author suggested that in the Say syndrome, the triad that includes vertebral anomalies and anal atresia might include either ectrodactyly of the thumb or radial polydactyly as the third component.

Quan and Smith[4] described a group of patients with five associated defects: Vertebral defects, Anal atresia, TracheoEsophageal fistula, Renal anomalies and Radi-al dysplasia. They suggested the acronym VATER to summarize these associated anomalies. They pointed out that the importance of the association is that any one or two of these defects should alert the physician to examine the patient carefully for the other frequently associated defects.

Imperforate anus is commonly associated with esophageal atresia and also is commonly associated with renal anomalies. Quan and Smith[4] thought that they were not dealing with a discrete single etiology syndrome but, rather, that there was a nonrandom tendency for the defects to associate. All seven of their patients had vertebral anomalies and a serious intestinal malformation, but only one of them had all of the VATER components. Two had cardiac defects.

DIFFERENTIAL DIAGNOSIS

Congenital heart disease is seen with some regularity in the VATER association. This had led to confusion of the VATER association with the Holt-Oram syndrome (see chapter, this section), in which atrial or ventricular septal defect is associated with absent or rudimentary thumbs and absent radii or with phocomelia. The Holt-Oram syndrome is transmitted as an autosomal dominant. Sporadic cases of rudimentary or absent thumb and absent radius with cardiac defects should be examined closely for the defects related to the VATER association. This anomaly also should be differentiated from the Fanconi anemia (see chapter, this section), in which a rudimentary or absent thumb or absent radius is associated with short stature, renal anomalies, pancytopenia, chromosomal breakage and tendencies toward malignancy. Among patients with imperforate anus, as many as 3–5% may have associated hand or other defects of the VATER association.[5] The anal malformation in this syndrome may show considerable variation from mild to complex.[5] Often a rectovaginal, rectovesical or rectourethral fistula is associated.

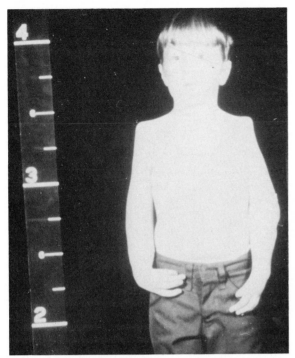

Fig. 1.—R. C. (Case Report). A 12-year-old boy with the VATER anomaly. The narrow hypoplastic shoulders and club hands are evident.

There is a striking similarity between the abnormalities of the VATER association and the malformations seen in the dominantly inherited hemimelia of mice in which absent or supernumerary digits and vertebral anomalies are associated with imperforate anus.[6] These mice, like at least one patient with the VATER association, have absence of the spleen.

Fig. 2.—R. C. The shoulders were flattened in the deltoid area. Both humeri were dislocated and the clavicles and acromion protruded above.

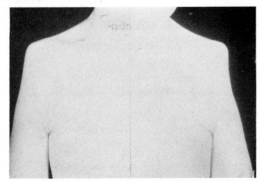

*We are indebted to Dr. Michael Sherman of the U.S.A.F. Hospital, Vandenberg Air Force Base, for the opportunity to study and report on this patient.

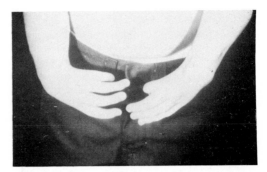

Fig. 3.—R. C. The hands illustrate the extreme radial deviation of a patient with absent radii. He had no thumbs and only four fingers. The camptodactyly of the index fingers had been produced surgically.

CASE REPORT*

R. C., a 13-year-old boy with multiple congenital malformations, was referred for diagnosis and evaluation. He was the product of a normal pregnancy and delivery. There was no history of exposure to drugs, radiation or infections during gestation. There was no family history of congenital defects. Birth weight was 7 lb, 8 oz, and length 19 inches. At birth he was found to have an imperforate anus. The defect was a web, which was opened at that time, and a supernumerary toe next to the left great toe was removed surgically. He also was noted

Fig. 4.—R. C. Roentgenograms of the neck illustrate fusion of C2 and C3.

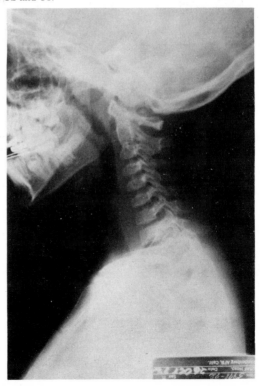

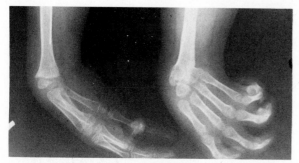

Fig. 5.–R. C. Roentgenograms of the upper extremities show bilaterally absent radii and thumbs.

to have bilaterally absent thumbs, bilateral radial aplasia and clubhands.

In infancy he had had a bout of pyelonephritis, and during childhood he had had chronic hay fever, many upper respiratory tract infections and two episodes of pneumonia. He was found to have shortness of stature, although at the time of examination he was growing at a nearly normal rate and was entering puberty. His intelligence was normal, and he was successfully attending the eighth grade.

Physical examination revealed a small, pleasant boy (Fig. 1). His height was 50 inches, which gave him a height age of 8 years at a chronologic age of 13. His weight was 60 lb. The neck was short. The lungs were clear and the heart appeared normal. Early puberal development was evident.

The shoulders (Fig. 2) were striking in that the distal ends of the clavicles appeared to protrude and the head of the humerus was dislocated bilaterally. The deltoid and biceps were extremely hypoplastic, limiting abduction. The elbows were limited in their range of flexion and extension motion. Pronation and supination were not possible. There were bilaterally absent radii, with the severe radial deviation known as club hands (Fig. 3). On the right hand there was an absent thumb, a hypoplastic index finger and partial syndactyly of the second and third fingers. On the left hand the thumb also was absent.

Roentgenographic study revealed congenital fusion of C2 and C3 (Fig. 4) and a mild S1/S2 sacral anomaly and bilateral coxa valga. There were hypoplastic humeri, with dislocated capital humeral epiphyses and congenital absence of both radii and thumbs (Fig. 5). The patient had cross-fused renal ectopia on the right (Fig. 6).

Complete blood count was normal, as was the bone marrow. Chromosome analysis was essentially normal. Fluorescent analysis revealed a prominent short arm and/or satellites on chromosome 15 (15 p+ or 15 ps). The minor variant also appeared to be present in metaphases from the father. Minor variants of this type are frequently seen in normal individuals, and in this patient it was interpreted as a coincidental finding, unrelated to his congenital abnormalities.

GENETICS

So far most cases have been sporadic. The resemblance of this syndrome to the hemimelia of the mutant mice suggests a genetic basis for the syndrome. Fuhrmann's[2] family suggests the possibility of a dominant with variable penetrance. More recently Say and colleagues[7] have observed a patient with the syn-

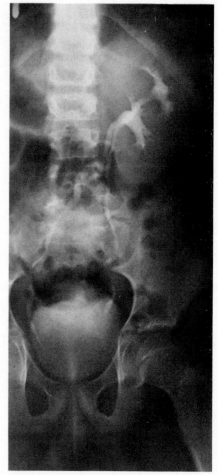

Fig. 6.–R. C. Intravenous pyelogram revealed fused, crossed renal ectopia. The two collecting systems were drained by separate ureters, which can be seen entering the bladder in the usual bilateral fashion.

drome whose grandmother had a hypoplastic thumb and vertebral anomalies.

TREATMENT

Orthopedic surgery may be helpful in the management of the deformities. Surgery is of course indicated for imperforate anus or tracheoesophageal fistula.

REFERENCES

1. Say, B., and Gerald, P. S.: A new polydactyly/imperforate-anus/vertebral-anomalies syndrome? (Letters to the Editor), Lancet 2:688, 1968.
2. Fuhrmann, W.: A new polydactyly/imperforate-anus/vertebral-anomalies syndrome? (Letters to the Editor), Lancet 2:918, 1968.

3. Tünte, W.: A new polydactyly/imperforate-anus/vertebral-anomalies syndrome? (Letters to the Editor), Lancet 2: 1081, 1968.

4. Quan, L., and Smith, M. D.: The VATER association: Vertebral defects, Anal atresia, T-E fistula with esophageal atresia, Radial and Renal dysplasia: A spectrum of associated defects, J. Pediatr. 82:104, 1973.

5. Say, B., Balci, S., Pirnar, T., and Tunçbilek, E.: A new syndrome of dysmorphogenesis: Imperforate-anus associated with poly-oligodactyly and skeletal (mainly vertebral) anomalies, Acta Paediatr. Scand. 60:197, 1971.

6. Searle, A. G.: The genetics and morphology of two "luxoid" mutants in the house mouse, Genet. Res. 5:171, 1964.

7. Say, B., Balci, S., Pinar, T., and Hiçsönmez, A.: Imperforate anus/polydactyly/vertebral anomalies syndrome: A hereditary trait? (Letters to the Editor), J. Pediatr. 79:1033, 1971.

Dermatoglyphics

DERMATOGLYPHICS*

Joan Schindeler Davies

Several books have been written on the subject of dermatoglyphics.[1-4] In general these writings are complex and, because of the complexity, many texts tend to intimidate those who attempt to use them without a great deal of experience. On the other hand, some of the articles that have appeared on the subject in medical journals are wordy and not well enough illustrated to be useful to the physician trying to make a diagnosis. This chapter is intended as a practical approach to dermatoglyphic analysis for physicians and others interested in using this technic in the assessment of human malformations.

The study of dermatoglyphics can trace its beginnings to the late 17th century and a physician named Nehemiah Grew, who described the "innumerable little ridges of equal bigness." The term dermatoglyphics was coined by the anatomist Harold Cummins in 1926.

Frances Galton,[2] a cousin of Charles Darwin, set up the first scientific classification of finger skin patterns in 1892. His methods and classification are those in use today. Skin patterns are produced by a system of parallel lines called ridges with depressions between them known as furrows. The formation of ridges begins early in fetal life and, by 18 to 20 weeks, they are completely formed. Once formed, the patterns are permanent. The ridges enlarge with physical growth but the patterns do not change.

In 1936 Cummins[5] published his work on the dermatoglyphic findings in children with the Down syndrome. In 1959 it was discovered that these children had a chromosomal abnormality. The association of abnormal chromosomes and abnormal dermatoglyphics in this syndrome led investigators to look for associations between dermatoglyphics and other cytogenetic abnormalities. This search was expanded to other malformation syndromes with and without abnormal chromosomes. A number of clinical abnormalities have been found that show patterns of ridge changes characteristic of the abnormality.[6-9] Not all patients with a syndrome will individually show the classic ridge patterns known to occur in the syndrome. This is true of most features in any syndrome. The dermatoglyphic abnormalities may be considered a subtle malformation, one of the features of the syndrome. In this chapter the classic dermal ridge findings, and whenever possible some variants, will be presented.

The formation of the dermatoglyphic patterns is multifactorial. It is under the control of many genes. It is also true that environmental factors can affect the development of ridge patterns in utero.

The use of dermatoglyphics in clinical medicine is facilitated by the fact that the patterns are easily available for study. They can be observed and easily recorded. Examination of the ridges is noninvasive for the patient, and children usually enjoy it.

Dermatoglyphics also may be used to help determine the zygosity of twins. In monozygotic twins the main characteristics are similar, whereas in the dizygotic the dermatoglyphs are no more similar than in siblings. Dermatoglyphics may suggest a diagnosis when the other manifestations of a condition are not apparent. A major problem in the clinical use of dermatoglyphics is that, often in the newborn infant, the ridges are poorly formed and it may be difficult to see the patterns; this is especially true in premature babies.

Dermatoglyphs do not include the flexion creases of the palms, soles and digits; however, the flexion creases usually are examined at the same time.[10] For example, the simian line occurs frequently in many syndromes. The simian line is a well-known characteristic of the Down syndrome. It also occurs with frequency in other syndromes and in some normal individuals. It is important not to overdiagnose the Down syndrome in the presence only of a simian line. Although the simian line occurs in the Down syndrome, the ridge patterns are of much more diagnostic value. The simian line alone without dermatoglyphic findings and without clinical features is a minor malformation and by itself of no significance. In the presence of other minor malformations, it may be significant and suggest a search for a major malformation.

TECHNICS IN DERMATOGLYPHICS

Finger ridges have been recorded for many centuries. Initially wax was used to make a permanent record of these patterns. Today there are a number of more satisfactory methods.

The ridges may be viewed by the naked eye provided there is proper light and the ridges are well developed. Often a 2.5× magnifying glass is helpful.

*Supported by the Hayward Foundation

351

When observing the palms and soles, the fingers and toes should be pointing in an upward direction. This position facilitates identification of the pattern types since all graphic illustrations of dermatoglyphs are presented in this way.

Viewing dermal ridges of newborn infants can seldom be accomplished with the naked eye. We have found that a flash magnifier used by stamp collectors is most helpful. This is a small 5× magnifying glass attached to a flashlight base, holding two size C batteries and a light bulb. The illumination is particularly helpful. Another advantage of this flash magnifier is its size. The over-all length is only seven inches and it will fit easily into a coat pocket. We use the Selsi flash magnifier, which is made in Japan and is quite inexpensive. Naked eye or magnified observation of the patterns is enough for diagnostic purposes. If a record of the pattern is desired, prints may be taken.

The technics for taking good prints vary with the type of ink used. There are three types of ink in most common usage. The Searche Company manufactures an ink used by most police departments. This is an undesirable method for our purposes because it blackens the skin and generally is too difficult to control. The advantage of this ink is in its performance.

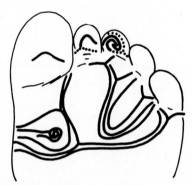

Fig. 2.—Graphic illustration of foot print.

Fig. 1.—Graphic illustration of hand print.

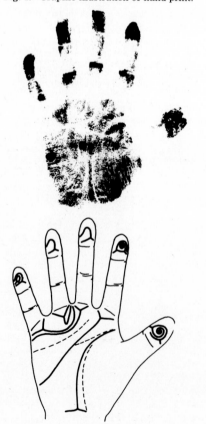

The ink most popular in clinical dermatoglyphic usage is manufactured by Faurot, Inc., of New York.* This ink is a clear liquid used with an applicator. The special paper is sensitized to the ink and leaves the hand only slightly, if at all, sticky. These prints fade in about 4–5 years.

Another fine printing material is made by Hollister of Chicago. This is a dry printer which, when used with their glossy paper, produces an excellent print. Hollister also produces the Disposable Footprinter, which gives a very black print with well-defined ridges. Both of these printers are used for palms and soles.

In the newborn infant the palms and soles should be wiped with an alcohol sponge before applying the ink. This removes any vernix, blanket fuzz or dead skin. Using the Hollister dry printer, rub gently along the hallucal area of the foot, holding the toes back. When using the Hollister Disposable Footprinter, merely touch the plate to the part that is to be printed. Without too much pressure, roll the glossy side of the paper from the tibial border to the fibular border so as not to miss any triradii. You will find that the Hollister dry printer will need to be scraped with a single edge razor

*299 Broadway, New York, New York 10007.

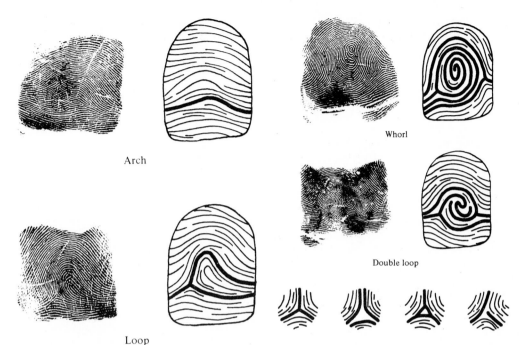

Arch

Whorl

Double loop

Loop

Fig. 3. — Five types of digital prints: the *arch* with no triradii, the *loop* (ulnar or radial) with one triradius and the *whorl* and the *double loop*, each with two triradii. A *triradius* is the center, or junction, of three streams of ridges. *Lower right*, four examples of triradii. Note that in some cases it is an arbitrary point.

blade from time to time to remove grease left by the skin. Sometimes heating the plate in front of a space heater will facilitate printing.

You should ink the entire hand and attempt to print it, if only to illustrate the over-all size and shape of the hand. It is rare that all digital and palmar triradii can be printed at one time. Re-ink only the palmar area, and be certain to cover the triradius under each digit (the interdigital area). Press gently with the paper. This process may need repeating to assure that all triradii are present on one print. Next, ink each digit and roll the paper across the fingertip to print, remembering to label the thumb 1, then fingers 2, 3, 4 and 5.

It should be remembered that obtaining readable prints of newborn infants is a time-consuming job and cannot be rushed. When printing older children and adults, the process is basically the same, although simplified.

The Faurot method, because of its cleanness, is the most popular method, and materials include an applicator, ink and sensitized paper. With an eye dropper apply the ink to the applicator until it is damp. This will probably last through three or four printings. Ink the hand, being certain not to miss the interdigital and lower palm areas.

Place the paper, glossy side up, on a thin piece of foam rubber. For the best printing, hold the paper with the foam rubber flat on your right palm. Place the patient's hand in the center of the paper. Press the fingers of your right hand into the palm of the patient to obtain a complete palm print. Print the fingertips separately, remembering to number them.

The sole and toes are inked the same way as the palm. The toes will need to be printed separately. They should be held back with the right hand while the paper is pressed under them, then up. The reason for this is that the triradii lie close to the crease on the underside of the toe.

CLASSIFICATION

NAME	LEFT HAND 5 4 3 2 1	RIGHT HAND 1 2 3 4 5	ATD ANGLE L	R	HYPO-THENAR L	R	THENAR L	R	HALLUCAL L	R	COMMENTS

The most important area of the sole is the hallucal area. Remember to roll the paper starting up on the tibial side. The *f* triradius often lies far up on this border.

When readable prints have been taken, the classifications may vary from a very simple one, which is presented here, to those that have been computerized.

Sometimes it is impossible to obtain prints, as in the case of a premature infant in an incubator or when there is hypoplasia or extremely dry skin. In this case, by using hand and foot outlines, you can draw in the type of print as you view it, using a magnifying glass or flash magnifier.

A simple classification is shown on page 353. Because of the difficulty in reproducing a clear original palm or hand print, a simple tracing procedure has been developed for illustrating the important areas of the dermatoglyphics (Figs. 1 and 2).

CLASSIFICATION OF DERMATOGLYPHICS

An international classification of ridge patterns was established in September 1967 in London, England.[9]

Fig. 4.—Classification of hand print and main lines. *Main lines* (labeled *A, B, C* and *D* for the triradii where they originate) are those that leave a triradius area and travel across or down the palm to or through the *hypothenar* area. *T = thenar*.

To understand dermatoglyphic classification, a knowledge of what constitutes a triradius is of fundamental importance. A triradius is an area or point at which ridges converge. Dermatoglyphic classification is based upon the identification of triradii (Fig. 3).

There are five digital patterns on the fingers and toes (see Fig. 3): the *arch* with no triradii, the *loop* (either ulnar or radial) with one triradius and the *whorl* and the *double loop*, each with two triradii. Double loops are considered to be a variant of a whorl pattern and may be classified as such. There are numerous other patterns that are variants of the five basic patterns.

The palm is divided into three large areas (Fig. 4): the *interdigital* area lies below and between the digits, the *thenar* area is at the base of the thumb and the *hypothenar* area is on the ulnar side of the palm.

There is a triradius at the base of each finger, with the exception of the thumb. Starting with the index finger, this triradius is labeled the *a* triradius, the third finger triradius is called the *b*, next the *c* and then the *d* (see Fig. 4). Occasionally there is a second triradius; this is classified with a prime mark and the letter from the triradius that it lies close to, for example, a', b', etc. The line that leaves the triradius area (made up

Fig. 5.—Triradii and *atd* angle. **Top,** double loop with two triradii. **Bottom,** the *atd* angle is subtended by lines drawn from the *t* triradius to the *a* and *d* triradii.

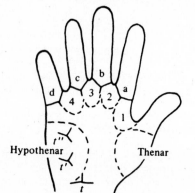

1, 2, 3, 4 = Interdigital areas
a, b, c, d = Digital triradii
t, t', t'' = Axial triradii

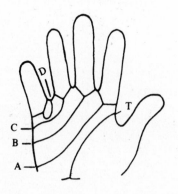

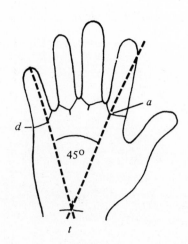

of the three sets of intersecting ridges) and travels across or down the palm carries the letter designated for the triradius where it began: the "main line," as it is called, is labeled with a capital letter: *A, B, C* and *D*. The main lines run in a horizontal direction on the palm (see Fig. 4).

The area between digital triradii is called interdigital: I, between the thumb and index finger; II, between the second and third digits; III, between the third and fourth digits; and IV, between the fourth and fifth digits (see Fig. 4). The most common pattern in these areas is a small loop opening in a distal direction. Whorls are rare.

An *ab* ridge count may be called for in some instances, i.e., the Turner syndrome, the Klinefelter syndrome. This count is a good index of pattern intensity and is independent of age. In making the ridge count, count in a straight line from triradius *a* to *b*, including any ridge that the line crosses. The count is reported as summed for both hands. The normals, as reported by Pons,[6] are: male 82, female 84.

The *thenar* area lies at the base of the thumb. The ridge patterns of this area are classified as loop or positive. The pattern, when it is not a loop, tends to become very complex in this area, and therefore no classification name, letter or numeral was given to it at the international conference.

The *hypothenar* area lies on the ulnar portion of the palm extending into the center. There are four primary types of ridge patterns in this area: whorls, loops, arches and S patterns. Another important triradius occurs in the hypothenar area at the base of the palm. This is called the *t* triradius (axial triradius). Its position on the palm is of important clinical significance. By drawing a line from the *a* and *d* triradius to the *t* triradius, the *atd* angle is obtained (Fig. 5). The angle is measured by protractor. When there is an *a* plus *a'* triradius, the most lateral *a* is used, and in the case of the *d* triradius, the most medial *d* is used. The *t* may move in a distal direction into the center of the palm; when this occurs another triradius usually is present at the base of the palm. Three positions have been designated for the *t* triradius with corresponding angles: *t* = 45 degrees, *t'* = 51 degrees and *t''* = 56 degrees and over (see Fig. 4). The most distal *t* always is used for the *atd* angle. The position of this distal *t* will give an estimate of the *atd* angle without having to draw it.

The *atd* angle has proved to be a most useful measurement for normal and abnormal palm patterns. The angle does change with palmar growth, but it is sufficient to use a value of 45 degrees as a normal *atd* angle. An angle less than 45 degrees also is normal.

The patterns on the toes are basically the same as on the fingers (Fig. 6). The arch with no triradius, the loop (either *tibial* or *fibular*) with one triradius and the

Fig. 6.—Classification of sole prints. The borders are *tibial* and *fibular*.

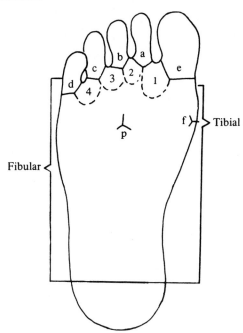

1, 2, 3, 4   = Interdigital areas
e, a, b, c, d   = Digital triradii
f, p   = Hallucal triradii

Fig. 7.—**Top,** example of an *ab* ridge count. The ridge count (**bottom**) is done under low magnification. The ridge count consists of the number of ridges between the triradius and the point of core, or center of the pattern. Neither the triradius nor the final ridge when it forms the center of the pattern is counted (**lower right**). In the case of whorl, both sides of the pattern are counted (**lower left**) and the higher count is reported for a total ridge count. When an absolute ridge count is desired, both counts are reported. In the case of a simple arch, there is no triradius and therefore no ridge count. The normal mean ridge count of 10 digits: males 145, females 126.

whorl and double loop, each with two triradii. There is a triradius at the base of each toe. Under the large toe it is designated as *e*, second toe *a*, third *b*, fourth *c* and fifth *d*. There is a triradius on the *tibial* border called *f*, and one in the center of the foot called *p*.

The interdigital areas fall between the digital areas, as on the palm: I, between the large and second toes; II, between the second and third toes; III, between the third and fourth toes and IV, between the fourth and fifth toes.

Classification of the entire sole is both complex and confusing. The sole has one area that will be classified in this text. This is the large area under the big toe and lying on the tibial border. It is called the hallucal area. Patterns in this area resemble hypothenar patterns.

Ridge counting (Fig. 7) often is useful in characterizing malformation syndromes, especially those involving the sex chromosomes. The *ab* ridge count is useful. A total ridge count is the sum of the counts for all 10 digits. The count for a digit is the number of ridges between the triradius and the point of core or the center of the pattern. Neither the triradius nor the final ridge where it forms the center of the pattern is counted. In the case of a whorl, both sides of the pattern are counted (see Fig. 7), and the higher ridge count is reported; when an "absolute" ridge count is desired, both sides are reported. By contrast, in an arch there is no triradius and hence a count of zero. The mean total counts in normal individuals are: males 145; females 126.

TABLE 1.—DERMATOGLYPHICS OF MAJOR ABNORMALITIES IN COMPENDIUM (FIGS. 8–29)[11-34]

Increase in digital whorls		Laurence-Moon-Biedl
		Smith-Lemli-Opitz syndrome
		Rubella
		G-deletion II
		Turner syndrome (XO)
		Holt-Oram syndrome
Increase in digital arches		Trisomy C
		Trisomy D
		Trisomy E
		Klinefelter syndrome (XXY)
		Klinefelter variant (XXYY)
		Cri-du-chat syndrome
Increase in digital ulnar loops		Down syndrome
Radial loops on digits other than digit 2		de Lange syndrome
		Laurence-Moon-Biedl syndrome
		Down syndrome
		G-deletion I
Elevated *t* triradius	*t'*	Hypercalcemia
		Smith-Lemli-Opitz syndrome
		de Lange syndrome
	t''	Arthrogryposis
		G-deletion II
		Turner syndrome
	t'''	Trisomy D
		Trisomy G
		Cri-du-chat syndrome
		Rubella
Thenar pattern		Trisomy C
		de Lange syndrome
		Rubinstein-Taybi syndrome
		Cri-du-chat syndrome
Vertical alignment of main lines		Cerebral gigantism
		Arthrogryposis
Elevated ridge count		Turner syndrome (XO)
		Cerebral gigantism
Low ridge count		Klinefelter syndrome (XXY)
		Trisomy D
		Trisomy E
Main line *A* exit in thenar area		Turner syndrome (XO)
		Cerebral gigantism
		Arthrogryposis
Missing *c* triradius		G-deletion II
		Turner syndrome (XO)

The following syndromes show either no dermatoglyphic changes or findings too subtle to be useful in diagnosis:

Cancer	Neurofibromatosis
Cat eye	Parkinson's disease
Cleft palate	Phenylketonuria
Deletion of no. 4 chromosome	Prader-Willi
Deletion of short and long	Schizophrenia
arms of no. 18 chromosome	Silver
Huntington's chorea	Tay-Sachs
Leukemia	Tuberous sclerosis
Mucopolysaccharidoses	XYY

Figures 8–13 show dermatoglyphs in autosomal chromosome aberrations, Figures 14–17 dermatoglyphs in sex chromosome aberrations and Figures 18–29 dermatoglyphs in other syndromes.

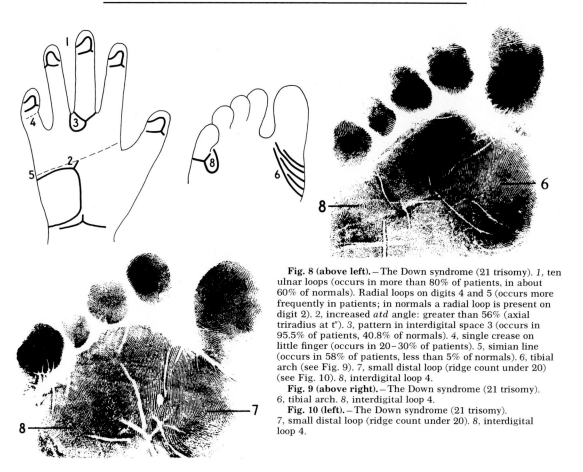

Fig. 8 (above left). – The Down syndrome (21 trisomy). *1*, ten ulnar loops (occurs in more than 80% of patients, in about 60% of normals). Radial loops on digits 4 and 5 (occurs more frequently in patients; in normals a radial loop is present on digit 2). *2*, increased *atd* angle: greater than 56% (axial triradius at *t″*). *3*, pattern in interdigital space 3 (occurs in 95.5% of patients, 40.8% of normals). *4*, single crease on little finger (occurs in 20–30% of patients). *5*, simian line (occurs in 58% of patients, less than 5% of normals). *6*, tibial arch (see Fig. 9). *7*, small distal loop (ridge count under 20) (see Fig. 10). *8*, interdigital loop 4.

Fig. 9 (above right). – The Down syndrome (21 trisomy). *6*, tibial arch. *8*, interdigital loop 4.

Fig. 10 (left). – The Down syndrome (21 trisomy). *7*, small distal loop (ridge count under 20). *8*, interdigital loop 4.

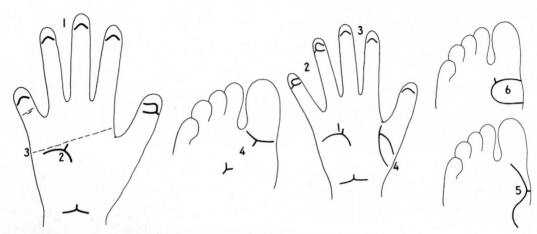

Fig. 11 (left). – 18 Trisomy (Edwards syndrome). *1*, increased arches (usually 6 or more, sometimes 10; only about 2% of normals have 6 or more arches). *2*, increased *atd* angle (axial triradius at *t″*). *3*, simian line. *4*, presence of *e* or *p* triradius; absence of *f* triradius in soles.

Fig. 12 (right). – 13 trisomy (the Patau syndrome). *1*, increased *atd* angle (77% of patients have marked increases in angle; mean sum of both *atd* angles varies between 186 degrees and 196 degrees; 93–98 degrees in normals). *2*, radial loops on digits 4 and 5 (rare in general population). *3*, increased arches (19% of digits in 13 trisomy, 5% in normals). *4*, patterns in thenar and interdigital area 1 (occurs in 50% of hands). *5*, arch fibular (S pattern) with an *f* triradius and elongated fibular loop related to *e* triradius, *6*, tibial loop.

Fig. 13 (left). – Trisomy C. *1*, arches increased, 2, thenar pattern.

Fig. 14 (right). – Cat cry syndrome (deletion of chromosome no. 5). *1*, increased arches. *2*, thenar pattern. *3*, increased *atd* angle. *4*, simian line.

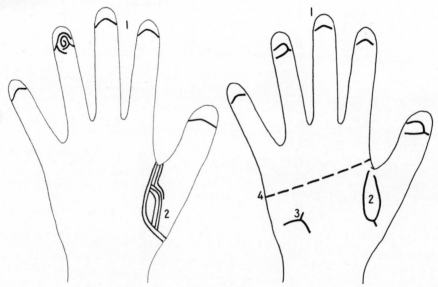

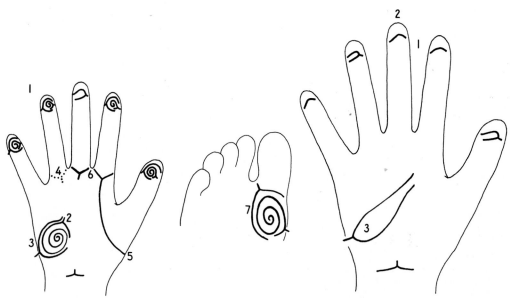

Fig. 15 (left). – The Turner syndrome (XO karyotype). *1*, increased whorls and large loops (ridge count in males increased: 166–178; general population of females: 126). *2*, increased *atd* angle (axial triradius at *t″*). *3*, large hypothenar pattern (occurs in 66% of patients, 12% of normals). *4*, missing *c* triradius. *5*, main line *A* exiting in thenar area. *6*, increased *ab* ridge count. *7*, frequent large whorls and distal loops.

Fig. 16 (right). – The Klinefelter syndrome (XXY karyotype). *1*, increased arches. *2*, mean total ridge count reduced to 114 (normal male: 145). *3*, hypothenar pattern (occurs in 36% of patients, 25% of normals).

Fig. 17. – Klinefelter variant (XXYY karyotype). *1*, increased arches (occurs in 17%; mean ridge count reduced to 90–106). *2*, increased *atd* angle. Axial triradius associated with either arch-carpal (*A*) or loop-radial (*B*) hypothenar pattern.

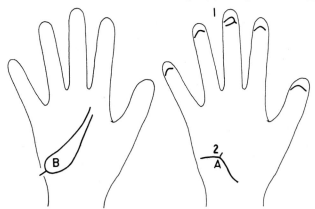

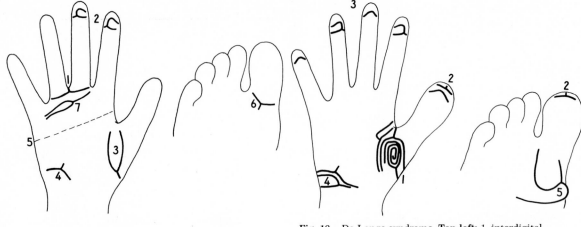

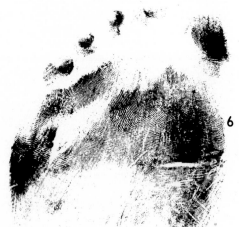

Fig. 18.—De Lange syndrome. **Top left:** *1*, interdigital triradius, usually in interspace 3. *2*, radial loops on digits 1 and 2. *3*, large loop in thenar area. *4*, increased *atd* angle (axial triradius at *t'*). *5*, simian line. *6*, open field with *e* triradius. *7*, transverse interdigital loop. **Bottom:** *6*, open field with *e* triradius.

Fig. 19 (**top right**).—Rubinstein-Taybi syndrome. *1*, large, complex thenar pattern. *2*, extra triradius at tip of thumb or great toe. *3*, reduced total ridge count. *4*, ulnar loop in hypothenar area associated with increased *atd* angle (axial triradius at *t'*). *5*, distal loop with *f* triradius displaced toward tibial border of foot and associated with loop opening toward fibular border.

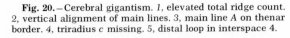

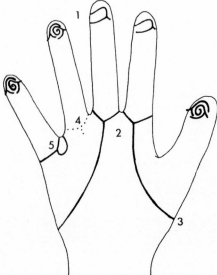

Fig. 20.—Cerebral gigantism. *1*, elevated total ridge count. *2*, vertical alignment of main lines. *3*, main line *A* on thenar border. *4*, triradius *c* missing. *5*, distal loop in interspace 4.

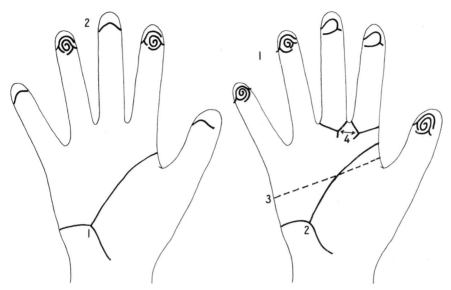

Fig. 21 (left). — Smith-Lemli-Opitz syndrome. *1*, elevated *t* triradius (*t'*). *2*, increased whorls and arches; decreased ulnar loops.
Fig. 22 (right). — Congenital rubella. *1*, increased whorls. *2*, elevated *t* triradius (*t'*). *3*, simian line. *4*, decreased *ab* ridge count.

Fig. 23 (left). — Laurence-Moon-Biedl syndrome. *1*, increased whorls. *2*, radial loops on digits 2 and 3. *3*, extra triradius in patients with polydactyly, sometimes associated with whorl pattern, in interspace 4. *4*, elevated *t* triradius (*t''*) generally in patients without polydactyly.
Fig. 24 (right). — Hypercalcemia. *1*, elevated *t* triradius (*t'*). *2*, small ulnar loop in hypothenar area. *3*, distal loop in interspace 4.

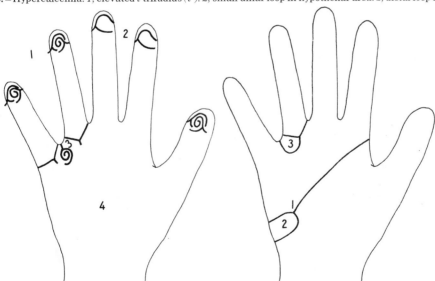

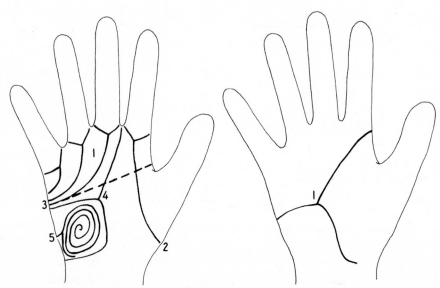

Fig. 25 (left). — Arthrogryposis (severe). *1*, vertical alignment of main lines. *2*, main line *A* present in thenar area. *3*, simian line on both palms. *4*, elevated *t* triradius (t″). *5*, whorls on hypothenar area.

Fig. 26 (right). — Ectodermal dysplasia. *1*, elevated *atd* angle over 60 degrees in affected individuals and mothers.

Fig. 27 (left). — G-deletion I. *1*, increased radial loops. *2*, triradius *c* missing.

Fig. 28 (right). — G-deletion II. *1*, increased digital whorls. *2*, elevated *atd* angle. *3*, complex hypothenar area. *4*, triradius *c* missing or disturbed.

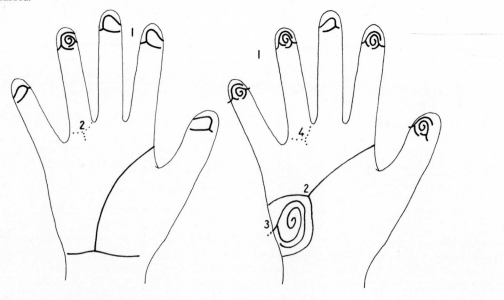

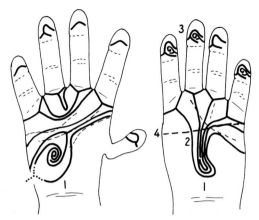

Fig. 29. – Holt-Oram syndrome. *1*, missing axial (*t*) triradius. *2*, distorted main lines due to missing axial triradius. *3*, increased digital whorl patterns. *4*, simian line.

REFERENCES

1. Elbualy, M. S., and Schindeler, J. D.: *Handbook of Clinical Dermatoglyphics* (Miami: University of Miami Press, 1971).
2. Galton, F.: *Finger Prints* (London: Macmillan Co., 1892).
3. Cummins, H., and Mildo, D.: *Finger Prints, Palms and Soles* (Philadelphia: Blakiston Co., 1943).
4. Holt, S. B.: *The Genetics of Dermal Ridges* (Springfield, Ill.: Charles C Thomas, Publisher, 1968).
5. Cummins, H.: Dermatoglyphic stigmata in mongoloid imbeciles, Anat. Rec. 64:11, 1936.
6. Pons, J.: Quantitative genetics of palmar dermatoglyphics, Am. J. Hum. Genet. 11:252, 1959.
7. Alter, M.: *Dermatoglyphics in Birth Defects*, Birth Defects: Original Article Series, Vol. V, No. 3 (New York: The National Foundation-March of Dimes, 1969).
8. Alter, M.: Variation in palmar creases, Am. J. Dis. Child. 120:424, 1970.
9. Penrose, L. S.: *Memorandum on Dermatoglyphic Nomenclature*, Birth Defects: Original Article Series, Vol. 4, No. 3 (New York: The National Foundation-March of Dimes, 1968).
10. Popich, G. A., and Smith, D. W.: The genesis and significance of digital and palmar hand creases: Preliminary report, J. Pediatr. 77:1017, 1970.
11. Penrose, L. S., and Smith, G. F.: Down's Anomaly (Boston: Little, Brown and Co., 1966).
12. Ford Walker, N.: The use of dermal configurations in the diagnosis of mongolism, J. Pediatr. 50:19, 1957.
13. Smith, G. F., Ridler, M. A. C., and Bat-Miriam, M.: Dermal patterns on the fingers and toes in mongolism, J. Ment. Def. Res. 10:105, 1966.
14. Smith, G. F.: Dermatoglyphic patterns on the fourth inter-digital area of the sole in Down's syndrome, J. Ment. Def. Res. 8:125, 1964.
15. Uchida, I. A., Patau, K., and Smith, D. W.: The dermal pattern of the new autosomal trisomy syndromes, Am. J. Dis. Child. 102:588, 1961.
16. Berg, J. M., Delhanty, J. D. A., Faunch, J. A., and Ridler, M. A. C.: Partial deletion of short arm of a chromosome of the 4–5 group (Denver) in an adult male, J. Ment. Defic. Res. 9:219, 1965.
17. Warburton, D., and Miller, O. J.: Dermatoglyphic features of patients with a partial short arm deletion of a B-group chromosome, Ann. Hum. Genet. 31:189, 1967.
18. Wertelecki, W., and Gerald, P.: Clinical and chromosomal studies of the 18q- syndrome, J. Pediatr. 78:1, 44, 1971.
19. Plato, C., Wertelecki, W., Gerald, P., and Niswander, J.: Dermatoglyphics in the 18q- syndrome, Pediatr. Res. 5:64, 1971.
20. Schindeler, J. D., and Warren, R. J.: Dermatoglyphics in the G- deletion syndromes, J. Ment. Def. Res. 17:149, 1973.
21. Hunter, H.: Finger and palm prints in chromatin-positive males, J. Med. Gent. 5:112, 1968.
22. Borgaonkar, D. W., Murdock, J. L., McKusick, V. A., Borkorof, S. P., and Money, J. W.: The YY Syndrome, Lancet 2: 461, 1968.
23. Bartlett, D. J., *et al.:* Chromosomes of male patients in a security prison, Nature 219:351, 1968.
24. Telfer, M. A., Baker, D., Clark, G. R., and Richardson, C. E.: Incidence of gross chromosomal errors among tall criminal American men, Science 159:1249, 1968.
25. Smith, G. F.: A study of the dermatoglyphs in the deLange syndrome, J. Ment. Def. Res. 10:241, 1966.
26. Milunsky, M. B., Cowie, V. A., and Donoghue, E.: Cerebral gigantism in childhood, Pediatrics 40:3, 395, 1967.
27. Chakanovskis, J. E., and Sutherland, G. R.: The Smith-Lemli-Opitz Syndrome in a profoundly retarded epileptic boy, J. Ment. Defic. Res. 15:153, 1971.
28. Achs, R., Harper, R., and Siegel, M.: Unusual dermatoglyphic findings associated with Rubella embryopathy, N. Eng. J. Med. 274:148, 1966.
29. Atasu, M., Balci, S., Tuncbilek, E., and Burhan, S.: Dermatoglyphic findings in Laurence-Moon-Biedl Syndrome, Lancet 2:98, 1973.
30. Bejar, R., Shear, C., Schindeler, J. D., Hernandez, F. A., and Smith, G. F.: Mental retardation, unusual facial appearance and abnormalities of the great vessels (Normocalcemic stage of idiopathic infantile hypercalcemia), J. Ment. Defic. Res. 14:16, 1970.
31. Hirsh, W., and Tonnis, D.: Die Bedentung de Dermatoglyphen and Hautfurchen fur die abgrenzung der Arthromyogrypposis multiplex congenita, Arch. Orthop. Unfallchir. 70:169, 1971.
32. Verbou, J.: Hypohidrotic (or anhidrotic) ectodermal dysplasia – an appraisal of diagnostic methods, Br. J. Dermatol. 83:341, 1970.
33. Verbou, Julian: Dermatoglyphs in leukemia, J. Med. Genet. 7:125, 1970.
34. Rosner, F., and Aberfeld, D. C.: Dermatoglyphics in the Holt-Oram Syndrome, Arch. Intern. Med. 126:1010, 1970.

Immunologic Problems

ATAXIA TELANGIECTASIA
Louis-Bar Syndrome, Syndrome of Syllaba-Henner

Cardinal Clinical Features

Progressive cerebellar ataxia, ocular and cutaneous telangiectasia, frequent sinopulmonary infections, decreased IgA and IgE, chromosomal breakage.

Clinical Picture

The syndrome was clearly described by Syllaba and Henner[1] in 1926. Louis-Bar[2] reported her patient in 1941. The major clinical manifestations of the syndrome have been summarized by Boder and Sedgwick[3, 4] on the basis of a sizable experience. Patients with ataxia telangiectasia appear normal at birth. The development of cerebellar ataxia usually is the first sign of the disorder. It tends to begin about the time that the child starts to walk and may be evident in a clumsiness and unsteadiness of gait (Fig. 1). Ataxia is progressive and may be evident on sitting and standing as well as on walking. Ultimately patients are confined to bed or a wheelchair. The speech is slurred or dysarthric. They drool constantly, and they have an expressionless, masklike face. Their disposition has been described as equable.[3] Nystagmus is common. These patients also may have involuntary movements, choreoathetosis, myoclonic jerks and intention tremors. Motor disability is progressive. Unusual apraxic eye movements may simulate ophthalmoplegia and usually are associated with blinking. The muscles become wasted and atrophic, particularly distally, and deep tendon reflexes are absent. A loss of vibration and position sense after 10–15 years points to posterior column abnormalities.[5]

The neuropathology includes atrophy of the cerebellar cortex, loss of Purkinje cells and demyelination of the posterior columns of the spinal cord. These patients are not retarded, but the IQ scores tend to drop as the ataxia progresses. Longitudinal growth is significantly retarded.

Telangiectases of the bulbar conjunctiva usually are manifest by 4–6 years of age. Rarely they may be present at birth.[5] Telangiectasia also occurs on the ears, in a butterfly distribution over the face and in the nasal septum. The dorsa of the hands and feet usually are involved, as are the antecubital and popliteal areas. The telangiectases are dilated venules. These and other skin changes increase on exposure to the sun.[4] The skin also may in time develop areas of atrophy, which give it a hidebound or sclerodermatous appearance. There are mottled patterns of hypo- and hyperpigmentation. Café-au-lait macules may resemble freckles or ephelides that follow actinic damage.[4] Depigmented spots also are seen. The skin is prone to develop cancers. The hair becomes gray prematurely.

Patients with this syndrome are unusually susceptible to infection. They may have chronic rhinorrhea. Recurrent infections of the lungs or sinuses may begin in infancy. Later they may develop chronic bronchitis. Cutaneous infections also are a problem. Most patients have impetigo,[4] and some have warts. Extensive candidiasis and coccidioidomycosis[4] have been reported. Eczema or seborrheic dermatitis is very commonly seen[4]; this relates to an immunodeficiency[6] that is a regular feature of the syndrome. The pattern of deficiency of circulating immunoglobulin is variable, but there usually is a deficiency of IgA; it may be undetectable or normal. IgG usually is normal, but occasional patients have low levels. IgE usually is low.[7] Secretory IgA also is deficient in patients in whom serum levels are low. Later they develop chronic bronchitis, bronchiectasis, clubbing of the fingers and pulmonary insufficiency. There also may be deficient delayed hypersensitivity. A decrease in peripheral small lymphocytes and lymphocyte depletion in lymph nodes suggest an absent or dysplastic thymus. Lateral roentgenograms of the neck may reveal little or no thymus or adenoids. Elevated platelet counts have been seen in some patients. Impaired rejection of a skin homograft may be present. A reduced lymphocyte transformation response to phytohemagglutinin is the rule.[5] Thymic dysplasia has been documented at autopsy in a number of cases.[3, 5, 6, 8] Plasma cells may be deficient in the bone marrow.[6]

Autoimmune phenomena are common in this disorder. There may be circulating antibodies against thyroid, smooth muscle and parietal cells. Endocrine disorders often are associated with ataxia telangiectasia.

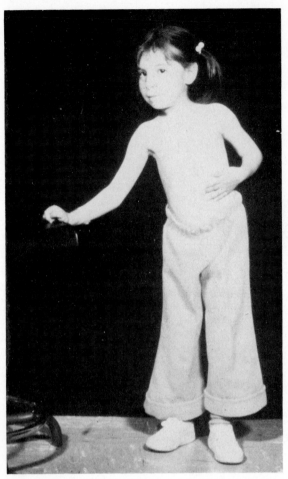

Fig. 1.—A. H. A patient with ataxia telangiectasia, illustrating the ataxia.

Diabetes mellitus has been reported in a large number of patients with this syndrome.[5] The diabetes was characterized by hyperglycemia, rarely with glycosuria and with a resistance to the development of ketosis. Abnormal liver function tests are commonly encountered.[5, 9] Biopsy of the liver may reveal mild fatty infiltration and chronic inflammation. Shortness of stature is common, but the growth hormone response to arginine is normal. Female patients with ataxia telangiectasia frequently lack ovarian tissue or have agenesis of the ovarian follicles. The ovaries may be absent, and there is a higher than average incidence of ovarian dysgerminoma. Males may have abnormal spermatogenesis or even testicular hypoplasia.

Patients with this syndrome have been found to have chromosomal breakage.[10, 11] In addition to gaps and breaks, there are rearrangements. It is thought that these chromosome changes predispose the patient to malignancy. Lymphoreticular malignancies are common and include the Hodgkin disease, reticu-

lum cell sarcoma, lymphoma and leukemia. Other tumors seen have been astrocytoma, cerebellar medulloblastoma, glioma and ovarian dysgerminoma. Adenocarcinoma of the stomach was reported in two sisters with the disorder.[12] Senile keratosis was discovered in a 23-year-old woman with ataxia telangiectasia. Multiple basal cell carcinomas were found in another patient.[4]

DIFFERENTIAL DIAGNOSIS

The differential diagnosis of ataxia telangiectasia includes the Bloom syndrome and the Fanconi anemia (see section on Thumb and Radial Dysplasias). All of these syndromes share chromosomal instability. There is consequent chromosomal breakage and associated malignancy, and retardation of growth. Hartnup disease is characterized by cerebellar ataxia, emotional instability and a pellagra-like, light-sensitive skin condition. In Hartnup disease there is a defect in the transport of amino acids, especially tryptophan in the intestines and kidneys. In the Cockayne syndrome there is short stature and intrauterine growth retardation, but patients also have pigmentary degeneration of the retina and optic atrophy. These patients may have photosensitive skin and progressive ataxia as well. The Rothmund-Thomson syndrome (see section on Dwarfism Syndromes with Premature Aging) and the deSanctis-Cacchione syndrome (see chapter on Xeroderma Pigmentosum in section on Neurocutaneous Diseases) have some features in common with ataxia telangiectasia.

Fig. 2.—R. S. (Case Report). An 8-year-old girl with ataxia telangiectasia, demonstrating conjunctival telangiectases. Also readily evident is the pinched, atrophic appearance around the nose. On closer inspection it is possible to see the clusters of telangiectases along the nose adjacent to the eyes. (Courtesy of Dr. Nancy Fawcett of the University of Miami, Florida.)

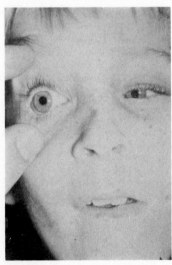

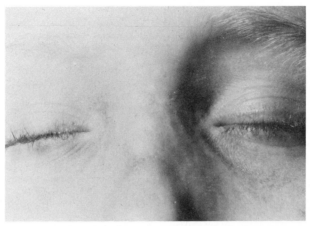

Fig. 3. – R. S. Scarring and atrophy of the skin over the bridge of the nose. Very fine telangiectases are seen along the side of the nose next to the eye. (Courtesy of Dr. J. A. McCrary of the University of Texas and reprinted from his *Pediatric Oculo-Neural Disease Case Studies* (Flushing, N. Y.: Medical Examination Co. In press).

CASE REPORT*

R. S., an 8½-year-old white girl, was admitted to the Clinical Research Unit of the University of Miami, Florida, because of progressive difficulty in walking. She was from Spanish Wells, a tiny island in the Bahamas, with a population of about 1,000 having a known propensity for inbred abnormalities.

The patient was the product of a second pregnancy of a 29-year-old mother. The first pregnancy resulted in a spontaneous miscarriage. This girl was born at home after 9 months of pregnancy, the birth weight being 8 lb, 1 oz. A small hypopigmented lesion was noted on the right leg at birth. She gained weight very slowly but was thought to be normal in development. She walked around the crib holding on at about 7 months and climbed out of the crib at 1 year. At 1 year of age it was noted that she had difficulty with gait and seemed to tire easily in the legs. She remained unsteady, tending to sway and fall, frequently sustaining facial trauma. She was able to walk alone between 2 and 3 years of age. At 4 years she was evaluated by a physician, who noted ataxia and telangiectasia. She had pneumonia, repeated bronchitis and a tonsillectomy and adenoidectomy at the age of 3 years.

Since about age 4 years, vascular changes had been noted in the conjunctivae, around the eyes, around the neck and in the popliteal areas. These continued to increase. She was able to walk to school, which she began at the age of 5½ years, but in subsequent years she had had progressively less ability to walk and more weakness. She began to encounter difficulty in writing and in speaking.

Her father died at the age of 33 years of an intracranial hemorrhage, possibly related to deep-sea diving.

Physical examination revealed a quiet, cooperative, frail, thin girl. Her height was 45¼ inches, and her weight 33 lb. Head circumference was 49.5 cm. She sat and walked with support but was unsteady and only attempted activities that she knew she could complete. She understood questions but responded in slow, slurred, hesitating speech. Her head slumped forward and the facies was mask-like; facial skin

was tight and similar to scleroderma. There was increased pigmentation around the eyes and small dilated vessels in the skin (Figs. 2 and 3). Small bright red blood vessels were apparent in the bulbar conjunctivae, in the pinnae, on the lips, around the neck, on the antecubital fossae, on the dorsa of the hands and on the popliteal fossae. There were areas of increased pigmentation on the medial aspects of the thighs. A hypopigmented area on the right lower leg had been present since birth. The patient was generally very thin and lacked muscle mass. She had blond hair, and the skin changes were definitely worse in areas exposed to the sun. The eyes showed slow nystagmoid movements and some inability of peripheral gaze. Telangiectatic lesions were present on the lids and conjunctivae. The vessels seemed to be more marked in the bulbar area and were very tortuous (Fig. 4). An alternate cover test revealed esotropia. She preferred to fix with the right eye. She had ataxic movements of the eyes and horizontal nystagmus. There was a notable absence of lymphoid tissue in the

Fig. 4. – R. S. Dilated tortuous vessels of the eye. (Courtesy of Dr. J. A. McCrary of the University of Texas and reprinted with permission from his *Pediatric Oculo-Neural Disease Case Studies*, (Flushing, N. Y.: Medical Examination Co. In press.)

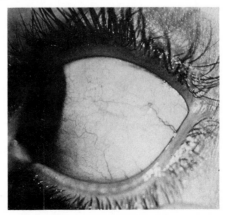

*We are indebted to Dr. Nancy Fawcett of the University of Miami, Florida, for the details of this patient.

pharynx, and there were numerous rhonchi and occasional wheezes bilaterally. The extremities were very thin. Gross truncal ataxia and decrease in deep tendon reflexes were also noted.

The hemoglobin was 14 gm/100 ml, RBC 5.10×10^6/cu mm and hematocrit 43. WBC was 8,000/cu mm with 71% segmented neutrophils, 1% staphylococci, 21% lymphocytes, 6% monocytes and 1% eosinophils. Skeletal survey revealed generalized demineralization of bone. There was a coxa valga deformity bilaterally. Bone age was 6–7 years. Roentgenograms of the chest revealed several confluent areas of increased density at the left base. The bone marrow was normocellular.

GENETICS

Ataxia telangiectasia is transmitted as an autosomal recessive. Several affected siblings with normal parents have been reported. The incidence of consanguinity is high in families with multiple affected offspring.

TREATMENT

Patients with this syndrome frequently die of respiratory failure or malignancy. Each should be managed appropriately. Repeated infusions of fresh plasma raise levels of IgA and may be clinically beneficial.[13]

REFERENCES

1. Syllaba, L., and Henner, K.: Contribution a l'indépendance de l'athétose double idiopathique et congénitale, Rev. Neurol. (Paris) 33:541, 1926.
2. Louis-Bar (Mme.): Sur un syndrome progressif comprenant des télangiectasies capillaries cutanées et conjonctivales symétriques, à disposition naevoïde et des troubles cérébelleux, Confina Neurol. 4:32, 1941.
3. Boder, E., and Sedgwick, R. P.: Ataxia-telangiectasia. A familial syndrome of progressive cerebellar ataxia, oculo-cutaneous telangiectasia and frequent pulmonary infection, Pediatrics 21:526, 1958.
4. Reed, W. B., Epstein, W. L., Boder, E., and Sedgwick, R.: Cutaneous manifestations of ataxia-telangiectasia, J.A.M.A. 195:746, 1966.
5. McFarlin, D. E., Strober, W., and Waldmann, T. A.: Ataxia-telangiectasia, Medicine 51:281, 1972.
6. Peterson, R. D. A., Kelly, W. D., and Good, R. A.: Ataxia-telangiectasia. Its association with a defective thymus, immunological-deficiency disease, and malignancy, Lancet 6:1189, 1964.
7. Polmar, S. H., Waldmann, T. A., Balestra, S. T., Jost, M. D., and Terry, W. D.: Immunoglobulin E in immunologic deficiency diseases. I. Relation of IgE and IgA to respiratory tract disease in isolated IgE deficiency, IgA deficiency, and ataxia telangiectasia, J. Clin. Invest. 51:326, 1972.
8. Peterson, R. D. A., Cooper, M. D., and Good, R. A.: Lymphoid tissue abnormalities associated with ataxia-telangiectasia, Am. J. Med. 41:342, 1966.
9. Schalch, D. S., McFarlin, D. E., and Barlow, M. H.: An unusual form of diabetes in ataxia-telangiectasia, N. Engl. J. Med. 282:1396, 1969.
10. German, J.: Oncogenic implications of chromosomal instability, Hospital Pract. 8:93, 1973.
11. Harden, D. G.: Ataxia telangiectasia: Cytogenetic and Cancer Aspects, in German, J. (ed.): *Chromosomes and Cancer* (New York: John Wiley & Sons, Inc., 1974), pp. 619–636.
12. Haerer, A. F., Jackson, J. F., and Evers, C. G.: Ataxia-telangiectasia with gastric adenocarcinoma, J.A.M.A. 210:1884, 1969.
13. Ammann, A. J., Good, R. A., Bier, D., and Fudenberg, H. H.: Long-term plasma infusions in a patient with ataxia-telangiectasia and deficient IgA and IgE, Pediatrics 44:672, 1969.
14. Goodman, W. N., Cooper, W. C., Kessler, G. B., Fischer, M. S., and Gardner, M. B.: Ataxia-telangiectasia. A report of two cases in siblings presenting a picture of progressive spinal muscular atrophy, Bull. Los Angeles Neurol. Soc. 34:1, 1969.

Craniosynostoses

APERT SYNDROME
Acrocephalosyndactyly

CARDINAL CLINICAL FEATURES

Acrocephaly due to synostosis of the coronal suture and syndactyly, which is symmetrical and involves all four extremities in a mitten- or stocking-like deformity.

CLINICAL PICTURE

The syndrome has been recognized at least since 1894 when Wheaton[1] reported two unrelated patients. The eponymic designation of the syndrome followed Apert's summary in 1906 of its characteristics in nine patients.[2]

Craniofacial Deformities

Patients with this syndrome have a very similar appearance, largely because of the changes induced in the head and face by asymmetrical growth in the presence of synostosis of the coronal suture (Fig. 1). The skull is shortened in anterior posterior diameter and lengthened vertically, producing the characteristic acrocephalic appearance. Obliteration of the cranial sutures occurs early, and coronal synostosis may be found at birth. The head is short and broad, the cranial vault high. There is frontal bossing. The occiput is flat. A deep transverse groove is sometimes seen above the supra-orbital ridges in the prominent forehead. Intracranial pressure is chronically increased, and digital markings become prominent roentgenographically.

The face usually looks flattened or pushed in. The bridge of the nose is depressed. The face may be asymmetrical. The maxillae often are hypoplastic and the mandible prominent, leading to a mild prognathism. There is ocular hypertelorism. The orbits are shallow and oblique laterally, giving rise to an antimongoloid slant and protrusion of the eyeballs. Strabismus is frequently marked. The nose is short and broad, often with a parrot-like configuration. The ears stand out from the head and appear large. The palate is high arched and narrow and the teeth crowded. Patients usually assume an open-mouthed appearance that is secondary to nasal obstruction. Cleft palate is present in some cases, especially posteriorly, but without an associated harelip.

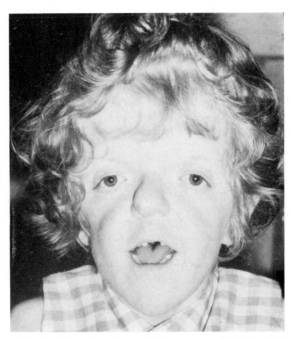

Fig. 1.—A female patient with the Apert syndrome. Craniofacial features are readily recognized in spite of her coiffure.

Skeletal Deformities

The syndactyly of this syndrome is the most marked encountered. It affects all four extremities symmetrically. The second, third and fourth fingers form a bony mass, a middigital mass, with a single common nail. The first and fifth digits are sometimes free and sometimes joined to this middigital mass, usually by soft tissues. This anomaly gives rise to a mitten-hand appearance (Figs. 2 and 3). Another way of looking at the hand[3] is as a spoon, the palm representing the concavity of the spoon. In the feet the second, third and fourth toes are joined by extensive soft tissue union. The first and fifth toes may be free or joined. Toenails usually are separate. The appearance of the feet has been described as sock foot (Fig. 4). A hallux varus deformity invariably is present.[4]

A variety of other anomalies may be seen. These in-

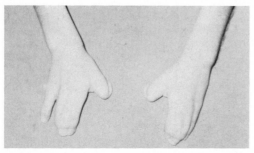

Fig. 2.–G. B. The hands illustrate the middigital mass and the mitten syndactyly.

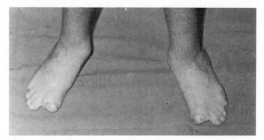

Fig. 4.–G. B. The syndactyly of the feet also was extreme. The great toes were broad.

clude contractures or ankylosis of major joints, especially the elbow, and synostosis of the radius. There may be bilateral absence of the radius and ulna. Winging of the scapula may be associated with deformity of the humeral head and glenoid. Diastasis of the symphysis pubis has been observed.

Bony fusion in the hands and feet is progressive.[5] It involves adjacent phalanges, often at their ends, and follows a definite pattern of increasing fusion of the tarsal and metatarsal bones. Bony outgrowths from the phalanges may represent a similar process. There is failure of normal tubulation of bones. Bone age often is retarded. Progressive fusion also is seen in the cervical spine. Pseudoarticulations develop in older patients between adjacent phalanges or metatarsals. An accessory bone of the foot, the os peroneum, seen in about 10% of normal individuals, is well developed.

Mental retardation is more common in this syndrome than in the general population. However, most children with the syndrome develop normally and have normal intelligence. The life span may be normal.

DIFFERENTIAL DIAGNOSIS

ACROCEPHALOPOLYSYNDACTYLY (ACPS).–It is clear that the ACPS syndromes are genetically distinct diseases from the Apert syndrome. There are a number of different phenotypic disorders.

Fig. 3.–G. B. Close-up of the hand.

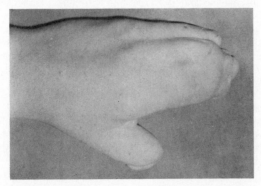

Among these are an autosomal dominant disorder reported by Noack[6] in a father and daughter. They had relatively mild acrocephaly and normal intelligence. Very broad great toes could be seen roentgenographically to contain two digits. The hands showed brachydactyly without syndactyly or polydactyly. This disorder has been listed by McKusick[7] as ACPS, type I.

The Carpenter syndrome (see chapter, this section) has been listed as ACPS, type II. It is thought to be transmitted as an autosomal recessive. It is characterized by oxycephaly with a triangular, pointed head and often with large areas of bony defect at the base of the triangle. Brachysyndactyly of the hands and preaxial polydactyly and syndactyly of the feet are characteristic. Patients also have unusual facies, with depressed nasal bridge and upturned nasal tip so that the nostrils face outward, as well as congenital heart disease and abdominal hernias.

A syndrome we believe to be distinct from that described by Carpenter was reported in one patient by Temtamy. The extremities were like those of Carpenter's, but the skull was acrocephalic in configuration, much like that in the Apert syndrome. In the chapter on the polydactyly-syndactyly syndrome with oxycephaly, we describe a patient with ACPS identical to that described by Temtamy, but with an additional anomaly around the knee as well as defects of the skin. This is known as ACPS, type III.

Norvig[8] also reported a distinct disorder in a mentally retarded brother and sister with mild acrocephaly, brachysyndactyly of the hands and polydactyly and syndactyly of the feet, along with displaced patellae and spina bifida occulta.

ACROCEPHALOSYNDACTYLY, TYPE II (APERT-CROUZON DISEASE OR VOGT CEPHALODACTYLY).–Vogt[9] described patients with hand and foot malformations identical to those in the Apert syndrome with the facial appearance of the Crouzon disease, in which there were extremely hypoplastic maxillae.

ACROCEPHALOSYNDACTYLY, TYPE III (WAARDENBURG TYPE).–Waardenburg[10] described a pedigree in which there were four generations of patients with acrocephaly, orbital and facial abnormalities and brachydactyly with soft tissue syndactyly.

ACROCEPHALOSYNDACTYLY, TYPE IV (PFEIFFER TYPE).–Pfeiffer[11] studied patients in three generations of a family in which there was autosomal dominant transmission of acrocephaly. The proximal phalanx of the thumbs and distal phalanx of the index fingers were dislocated radially. Double ossification centers were present in the proximal phalanx of each great toe. Radioulnar and radiohumeral synostoses were present.

It is not completely clear that these acrocephalosyndactylies are distinct from the Apert syndrome. There certainly is heterogeneity. This could represent variation in one disease or a number of diseases.

THE LAURENCE-MOON-BARDET-BIEDL SYNDROME. — In this syndrome oxycephaly is associated with mental retardation, retinitis pigmentosa, obesity and hypogonadism (see chapter, this section).

CASE REPORT

The patient, G. B., was a 17-year-old white male, a patient at Fairview State Hospital, Costa Mesa. He was the product of an 8-month gestation. Birth weight was 6 lb, 8 oz. The infant was found to have congenital anomalies of the skull and extremities characteristic of the Apert syndrome.

At 17 years of age the skull was observed to be short in the anteroposterior diameter and lengthened in the vertical diameter. The patient had a high-arched, narrow palate with crowded teeth and an open mouth. There was no cleft palate. He had prognathism and ocular hypertelorism. The eyes were prominent bilaterally because of the shallow orbits, and there was an antimongoloid slant. He had a left corneal opacity, and the left pupil did not react to light or accommodate.

The extremities illustrated the classic deformity of the disease. There was bilaterally symmetrical syndactyly. The metacarpals and phalanges were fused into a middigital mass involving the second, third and fourth digits. The first and fifth digits were joined to this mass. There was a single common nail. In the feet there was a complete sock-foot deformity of the toes. The toenails were separate.

The patient also had spastic paraplegia. He was markedly retarded, was blind and had no speech. He was able to sit in a wheel chair.

GENETICS

The genetics of the Apert syndrome has been studied by Blank.[12] It is transmitted as an autosomal dominant. Parent to child transmission has been noted. There is a marked parental age effect in which the incidence of sporadic cases increases with the age of the father. This would suggest a problem in the copying of genes at cell division. The numbers of divisions are few in female germ cells but relatively many in the male and continue throughout reproductive life. The modal ages of the parents in Blank's series were 36 (35–39) for the father and 32 (30–34) for the mother. These modal ages were 10 and 5 years older, respectively, than those of the general population.

TREATMENT

Surgery for the correction of the craniostenosis should be done early in life. This should be effective in management of the cosmetic problem and should prevent brain damage from craniosynostosis. There is controversy concerning effectiveness. Mental retardation may occur nevertheless. The syndactyly is correctable surgically.

REFERENCES

1. Wheaton, W. S.: Two specimens of congenital cranial deformity in infants associated with fusion of the fingers and toes, Trans. Pathol. Soc. (London) 45:238, 1894.
2. Apert, M. E.: De l'acrocéphalosyndactylie, Bull. Soc. Med. Hop. Paris, 23:1310, 1906.
3. Park, E. A., and Powers, G. F.: Acrocephaly and scaphocephaly with symmetrically distributed malformations of the extremities. A study of the so-called acrocephalosyndactylism, Am. J. Dis. Child. 20:235, 1920.
4. Dunn, F. H.: Apert's acrocephalosyndactylism, Radiology 78:738, 1962.
5. Schauerte, E. W., and St. Aubin, P. M.: Progressive synostosis in Apert's syndrome (acrocephalosyndactyly), Am. J. Roentgenol. Radium Ther. Nucl. Med. 97:67, 1966.
6. Noack, M.: Ein Beitrag zum Krankheitsbild der Akrozephalo-syndaktylie (Apert), Arch. Kinderheilkd. 160:168, 1959.
7. McKusick, V. A.: *Mendelian Inheritance in Man, Catalogs of Autosomal Dominant, Autosomal Recessive and X-linked Phenotypes* (2d ed.; Baltimore: Johns Hopkins Press, 1968), p. 5.
8. Norvig, J.: To Tilfaelde af Akrocephalosyndaktyli hos Soskende, Hospitalstidende 72:165, 1929.
9. Vogt, A., Dyskephalie (dysostosis craniofacialis maladie de Crouzon 1912) und eine neuartige Kombination dieser Krankheit mit Syndaktylie der 4 Extremitaen, Klin. Monatsbl. Augenheilkd. 90:441, 1933.
10. Waardenburg, P. J., Franceschetti, A., and Klein, O.: *Genetics and Ophthalmology* (Springfield, Ill.: Charles C Thomas, 1961). Vol. I, p. 320.
11. Pfeiffer, R. A.: Dominant erbliche Acrocephalosyndaktylie, Z. Kinderheilkd. 90:301, 1964.
12. Blank, C. E.: Apert's syndrome, a type of acrocephalosyndactyly. Observations on a British series of thirty-nine cases, Ann. Hum. Genet. 24:151, 1960.

POLYDACTYLY-SYNDACTYLY SYNDROME WITH OXYCEPHALY AND OTHER ANOMALIES
Polydactyly-Syndactyly Type II, Acrocephalopolysyndactyly Type III

CARDINAL CLINICAL FEATURES

Cranial synostosis, acrocephaly, unusual facies, polydactyly-syndactyly of the toes, bowed femur, hypoplastic tibiae, displacement of the fibulae, congenital heart disease, inguinal hernia.

Fig. 1.—D. B. (Case Report). Illustrated are the overall appearance, the acrocephaly and facial configuration, the short fat legs, and the special crutches with which the patient learned to walk.

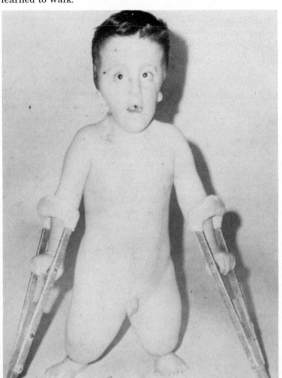

CLINICAL MANIFESTATIONS

A number of conditions present with polydactyly and syndactyly. Among the most striking of these is the Carpenter syndrome[1] (see chapter, this section). Involved patients have congenital synostosis of all of the cranial sutures. As a result, they develop acrocephaly and an unusual appearance similar to that of the patient illustrated in Figure 1. They also have brachysyndactyly of the fingers and polydactyly and syndactyly of the toes, as well as mental retardation, hypogonadism and obesity. These patients do not have long

Fig. 2.—D. B. Roentgenogram of the lower extremity illustrating hypoplasia of the tibia and deformity and displacement of the fibula.

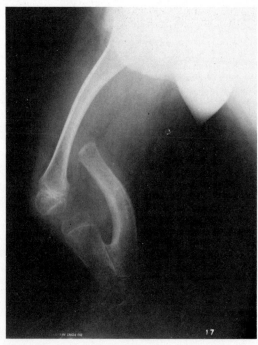

bone anomalies. A much more bizarre cranial anomaly usually characterizes the acrocephaly of the Apert syndrome,[2] and the syndactyly of that syndrome usually is very severe. Most patients are severely retarded mentally.

A family was recently reported by Eaton and McKusick[3] with what they called a polydactyly syndactyly syndrome. A father, both of his daughters and a granddaughter were involved. In addition to polydactyly and syndactyly involving the toes, they had striking malformations around the knees identical to those illustrated in Figure 2. The femora were short and the tibiae hypoplastic and malformed. The fibulae were subluxed posteriorly. These patients had no other anomalies, e.g., of the skull or elsewhere.

A syndrome has been described by Laurin et al.[4] in which polydactyly and syndactyly of the feet and mitten syndactyly of the hands and supernumerary fingers were associated with bilaterally absent radius and tibia, with duplication of the ulna and fibula. There was bilateral telescoping at the knee and elbow.

None of these patients had anomalies of the ear or heart, as described in the Case Report in this chapter. None had scars or clefts in the neck.

CASE REPORT

D. B., a boy, was the second child of a 35-year-old mother and a 40-year-old father. A 24-year-old sibling was well. He was the product of an 8-month gestation. Multiple anomalies were noted at birth, including an unusual head shape and absence of sutures. The first year was very difficult. The patient had noisy, difficult breathing and many serious cyanotic attacks, necessitating hospitalization. The parents kept oxygen in the home. He also was difficult to feed. Frequent respiratory infections and pneumonia were recurrent through the seventh year. A hernia was noted at 6 years. Development was slow. The child sat at 2 years, learned to walk in parallel bars at 4 years and at 5 learned to stand up and walk with the aid of special crutches made by his father (see Fig. 1). When evaluated, the child was 8 years old and attending a school for handicapped. His IQ was around 90.

Examination revealed the abnormality in the shape of the skull. The calvaria was large in circumference (58 cm) and the face small. The acrocephalic appearance was that of premature synostosis of all of the cranial sutures. He had bilateral patches of alopecia with atrophic skin above the ears, a coin-shaped lesion about 3 cm in diameter on the left and a somewhat larger, less regular lesion on the right. The ears were low set, the pinna deformed and there was an ear tag. The ear canal was normal on the left, with a normal tympanic membrane. On the right the canal was very small and very angulated so that it presented as a cleft, but it appeared otherwise normal. Hearing was normal. Over the left eyebrow was a small lesion resembling an inclusion cyst, which was present at birth and had been much larger. The orbits were shallow and the eyes protruding. The interorbital distance was small. The nose was elongated and beak-like and extended into a forehead that was flat and vertical in profile. He had a narrow, high-arched palate and thin tongue. The maxillae were shallow and the upper teeth crowded. The mandible ap-

peared large, overbiting the smaller maxilla. The patient had a short neck and low hairline. In the submental area linear scars present from birth radiated outward in triangular fashion, resembling remnants of clefts.

The nipples were widely spaced (16 cm); the chest circumference was 64 cm. There was a grade II systolic murmur and a fixed split second sound. There was a right inguinal hernia. The testes were undescended and the phallus somewhat small. The extremities were unusual. The arms were short, and in complete supination, the elbows displayed a wide carrying angle. The proximal interphalangeal joints of the fourth finger were fused bilaterally, and the fifth fingers were deviated radially at the distal phalanx. The thumbs took off proximally. A sixth digit had been removed from the right hand at birth. The legs were very short (see Fig. 1). There were linear dimples over the fibulae bilaterally. The feet were held in adduction, but there was no limitation of motion. The feet showed polysyndactyly with seven toes on the right and six toes on the left. On the right, the great toe appeared duplicated and there was syndactyly between these two toes. (Fig. 3). The next two toes appeared to be duplicated second toes and, again, there was syndactyly of soft tissues but no bony fusion. The next two toes appeared to be normal. The seventh toe was very broad and was syndactylous with its neighbor down to the terminal phalanx. On the left foot the great toe appeared to be duplicated, these toes being syndactylous. The next two toes appeared to be duplicated second toes and were syndactylous. The last two toes on this foot appeared normal.

The ECG revealed left axis deviation and incomplete right bundle branch block. There was left ventricular and atrial hypertrophy and right atrial hypertrophy. Cytogenetic analysis revealed a normal karyotype.

Roentgenographic examination revealed the multiple anomalies. There was absence of sutures, oxycephaly and a marked disproportion between the large cranium and small face. There were signs of chronic increased intracranial pressure, with a flattened, J-shaped sella turcica. The orbital roofs were elevated and the interorbital distance reduced. The mandible was hypoplastic but relatively larger than the maxilla and other facial bones. There was bilateral coxa valga and lateral bowing of both femora, which were short. The tibiae were hypoplastic and malformed bilaterally. The bowed right fibula overrode the right femur laterally and posteriorly. The left fibula overrode the left femur posteriorly. The metatarsals and tarsals were malformed.

COMMENT.—This patient presented with a unique syndrome of malformations. However, many of his clinical features also are seen in other syndromes.

Fig. 3.—D. B. Polydactyly and syndactyly of the foot.

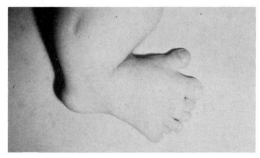

The extremities are virtually identical to those described by Eaton and McKusick[3] in four persons in three generations. The abnormality around the knee with hypoplastic tibiae and displacement of fibulae as well as the polydactyly and syndactyly of the toes are striking malformations. They were entirely similar in those described[3] and in our patient.

In many other ways the patient resembled those with the Carpenter syndrome as described by Temtamy[1]. Craniosynostosis with acrocephaly and similar facial appearance are seen in that syndrome along with brachysyndactyly of the fingers and polydactyly and syndactyly of the toes. A number of other features of that syndrome are lacking. Acrocephalopolysyndactyly (ACPS) differs from the Apert syndrome or acrocephalosyndactyly in the presence of polydactyly. Among the ACPS syndromes, Noack[5] described a dominant syndrome in which acrocephaly and syndactyly occurred with enlarged thumbs and enlarged, duplicated great toes. This type of preaxial polydactyly also is seen in the Carpenter syndrome.

GENETICS

In a sporadic case with multiple congenital anomalies, the genetics are not clear. The polydactyly syndactyly syndrome is transmitted as an autosomal dominant. The Carpenter syndrome is an autosomal recessive. We assume that this condition is a dominant one and that the patient seen represented a new mutation.

TREATMENT

The craniosynostosis usually would be corrected surgically early in life.

REFERENCES

1. Temtamy, S. A.: Carpenter's syndrome, acrocephalopolysyndactyly, J. Pediatr. 69:111, 1966.
2. Blank, C. E.: Apert's syndrome (a type of acrocephalosyndactyly). Observations on a British series of 39 cases, Ann. Hum. Genet. 24:151, 1960.
3. Eaton, O., and McKusick, V. A.: A Seemingly Unique Polydactyly-Syndactyly Syndrome in Four Persons in Three Generations, in *The First Conference on the Clinical Delineation of Birth Defects. Part III, Limb Malformations,* Birth Defects: Original Article Series (New York: The National Foundation-March of Dimes, 1969), pp. 221–225.
4. Laurin, C. A., Favreu, J. C., and Labelle, P.: Bilateral absence of the radius and tibia with bilateral reduplication of the ulna and fibula, J. Bone and Joint Surg. [Am.] 46:137, 1969.
5. Noack, M.: Ein Beitrag zum Krankheitsbild der Akrozephalosyndaktylie (Apert), Arch. Kinderheilkd. 160:168, 1959.
6. Sakati, N., Nyhan, W. L., and Tisdale, W. K.: A new syndrome with acrocephalopolysyndactyly, cardiac disease, and distinctive defects of the ear, skin, and lower limbs, J. Pediatr. 79:104, 1971.

CARPENTER SYNDROME
Acrocephalopolysyndactyly Type II, Temtamy Syndrome

Cardinal Clinical Features

Acrocephaly, brachydactyly and syndactyly of the hands; preaxial polydactyly and syndactyly of the feet.

Clinical Features

The term Carpenter syndrome was first coined by Temtamy[1] in a report of a patient with congenital synostosis of all cranial sutures, leading to acrocephaly, who also had polysyndactyly of the feet and brachysyndactyly of the hands. She analyzed the data in 12 published cases and reasoned that this was a distinct syndrome. The first report was that of Carpenter[2] in 1901. We[3] have thought that the syndrome might more appropriately be called the Temtamy syndrome, as the sisters originally reported by Carpenter did have cranial sutures and had other differences from Temtamy's patient and others reported since.[4] However, the eponym is that most commonly used for what may still be a family of syndromes.

Patients with the syndrome characteristically have premature closure of the sutures, particularly the sagittal sutures. This produces a tower skull, or acrocephaly. They have an unusual triangular appearance to the face. The nasal bridge is flat, and there may be a slight antimongoloid slant of the eyes. The inner canthi are laterally displaced and there may be epicanthal folds. Micrognathia and a high-arched palate are typical, and the ears are low set.

Brachysyndactyly of the digits of the hand varies in expression from mild soft tissue syndactyly of the third and fourth fingers to soft tissue syndactyly of these fingers that reaches almost to the nails. No bony fusions are seen. Some of the short fingers may have a single flexion crease. The fingers may be deformed— particularly an ulnar deviation of the radial fingers and a radial deviation of the fourth and fifth fingers. The thumbs may be rather broad and there often is clinodactyly of the fifth finger.

All of the patients described have had preaxial polydactyly of the toes along with an impressive degree of syndactyly. The great toe or the second toe may be duplicated, and usually the medial toes are medially deviated. Associated anomalies may include spina bifida occulta, genu valgum, metatarsus varus and coxa valga. Lateral displacement of the patellae has been observed. Patients may have congenital heart disease. Hypogonadism in the male and obesity has been reported. Mental retardation may of course be present. Funduscopic examination may reveal papilledema or optic atrophy. There is no retinitis pigmentosa.

Roentgenographic features include craniosynostosis and hypertelorism, with shallow orbits. There is brachymesophalangy of the digits of the hands and broad or duplicated distal or proximal phalanges of the thumb. The extra metatarsal bones may be hypoplastic and may or may not be fused to the neighboring metatarsal. There is flaring of the iliac wings and there may be mild acetabular dysplasia.

Differential Diagnosis

The acrocephalopolysyndactylies (ACPS) are currently divided into at least three types. The Carpenter syndrome is known as ACPS type II. Type I is the syndrome described by Noack,[5] in which mild acrocephaly with normal mentality was associated with very broad great toes, syndactyly of the feet and brachydactyly with or without syndactyly of the hands. Noack reported a 43-year-old father and his 11-month-old daughter who had acrocephaly, brachydactyly, an enlarged thumb and preaxial polysyndactyly of the feet with duplication of the great toes. In the Carpenter syndrome, patients have regularly been mentally retarded, and inheritance is recessive. In ACPS type III,[3, 4] (see chapter on the polydactyly-syndactyly syndrome with oxycephaly, this section) cranial synostosis, brachydactyly of the hand and polysyndactyly of the feet were associated with an unusual malformation around the knees. This case was sporadic and the parents normal, but increased paternal age at the time of birth suggested that it is dominant.

The Apert syndrome also is characterized by acrocephaly and syndactyly (see chapter, this section). It

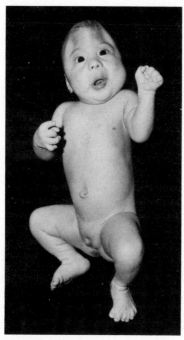

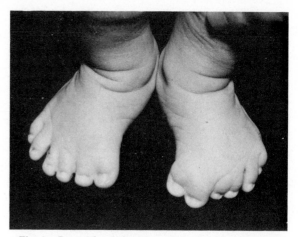

Fig. 3.—Preaxial polydactyly and syndactyly of the toes. (Courtesy of Drs. V. M. Der Kaloustian, A. A. Sinno and S. I. Nassar and reprinted with permission of Am. J. Dis. Child.[4])

Fig. 1.—R. S. (Case Report). A 2-month-old infant with the Carpenter syndrome. The acrocephaly and unusual facies are apparent, as well as the brachydactyly of the hand and preaxial polysyndactyly of the feet. (Courtesy of Drs. V. M. Der Kaloustian, A. A. Sinno and S. I. Nassar and reprinted with permission of Am. J. Dis. Child.[4])

Fig. 2.—R. S. The forehead was very high. The malformations of the hand illustrated include brachydactyly and syndactyly of the third and fourth fingers, ulnar deviation of the third finger and radial deviation of the fourth finger. (Courtesy of Drs. V. M. Der Kaloustian, A. A. Sinno and S. I. Nassar and reprinted with permission of Am. J. Dis. Child.[4])

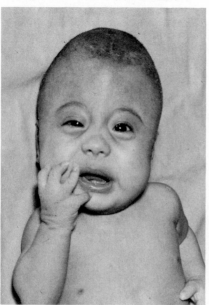

is a genetic disorder transmitted as an autosomal dominant. These patients have severe syndactyly of the hands and feet with mitten hands and sock feet. They do not have polydactyly. Patients with the Laurence-Moon-Biedl syndrome have by contrast postaxial polydactyly of the feet. They also have postaxial polydactyly of the hands, as well as obesity, hypogonadism, mental retardation and retinitis pigmentosa.

CASE REPORT*

R. S. was a 2-month-old Arab Palestinian female, the product of a normal pregnancy and delivery. She was the second child of healthy young parents who were first cousins. There was no history of exposure to illness, x-irradiation or drugs during pregnancy. The first child was born prematurely and died on the fifth day of life, but no anomalies were noted.

Physical examination revealed a skull with an increased vertical diameter and a triangular appearance of the face and head (Figs. 1 and 2). There was a prominent median ridge of bone. She had hypertelorism, a slight antimongoloid slant, and a flat nasal bridge with lateral displacement of inner canthi. The cheeks were broad and the ears low set. There was a high-arched palate and a hypoplastic mandible. The hands were short and stubby. There was partial syndactyly of the third and fourth fingers. There was ulnar deviation of the distal phalanx of the third finger, radial deviation of the distal phalanx of the fourth finger and clinodactyly of the fifth finger. The patient had preaxial polysyndactyly of both feet (Fig. 3). There was no retinitis pigmentosa and no heart disease.

X-ray examination revealed closed sagittal sutures, partially closed coronal sutures and shallow orbits

*We are grateful to Drs. V. M. Der Kaloustian, A. A. Sinno and S. I. Nassar of the American University of Beirut in Lebanon for details of this Case Report and to Am. J. Dis. Child.[4] for permission to reprint.

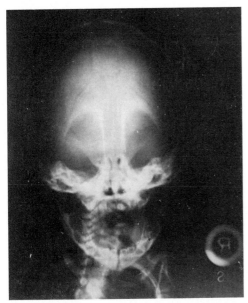

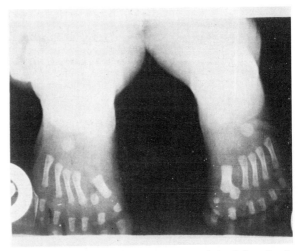

Fig. 5.—Roentgenogram of the feet revealed polydactyly of the great toes with hypoplastic extra metatarsals. (Courtesy of Drs. V. M. Der Kaloustian, A. A. Sinno and S. I. Nassar and reprinted with permission of Am. J. Dis. Child.[4])

Fig. 4.—Roentgenogram of the skull illustrates the cranio-synostosis and unusual orbits. (Courtesy of Drs. V. M. Der Kaloustian, A. A. Sinno and S. I. Nassar and reprinted with permission of Am. J. Dis. Child.[4])

(Fig. 4). In the pelvis, flaring of the iliac wings was prominent (Fig. 5). There were duplication of the great toe, metatarsus varus and incomplete formation of the extra metatarsal; this was fused on the left with the adjacent metatarsal (Fig. 6). The patient underwent a craniectomy to create open sutures and remove the ridge of bony deformity of the forehead.

GENETICS

The Carpenter syndrome is a genetic disorder transmitted by an autosomal recessive gene. Consanguinity is high among parents.

TREATMENT

Early correction of the craniosynostosis may prevent mental retardation as well as improve the appearance of the infant. Surgical repair of the polydactyly and syndactyly may have cosmetic value. Removal of the extra toe may allow the patient to wear regular shoes.

REFERENCES

1. Temtamy, S. A.: Carpenter's syndrome: Acrocephalopoly-syndactyly. An autosomal recessive syndrome, J. Pediatr. 69:111, 1966.
2. Carpenter, G.: Two sisters showing malformations of the skull and other congenital abnormalities, Rep. Soc. Study Dis. Child. 1:110, 1901.

Fig. 6.—Roentgenogram of the pelvis. Flaring of the iliac wings was prominent. (Courtesy of Drs. V. M. Der Kaloustian, A. A. Sinno and S. I. Nassar and reprinted with permission of Am. J. Dis. Child.[4])

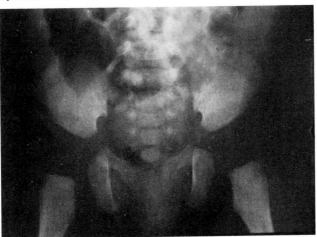

3. Sakati, N., Nyhan, W. L., and Tisdale, W. K.: A new syndrome with acrocephalopolysyndactyly, cardiac disease, and distinctive defects of the ear, skin and lower limbs, J. Pediatr. 79:104, 1971.
4. Der Kaloustian, V. M., Sinnó A. A., and Nassar, S. I.: Acrocephalopolysyndactyly type II Carpenter's syndrome, Am. J. Dis. Child. 124:716, 1972.
5. Noack, M.: Ein Beitrag zum Krankheitsbild der Akrozephalosyndaktylie, Arch. Kinderheilkd. 160:168, 1959.
6. Palcios, E., and Schimke, R. N.: Craniosynostosis – syndac-
tylism, Am. J. Roentgenol. Radium Ther. Nucl. Med. 106: 144, 1969.
7. Blank, C. E.: Apert's syndrome – type of acrocephalosyndactyly observation on a British series of 39 cases, Ann. Hum. Genet. 24:151, 1960.
8. Laurence, J. Z., and Moon, R. C.: Four cases of retinitis pigmentosa occurring in the same family and accompanied by general imperfection of development, Ophthalmol. Rev. 2:32, 1865.

CROUZON DISEASE
Craniofacial Dysostosis

CARDINAL CLINICAL FEATURES

Craniosynostosis, maxillary hypoplasia, hypertelorism, shallow orbits, proptosis of the eyes.

CLINICAL PICTURE

The syndrome was first described by Crouzon[1] in 1912 in a report of a 29-year-old mother and her 3-year-old son with an unusual combination of malformations of the skull and face. The patients have been described as having a frog-like appearance. The skull defect is a consequence of craniosynostosis. Premature fusion of the coronal suture is characteristic. The sagittal and lambdoidal also may be involved. The shape of the head depends on which sutures close prematurely, and when.[2] Oxycephaly results from clo-

Fig. 1.—A patient with the Crouzon disease. She had a craniofacial dysostosis with an increased transverse diameter of the skull, maxillary hypoplasia and shallow orbits with proptosis of the eyes. She had a marked antimongoloid slant. (Courtesy of Drs. Douglas Cunningham and Geoffrey MacPherson of the U.S. Naval Hospital, San Diego.)

Fig. 2.—Same patient as shown in Figure 1. In profile the skull could be seen to be shortened in the anteroposterior diameter. Her deafness is illustrated by the hearing aid. (Courtesy of Drs. Douglas Cunningham and Geoffrey MacPherson of the U.S. Naval Hospital, San Diego.)

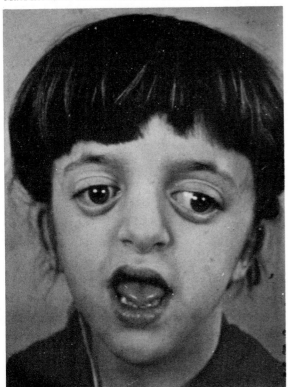

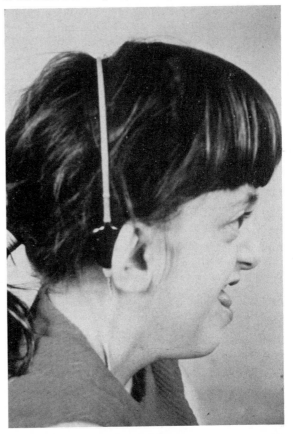

sure of all of the cranial sutures. There is regularly an increase in the transverse diameter of the skull and shortening of the anterior posterior diameter of the skull, leading to brachycephaly (Figs. 1 and 2). Scaphocephaly or trigonocephaly was the appearance in the original patients of Crouzon[1] in whom there was a large bony lump with a truncated top in the frontal bone. A hard protrusion in the region of the anterior fontanel is common, as is a palpable ridge over the sagittal suture. The occiput usually is flat. What has been called the Kleeblattschadel syndrome, or more appropriately the Kleeblattschadel anomaly, or cloverleaf skull may be seen in the Crouzon disease.[3] Increased intracranial pressure may result from premature closure of the sutures. This may lead to papilledema, sixth nerve palsy and strabismus, compression of the optic nerve, optic atrophy and blindness. Intellectual development may be impaired, and hydrocephalus has been reported.

The characteristic facial defect is hypoplasia of the maxilla (see Fig. 1). The zygomatic arches also are usually hypoplastic, and these changes lead to midfacial hypoplasia, shortness of the upper lip and apparent mandibular prognathism. A high-arched palate is common. The roofs of the orbits are shortened as a result of the shortened anterior fossa caused by coronal synostosis, and the floors of the orbits are shortened as a consequence of the maxillary hypoplasia. Consequently, there are very shallow orbits and the eyes protrude (see Fig. 1). In addition to the appearance of exophthalmos, there is regular hypertelorism and occasionally an antimongoloid slant. There may be a divergent strabismus.[4] Some patients have a beaked, parrot-like nose.

Occasional defects seen in the syndrome are septal displacement, which may produce nasal obstruction. Choanal atresia has been described. Oral anomalies occasionally seen include partial anodontia, crowding of the teeth and unerupted teeth. Cor pulmonale has been reported[5] in the Crouzon disease and interpreted to be a result of narrowing of the upper airway.

Deafness is sometimes part of the syndrome. There may be bilateral atresia of the auditory canals, as well as various anomalies of the middle ear. The ears may be low set, and the lobes may be abnormally large. Ankylosis of the elbow has been reported. Broad thumbs were described in one patient who also had partial syndactyly of the third and fourth toes.[2]

Roentgenographic findings include premature closure of the sutures, hypoplasia of the maxilla and unerupted teeth. There may be increased digital markings in the skull.

The syndrome should be differentiated from the Apert syndrome, or acrocephalosyndactyly (see chapter, this section), in which craniosynostosis is associated with mitten syndactyly of the hands and feet and ankylosis of joints, and from the ACPS syndromes (see chapters on the Carpenter syndrome and on the poly-

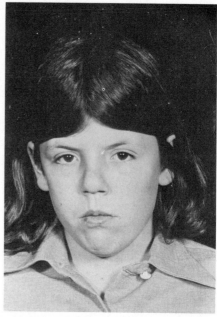

Fig. 3. – D. R. W. (Case Report). A 10-year-old girl with the Crouzon disease. She was treated surgically for coronal synostosis at 1 month of age and at 5 years of age for sagittal craniosynostosis. (Courtesy of Dr. Paul Schultz of the University of California, San Diego.)

Fig. 4. – D. L. W. The mother of D. R. W. She, too, had had craniosynostosis. (Courtesy Dr. Paul Schultz of the University of California, San Diego.)

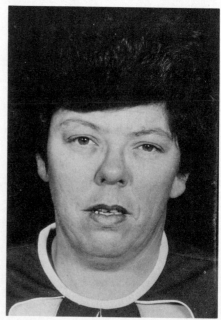

Fig. 5.—A. H. Grandmother of D. R. W., who also had had craniosynostosis. The brachycephaly was prominent. (Courtesy of Dr. Paul Schultz of the University of California, San Diego.)

dactyly syndactyly syndrome with oxycephaly, this section).

CASE REPORT*

D. R. W., a 10-year-old girl, was evaluated in the Pediatric Neurology Clinic at University Hospital because of headaches. The patient was born uneventfully, but brachycephaly was apparent in early infancy. A craniectomy was performed at about 6 weeks of age for coronal craniosynostosis. The patient was readmitted to hospital at 5½ years of age, and another craniectomy was performed for craniosynostosis of the sagittal suture. The patient was admitted at 6 years of age for what was thought to be a cerebral concussion. At 10 years of age she was admitted for 3 days because of headaches, dizziness and vomiting. An EEG 1 year previously was said to have been markedly abnormal, and the patient was started on treatment with Dilantin.

The family history was of considerable interest in that the patient's mother had right renal agenesis and some difficulty in hearing. The mother's first pregnancy resulted in a miscarriage of dizygotic twins in the fourth month of gestation. The

*We are indebted to Dr. Paul Schultz of the University of California San Diego for the history of this patient and for permission to report on her.

patient's only sibling was in good health. Detailed family history was otherwise negative.

Physical examination showed an alert, cooperative girl with normal intelligence. There was brachycephaly, and the head circumference was 51½ cm. There was no proptosis. The pupils were equal, round and reactive to light. Funduscopic examination was normal. Horizontal and vertical nystagmus was present, possibly related to Dilantin therapy. She had a high-arched palate. The patient's physiognomy was that of the Crouzon disease (Fig. 3) and was strikingly similar to that of her mother (Fig. 4) and maternal grandmother (Fig. 5).

GENETICS

The syndrome is transmitted in an autosomal dominant pattern of inheritance. The degree of expression may be variable. It is possible that the Crouzon disease is genetically heterogeneous, but it is clear that variation in expression may be seen within a single kindred.[6] Sporadic cases are the result of new mutation. Autosomal recessive forms of craniosynostosis may occur.[7]

TREATMENT

Early craniectomy is recommended and may have to be repeated during the period of brain growth.[2] Orbital decompression also may be helpful. Cosmetic facial surgery may be of considerable esthetic benefit.[8]

REFERENCES

1. Crouzon, O.: Dysostose cranio-facial héréditaire, Bull. Soc. Med. Hop. Paris 33:545, 1912.
2. Kushner, A., Alexander, E., Jr., David, C. H., Jr., Kelly, D. L., Jr., and Kushner, A. H.: Crouzon's disease (craniofacial dysostosis): Modern diagnosis and treatment, J. Neurosurg. 37:434, 1972.
3. Hall, B. D., Smith, D. W., and Shiller, J. G.: Kleeblattschadel (clover leaf) syndrome: Severe form of Crouzon's disease, J. Pediatr. 80:526, 1972.
4. Flippen, J. H., Jr.: Cranio-facial dysostosis of Crouzon: Report of a case in which the malformation occurred in four generations, Pediatrics 5:90, 1950.
5. Don, N., and Siggers, D. C.: Cor pulmonale in Crouzon's disease, Arch. Dis. Child. 46:394, 1971.
6. Shiller, J. G.: Craniofacial dysostosis of Crouzon: A case report and pedigree with emphasis on heredity, Pediatrics 23:107, 1959.
7. Cross, H. E., and Opitz, J. M.: Craniosynostosis in the Amish, J. Pediatr. 75:1037, 1969.
8. Fleming, J. P.: Improvement of the face in Crouzon's disease by conservative operations, Plast. Reconstr. Surg. 47:560, 1971.

LAURENCE-MOON-BIEDL SYNDROME
Laurence-Moon-Bardet-Biedl Syndrome

Cardinal Clinical Features

Polydactyly, mental retardation, obesity, retinitis pigmentosa, hypogenitalism.

Clinical Picture

The syndrome was first described by Laurence and Moon in 1866[1] in a report of four siblings with retinitis pigmentosa, mental retardation, obesity, shortness of stature and, in the three males, hypogonadism. Bardet[2] in 1920 and Biedl[3] in 1922 added polydactyly to the entity and thus established the complete syndrome.

Retinitis pigmentosa is the most incapacitating aspect of the syndrome. It is not recognized at birth but appears later, and it is progressive. One of the earliest symptoms is night blindness. Nystagmus is another. The patient may complain of decreased visual acuity, especially at night or in artificial light. He may look to one side of an object rather than straight at it. Often the problem is appreciated for the first time when the child goes to school. Blindness has developed by the time the patient is 30 years of age. The evolution of blindness in this syndrome is more rapid than in simple retinitis pigmentosa.[4] Optic atrophy is regularly observed at this stage. The pigment is deep black in color and is widely scattered, particularly in the peripheral fundus. It may occur in flakes, streaks or spots, with lines extending from them like bone corpuscles. Most commonly, pigment is finely scattered. Macular degeneration may be seen. Tapetoretinal degeneration has been seen in several patients without pigmentary retinopathy. The electroretinogram may be very useful in establishing the presence of this retinal defect.[4] Other ocular anomalies that may be associated with the syndrome include microphthalmia, anophthalmia, colobomas of the choroid or iris, and aniridia.

The mental defect is seen in most patients. It may be severe but usually is not. Most patients are thought of as backward, or subnormal, unable to learn in school. Doubtless the visual defect contributes to this. Some patients have been described in whom intelligence is normal. Obesity, a rather constant feature, is recognized in infancy and is most marked on the trunk, abdomen and thighs (truncal and rhizomelic). It increases with age, especially after puberty. In some patients obesity is the presenting complaint. The weight of these patients is generally well above the mean for their age.[4] By the time they have reached the age of puberty, most patients are well below the average height for their age, and this makes the adiposity even more impressive.

Polydactyly is a characteristic component of the syndrome. It is postaxial polydactyly, occurring on the ulnar or fibular sides of the extremities. Polydactyly may be seen in one or more extremities. Sometimes the accessory digit is hypoplastic. The most frequent expression is a supernumerary sixth digit. Occasionally, syndactyly is associated, usually confined to the second and third toes, or to the extra digit and its neighbor. Brachydactyly also may be present.

Hypogenitalism has been more commonly reported in males than in females. However, in children, genital abnormalities are more readily recognized in males, and most reported females have been inadequately studied endocrinologically. Male patients have been described to have a small penis, hypoplastic or bifid scrotum and undescended or hypoplastic testes. No male patient is recorded as having become a father. Hypogenitalism is less clearly delineated in the female, and some women with certain features of the syndrome have borne normal children.

Associated cerebral defects may include ataxia, chorea or spasticity. Other associated anomalies described have been imperforate anus and congenital heart disease, including patent ductus arteriosus, pulmonic stenosis, hypoplasia of the aorta, ventricular septal defect and atrial septal defect.

Deafness has been reported in a number of cases, but there is some question as to whether these patients may represent another syndrome in which retinitis pigmentosa and deafness are associated. Renal anomalies are clearly a part of the syndrome. They include renal hypoplasia, hydronephrosis and cystic kidneys. Multiple structural anomalies of the urinary tract have been described.[5]

DIFFERENTIAL DIAGNOSIS

The syndrome should probably be differentiated from the Biemond syndrome,[4] in which coloboma of the iris is associated with obesity, hypogenitalism, polydactyly and mental retardation. Patients with this syndrome do not have retinitis pigmentosa. The Carpenter or Temtamy syndrome (see chapter, this section) is characterized by acrocephaly, polysyndactyly, mental retardation and hypogenitalism. Again, there is no retinitis pigmentosa.

CASE REPORTS*

CASE 1.—W. B. was a 13-year-old boy when he was admitted to the Clinical Research Unit of the University of Miami, Florida, for evaluation of marked obesity. He also was known to have impaired intellectual ability. His IQ was between 70 and 80. He had reduced visual acuity, uncorrectable past 20/50 in each eye. He had had two operations for esotropia. Hearing had been formally tested and found to be normal.

His weight at birth was 8 lb, 8 oz. There was an extra digit on each hand and each foot that were removed during the first week of life. The patient was obese as an infant and this continued. All weights recorded were above the 90th percentile. The height was in about the 84th percentile through the first

Fig. 1.—W. B. (Case 1). A 13-year-old boy with the Laurence-Moon-Bardet-Biedl syndrome. His obesity and hypogonadism are evident. (Courtesy of Drs. Carol Shear and Douglas Sandberg of the University of Miami, Florida.)

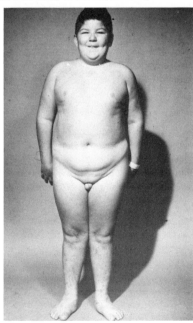

*We are grateful to Drs. Carol Shear and Douglas Sandberg of the University of Miami, Florida, for providing us with the details from the clinical records of these patients.

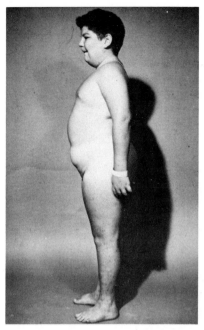

Fig. 2.—W. B. The obesity is particularly evident in profile. (Courtesy of Drs. Carol Shear and Douglas Sandberg of the University of Miami, Florida.)

year, falling to the third percentile by 13 years. The penis and testes were noted to be small and there was no pubic or axillary hair.

Physical examination revealed a markedly obese white male (Figs. 1 and 2). The weight of 137 lb, 7 oz was in the 95th percentile. The height was 55 inches, well below the third percentile, and height age was 9. The head circumference was 53 cm, which is within 2 SD of the mean. The ratio of upper to lower segments was 1.2, which was high for the boy's age. There was a mild bilateral esotropia, as well as a distinct, abnormal, stippled retinal pigmentation most marked peripherally. There was a moderately high palatal arch. The genitalia were small (see Fig. 1): the testes measured 1 × 0.5 cm, and

Fig. 3.—W. B. The great toe seemed somewhat broad. The feet were short, the toes particularly so. (Courtesy of Drs. Carol Shear and Douglas Sandberg of the University of Miami, Florida.)

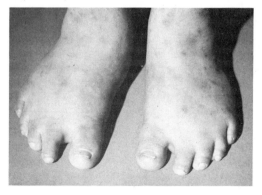

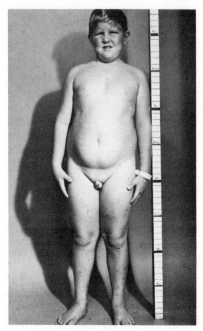

Fig. 4. – D. B. (Case 2). The 8-year-old brother of W. B. His obesity and hypogonadism also are evident. (Courtesy of Drs. Carol Shear and Douglas Sandberg of the University of Miami, Florida.)

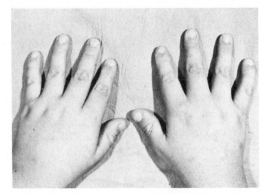

Fig. 5. – D. B. The hands appeared short, and the distal phalanges were particularly small. (Courtesy of Drs. Carol Shear and Douglas Sandberg of the University of Miami, Florida.)

the penis 1.5 × 0.5 inches. There was no pubic or axillary hair. There were bilateral scars on the lateral aspects of his hands and feet, which were short and had particularly short distal phalanges (Fig. 3).

An intravenous pyelogram revealed some dilatation of the calices. A renal scan was negative. The urine could not be concentrated further than 1.014, and an Addis count revealed an increased number of casts, but no other abnormalities. The bone age was 11 years. He had a normal response to Metopirone and no urinary gonadotropins.

CASE 2. – D. B. was the 8-year-old brother of W. B. He too had decreased vision, uncorrectable past 20/50. His IQ was 70 and he was attending a special class in school. He had always been obese.

Physical examination revealed a somewhat obese white male (Fig. 4). His weight was 103 lb, which was well above the 97th percentile for his age and at the 50th percentile for a boy of 14 years. The height was 53.75 inches, which was at the 84th percentile. The ratio of upper to lower segments was 1.15, which was higher than normal. The head circumference was 53.3 cm. The boy had clear-cut retinitis pigmentosa, which appeared to be at an earlier stage than that of the brother. The genitalia measured approximately the same as those of the brother. The hands and feet were short and the distal phalanges quite short (Fig. 5).

An intravenous pyelogram showed some clubbing of the calices.

GENETICS

The disorder as described in the literature may well be heterogeneous. Nevertheless, it is clear that the classic syndrome is inherited in an autosomal recessive pattern. Several involved children in families with normal parents have been described repeatedly.[6, 7]

Furthermore, there is high incidence of parental consanguinity among reported families.[4, 8]

TREATMENT

None.

REFERENCES

1. Laurence, J. Z., and Moon, R. C.: Four cases of "retinitis pigmentosa" occurring in the same family and accompanied by general imperfections of development, Ophthalmol. Rev. (London) 2:32, 1866.
2. Bardet, G.: Sur un syndrome d'obésité congénitale avec polydactylie et retinité pigmentaire, Paris, 1920.
3. Biedl, A.: Ein Geschwisterpaarmit adiposo-genitaler Dystrophie, Dtsch. Med. Wochnschr. 48:1630, 1922.
4. Klein, D., and Ammann, F.: The syndrome Laurence-Moon-Bardet-Biedl and allied diseases in Switzerland, J. Neurol. Sci. 9:479, 1969.
5. Nadjmi, B., Flanagan, M. J., and Christian, J. R.: Laurence-Moon-Biedl syndrome, Am. J. Dis. Child. 117:352, 1969.
6. Blume, L. J., and Kniker, W. J.: Laurence-Moon-Bardet-Biedl syndrome. Review of the literature and report of 5 cases including a family group with three affected males, Tex. Rep. Biol. Med. 17:391, 1959.
7. Walsh, F. B., and Hoyt, W. F.: The Laurence-Moon-Biedl Syndrome, in Clinical Neuro-Ophthalmology, (3d ed.; Baltimore: Williams & Wilkins Co., 1969), pp. 898 – 900.
8. Beil, J.: The Laurence-Moon Syndrome, in Penrose, L. S. (ed.): Treasury of Human Inheritance (London: Cambridge University Press, 1958), Vol. 5, pp. 51 – 69.
9. Bauman, M. L., and Hogan, G. R.: Laurence-Moon-Biedl syndrome. Report of two unrelated children less than three years of age, Am. J. Dis. Child. 126:119, 1973.
10. Reinfrank, R. F., and Nichols, F. L.: Hypogonadotrophic hypogonadism in the Laurence-Moon syndrome, J. Clin. Endocrinol. Metab. 24:48, 1964.

POLYDACTYLY-SYNDACTYLY SYNDROME WITH TRIPHALANGEAL THUMBS
Eaton-McKusick Syndrome, Polydactyly-Syndactyly Type I

CARDINAL CLINICAL FEATURES

Five triphalangeal digits on the hands with or without a sixth rudimentary digit, preaxial polydactyly of the feet and hypoplastic halluces; malformation of the lower leg involving a markedly hypoplastic tibia, absent patella and a thick, curved fibula dislocated on the femur.

CLINICAL PICTURE

This syndrome was first described by Eaton and McKusick[1] in four individuals in three generations. All four had virtually identical hand and foot malformations, and three had the malformation of the lower extremity. We have now seen this syndrome in four individuals in three generations, only one of which had the malformation of the lower extremity.

This malformation is unique. A major component is hypoplasia of the tibia, which may at first appear to be congenitally absent. Later it ossifies and develops slowly, always remaining markedly hypoplastic. The patella is absent. The fibula is thickened and curved and is displaced on the femur posteriorly, laterally or both. Motion at the knees is limited, but these patients learn to walk and usually have no problem getting about.

The hands are characteristically without recognizable thumbs. Instead they have five finger-like triphalangeal digits. The thenar area has a hypothenar appearance. There may be a sixth rudimentary digit, usually only a distal portion originating from the lateral side of the fifth digit and syndactylous with it, and usually with no bones. Mild membranous syndactyly was present in the hands of our patients, but one of Eaton and McKusick's[1] patients had third and fourth fingers that were completely webbed and another had six triphalangeal fingers, the ulnar two of which had extensive soft tissue syndactyly. That patient also had a pedunculated postminimus on the left hand.

Polydactyly of the feet was observed regularly. The number of toes per foot was variable, ranging from six to nine. The pattern always was one of preaxial polydactyly. The lateral four toes appeared normal. The next toe was sometimes not recognizable as a hallux, although often it was. Nearly always it was hypoplastic, and the hypoplasia usually involved the metatarsal as well. The extra digits were variable in size and number and sometimes were triphalangeal.

The syndrome is distinct. An identical anomaly of the lower extremities was described by Sakati et al.[2] in a patient with acrocephalopolysyndactyly (ACPS) (see chapter on the polydactyly-syndactyly syndrome with oxycephaly).

CASE REPORT

Our index patient was a male infant referred because of an abnormality around the knees.

The baby was born with what appeared to be five fingers on

Fig. 1.—(Case Report). Infant with the polydactyly-syndactyly syndrome. The hands appeared to have five fingers in which there was digitalization of the thumbs.

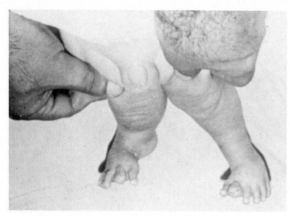

Fig. 2.–Lower extremities of same infant. The legs were short and had an extra skin fold. The knees were in valgus. There were six toes on each foot. The second toe looked like a hallux.

each hand, but no thumbs (Fig. 1). An extra digit on the right hand had been removed surgically. The legs were short and had an extra skin fold (Fig. 2). The knees were in valgus. There were six toes on each foot in a pattern of preaxial polydactyly. The second toe looked like a hallux and the four toes lateral to it appeared normal.

Roentgenograms of the hand revealed five normal-appearing triphalangeal fingers (Fig. 3). Three carpal bones were present in a row. Roentgenograms of the feet showed both to be in adduction. On the right there was a hallux with a hypoplastic metatarsal and a preaxial extra digit with three phalanges. The four lateral toes appeared normal. On the left, the appearance was similar, but in addition the hallux had a duplicated distal phalanx. Four tarsal bones were evident bilaterally. The tibia on the right was simply an ossification center (Fig. 4); on the left, the tibia was dysplastic. Bilaterally the fibula was bowed and thickened. No patella was evident, and the fibula was dislocated proximally and laterally on the femur.

Family history revealed that the mother had hands and feet similar to those of her infant (Figs. 5 and 6), as did a maternal aunt and the maternal grandfather. None of these individuals had the abnormality of the lower extremities. The grandfather

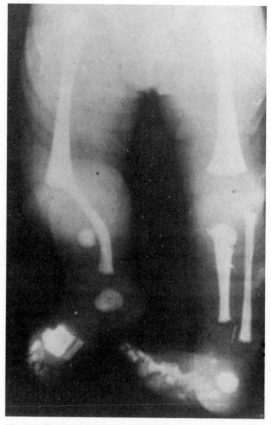

Fig. 4.–Roentgenogram of the lower extremities, same infant. The right leg had a bowed and proximally displaced fibula. The hypoplastic tibia was just an ossification center. In the left leg the tibia was dysplastic.

had had six healthy siblings in whom there were no anomalies. His 48-year-old wife was well. Their first pregnancy had yielded twin girls who died very early in life and were reliably reported to have had the syndrome. The next child was an uninvolved male who had drowned at 12 years of age. The

Fig. 3.–Roentgenograms of hands of this infant. (The right hand is on the *left* of the photograph.) On both sides there were five metacarpals. Each finger was triphalangeal and none looked like a thumb. Three carpal bones were evident.

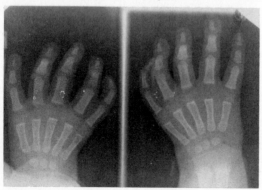

Fig. 5.–Hands of the mother of the infant shown in Figures 1–4. There were six fingers on each hand. There was membranous syndactyly between the first and second fingers, neither of which had the appearance of a thumb. However, the distal phalanx of the first digit had a rudimentary duplication, including a nail. There was a milder syndactyly between the fourth and fifth fingers. The sixth fingers were rudimentary and fused to the fifth.

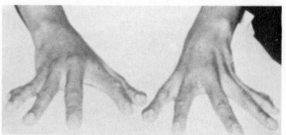

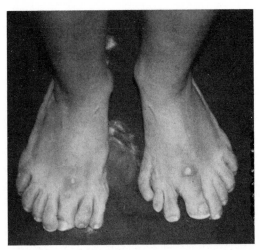

Fig. 6. – Feet of the mother of same infant. The left foot had eight toes. This foot had preaxial polydactyly, in which there were three toes to the right of the great toe. From the great toe laterally there were four normal-appearing toes. The right foot was different: there were seven toes. The hallux appeared to be the third toe. There were two preaxial extra toes, the second of which was widened terminally.

mother of the patient was the fourth child. The fifth was a healthy, uninvolved female now 19 years of age; the sixth was a 17-year-old healthy male, the seventh was a 15-year-old healthy male; and the eighth was the involved aunt.

GENETICS

The disorder is transmitted as an autosomal dominant.[3] In each of the families reported, expression of the abnormalities of the hands and feet was very similar and present in all of the patients. The tibial malformation was not regularly seen, and in each family it spared the oldest generation. It is possible that, once expressed in a generation, it will be continued in the next.

TREATMENT

Most patients require no treatment. Many will want postminimi removed. Extensive syndactyly may be treated surgically.

REFERENCES

1. Eaton, O., and McKusick, V. A.: A Seemingly Unique Polydactyly Syndrome in Four Persons in Three Generations, in *The First Conference on the Clinical Delineation of Birth Defects. Part III, Limb Malformations*, Birth Defects: Original Article Series (New York: The National Foundation-March of Dimes, 1969), p. 221.
2. Sakati, N., Nyhan, W. L., and Tisdale, W. K.: A new syndrome with acrocephalopolysyndactyly, cardiac disease, and distinctive defects of the ear, skin, and lower limbs, J. Pediatr. 79:104, 1971.
3. Yujnovsky, O., Ayala, D., Vincitoria, A., Sakati, N., and Nyhan, W. L.: A syndrome of polydactyly-syndactyly and triphalangeal thumbs in three generations, Clin. Genet. 6:51, 1974.
4. Temtamy, S., and McKusick, V. A.: Synopsis of Hand Malformations with Particular Emphasis on Genetic Factors, in *The First Conference on the Clinical Delineation of Birth Defects. Part III, Limb Malformations*, Birth Defects: Original Article Series (New York: The National Foundation-March of Dimes, 1969), p. 125.

Single Syndromic Malformations

CONGENITAL ABSENCE OF ABDOMINAL MUSCULATURE
Prune Belly Syndrome, Triad Syndrome

CARDINAL CLINICAL FEATURES

Absence of the abdominal muscles, undescended testes, urinary tract anomalies.

CLINICAL PICTURE

The syndrome was first reported in 1895 by Parker.[1] Most reported patients have been male, although the disorder has occurred rarely in females. Females with the disorder have tended not to have urinary tract abnormalities. The characteristic alerting feature of the syndrome is congenital hypoplasia of the abdominal muscle, which may be a complete or partial absence and causes the skin of the abdomen to appear wrinkled and flabby. It is this appearance that is somewhat reminiscent of a dried prune. The abdominal wall is flaccid. Intestinal patterns of loops and peristalsis may be observed, and abdominal organs are readily palpated.

Renal anomalies are regularly associated with the syndrome (Fig. 1). They usually take the form of hydronephrosis, along with elongated, tortuous, extremely dilated ureters and a large bladder. Some patients have had obstruction at the vesical neck or a hypoplastic-to-absent urethra. Occasionally a megalourethra has been seen, and some patients have had a persistent or patent urachus. Dysplasia of the renal parenchyma has been observed, as well as hypoplastic or cystic kidneys. The high morbidity and mortality in this syndrome is due to the urinary tract abnormalities. The mortality has approximated 50%, and death is due most commonly to urinary infection. A weak cough may facilitate pulmonary infection.

Undescended testes are common among males with this syndrome; almost without exception the testes are found intra-abdominally. Other associated anomalies include talipes equinovarus, dislocation of the hips, gastrointestinal anomalies such as malrotation, and congenital heart disease.

The syndrome is thought to be due to a defect early in embryogenesis, probably during the sixth to tenth

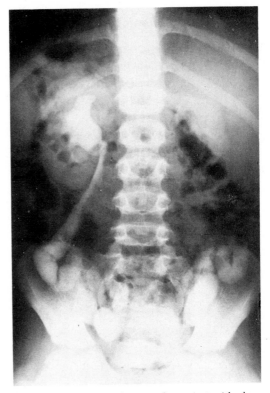

Fig. 1.—Intravenous pyelogram of a patient with absence of abdominal muscles. He had bilateral ureters and dilatation of the renal calices. (Courtesy of Dr. George Kaplan of the University of California San Diego.)

weeks of fetal life. Smith[2] has postulated that a localized defect in the early mesoderm at about the 23d day of fetal life would lead to defects in three areas of the segmenting mesoderm and ultimately to agenesis or hypoplasia of the abdominal muscle, abnormalities of the kidneys and dilated ureters and bladder. Myotubular embryonic muscle has been found in the abdomen of a patient with this syndrome.[3]

Intelligence in these patients is in the normal range.

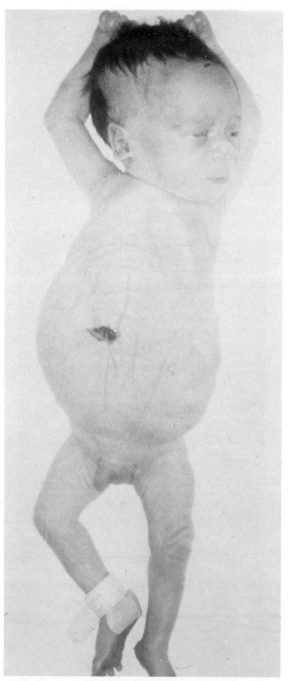

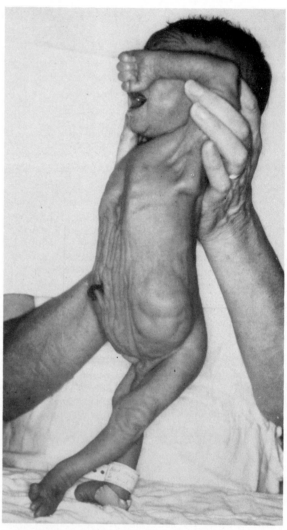

Fig. 3.—Baby B. In the side view the dilated loops of the ureter are clearly outlined. (Courtesy of Drs. David Allan and Joseph Joyner, San Diego.)

Fig. 2.—Baby B. (Case Report). A newborn infant with congenital absence of the abdominal musculature. The abdomen was large and the skin loose and wrinkled. This appearance has been reminiscent to some of the surface of a prune. (Courtesy of Drs. David Allan and Joseph Joyner, San Diego.)

CASE REPORT

Baby B. was the product of a normal pregnancy and premature delivery at 36 weeks to a gravida III, para II mother. His birth weight was 4 lb, 4 oz.

Physical examination revealed a striking lax, wrinkled abdomen (Figs. 2 and 3). The abdominal organs were easily palpable through the very soft wall. The testes were not palpable (Fig. 4). Following the diagnosis an intravenous pyelogram was done within the first 36 hours of life and revealed bilateral hydronephrosis and hydroureters, a dilated bladder and a patent urachus. Ureterostomy was done on the left side, and a nephrostomy on the right. The infant's renal function appeared adequate, but at 6 weeks of age he developed an Aerobacter sepsis and died.

GENETICS

This disorder appears to be nongenetic. The cause is not known. The disease has been reported in twins and in siblings[6]; one of the affected twins also had an omphalocele. More often in the case of twins, only one of the pair has been involved.

Fig. 4.—Baby B. Close-up of the abdomen. The cryptorchidism is illustrated. (Courtesy of Drs. David Allan and Joseph Joyner, San Diego.)

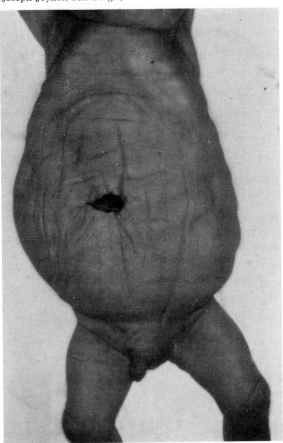

TREATMENT

The prognosis for babies with this syndrome depends on the degree of renal insufficiency and the drainage of the urinary tract. Prompt and aggressive surgery is thought by most to be indicated to establish drainage. The early use of nephrostomy drainage has been emphasized, along with renal biopsy to evaluate the prognosis for renal function.[8] At the same time, it is recognized that demonstrable organic obstruction is the exception. Infection always is a concern. The abdomen itself is no problem. Most parents prefer the cosmetic appearance obtainable with a soft abdominal binder, and a diaper is suitable for this purpose. A corset may assist with defecation. The defect does not interfere with standing or walking. Ultimately the cryptorchidism may require surgery.

REFERENCES

1. Parker, R. W.: Absence of abdominal musculature in an infant, Lancet 1:1252, 1895.
2. Smith, D. W.: Single Syndromic Malformations Resulting in Secondary Defects, in *Recognizable Patterns of Human Malformation* (Philadelphia: W. B. Saunders, Co., 1970, pp. 5–13.
3. O'Kell, R. T.: Embryonic abdominal musculature associated with anomalies of the genitourinary and gastrointestinal systems, Am. J. Obstet. Gynecol. 105:1283, 1969.
4. Silverman, F. N., and Huang, N.: Congenital absence of the abdominal muscles, Am. J. Dis. Child. 80:91, 1950.
5. Cremin, B. J.: The urinary tract anomalies associated with agenesis of the abdominal walls, Br. J. Radiol. 44:767, 1971.
6. Petersen, D. S., Fish, L., and Cass, A. S.: Twins with congenital deficiency of abdominal musculature, J. Urol. 107:670, 1972.
7. Harley, L. M., Chen, Y., and Rattner, W. H.: Prune belly syndrome, J. Urol. 108:1974, 1972.
8. Waldbaum, R. S., and Marshall, V. F.: The prune belly syndrome: A diagnostic therapeutic plan, J. Urol. 103:668, 1970.
9. Eagle, J. F., Jr., and Barrett, G. S.: Congenital deficiency of abdominal musculature with associated genitourinary abnormalities: A syndrome, Pediatrics 6:721, 1950.

EXSTROPHY OF THE CLOACA

CARDINAL CLINICAL FEATURES

Exstrophy of the bladder, omphalocele, prolapse of the intestine, imperforate anus, lumbosacral myelocele.

CLINICAL PICTURE

Cloacal exstrophy is a rare congenital malformation and a distinctive morphologic syndrome. It has been classified by Smith[1] as being due to a single localized defect in mesodermal morphogenesis. One would expect the defect to have occurred at 4–5 weeks of gestation. However, Welch[2] has said that no single embryologic event satisfactorily explains this anomaly.

The essential feature of the syndrome is a failure of cloacal septation, which results in a persistent cloaca into which the ureters and a rudimentary hindgut open. As there is no cloacal membrane, this structure is completely exstrophic. Thus, exstrophy of the bladder produces a heart-shaped mucosal field and anorectal agenesis with an imperforate anus. The pubic rami do not fuse. There usually is an omphalocele. In about half the patients there is a prolapse of the intestine into the exstrophied area. This prolapse sometimes produces a striking appearance known as the elephant-trunk deformity. There may be one to four intestinal openings on the exstrophied surface. The superior opening usually represents the terminal ileum. There is incomplete development of the lumbosacral vertebrae and herniation of a markedly dilated central canal through it. This looks like a meningomyelocele but is a lumbosacral myelocele or hydromelia. It is soft, cystic and covered with skin. Information on 14 patients with this syndrome was collected by Keith in 1908.[4]

The genital tubercles fail to fuse. In the male the external genitalia may be absent, and cryptorchidism usually is present. He may have unilateral or bilateral germinal aplasia or a short, bifid penis with epispadias. In the female the two halves of the müllerian system fail to fuse in the midline, resulting in a completely bifid uterus and a short, bifid or atretic vagina. Vaginal atresia may lead to hydrocolpos. In both sexes renal anomalies occur in association with the syndrome. They include hydronephrosis, hydroureter and dysplastic kidneys. Various forms of renal or ureteral duplication may occur.

These patients may have associated neurologic manifestations, and diastomyelia and hydrocephalus sometimes develop, with or without an Arnold-Chiari malformation. Occasionally seen are talipes equinovarus or other anomalies of the lower limbs such as absence of bones in the foot or ankle.

An ileocecal fistula may open into the bladder surface. The ileum, which is prolapsed through the vesicointestinal fistula, may become gangrenous. A double appendix has been reported, as has colonic atresia. Most of the patients have a single umbilical artery.

It is striking that all of the anomalies in this complex syndrome are limited to the area below the diaphragm. Hydrocephalus, when it occurs, is secondary to the lower cord malformation. This type of localization is reminiscent of malformations seen in mosaic flies, which involve an entire section of the body, a segment or half. It is in this sense reminiscent of the various hypertrophies and other unilateral diseases.

CASE REPORT

A. C. was a 4-hour-old white female when admitted to University Hospital because of multiple congenital anomalies. She was the product of a 9-month uncomplicated pregnancy to a 28-year-old gravida I, para I female in good health. Delivery

Fig. 1.–A. C. (Case Report). Newborn girl with cloacal exstrophy syndrome. The ventral and dorsal defects are both evident in this view. There was ventrally an omphalocele, prolapsed bowel and exstrophy of the bladder. The anus was not perforate.

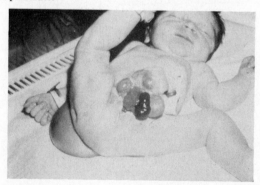

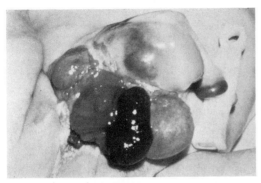

Fig. 2.—A. C. Close-up of the omphalocele and exstrophy of the bladder. The possibility of compromised circulation is evident from the black color of the prolapsed intestine.

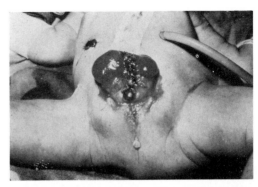

Fig. 4.—A. C. After surgical treatment, the bladder remained exstrophied. In this view the imperforate anus is evident and the widely separated pubic rami are visible.

was by cesarean section because of the presence of a large mass in the baby's lumbosacral area. At birth the baby was alert and vigorous. She was noted to have a large omphalocele, an imperforate anus, ileal prolapse and exstrophy of the bladder (Figs. 1 and 2). Only one umbilical artery was evident and two appendices were attached to the cecum. Other anomalies included the absence of part of the transverse colon, the left hemicolon and the ligament of Treitz. She had a bifid uterus and vaginal atresia. The left fallopian tube was not attached to the uterus. She also had a large lumbosacral myelocele (Fig. 3).

Roentgenographic examination revealed a large lumbosacral spina bifida and widely spread pubic rami. Chromosome analysis was normal. On the day of birth, laparotomy was done to repair the omphalocele. Colostomy and gastrostomy also were done at this time and the ileal prolapse reduced. Intravenous urogram revealed both kidneys to visualize well. There was no ureteral obstruction.

At 2½ months of age the child was doing well and gaining weight. She was hospitalized again to repair the lumbosacral

Fig. 3.—A. C. Hydromelia. Lumbosacral spina bifida with a protruding and dilated central canal of the spinal cord.

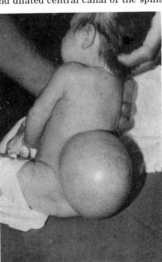

defect (Fig. 4). The mass was removed and the dura closed over the cord. The lower extremities were noted to move well.

GENETICS

Most cases of cloacal exstrophy are sporadic. None has been shown to be genetic in origin. Smith has proposed that most abnormalities following a single localized defect in morphogenesis are due to polygenic factors.

TREATMENT

Surgical treatment has been carried out with limited effectiveness. The prognosis may be hopeless in a significant number of these infants, particularly regarding the problem of urinary and fecal incontinence. An omphalocele requires prompt treatment. Some of these can be closed primarily, some in stages using plastic mesh, and the largest are treated by periodic application of mercurochrome. Acute surgical intervention may be required in a patient with symptoms and signs of intestinal obstruction resulting from a midgut volvulus due to malrotation. Strangulation of the prolapsed ileum also may represent a surgical emergency. Neurosurgical repair of myelocele or hydromelia usually is carried out, as in the patient discussed in the Case Report. Primary repair of the exstrophy of the bladder usually is recommended to be done at 12–18 months of age. Orthopedic surgery may be required to correct malformations of the lower limbs. As reported in Welch's[2] summary of 18 cases of this syndrome, 15 patients died of various surgical attempts at correction of one or another of the defects. Some surgical successes have been reported.[5]

REFERENCES

1. Smith, D. W.: Recognizable Patterns of Human Malformation, in *Major Problems in Clinical Pediatrics* (Philadelphia: W. B. Saunders Co., 1970), Vol. VII, p. 6.

2. Welch, K. J.: Cloacal Exstrophy (Vesicointestinal Fissure), in Mustard, W. T., *et al.* (eds.): *Pediatric Surgery* (2d ed.; Chicago: Year Book Medical Publishers, Inc., 1969), Vol. II, p. 1344.
3. Gough, M. H.: Anorectal agenesis with persistence of the cloaca, Proc. R. Soc. Med. 52:886, 1959.
4. Keith, A.: Three demonstrations of malformations of the hind end of the body, Br. Med. J. 2:1856, 1908.
5. Zwiren, G. L., and Patterson, J. H.: Exstrophy of the cloaca. Report of a case treated surgically, Pediatrics 35:687, 1965.
6. Beckwith, T. B.: The congenitally malformed. VII. Exstrophy of the bladder and cloacal exstrophy, Northwest Med. 65:496, 1966.
7. Swan, H., and Christensen, S. P.: Exstrophy of the cloaca, Pediatrics 12:645, 1953.

CYCLOPIA

CARDINAL CLINICAL FEATURES

Single orbital fossa with varying degrees of fusion presenting one or two eyes, a proboscis-like appendage above, holoprosencephaly.

CLINICAL PICTURE

Many of us first read of this deformity in Homer's *Odyssey*. It is clear from this and other references that this malformation has been around about as long as man. Cyclopia was referred to as early as the cuneiform tables from Chaldea. However, early descriptions of the cyclops were of fierce beings against whom Ulysses fought and who were destroyed by Apollo because of their responsibility for the death of Aesculapius, the mythological god of medicine. These characterizations were a long way from the reality of a defect that usually leads to stillbirth or death after a few hours of life.

The cyclops malformation is characterized by a single orbital fossa located in the area normally occupied by the bridge of the nose (Fig. 1). The socket usually is diamond shaped, with open angles extending beyond the globe. There may be all degrees of fusion, from two eyes in the same orbital cavity, to a single globe with two corneas and two pupils to a single eye. There may be one or no optic nerves. A nose-like proboscis is found above the eye (see Fig. 1). It is covered by normal skin and may contain cartilage but does not contain bone. The structure usually has a single external orifice that ends blindly at bone. The nasopharynx does not connect externally with the face.

The cyclopian deformity is the most severe of the holoprosencephalies or median faciocerebral dysplasias.[1, 2]

In these disorders (also known as arhinencephaly or alobar holoprosencephaly) median dysplasia of the face and aplasia of the olfactory bulbs are associated with defective midline cleavage of the embryonic forebrain into mirror image hemispheres. In cyclopia there usually is a single ventricle and prosencephalon along with absence of the corpus callosum, chiasma and septum pellucidum. The thalami are fused. The second and fourth pairs of cranial nerves usually are absent. Other central nervous system abnormalities that

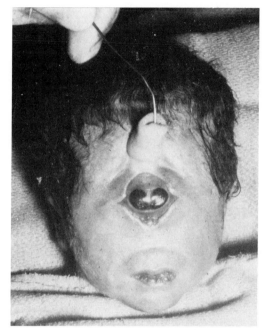

Fig. 1.—Patient with cyclopia. (Courtesy of Dr. David Allan, University of California and Mercy Hospital, San Diego.)

may be seen include microcephaly, or occasionally hydrocephaly, spina bifida, anterior encephalocele, acrania and anencephaly.

There may be a variety of oral manifestations.[3] The mouth generally is triangular because of the small size of the upper lip and normal lower lip. The philtrum and frenulum of the upper lip are missing. The ears are often low set, malformed or absent. Some of the facial bones usually are missing such as the ethmoids, turbinates and nasal bones, as well as portions of the malar bones. There may be a cleft palate or bifid uvula.

A related severe defect is cebocephaly, in which there is pronounced ocular hypotelorism with two eyes very close to each other and a rudimentary nose with a single nostril that seldom connects with the nasopharynx. The least severe defect in this series of disorders is simple arhinencephaly in which there is absence of the corpus callosum. Cyclopia with unilateral or bi-

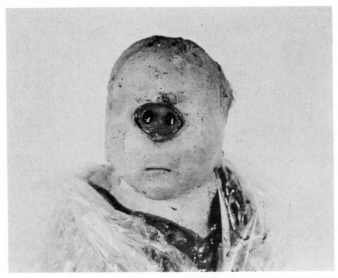

Fig. 2.–Baby S. (Case 1). There were two corneas, pupils and irises in a single globe. The orbital fossa was triangular.

lateral abnormality of the ear[4] may be associated with agnathia and has been called cyclopia hypognathus.

The pituitary may be absent and associated with aplasia or hypoplasia of the adrenal or thyroid. There also may be anomalies of the lung, cryptorchidism, testicular aplasia or hypospadias.

CASE REPORTS

CASE 1.– Baby S. was born at University Hospital to a 22-year-old black woman after 40 weeks of uneventful gestation. There was no exposure to any substance that might be considered teratogenic, and there was no family history of a similar condition. At birth the infant showed the cyclops malformation. He survived only a few minutes.

On examination there was a centrally placed orbital fossa, which was diamond shaped and contained a single eyeball with double corneas and pupils (Fig. 2). A proboscis-like appendage was located above the eye on the lower part of the forehead (Fig. 3). The infant was microcephalic; the head circumference was 12 cm. The mouth was triangular. The palate was high arched and had bilateral grooves adjacent to the midpalatal suture, but there was no cleft of the palate or lip. The ears were low set and the neck short.

He had hypospadias and cryptorchidism. Examination of the placenta revealed a pronounced amnion nodosum (Fig. 4). This finding suggested the possibility of agenesis of the kidney, as in the Potter syndrome (see chapter, this section). This point could not be clarified because a postmortem examination could not be done.

CASE 2.– Baby R. was the product of an 8-month, uneventful gestation. The birth weight was 3 lb, 11 oz. The cyclops deformity was immediately evident. She had a poor Apgar score of 3 at 5 minutes. At 1½ hours of age, she died. There was no history of exposure to possible teratogens. Three siblings were normal.

Autopsy revealed a skull flattened posteriorly and shortened

Fig. 3.– Baby S. Profile illustrating the position of the rudimentary proboscis above the eye. The ears·were low set.

Fig. 4.– Baby S. Amnion nodosum.

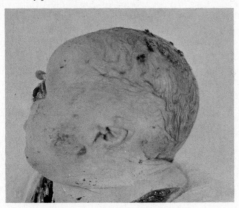

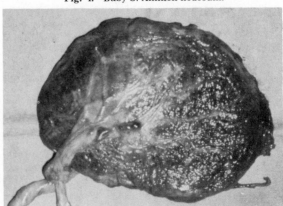

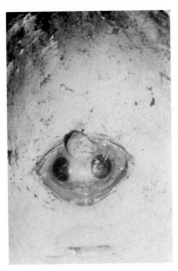

Fig. 5.—Baby R. (Case 2). Cyclopia.

in its anteroposterior dimension. There was a single eye located in the midline. Directly above the eye was a 2-cm pedunculated proboscis (Fig. 5). The heart revealed the right atrium to be somewhat large and an atrial septal defect to be present. The pulmonic valve was hypoplastic and bicuspid. The right ventricle was large. The right middle lobe of the lung was absent. The adrenal glands were hypoplastic. There was no sella turcica or pituitary. There was a single optic tract, which ran in the midline to the single orbit. On dissection of the eye, a lacrimal gland was found on either side of the single globe. The globe appeared to have a normal iris.

Examination of the brain revealed a rudimentary cerebral cortex without hemispherical division. There was a single, large ventricle but no corpus callosum, septum pellucidum or olfactory bulbs. Microscopic examination of the cortex revealed subependymal and perivascular calcifications. Cortical layers were absent, and nerve cells were clustered in groups and streaks. There was glial overgrowth in the subpial region. Cytogenetic analysis revealed that the baby had a normal karyotype.

GENETICS

Cyclopia has been observed rarely but definitely in identical twins. In fact, this abnormality of cleavage or fusion has been suggested to be related to twinning. Cyclopia has been reported on one side of the face, with normal structures on the other side.[5] François[6] suggested that cyclopia is a lethal autosomal recessive trait. This is probably generally the case, but the etiology of the defect may well be heterogeneous. Nitowsky[7]

reported a patient with cyclopia in whom there was a deletion of the short arm of chromosome no. 18. This is not the usual result with this deletion, but another cyclops with a similar karyotype has been reported. On the other hand, Cohen[8] observed mosaicism in a patient with cyclopia in which one cell line was monosomic for a G chromosome (21 or 22) and the other cell line was normal. A normal karyotype has been reported by others. Cyclopia and cebocephaly, or a variety of arhinencephalies, have been seen in the same sibship. Hereditary cyclopia and other manifestations of alobar holoprosencephaly have been documented in guinea pigs.[9] Cyclopia also has been produced by teratogenic agents including lithium chloride, magnesium chloride and vitamin A. Ingestion of the plant *Veratrum californeum* during pregnancy can produce cyclopia in sheep.[10]

TREATMENT

None is known. Children with cyclopia generally are stillborn. Although there are rare exceptions, usually those born alive survive only a few minutes.

REFERENCES

1. Yakovlev, P. I.: Pathoarchitectonic studies of cerebral malformations. III. Arhinencephalies (Holotelencephalies), J. Neuropathol. Exp. Neurol. 18:22, 1959.
2. DeMyer, W., and Zeman, W.: Alobar holoprosencephaly (arhinencephaly) with median cleft lip and palate: Clinical, electroencephalographic and nosologic considerations, Confin. Neurol., 23:1, 1963.
3. Sedano, H. O., Rosario, A., and Gorlin, R. J.: The oral manifestations of cyclopia, Oral Surg. 16:823, 1963.
4. Hagens, E. W.: Unilateral malformation of the ear associated with cyclopia, Arch. Otolaryngol. 18:332, 1933.
5. Meeker, L. H., and Aebli, R.: Cyclopean eye and lateral proboscis with normal one-half face, Arch. Ophthalmol. 38:159, 1947.
6. François, J.: *Heredity in Ophthalmology* (St. Louis: C. V. Mosby Co., 1961), pp. 173–176.
7. Nitowsky, H. N., Sindhvananda, N., Konigsberg, U. R., and Weinberg, T.: Partial 18 monosomy in the cyclops malformation, Pediatrics 37:260, 1966.
8. Cohen, M. M.: Chromosomal mosaicism associated with a case of cyclopia, J. Pediatr. 69:793, 1966.
9. Wright, S.: The genetics of vital characters of the guinea pig, J. Cell. Comp. Physiol. [Suppl.] 56:123, 1960.
10. Binns, W., James, L. F., Shupe, J. L., and Thacker, E. J.: Cyclopian-type malformation in lambs, Arch. Environ. Health 5:109, 1962.

SIRENOMELIA
Mermaid Syndrome,
Sympodia,
Caudal Regression Syndrome

CARDINAL CLINICAL FEATURES

Massive abnormality of the lower half of the body, with absence of the anus and external genitalia and fusion of the lower extremities throughout their length.

CLINICAL PICTURE

Sirenomelia is a rare malformation, which was recognized many centuries ago and recorded in the literature as early as the first century A.D. Homer's Greek sirens were women who sang and lured ships' crews to their deaths in rocky channels, and the mermaid as a water nymph temptress is found in the seafarers' legends of many countries. Early responses to the birth of a baby whose limbs resembled these mythical creatures were often to blame the misfortune on the mother's cohabiting with mermen or incubi, or being frightened in her dreams by a triton.[1] By the 18th century reports became more factual, but the fascination of legend may still have influenced the reporting, as some 190 instances of this disorder have been recounted.

This malformation is thought to occur during the first 3 weeks of fetal life and to be due to a wedge-shaped defect of the posterior axis mesoderm, which permits fusion of the early limb buds at their fibular margins and absent or incomplete caudal development.[2] Involved infants are stillborn or die within a few hours after birth. They usually have a single umbilical artery that arises from the aorta instead of from the hypogastric artery.

Patients commonly have lumbar or other vertebral defects such as hemivertebrae or segmental vertebrae, and the sacrum may be absent. Agenesis of the kidneys, ureters and bladder is frequent. Rarely, a rudimentary or horseshoe kidney may be present, or abortively developed renal elements may be seen. Oligohydramnios and Potter facies (see chapter this section)[3] and the ears are characteristically large and low set and the nose flattened.

Hypertelorism and epicanthal folds may be present, and a typical semicircular fold usually is seen below each eye.

Most of the reported cases of sirenomelia have been males. They usually have no external genitalia and, although the gonads are always present, the wolffian or müllerian ducts are absent. There usually is an imperforate anus and rectal atresia.

The fused lower extremity may take a number of forms. In general, there is a lack of the normal inward rotation so that the fibular sides of the legs are fused. There may be complete, separate extremities, starting with two femora within a single extremity mass and ending with two feet, which may be distinct. Fusion may yield a single femur, and there may be varying degrees of distal hypoplasia, so that there may be one foot or none at all. The lower leg may be a short, irregular stump. The pelvic bones may be a coalesced mass, or there may be absence of individual pelvic bones.

The fibular fusion and lack of inward rotation leads to a reversal of the usual positions so that the soles of the feet and the popliteal regions or the posterior surfaces of the legs face forward. The rotation gives the feet, when present, a fishtail appearance. The major muscle groups and innervation of the lower extremity are present. The position of the fused limb demonstrates clearly the position of the early-developing limb buds.

Hypoplasia of the lungs may occur in this condition, as it does in the Potter syndrome of renal agenesis. Thirteen ribs have been reported,[4] as have been ten.[5] Cardiac malformations have rarely been seen. Tracheoesophageal fistula has been observed.[6]

CASE REPORT*

Baby N. was a stillborn female infant, delivered prematurely of a 21-year-old, gravida I mother who had had no

*We are indebted to Drs. Harold Koenig and K. C. Campbell of the U.S. Naval Hospital, San Diego, for permission to describe this patient.

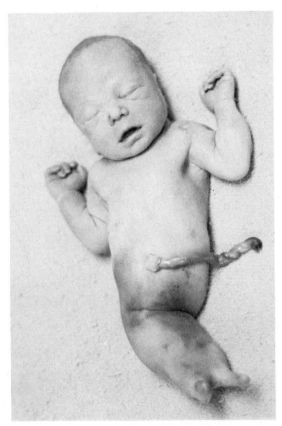

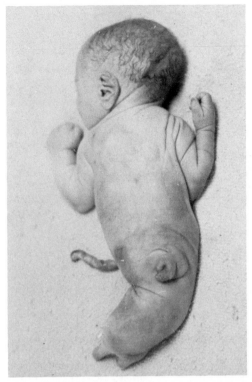

Fig. 1.—Baby N. (Case Report). An infant with sirenomelia. The Potter facies is indicated by the characteristic semilunar creases under the eyes. The genitalia were absent and the lower extremities fused to give a mermaid-like appearance. (Courtesy of Drs. Harold M. Koenig and K. C. Campbell of the U.S. Naval Hospital, San Diego.)

Fig. 2.—Baby N. In this view the large ear is evident. There was no anus. A prominent set of skin folds was present in the anal area. The lower extremities were smoothly rounded, terminating in two toe-like buds. (Courtesy of Drs. Harold M. Koenig and K. C. Campbell of the U.S. Naval Hospital, San Diego.)

complications of pregnancy. No history could be obtained of prior anomalies in the family. Birth weight was 2 lb, 12 oz.

The upper portion of the body appeared to be normal. The lower extremities were fused throughout their length and presented a stump-like projection 9 cm in length and 15 cm in diameter that terminated in two small, toe-like buds (Figs. 1 and 2). Roentgenograms showed the presence of a single femur with two small, separate distal bones (Fig. 3). The extremity appeared to be hinged at the pelvis, although only anterior flexion was possible. Roentgenogram of the chest revealed 13 ribs bilaterally.

No midline genital structures were present anteriorly. There was no anus, but a small, blind opening at the terminal end of the sacrum. About 1 cm distal to this was a small, fleshy protrusion of skin and fatty tissue about 1.5 cm in length. Superior to this protrusion was another small opening.

Autopsy revealed hypoplastic lungs. Kidneys, ureters, bladder and urethra were absent. The descending colon terminated in a blind pouch hanging freely in the mesentery. No anus could be identified. The infant was considered to be female because of the presence of bilateral structures in the pelvis that resembled ovaries, and histologic examination confirmed this observation. The ovaries were attached to the distal portions of what appeared to be fallopian tubes. There were no proximal portions of these tubes and no uterus or vagina. The

fleshy protrusion over the posterior pelvis was dissected; the lower opening led to a small, cord-like structure, 0.2 cm in diameter, which had a small lumen that extended through an opening into the anterior fossa of the pelvis. Sections of the umbilical cord revealed the presence of a single artery and a single vein.

GENETICS

Sirenomelia is thought to be due to a single syndromic defect early in fetal life. It occurs in one in 60,000 births. There is a 2.7:1 male preponderance. The syndrome occurs sporadically. It may be nongenetic. The risk of recurrence appears to be negligible. One instance of involvement of identical twins has been reported, but there have been no other familial cases reported. An analogous condition in the mouse appears to be polygenic or multifactorial in origin.[4]

REFERENCES

1. Kampmeier, O. F.: On sireniform monsters, with a consideration of the causation and the predominance of the male sex among them, Anat. Rec. 34:365, 1927.

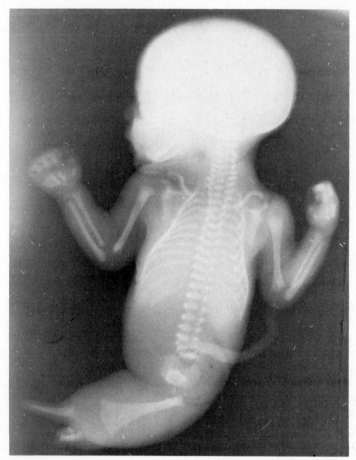

Fig. 3. — Baby N. Roentgenographic examination revealed that both femora were fused into a single bone with a marked distal widening. The protruding buds were hypoplastic lower legs with hypoplastic tibiae or fibulae. (Courtesy of Drs. Harold M. Koenig and K. C. Campbell of the U.S. Naval Hospital, San Diego.)

2. Smith, D. W.: *Recognizable Patterns of Human Malformation* (Philadelphia: W. B. Saunders Co., 1970), p. 5–11.
3. Bearn, J. G.: The association of sirenomelia with Potter's syndrome, Arch. Dis. Child. 35:254, 1960.
4. Crawfurd, M. d'A., Ismail, S. R., and Wigglesworth, J. S.: A monopodal sireniform monster with dermatoglyphic and cytogenetic studies, J. Med. Genet. 3:212, 1966.

5. Newbill, H. P.: Sirenomelian monster (sympus), an anatomic presentation, Am. J. Dis. Child. 62:1233, 1941.
6. Williams, H. I.: Sympodia, Arch. Pathol. 74:472, 1961.
7. Duhamel, B.: From the mermaid to anal imperforation: The syndrome of caudal regression, Arch. Dis. Child. 36:152, 1960.

POTTER SYNDROME
Bilateral Renal Agenesis

CARDINAL CLINICAL FEATURES

Bilateral renal agenesis; low-set ears with proportionately less cartilage; deep, semicircular infraorbital skin fold; increased distance between the inner canthi; broadening and flattening of the nose; small receding mandible; pulmonary hypoplasia; oligohydramnios; amnion nodosum.

CLINICAL PICTURE

Infants with the Potter syndrome can be recognized at the time of birth by the presence of a number of striking characteristics. These findings were initially observed in patients with complete renal agenesis.[1] They also may occur with massive polycystic changes in the kidney (Fig. 1) or other major renal anomalies.[2] These infants have strikingly similar facies. They often are first considered because of the ears, which are extremely low set (Fig. 2) and usually lack cartilage. There is an increase in the width between the eyes, and an unusually prominent skin fold appears under

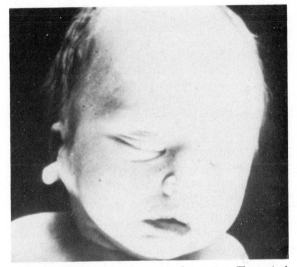

Fig. 2.—The ears were low set in this patient. The typical semilunar folds are seen under each eye. The skin was wrinkled and senile in appearance.

Fig. 1.—Polycystic kidneys in the Potter syndrome; the renal lesion usually is agenesis.

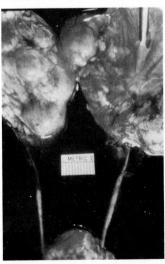

the eyes, starting at the inner canthus and forming a deep semicircle (Fig. 3). In profile the chin is seen to be receding. The skin often is dry, loose or wrinkled (see Fig. 2). The over-all appearance suggests premature senility. The nose is flattened and slightly broadened.

The hands are broad and spade-like, and the fingers appear relatively short in relation to the width of the hand (Fig. 4). Some form of talipes often is found in the feet.

Oligohydramnios is uniformly recorded when it is looked for. It is not a very useful alerting manifestation, as it is often not noted. Amnion nodosum is a better sign and appears uniformly in this syndrome.[3] In amnion nodosum (Fig. 5) there are many 1–2 mm yellowish nodules, which look something like miliary tubercles. They lie on the fetal side of the amnion. On section they are masses of keratinized squames imbedded in the amniotic mesoderm of the fetus. Antenatal films may reveal extreme flexion of the spine of the fetus. This finding may occur with fetal death in utero from any cause, but it is seen with live babies in the presence of oligohydramnios. They are very often

407

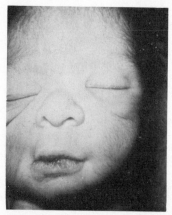

Fig. 3. – The characteristic facies is illustrated with a broad nasal root, micrognathia and typical semicircular infraorbital lines. (Courtesy of Dr. Louis Gluck of the University of California San Diego.)

born by breech, which also is probably a result of oligohydramnios. The lungs of these infants are always hypoplastic.[4]

Female pseudohermaphroditism and other forms of genital maldevelopment are common in renal agenesis.[5] Imperforate anus (Fig. 6) also is common and may occur in the infant with genital defects. Probably the most extreme malformation is the sirenomelus deformity, in which there is a single fused lower extremity. All of these infants have genital abnormalities. Bilateral renal agenesis occurs more frequently in the male by a factor of about 3 to 1. None of the females observed by Carpentier and Potter[5] had an entirely normal genital system. In most, the uterus and vagina were absent. The fallopian tubes more often were normal. In the female pseudohermaphrodites there usually was a well-formed penile structure and often a centrally placed penile urethra, along with no vagina or uterus. These deformities have been documented to occur in infants with female sex chromatin and with two X chromosomes.[6] These findings suggest the possibility of an abnormal androgenic influence on these fetuses.

Fig. 4. – Hands of the patient with the Potter syndrome are short and broad.

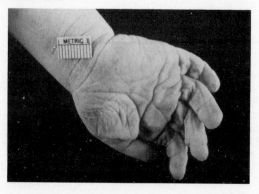

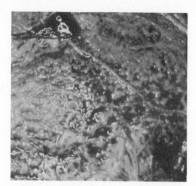

Fig. 5. – Amnion nodosum. (Courtesy of Dr. Louis Gluck of the University of California San Diego.

In the males abdominal testes are common, as well as a number of abnormalities of the external genitalia.

CASE REPORT

Baby boy R. was born by breech extraction at University Hospital after 34 weeks of gestation. Multiple anomalies were evident at birth that were consistent with the Potter syndrome. He did not breathe spontaneously and had a heart rate of 60. He was intubated, but his respiratory pattern was very poor, and he died about 2 hours after birth.

His facial characteristics included very low-set ears and widely separated eyes, with prominent folds arising at the inner canthus running downward and laterally from each eye. The nose was flattened, the chin receding. The genitalia were male, the testes undescended.

Autopsy revealed bilateral renal agenesis. There also was agenesis of the ureters, bladder and spleen. The lungs were hypoplastic (Fig. 7). The right lung had two small lobes, the left lung one lobe. The stomach was very small and consisted of a short narrow tube (Fig. 8).

GENETICS

The association of maldevelopment of the urinary tract and displacement of the ears could result from linked genes or a common teratogenic injury at an appropriate stage of differentiation. However, many of the features of this syndrome appear to be a consequence of the oligohydramnios that follows the ab-

Fig. 6. – The Potter syndrome. Imperforate anus. (Courtesy of Dr. Louis Gluck of the University of California San Diego.)

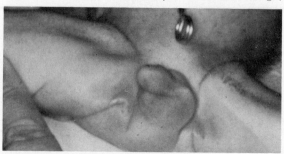

Fig. 7.—Baby boy R. (Case Report). Hypoplastic lungs.

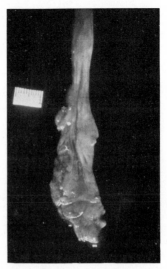

Fig. 8.—Baby boy R. Hypoplastic stomach.

sence of renal function and fetal urine. Cytogenetic analysis has been normal on all of those tested.[7]

TREATMENT

None. Affected infants are stillborn or live only very briefly.

REFERENCES

1. Potter, E. L.: Facial characteristics of infants with bilateral renal agenesis, Am. J. Obstet. Gynecol. 51:885, 1946.
2. Bain, A. D., and Scott, J. S.: Renal agenesis and severe urinary tract dysplasia, Br. Med. J. 5176:841, 1960.
3. Landing, B. H.: Amnion nodosum. A lesion of the placenta apparently associated with deficient secretion of fetal urine, Am. J. Obstet. Gynecol. 60:1339, 1950.
4. Potter, E. L.: Bilateral renal agenesis, J. Pediatr. 29:68, 1946.
5. Carpentier, P. J., and Potter, E. L.: Nuclear sex and genital malformation in 48 cases of renal agenesis, with special reference to nonspecific female pseudohermaphroditism, Am. J. Obstet. Gynecol. 78:235, 1959.
6. Schlegel, R. J., Neu, R. L., Carneiro-Leao, J., and Gardner, L. I.: An XX sex chromosome complement in an infant having male-type external genitals, renal agenesis, and other anomalies, J. Pediatr. 69:812, 1966.
7. Passarge, E., and Sutherland, J. M.: Potter's syndrome, Am. J. Dis. Child. 109:80, 1965.

CRANIOFACIAL AND LIMB DEFECTS SECONDARY TO ABERRANT TISSUE BANDS

Kenneth Lyons Jones, Jr.

CARDINAL CLINICAL FEATURES

Facial clefts, encephalocele, limb defects, aberrant tissue bands.

CLINICAL PICTURE

The earliest report of this pattern of malformation was published by Portal[1] in 1685. Torpin[2] accumulated reports of a number of affected individuals in 1968 and discussed the etiology of this disorder. More recently, the clinical spectrum has been delineated by Jones *et al.*[3]

Fig. 1.—(Case Report). Stillborn child with craniofacial and limb defects secondary to amniotic bands. (Courtesy of Dr. James MacLaggan of San Diego.)

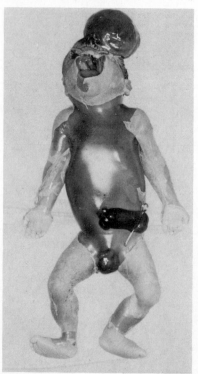

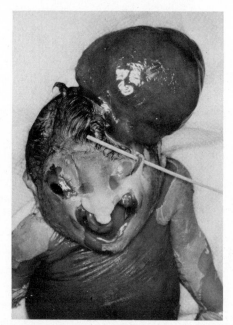

Fig. 2.—The aberrant tissue band, elevated by the probe inserted at the base of the asymmetrical anterior encephalocele. There were bilateral clefts of the lip and distorted palpebral fissures. (Courtesy of Dr. James MacLaggan of San Diego.)

All reported patients have had severe microcephaly and a deficiency of the anterior cranium. The majority have had asymmetrical encephaloceles, but anencephaly also has been observed. Short distorted palpebral fissures are common. Microphthalmia has been reported. All patients have had nasal deformities; in some cases this has been manifested by a unilateral proboscis. Facial clefts are common, and they do not necessarily conform to the normal planes of closure of the facial processes. Aberrant facial tissue bands have varied from filamentous strands, to bands of tissue 1–2 mm wide to fusiform peduncles up to 20 mm in length. These may lead to tremendous distortions in the appearance of the face.

Limb defects usually consist of amputation or constriction. Pseudosyndactyly may be seen in which a distal band of tissue holds two digits together. In other patients a combination of amputation and constriction leads to a syndactylous, disorganized mass of digits. The bands themselves may or may not be present on the limbs. Clubfoot has been reported in three patients. Two have had gastroschisis.

CASE REPORT*

The child was stillborn after a 32-week uncomplicated pregnancy to a 21-year-old primigravida. Fetal heart tones were lost 6 hours prior to delivery. Birth weight was 3 lb, 6½ oz and birth length was 17 inches.

Physical examination revealed an anterior encephalocele (Fig. 1) to the left of midline with lack of complete development of the anterior cranium. A 20-mm long fusiform peduncle protruded from the right base of the encephalocele. There was bilateral microphthalmia and clefts of both upper eyelids. A thick tissue band connected inferiorly to the skin below the left eye and superiorly to the base of the encephalocele (Figs. 2 & 3). There was bilateral cleft lip and palate. Amputations and constriction defects (Fig. 4) with bands encircling some fingers were present.

GENETICS

The combination of findings in this syndrome suggests that the craniofacial and limb defects are secon-

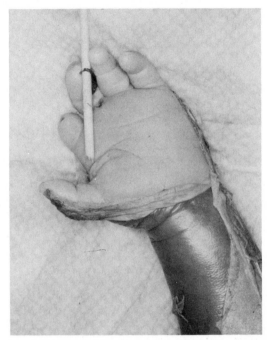

Fig. 4.—The index finger was partially amputated. An aberrant tissue band between the third and fourth fingers is illustrated by the probe. (Courtesy of Dr. James MacLaggan of San Diego.)

dary to the disruptive forces of the aberrant tissue bands.

All patients with this disorder have represented sporadic occurrences in otherwise normal families. With rare exception, involvement of limbs alone by tissue bands also has been sporadic. Thus, the recurrence risk appears to be negligible.

The degree of craniofacial anomaly would lead one to anticipate an early death for infants with this disorder. However, a number of affected children have survived past the newborn period.

Fig. 3.—Close-up of the aberrant band. (Courtesy of Dr. James MacLaggan of San Diego.)

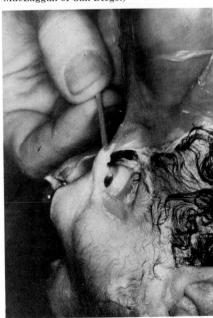

*We are indebted to Dr. James MacLaggan of San Diego for permission to describe this patient.

TREATMENT

None.

REFERENCES

1. Portal, P.: La Pratique des Accouchements, Paris, 1685.
2. Torpin, R.: *Fetal Malformations Caused by Amnion Rupture During Gestation* (Springfield, Ill.: Charles C Thomas, 1968).
3. Jones, K. L., Smith, D. W., Hall, B. D., Hall, J. Q., Ebbin, A. J., Massoud, H., and Golbus, M. S.: A pattern of craniofacial and limb defects secondary to aberrant tissue bands, J. Pediatr. 84:90, 1974.

Syndromes with Skeletal Dysmorphogenesis

OSTEOPETROSIS
Albers-Schönberg Disease,
Marble Bone Disease,
Infantile Malignant Form of Osteopetrosis

CARDINAL CLINICAL FEATURES

Thick sclerotic, brittle bones; multiple fractures; pancytopenia; blindness; deafness.

CLINICAL PICTURE

Osteopetrosis was first described by Albers-Schönberg,[1] a German radiologist, in 1904. He described a 26-year-old man with generalized skeletal sclerosis and multiple fractures. The name osteopetrosis was first given to the disease by Karshner[2] in a review of the literature in 1926, by which time only 18 patients had been reported. There are two major forms of osteopetrosis, a benign form transmitted by a single autosomal dominant gene, and a severe, malignant form transmitted as an autosomal recessive.[3] The original patient with Albers-Schönberg disease must have had the former.

The malignant form of osteopetrosis occurs in infancy and has been diagnosed in utero. These babies have thick sclerotic bones. The bones also are brittle or fragile and fracture easily. Fractures are most common in the long bones, ribs and pelvis. The bones also are unusually susceptible to osteomyelitis. Osteomyelitis of the mandible is a particularly common feature of the disease, and the maxilla may be infected as well. Dentition often is delayed, presumably because of the sclerotic bone, which surrounds the tooth buds. The teeth also have enamel hypoplasia or defective dentine and are especially susceptible to caries.

These patients have characteristic facies. The head often is large and there is frontal bossing (Fig. 1). The face is broad and the nasal bridge depressed. The eyes appear prominent and widely separated.[2] Thickening of the base of the skull leads to a narrowing of all of the foramina. Hydrocephalus is a regular consequence. Pressure on the cranial nerves leads to cranial nerve palsies.[4] Blindness may be an early symptom of the disease and may result from optic atrophy. Central retinal degeneration also has been reported.[5] In these cases optic nerve compression appeared not to be a factor, so that there are at least two causes of blindness in this disease. Other ocular manifestations include nystagmus, exophthalmos, ptosis and strabismus. Deafness is common and is thought to be labyrinthine in origin.

Anemia is relentless in this disease and may be the cause of death. It is due to the narrowing of the marrow cavities by the osteosclerotic process. There may be pancytopenia, but the anemia usually is the most prominent feature. Thrombocytopenia may be prominent. These patients develop extensive extramedullary hematopoiesis. There usually is major enlargement of the liver and spleen, and there may be lymph node enlargement. Furthermore, leukemoid reactions have been observed in a number of infants with this disor-

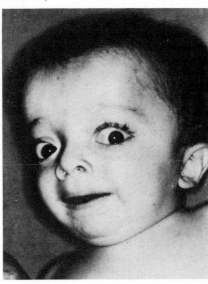

Fig. 1.—A patient with osteopetrosis, illustrating the large head with frontal bossing, wide-set prominent eyes and flattened nose. (Courtesy of Dr. Rodrigo Loria-Cortes of San Jose, Costa Rica.)

der. As the disease progresses, a hemolytic component may develop because of hypersplenism.[6]

Roentgenograms reveal a generalized sclerosis of all of the bones.[7] The base of the skull is especially thick and dense, as is the orbital roof. This makes the skull film look as if it is wearing a small mask of the type affected in the opera or by Batman (Fig. 2). Club-like thickening of the posterior clinoid has regularly been described.[2] This probably is simply a part of the generalized process in which all angular bony prominences are smoothed and rounded.[2] The optic canals are reduced in size. The ribs are thick. The long bones may be club shaped. The bones of the extremities, particularly the metacarpals, may reveal a "bone within bone" appearance, which is pathognomonic of this disease. Metaphyseal bands of increased and decreased density produce characteristic striations. The vertebrae may have a characteristic look, with pronounced anterior notching and a "sandwich" appearance, resulting from sclerosis of the upper and lower plates and an area of lesser density in between.

Laboratory findings include normal concentrations of calcium, phosphate and alkaline phosphatase. Balance studies have shown an excessive absorption of calcium.[8]

The cause of osteopetrosis is not known. The condi-

tion could result from a defect in bone resorption. There is a syndrome in gray-lethal mice in which osteopetrosis is associated with increased amounts of a calcitonin-like hypocalcemic factor in the plasma.[9] Calcitonin does inhibit bone resorption. These mice have increased numbers of thyroid parafollicular cells,[10, 11] which produce calcitonin. These observations have suggested that overproduction of calcitonin may play a role in the causation of osteopetrosis in man.

DIFFERENTIAL DIAGNOSIS

The benign, dominant form of osteopetrosis[3] has a considerable variety of expression. It may be asymptomatic, or may present with fractures. As many as 33 fractures have been observed in one patient.[3] Osteomyelitis is a common complication and mandibular osteomyelitis is a particular problem. Optic atrophy is rare in these patients but has been reported. Palsies of cranial nerves II, III and VII have been seen in about 15% of patients, as has frontal bossing. A smaller percentage of patients may appear exophthalmic. They do not have hepatosplenomegaly. On the other hand, the

Fig. 2.—Roentgenogram of the skull of a patient with osteopetrosis, illustrating the mask-like appearance of the dense bone around the eyes. (Courtesy of Dr. Rodrigo Loria-Cortes of San Jose, Costa Rica.)

Fig. 3.—B. L. (Case Report). A 4-month-old baby with osteopetrosis. He was pale, hypotonic and weak. The forehead was very prominent, giving the face a relatively small appearance. Actually, the head circumference was in the 75th percentile, and over-all the baby was below the third percentile for his age; thus, he was really macrocephalic. The abdomen was protuberant.

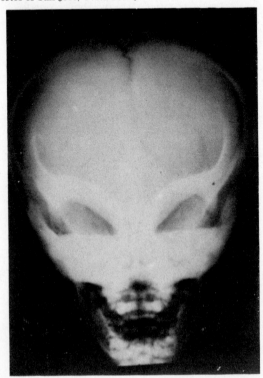

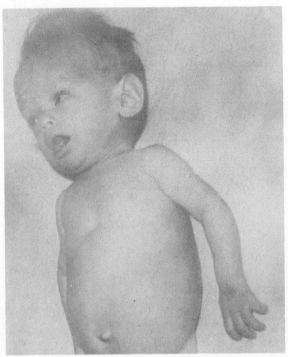

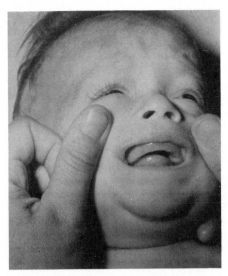

Fig. 4.–B. L. Close-up of face and mouth. Patients with this disease always have marked thickening and sclerosis of the maxilla and mandible. This causes problems in eruption of teeth.

roentgenographic picture may be indistinguishable from the infantile form. Bone pain is common. Anemia may be prominent but usually is not. The level of acid phosphatase may be elevated.

Pyknodysostosis (see chapter, this section) also is characterized by dense bones, prone to multiple frac-

Fig. 5.–B. L. Roentgenogram of the lower extremities. The bones were extremely dense. This is little or no evidence of a marrow cavity. In addition, at the ends of the long bones rather subtle horizontal areas of metaphyseal translucency can be seen.

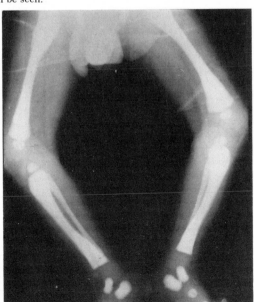

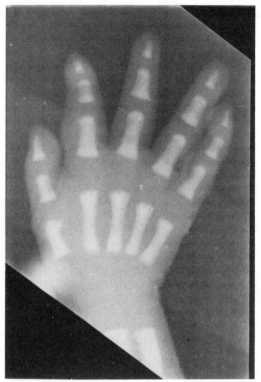

Fig. 6.–B. L. Roentgenogram of the hands. The dense bones are striking here, as elsewhere. The phalanges give the characteristic appearance of bone within bone, which is seen only in this condition.

tures. These patients have resorption of the distal ends of the phalangeal tufts and dysplastic clavicles. Macrocephaly is associated with wide-open fontanels and sutures. The mandibular angles are obtuse.

Generalized sclerosis of bone also may be seen in acute lymphocytic or myelocytic leukemia. The disease also should be differentiated from the Engelman disease or progressive diaphyseal dysplasia, which is characterized by leg pains, a waddling gait and fusiform enlargement of the diaphyseal areas of the long bones because of increased density of the bony cortex in these areas. The vertebrae and ribs of these patients are normal. This disorder is transmitted as an autosomal dominant.

CASE REPORT*

B. L., a 4-month-old white male, was seen at the U. S. Naval Hospital Walk-In Clinic with a history of vomiting and diarrhea for 12–24 hours. He was the product of a full-term pregnancy and had a birth weight of 6 lb. Head size at birth was in the 75th percentile, but the length was below the third percen-

*We are indebted to Dr. A. L. Lightsey of the U. S. Naval Hospital, San Diego, for the details of this patient and permission to report him here.

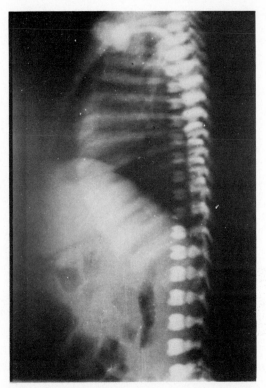

Fig. 7.—B. L. Roentgenographic appearance of the vertebrae. The anterior notching is striking, as is the adjacent linear translucency, which gives the vertebrae the appearance of a hamburger bun with a lucent hamburger in between. The bones were, of course, very dense.

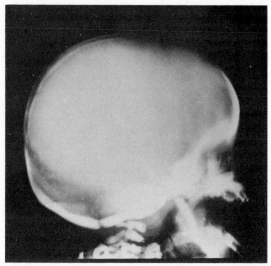

Fig. 8.—B. L. Roentgenogram of the skull. The base of the skull is very dense and thick. The orbital ridges are outlined by dense, opaque bone.

tile. The 17-year-old mother was well, as was the father. There was no consanguinity. An older sibling was in good health. The infant had always been a poor feeder and failed to thrive.

At 4 months he was very pale and had a protuberant abdomen (Fig. 3). He was hypotonic and had a weak cry. His weight was 8 lb, 3 oz, and his length 22 inches, both less than the third percentile. Frontal bossing was prominent. The pupils reacted to light but the discs were chalky white. There were marked thickening and sclerosis of the maxilla and mandible (Fig. 4). The liver was palpable two fingerbreadths below the costal margin. The spleen also was large and was palpable 6–7 cm below the left costal margin. There was mild wasting of the extremities. Motor development was grossly retarded. The patient could not lift his head in the prone position and did not respond to noise, although his gaze appeared to follow a moving light. He did not laugh or smile.

The hematocrit was 22%. The leukocyte count was 26,600/cu mm; there were 6% bands, 61% lymphocytes, 5% basophils and 26% nucleated red cells/100 nucleated cells. The reticulocyte count was 2.3%, and the platelet count 162,000/cu mm. The serum concentration of calcium was 8.9 and of phosphorus 7.2 mg/100 ml. On the fourth day after admission, the hemoglobin was 6.2 mg/100 ml. The reticulocyte count had risen to 9.3%, and the platelet count was 19,000/cu mm. Roentgenographic survey of the skeleton revealed the generalized dense sclerotic bones characteristic of osteopetrosis (Figs. 5–8). There were fractures of both humeri.

The patient was treated with blood transfusions and predni-

sone. At 18 months he was blind and deaf and had profound anemia, thrombocytopenia and hepatosplenomegaly.

Genetics

The infantile malignant type of osteopetrosis is transmitted as an autosomal recessive. Consanguinity is frequent.[4, 12, 13] The benign type is transmitted as an autosomal dominant.

Treatment

There is no really effective treatment. Repeated transfusions are necessary for anemia. Splenectomy may be useful in the presence of hypersplenism. Good oral hygiene is important. The possibility of osteomyelitis, particularly of the mandible, should be kept in mind and treated vigorously. Prednisone and a low calcium diet have been used.[4] Infants with the malignant form of the disease have been said rarely to survive the first year of life.[4] However, some exceptions have been reported. Death usually is due to anemia or sepsis.

REFERENCES

1. Albers-Schönberg, H.: Röntgenbilder einer seltenen Knochenerkrankung. Aerzlicher Verein Hamburg, 1904, Munch. Med. Wochenschr. 51:365, 1904.
2. Karshner, R. G.: Osteopetrosis, Am. J. Roentgenol. Radium Ther. Nucl. Med. 16:405, 1926.
3. Johnston, Jr., C. C., Lavy, N., Lord, T., Vellios, F., Merritt, A. D., and Deiss, W. P., Jr.: Osteopetrosis. A clinical, genetic, metabolic, and morphologic study of the dominantly inherited, benign form, Medicine 47:149, 1968.

4. Yu, J. S., Oates, R. K., Walsh, K. H., and Stuckey, S. J.: Osteopetrosis, Arch. Dis. Child. 46:257, 1971.
5. Keith, C. G.: Retinal atrophy in osteopetrosis, Arch. Ophthalmol. 79:234, 1968.
6. Gamsu, H., Lorber, J., and Rendle-Short, J.: Haemolytic anaemia in osteopetrosis, Arch. Dis. Child. 36:494, 1961.
7. Gwinn, J. L., and Lee, F. A.: Radiological case of the month, Am. J. Dis. Child. 124:91, 1972.
8. Dent, C. E., Smellie, J. M., and Watson, L.: Studies in osteopetrosis, Arch. Dis. Child. 40:7, 1965.
9. Murphy, H. M.: Calcitonin-like activity in the circulation of osteopetrotic grey-lethal mice, J. Endocrinol. 53:139, 1972.
10. Marks, S. C., Jr., and Walker, D. G.: The role of the parafollicular cell of the thyroid gland in the pathogenesis of congenital osteopetrosis in mice, Am. J. Anat. 126:299, 1969.
11. Walker, D. G.: The induction of osteopetrotic changes in hypophysectomized, thyroparathyroidectomized, and intact rats of various ages, Endocrinology 89:1389, 1971.
12. Hanhart, Von E.: Uber die Genetik der einfach-rezessiven Formen der Marmorknochenkrankheit und zwei entsprechende Stammbäume aus der Schweiz, Helv. Paediatr. Acta 2:113, 1948.
13. Farber, S., and Vawter, G. F.: Clinical Pathological Conf., The Children's Hospital Medical Center, Boston, Mass., J. Pediatr. 67:133, 1965.
14. Aasved, H.: Osteopetrosis from the ophthalmological point of view. A report of two cases, Acta Ophthalmol. 48:771, 1970.
15. Sjölin, S.: Studies on osteopetrosis. II. Investigations concerning the nature of the anaemia, Acta Paediatr. 48:529, 1959.
16. Verdy, M., Beaulieu, R., Demers, L., Sturtridge, W. C., Thomas, P., and Kumar, M. A.: Plasma calcitonin activity in a patient with thyroid medullary carcinoma and her children with osteopetrosis, J. Clin. Endocrinol. Metab. 32:216, 1971.
17. Ambs, E., Mingers, A.-M., Podgorski, J., and Weber, H.: Leukämieartige Krankheitsbilder mit blastenähnlichen Zellen im Säuglingsalter, Z. Kinderheilkd. 104:308, 1968.
18. Sofferman, R. A., Smith, R. O., and English, G. M.: Albers-Schönberg's disease (osteopetrosis). A case with osteomyelitis of the maxilla, Laryngoscope 81:36, 1971.

PYKNODYSOSTOSIS

Cardinal Clinical Features

Shortness of stature; frontal bossing, prominent parrot-like nose, an obtuse mandibular angle; dense brittle bones that are prone to fractures; failure of closure of sutures and fontanels; hypoplasia of distal phalanges; wide, flattened ends of fingers; dysplasia of nails.

Clinical Picture

Pyknodysostosis was first recognized as an entity by Maroteaux and Lamy in 1962.[1, 2] They coined the term "Pycnodysostose" to indicate the two elements of the disease. The first part of the word comes from the Greek πυκνος, which means dense, and describes the abnormal character of the bones; the dysostosis indicates the many malformations of the skeleton, cranium and fingers. It was these investigators who developed the theory[3] that Henri Toulouse-Lautrec, the French impressionist, had, in fact, had pyknodysostosis; it had previously been proposed that he had osteogenesis imperfecta. Toulouse-Lautrec had the characteristic short stature and short, square hands with short fingers. He also had a disproportionately large cranium and a small mandible and face. His beard could have been a cover-up for his receding chin, and his ubiquitous hat would have covered and might have been thought to protect an imperfectly ossified skull. He had had two fractures as a result of minor trauma. Furthermore, his parents were first cousins, a significant clue to an autosomal recessive disease.

Pyknodysostosis is a rather rare disease. Yet, in 1967 Elmore[4] was able to review 33 cases in the literature. Shortness of stature is one of the hallmarks of the syndrome. The adult height of patients ranges from 53–60 inches. The cranium is large. There is imperfect closure of the lambdoid and sagittal sutures, and the anterior and occipital fontanels remain open. Pyknodysostosis is one of the few conditions in which the fontanel remains open in an adult. There is frontal and occipital bossing. The face is small, the maxilla hypoplastic and the chin receding. The eyes may be prominent or appear proptosed. The nose is parrot-like. A variety of oral and dental anomalies are associated with pyknodysostosis. These include persistence of the deciduous teeth, so that a double row of teeth may be seen.[5] There may be partial anodontia.[6] Crowded irregular and malpositioned teeth are common and teeth may be missing. There is enamel hypoplasia. There may be premature or delayed eruption. The arch of the palate is high, and it may be grooved. Intelligence is normal.

The thorax is narrow and there may be a pectus excavatum. The extremities are short. The fingers and toes are short and blunt, and the skin is wrinkled. The nails are thin and brittle and appear to fold over the ends of the fingers.

Roentgenographic findings include a generalized increase in density of all the bones and thickening of the base of the skull. The sutures and fontanels remain wide open (Fig. 1). There is a wormian bone pattern over the parietal bone. Hypoplasia and lack of pneumatization of the paranasal sinuses are characteristic, as is an obtuse or straightened angle of the mandible. The ends of the clavicles are dysplastic, particularly at the acromial ends. Spondylolisthesis of the lumbar or sacral vertebrae is common. The acetabula are shallow and have increased angulation with the vertical plane of the body.[7] The medullary canals of the bones may still be distinguishable. These dense bones are brittle and fragile, and multiple fractures, especially in the lower extremities, are the rule. Fractures of the mandible or maxilla, especially during tooth extraction, are relatively common. Absence or poor development of the distal phalanges is a prominent feature, with partial or complete absence of bony structure in the index fingers (Fig. 2). These changes have been referred to as acro-osteolysis.[8]

Differential Diagnosis

The differential diagnosis of pyknodysostosis includes differentiation from cleidocranial dysostosis (see chapter, this section), in which there is partial or complete absence of the clavicles and very delayed closure of the sutures and fontanels. Cleidocranial dysostosis is transmitted as an autosomal dominant. Differentiation from osteopetrosis, or Albers-Schönberg disease (see chapter, this section) is relevant because of the density of the bones. Patients with pyknodysostosis never develop the anemia or evidence of extramedullary hematopoiesis characteristic of osteopetrosis. There is no cranial nerve involvement in

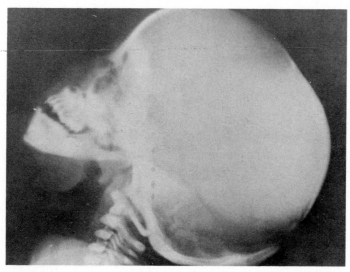

Fig. 1.—M. Z. (Case Report). Roentgenogram of the skull. The angle of the mandible is characteristically obtuse. In this case the angle was so wide that there appeared almost to be a straight line from the tip of the mandible to the upper mandibular joint. The base of the skull and occiput were densely sclerotic. The posterior and anterior fontanels were still open. (Courtesy of Drs. O. Yujnovsky, H. Viale and A. Vincitorio of Argentina.)

Fig. 2.—M. Z. Roentgenograms of the hands and feet reveal the dense sclerosis of bone. The distal phalanges show the acro-osteolysis that is classic for this syndrome. It is most prominent in the index fingers. (Courtesy of Drs. O. Yujnovsky, H. Viale and A. Vincitorio of Argentina.)

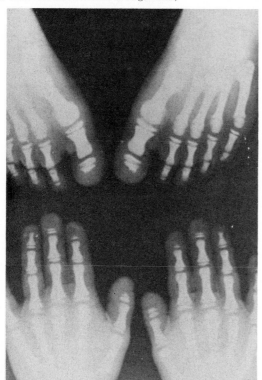

pyknodysostosis. Another disease with which this syndrome is sometimes confused is the Cheney syndrome, in which there are acro-osteolysis, multiple wormian bones and osteoporosis. Patients with this disease also have hypoplasia of the mandibular rami and collapsed vertebrae.[9] A syndrome reported by Young, Radebaugh and Rubin[10] is similar to pyknodysostosis. These patients have acro-osteolysis in which the distal phalanges, as well as the mandible and clavicles, are hypoplastic. The sagittal sutures are widened and open. In these patients the joints are limited in motion, and the skin atrophied. These patients also differ from those with pyknodysostosis in that they lack the characteristic density of the bones. Progeria shares with this syndrome the acro-osteolysis of the lateral ends of the clavicles, and the terminal phalanges.

Fig. 3.—M. Z. and F. Z. (Case Report). Two siblings with pyknodysostosis. There was frontal bossing and a prominent nose. (Courtesy of Drs. O. Yujnovsky, H. Viale and A. Vincitorio of Argentina.)

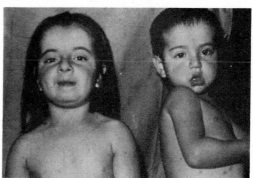

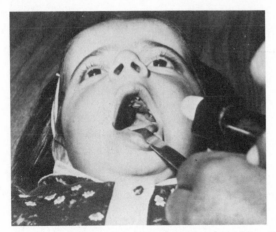

Fig. 4. – M. Z. The high-arched palate. (Courtesy of Drs. O. Yujnovsky, H. Viale and A. Vincitorio of Argentina.)

CASE REPORT*

M. Z. (female) and F. Z. (male) were siblings, both of whom had pyknodysostosis (Fig. 3). M. Z. was 8 years, 2 months old. Her height was 39½ inches and her weight 37½ lb. Height age was 4 years, and bone age 5½ years. She was a good student in the second grade. She presented with a fracture of the tibia at 8 years, 2 months.

Examination revealed micrognathia and a high-arched palate (Fig. 4), as well as a prominent nose and small face (Fig. 5). There was an abnormal implantation of the teeth (Fig. 6). The extremities were short and the fingers broad, with

Fig. 5. – M. Z. In the lateral view the prominent or parrot-like nose and small face are well demonstrated. The occipital prominence is obscured by the hair. The chin was small. (Courtesy of Drs. O. Yujnovsky, H. Viale and A Vincitorio of Argentina.)

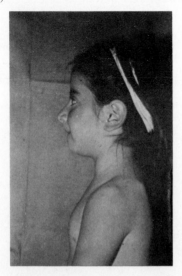

*We are indebted to Drs. O. Yujnovsky, Hercules Viale and A. Vincitorio for providing us with the details of their patients.

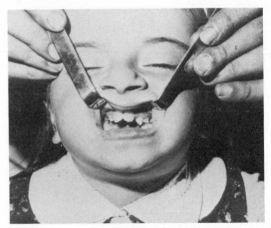

Fig. 6. – M. Z. The teeth were irregular in position and poorly aligned. They were hypoplastic, pointed and serrated. (Courtesy of Drs. O. Yujnovsky, H. Viale and A. Vincitorio of Argentina.)

dystrophic nails (Fig. 7). IQ was 79, which is borderline, and she came from a very poor environment.

F. Z., 2 years, 7 months old, presented with a tibial fracture. He looked exactly like his sister (see Fig. 1). The height was 32⅓ inches. Height age was 1½ years, as was the bone age. Fingers were broad but the toes looked normal. There was abnormal implantation of the teeth and a high-arched palate.

The mother was 47 and the father 52. Both were well, and there was no history of consanguinity. There were 8 normal siblings.

GENETICS

Pyknodysostosis is transmitted as an autosomal recessive disease. A number of families have been reported in which there was more than one affected sibling. Parental consanguinity has been prominent. In one family the disorder was present in a boy and his maternal uncle,[6] suggesting that these otherwise characteristic patients might represent an X-linked variant.

Fig. 7. – The hands, like the arms and legs, were short. The tips of the fingers over the distal phalanges were widened and flattened, giving a bulbous appearance of the ends of the fingers. The nails were soft, brittle and spoon shaped. (Courtesy of Drs. O. Yujnovsky, H. Viale and A. Vincitorio of Argentina.)

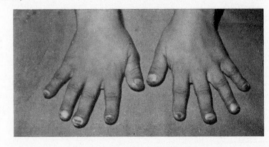

TREATMENT

None, except for the standard treatment of fractures. The prognosis for a normal life span is good.

REFERENCES

1. Maroteaux, P., et Lamy, M.: La pycnodysostose, Presse Med. 70:998, 1962.
2. Maroteaux, P., et Lamy, M.: Deux observations d'une affection osseuse condensante: La pycnodysostose, Arch. Fr. Pediatr. 19:267, 1962.
3. Maroteaux, P., and Lamy, M.: The malady of Toulouse-Lautrec, J.A.M.A. 191:715, 1965.
4. Elmore, S. M.: Pycnodysostosis: A review, J. Bone Joint Surg. [Am.] 49:153, 1967.
5. Sedano, H., Gorlin, R. J., and Anderson, V. E.: Pycnodysostosis, Am. J. Dis. Child. 116:70, 1968.
6. Shuler, S. E.: Pycnodysostosis, Arch. Dis. Child. 38:620, 1963.
7. Dusenberry, J. F., Jr., and Kane, J. J.: Pycnodysostosis, Am. J. Roentgenol. Radium Ther. Nucl. Med. 99:717, 1967.
8. Andrén, L., Dymling, J. F., Hogeman, K. E., and Wendeberg, B.: Osteopetrosis acro-osteolytica. A syndrome of osteopetrosis, acro-osteolysis and open sutures of the skull, Acta Chir. Scand. 124:496, 1962.
9. Dorst, J. P., and McKusick, V. A.: *Acro-osteolysis (Cheney Syndrome)*, Birth Defects: Original Article Series (New York: The National Foundation-March of Dimes, 1969), p. 215.
10. Young, L. W., Radebaugh, J. F., and Rubin, P.: *A New Syndrome Manifested by Mandibular Hypoplasia, Acro-osteolysis, Stiff Joints and Cutaneous Atrophy (Mandibuloacral Dysplasia) in Two Unrelated Boys*, Birth Defects: Original Article Series (New York: The National Foundation-March of Dimes, 1971), p. 291.

NAIL-PATELLA SYNDROME

Cardinal Clinical Features

Dystrophic nails, hypoplasia or absence of the patellae, dysplasia of the elbows, iliac horns, nephropathy.

Clinical Picture

The nail-patella syndrome, or osteo-onycho-dysplasia, consists of an association between dysplasia of the nails and a number of distinct skeletal abnormalities. Other organs, particularly the kidneys, may be involved.

The various features of this syndrome have gradually been recognized and assembled. In 1897 Little[1] reviewed 42 patients who had congenital absence or delayed development of the patella. He had information on an additional family in which 18 individuals over four generations had no patellae and no thumbnails. In the same year Mayer[2] reported a mentally retarded man with absent patellae and dysplastic thumbnails. Turner[3] in 1933 studied two large kindreds in which there was a dominant pattern of transmission. He pointed out that there was a triad of dystrophy of the nails, hypoplasia or absence of the patellae and arthrodysplasia of the elbows. Iliac horns were first recognized in 1946 by Fong[4] and associated with this syndrome by Thompson, Walker and Weens[5] in 1949. This makes up a tetrad. The association with this syndrome of a nephropathy presenting as chronic nephritis was published in 1950 by Hawkins and Smith.[6]

Onychodysplasia.—The nail changes, always symmetrical, are the most regularly observed abnormality in this syndrome. A spectrum of nail abnormalities ranges from longitudinal ridging to complete absence of the nails. The thumbnail is the one most frequently involved (Fig. 1). Often there is a graded decrease in severity from the thumb to the fifth finger. The toenails, always less affected than the fingernails, may be normal. Changes in the fingernails have been seen at birth. The deformities persist throughout life. The typical dysplastic nail has a normal-appearing base, but the nail is thinner than normal and disappears about halfway down the nail bed. This permits the fleshy portion of the finger or thumb to turn backward over the nail, often making it difficult to pick up pins or small objects. Less severe abnormalities include small or spoon nails, as well as an unusual softness and thinning. These nails crack, split or fray distally and often develop a brownish discoloration.

Patellar dysplasia.—The patellae may be completely absent, very small or only moderately hypoplastic (Figs. 2 and 3). The patella may be tripartite or polygonal. Recurrent dislocation of the patella laterally is frequent because of hypoplasia of the patellar tendons and it is particularly likely to occur on descending the stairs. The anterior aspect of the knee may present a flat or concave appearance when the legs are extended. Hyperextension easily produces an appearance of genu recurvatum. The knee joints are loose and permit more lateral motion than normal. The lateral femoral condyles often are hypoplastic and the medial femoral condyles prominent. The head of the fibula may be small. Subluxation of the knee joint occurs readily, and there frequently is early onset of osteoarthritis of the knee.

Dysplasia of the elbows.—This disorder must be considered in the differential diagnosis of those conditions with limited mobility at the elbows (Fig. 4). There may be an inability to extend, pronate or supinate fully. There may be an increased carrying angle. These problems usually result from hypoplasia of the radial head, which may be small and dislocated dorsally, or from bowing of the radial shaft. The internal condyles of the humerus often are unusually prominent.

Iliac horns.—The horns produce no symptoms or obvious deformity, but they are a striking manifesta-

Fig. 1.—Thumbs of a patient with the nail-patella syndrome. The nails were dystrophic and have both split.

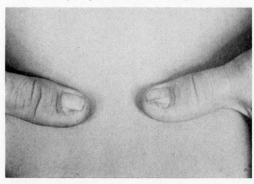

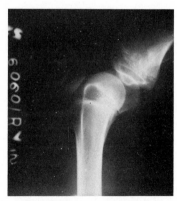

Fig. 2.—Lateral roentgenogram of the knee. The patella was hypoplastic.

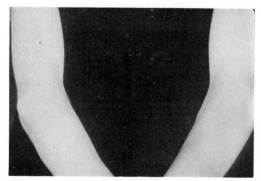

Fig. 4.—Limitation of motion at the elbows in a patient with the nail-patella syndrome.

tion (Figs. 5 and 6). They often can be palpated in the glutei. These bilateral, symmetrical protuberances arise from the posterior surface of the ileum. A cartilaginous, epiphysis-like cap may be seen even in an adult. It is often the appearance of these horns on a roentgenogram that leads a radiologist to suggest the diagnosis. Iliac horns are pathognomonic of the nail-patella syndrome; they appear not to occur in any other condition.

The incidence of these common manifestations of the nail-patella syndrome varies. In a series of 25 patients from five kindreds, Lucas and Opitz[7] found nail dysplasia in 20 and knee and elbow dysplasia in 15 each.

From a review of 62 published patients, Carbonara and Alpert[8] calculated an incidence of 98% nail anomalies, 90% elbow involvement and 92% involvement of the knees. Iliac horns were present in 81% and palpable in 71%.

RENAL DYSPLASIA.—Renal dysplasia is a much less common manifestation. It was seen in none of the pa-

Fig. 3.—Roentgenogram of the knee. Hypoplasia of the patella also can be appreciated in this view.

tients of Lucas and Opitz.[7] It is important to be alert to the possibility of this aspect of the syndrome, as it may be fatal.[9] On the other hand, death due to hereditary nephropathy in this syndrome is unusual. The renal lesions usually are asymptomatic, manifesting only proteinuria. Casts and occasional red or white cells may be found in the urine. The pathologic anatomy of the involved kidney is that of chronic glomerulonephritis. There is diffuse hyaline thickening of the glomerular basement membrane progressive to complete hyalinization of the glomeruli. Foam cells of the type observed in the Alport syndrome of hereditary nephritis with nerve deafness have never been seen in the nail-patella syndrome.

THE EYES (LESTER'S SIGN).—An abnormality in the pigmentation of the iris is seen in this syndrome.[10] A clover-leaf-shaped area of dark pigmentation occurs around the inner margin of the iris. This abnormality has been described in six of 13 patients examined.[8] The heterochromia is more readily seen in patients with blue eyes. Other anomalies include microcornea and cataracts.

OTHER ABNORMALITIES.—A considerable variety of other abnormalities is seen in this syndrome. Anoma-

Fig. 5.—In an adult patient such as this one the iliac horns are visible.

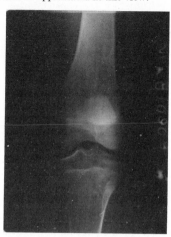

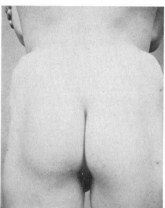

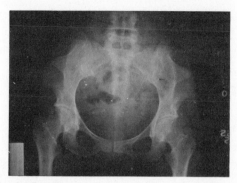

Fig. 6.—Roentgenogram of the pelvis of patient shown in Figure 5, illustrating the appearance of the iliac horns.

lies of the scapula, including thickening of the lateral border and hypoplasia of the glenoid, are common. Scoliosis, lordosis or even spina bifida is seen. Clubfoot has been described. These patients may have clinodactyly of the fifth fingers, laxity of the finger joints or flexion deformity of the fingers. Absence of the flexion creases over the distal interphalangeal joints has been described. Subluxation at the wrist,[11] as well as at the acromial end of the clavicle, has been observed. Multiple angiomatoses of the skin have been reported.[6] Mental retardation is not common but has been described.

CASE REPORT

A. H., a 16-year-old white male, was studied at the U. S. Naval Hospital at San Diego. He had a diagnosis of the nail-patella syndrome and associated nephropathy.

He had been noted to have dysplastic thumbnails at birth, but no other anomalies were observed at that time. At 8 years of age he had a right herniorrhaphy and orchiopexy for cryptorchidism. He was noted to have prominent iliac horns and proteinuria. An intravenous pyelogram was normal.

He was asymptomatic until 13 years of age when he developed dysuria. Urinalysis revealed many WBC and RBC, RBC casts, and a 4+ proteinuria. Audiogram was normal. The BUN and creatinine were normal. The 24-hour urinary excretion of protein was 6.1 gm.

Physical examination revealed visible iliac horns, dislocated and hypoplastic radial heads and apparently normal patellae. The 24-hour urinary excretion of protein was 2.75 gm.

A renal biopsy revealed a slight increase of mesangial cells and matrix, mild local thickening of the glomerular basement membrane, slight thickening of Bowman's capsule and nonfunctional glomeruli with proliferation, fibrosis and tubular atrophy. Immunofluorescent studies revealed an irregular deposition of IgG along the glomerular capillary walls and some mesangial areas and a local focal deposition of C′2 and C′4 in similar areas.

Histologic diagnosis was mild focal chronic membranous glomerulonephritis.

Study of the family revealed the mother to be similarly affected, except that she had normal nails. She had proteinuria and chronic urinary tract infection, but normal renal function. A maternal male first cousin had iliac horns, abnormal radial heads and dislocated patellae, but no nail changes or evidence of renal disease. The rest of this large family has been unavailable for study, but evaluation of photographs suggests that the maternal grandfather, a maternal uncle and possibly six of nine maternal siblings also have the disease. One maternal aunt died of kidney disease at 28 years of age.

GENETICS

The nail-patella syndrome is due to a dominant gene. Both sexes are affected in equal numbers. The gene is fully penetrant. Sporadic new mutation in association with increased paternal age has been reported.[7] The gene for the nail-patella syndrome has been found to be closely linked to the ABO blood group locus. Renwick and Lawler[12] established a 10% recombination frequency.

TREATMENT

Dislocations of the patella or at the joints may require orthopedic management.

REFERENCES

1. Little, E. M.: Congenital absence or delayed development of the patella, Lancet 2:781, 1897.
2. Mayer, H. N.: Congenital absence or delayed development of the patella, Lancet 2:1384, 1897.
3. Turner, J. W.: An hereditary arthrodysplasia associated with hereditary dystrophy of the nails, J.A.M.A. 100:882, 1933.
4. Fong, E. E.: "Iliac horns" (symmetrical bilateral central posterior iliac processes), Radiology 47:517, 1946.
5. Thompson, E. A., Walker, E. T., and Weens, H. S.: Iliac horns, an osseous manifestation of hereditary arthrodysplasia associated with dystrophy of the fingernails, Radiology 53:88, 1949.
6. Hawkins, C. F., and Smith, O. E.: Renal dysplasia in a family with multiple hereditary abnormalities including iliac horns, Lancet 1:803, 1950.
7. Lucas, G. L., and Opitz, J. M.: The nail-patella syndrome, J. Pediatr. 68:273, 1966.
8. Carbonara, P., and Alpert, M.: Hereditary Osteo-onychodysplasia (HOOD), Am. J. Med. Sci. 248:139, 1964.
9. Leahy, M. S.: The hereditary nephropathy of osteo-onychodysplasia, nail-patella syndrome, Am. J. Dis. Child. 112:237, 1966.
10. Lester, A. M.: A familial dyschondroplasia associated with anonychia and other deformities, Lancet 2:1519, 1936.
11. Karno, M. L.: The iliac horn syndrome, J. Bone Joint Surg. 44:1435, 1962.
12. Renwick, J. H., and Lawler, S. D.: Genetical linkage between the ABO and nail-patella loci, Ann. Hum. Genet. 191:312, 1955.
13. Clarke, C. A., McConnell, R. B., and Sheppard, P. M.: Data on linkage in man. The nail-patella syndrome, family S, Ann. Hum. Genet. 25:25, 1961.
14. Goodall, C. M.: Nail-patella syndrome coupled with blood group B in a New Zealand family, Ann. Hum. Genet. 26:243, 1963.
15. Similä, S., Vesa, L., and Wasz-Höckert, O.: Hereditary onycho-osteo dysplasia (the nail-patella syndrome) with a nephrosis-like renal disease in a newborn boy, Pediatrics 46:61, 1970.

MELNICK-NEEDLES SYNDROME

CARDINAL CLINICAL FEATURES

Severe bone dysplasia with bowing and S-like curvatures of the tubular bones and tortuous ribboned ribs, exophthalmos, hypertelorism, micrognathia, malalignment of the teeth.

CLINICAL PICTURE

Melnick and Needles[1] described two families in 1966 in which there were several patients with a dramatic dysplasia of bone in a number of generations. The facial features of the patients were characteristic (Fig. 1). They displayed a prominent appearance of the eyes, which resulted from exophthalmos and hypertelorism. Patients have micrognathia, a marked overbite and a receding chin. The teeth are poorly aligned. There is delayed closure of the anterior fontanel, and x-rays reveal sclerosis of the base of the skull. In the adult the cranium is thickened.

All of the bones are involved. The upper arms are abnormally short, especially when compared to the elongated forearms and hands. The scapulae are

Fig. 2.—(Case Report). Family portrait. The mother and two girls (S. R. on *left*) were clearly involved; the boy was normal. Variety of expression is evident in the severity of the pulmonary consequences only in S. R., whereas the appearance of the forearms and the face were striking in all three.

Fig. 1.—S. R. (Case Report). A 7-year-old girl with the Melnick-Needles syndrome. The tracheostomy indicates the potential severity of this condition. In addition, the facial appearance is characteristic. She had wide, prominent eyes and micrognathia. The chest deformity also is visible.

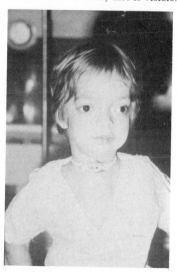

shortened. The arms turn away from the trunk and give an appearance of cubitus valgus. There is marked bowing of the radii and tibiae, causing S-like curvatures of these bones. Enlargement and flaring of the metaphyses are present on both ends of all of the long bones. The distal phalanges are short.

Coxa valga has been observed. The iliac bones are flattened in the area of the acetabulum and flared at the crest. Lumbar lordosis is common. Increased height of the lumbar and cervical vertebrae is characteristic of the syndrome. The vertebrae have anterior concavities, which produce a double-beaked appearance.

These patients have irregular constrictions and curvatures of the ribs that give these bones a wavy, ribboned appearance. This is one of the most striking roentgenographic features of the syndrome. The ribs are rather wide except near their attachment to the

427

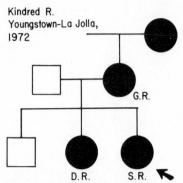

Fig. 3.—Pedigree of family shown in Figure 2. *Arrow* indicates S. R.

vertebrae, at which point they are abnormally narrow. There is cortical irregularity of the clavicle with medial flaring.

Hydronephrosis and hydroureter have been observed in one patient. Routine clinical laboratory determinations including calcium, phosphorus and alkaline phosphatase appear to be normal.

Melnick and Needles described a total of 18 patients with this syndrome. The youngest (Case Report) was diagnosed roentgenographically in utero at 6 months because of the classic appearance of the ribs and long bones. Coste, Maroteaux and Chouraki[2] described a 50-year-old woman with osseous abnormalities identical to those found in the patients described by Melnick and Needles. Their patient also experienced a spontaneous subluxation of the eye, which was successfully reduced. These authors referred to this entity as osteodysplasty. They also described another patient, 7 months old, in 1968.[3]

Fig. 4.—S. R. was small and had very narrow shoulders. The chest was striking. The abnormality of the forearms also is visible.

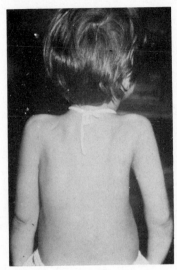

Fig. 5.—S. R. Posterior view of the chest indicating its narrowness, asymmetry and curvature.

Most of the patients described have had frequent respiratory infections. Frailness, shortness of stature and failure to thrive are common. The size of the thorax is disturbingly small: by 12 months it usually has fallen below the third percentile, and it may not grow significantly thereafter. The literature gives the impression that this is a benign condition. However, our patient (Case Report) had significant respiratory problems requiring tracheostomy and intensive care for survival. Ultimately, she died of the disease at 8 years of age. Another patient has died as a result of respiratory disease.

Fig. 6.—S. R. The forearm. There was a striking ulnar deviation of the hand and lower forearm. The entire forearm was bowed.

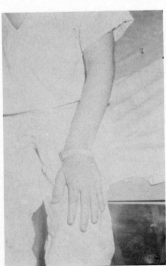

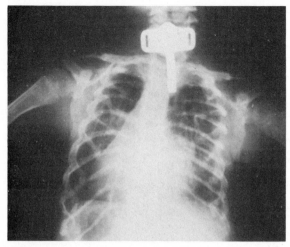

Fig. 7.—S. R. Roentgenogram of the chest. The ribs had the striking ribbon-like appearance that characterizes this syndrome.

This syndrome has clinical similarities to the Jeune syndrome, asphyxiating thoracic dystrophy. However, infants with the Jeune syndrome have respiratory difficulties early in infancy, whereas in this syndrome there is little early difficulty.

Case Report*

S. R., a white female, age 7 years, 10 months, was first seen at the U. S. Naval Hospital at San Diego. She was diagnosed in utero by Melnick and Needles and reported as a member of the second family in their original article.[1] Characteristic roentgenographic findings of the syndrome were present at 6 months of gestation.

She was delivered at full term and had a birth weight of 7 lb, 6 oz, and a length of 21 inches. She was found to have prominent eyes, hypertelorism and a receding chin. The anterior fontanel was large.

Exploration of the family history (Figs. 2 and 3) revealed that the 9-year-old sister of the patient had features and a clinical course similar to those of her sister but not so severe. An older brother was normal. The mother, aged 38, was 58 inches tall and weighed 98 lb. She had delivered all three children by cesarean section because of a narrow and triangular superior pelvic opening, also a characteristic of the syndrome. She had

*We are indebted to Dr. Harold Koenig of the U. S. Naval Hospital, San Diego, for permission to report the details of this patient.

all of the other features of the syndrome as well. Examination of a photograph of the grandmother indicated that she too had the disease.

Physical examination revealed a frail, underweight girl (see Fig. 1). She weighed 31 lb and had a height of 41 inches. Her height age was 4 years, 2 months, and her chronologic age 7 years, 10 months. The anterior fontanel was still open and measured 5×4 cm. Her eyes displayed an antimongoloid slant, hypertelorism and exophthalmos. There was a marked micrognathia and a severe overbite. The shoulders were very narrow, as was the chest (Figs. 4 and 5). The forearms and hands were abnormally long and turned away from the trunk because of significant ulnar deviation (Fig. 6). The legs were bowed. She had a severe lordosis.

Roentgenographic examination confirmed the presence of the characteristic osseous dysplasty of the syndrome (Fig. 7).

Throughout infancy and childhood the patient had repeated respiratory infections. At 14 months of age she had her first of many episodes of pneumonia, numbering 4–6 annually. At 6 years of age she developed respiratory failure, necessitating a tracheostomy, which never could be successfully removed. At 7 years of age she developed cor pulmonale. This was treated with diuretics and cardiac glycosides but was nevertheless progressive. She died from congestive heart failure and respiratory failure at 8 years of age.

Genetics

The Melnick-Needles syndrome is transmitted by a single autosomal dominant gene. Male to male transmission has been reported.

Treatment

There is currently no treatment for the bone dysplasia. Treatment for the respiratory and consequent cardiac problems deserves emphasis.

References

1. Melnick, J. C., and Needles, C. F.: An undiagnosed bone dysplasia. A 2 family study of 4 generations and 3 generations, Am. J. Roentgenol. Radium Ther. Nucl. Med. 97:39, 1966.
2. Coste, F., Maroteaux, P., and Chouraki, L.: Osteodysplasty (Melnick and Needles syndrome). Report of a case, Ann. Rheum. Dis. 27:360, 1968.
3. Maroteaux, P., Chouraki, L., and Coste, F.: L'ostéodysplastie (syndrome de Melnick et de Needles), Presse Med. 76:715, 1968.

CLEIDOCRANIAL DYSPLASIA
Cleidocranial Dysostosis

CARDINAL CLINICAL FEATURES

Absent or dysplastic clavicles, delayed closure of the cranial sutures and fontanels, dental dysplasia.

CLINICAL PICTURE

This syndrome has been generally known for some time. Early descriptions date from 1760,[1, 2] and the skull of a Neanderthal man has been reported to have the osseous and dental defects typical of the syndrome.[3] The classic paper was that of Marie and Sainton[4] in 1898. These authors recognized the disorder as a syndrome and its hereditary nature in the occurrence in a father and son, and in a mother and daughter, and gave it the name cleidocranial dysostosis. There may be a variety of other defects, but this term recognizes the prominent involvement of the clavicles and the membranous bones of the skull. As a generalized defect in the development of tissues programmed to calcify, it is more properly considered a dysplasia.[5]

Patients with cleidocranial dysplasia have a characteristic head in which there is a generalized delay in the development of membranous bone. This is most prominently manifest in the fontanels, which are persistently widely open into late childhood or even adulthood. Obvious pulsations may be seen over the area as late as 47 years of age. Even when it ultimately closes, the anterior fontanel leaves a deep depression or midline furrow. The head may be large. It is characteristically brachycephalic and has a demonstrable increase in its transverse diameter.[4] Frontal and parietal bossing may give the cranium a fetal appearance. The sutures also close very late. Numerous wormian bones may be found along the sutures. The facial bones may be underdeveloped, and the face may appear small in relation to the head. There may be hypoplasia or absence of the paranasal sinuses and mastoid air cells. The eyebrows often appear unusually raised and arched. Hypertelorism and a depressed nasal bridge may contribute to the facial appearance, and the palate may be high arched or, rarely, cleft.

The clavicle may be completely absent (Fig. 1). There may be virtually all degrees of dysplasia. The mildest is an absence of bone at the acromial end.

There alternatively may be an absence of the central part of the bone, with small stumps attached to the sternum and the acromion. Muscular defects in this region may accompany the osseous deficiencies. The defect in the clavicle often may be sufficiently minor that roentgen examination is necessary to confirm the diagnosis. Many patients may not know that there is anything wrong with their clavicles until it is pointed out by a physician. In the well-developed defect, the shoulders are narrow and drooping. The ability of these patients, as a consequence of the absence of the clavicles, to approximate their shoulders in front is unique.

Defective dental development is an integral feature of the syndrome. Dentition is regularly delayed. Some teeth may be totally absent. The teeth often are malformed, with cone shapes or accessory cusps. They become extensively carious. The deciduous teeth are commonly retained for years, and there may be supernumerary teeth. On the other hand, many patients will have lost all of their teeth by the age of 20 years. Dentigenous cysts are frequently seen.

A lack of mineralization or late ossification also is seen in other parts of the skeleton. This is especially true of the symphysis pubis, which may never fuse. Actually, ossification generally progresses in these patients, although very slowly, so that in the older patient the defects are less and less evident. Other skeletal defects include spina bifida occulta, cervical ribs, hypoplastic or deformed scapulae, genu valgum, coxa vara or valga and dislocation of the hips. The ossification of the proximal ends of the humeri and femora and other long bones may be irregular and may cause this syndrome to be confused with rickets. The distal phalanges may be short and hypoplastic, especially in the thumb and great toe. Because of the pelvic deformities, female patients may require cesarean section.

Intelligence usually is normal in these patients and they have a normal life span. Many patients with this syndrome have been quite short, but shortness of stature is not a necessary feature of the disease.

DIFFERENTIAL DIAGNOSIS

The disorder most commonly confused with cleidocranial dysplasia is pyknodysostosis (see chapter, this

Fig. 1.—Sculpture entitled *Cavalieri*, done in 1947 by Marino Marini, one of the giants of modern sculpture. The horse and its rider are a recurrent theme in Marini's work, and many of the riders appear to have a problem with their shoulders. The close approximation of the shoulders of this rider indicates that he has no clavicles. The ability to assume this position is the hallmark of cleidocranial dysostosis.

section), in which there are marked shortness of stature, wide open sutures and fontanels, an obtuse mandibular angle, dysplastic clavicles and partial or total absence of the distal phalanges. In pyknodysostosis the bones are dense, sclerotic and predisposed to fractures.

CASE REPORT

T. G., a boy, was admitted to the Jackson Memorial Hospital, Miami, Florida, with high fever and an area of inflammation on his arm. He was admitted because of the possibility of osteomyelitis. It turned out that the inflammatory reaction was a cellulitis, but a careful examination of the bones revealed certain striking abnormalities. It was possible to approximate his

shoulders because of the absence of clavicles (Fig. 2). This was documented by roentgenographic examination (Fig. 4). He also had defective ossification of the skull (Figs. 5 and 6). The anterior fontanel appeared clinically closed, but there was a permanent depression over the forehead in the midfrontal region (Fig. 3).

GENETICS

Cleidocranial dysplasia is transmitted by an autosomal dominant gene. Jackson[5] reported an interesting family in which there were at least 70 affected persons among 356 descendants of a Chinese man with the disorder who changed his name to Arnold, and became a Mohammedan in South Africa. This family was

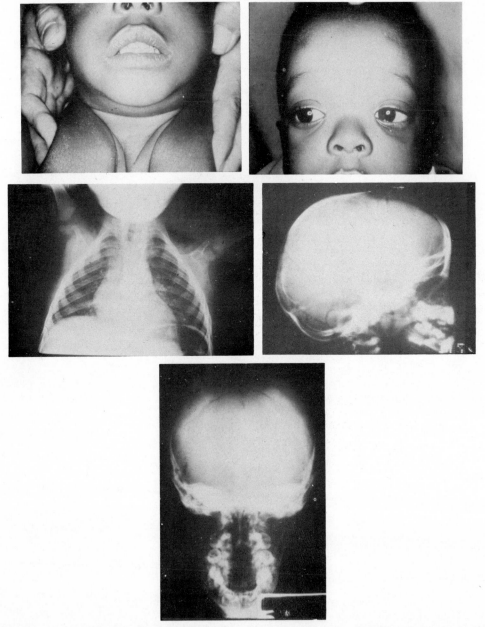

Fig. 2 (top left).—T. G. (Case Report.) A boy with cleidocranial dysplasia. He is shown in the classic position approximating the shoulders anteriorly, a position impossible for a person with clavicles.

Fig. 3 (top right).—T. G. The depression in the upper forehead is seen as a shadow. He also had prominent frontal bossing. Brachycephaly is a regular characteristic of the syndrome.

Fig. 4 (center left).—T. G. Roentgenogram of the chest. There were no clavicles.

Fig. 5 (center right).—T. G. Lateral roentgenogram of the skull illustrating widely separated sutures and wormian bones. The open frontal furrow extended forward from the anterior fontanel.

Fig. 6 (bottom).—T. G. Roentgenogram of the skull. The anterior and posterior fontanels were still open. Wormian bones are typical of this syndrome.

quite proud of the "Arnold head," so records of its presence were well kept. In this family affected parents studied had 70 affected children and 73 unaffected in perfect agreement with the mendelian ratio. Sporadic cases of this disorder due to new mutation also have been described.[6] There is a considerable variation in expression. An apparently unaffected parent of a patient may be found on x-ray to have evidence of the disease.

TREATMENT

The disorder is generally considered to be harmless, and it is a rather minor affliction. The dental problems are, on the other hand, considerable. The attentions of a dentist should be sought early.

REFERENCES

1. Terry, R. J.: Rudimentary clavicles and other abnormalities of the skeleton of a white woman, J. Anat. Physiol. 33:413, 1899.
2. Terry, R. J.: A skeleton with rudimentary clavicles, divided parietal bones and other anomalous conditions, Am. J. Anat. 1:509, 1902.
3. Greig, D. M.: Neanderthaloid skull presenting features of cleidocranial dysostosis, Edinburgh Med. J. 40:497, 1933.
4. Marie, P., and Sainton, P.: Sur la dysostose cleido-cranienne héréditaire (Rev. Neurol. 6:835, 1898), translated by Bick, E. M., Clin. Orthop. 58:5, 1968.
5. Jackson, W. P. U.: Osteo-dental dysplasia (Cleido-cranial dysostosis), Acta Med. Scand. 139:292, 1951.
6. Forland, M.: Cleidocranial dysostosis. A review of the syndrome and report of a sporadic case, with hereditary transmission, Am. J. Med. 33:792, 1962.
7. Lasker, G. W.: The inheritance of cleidocranial dysostosis, Hum. Biol. 18:103, 1946.
8. Anspach, W. E., and Huepel, R. C.: Familial cleidocranial dysostosis (cleidal dysostosis), Am. J. Dis. Child. 58:786, 1939.

LARSEN SYNDROME

CARDINAL CLINICAL FEATURES

Multiple congenital dislocations, especially at the knee; foot deformities; unusual facies; cleft palate. Mentality usually is normal.

CLINICAL PICTURE

The hallmark of the Larsen syndrome[1-3] is the occurrence of multiple congenital dislocations. Bilateral anterior dislocation of the tibiae on the femora is characteristic. The tibiae also may be bowed anteriorly. Patients also may have bilateral dislocation of the hip or elbow. They often have bilateral equinovarus or equinovalgus deformities of the foot. These patients often are mistaken for patients with arthrogryposis.

The fingers usually are long and cylindrical. The flattened facial appearance is created by wide spacing of the eyes, prominence of the forehead and a depressed bridge of the nose. Associated findings are deformities of the soft palate, cleft palate, cleft uvula or a midline groove of the uvula. Vertebral anomalies are common. In previously reported patients the central nervous system has been normal.

Latta and colleagues[4] recently have reported an additional patient and called attention to further characteristics including an unusual accessory bone posterior to the calcaneus. There may be transient diminished rigidity of the cartilage in the epiglottis and possibly in the trachea. The fingernails and metacarpals may be short. Extra creases may be seen on the fingers and palms.

CASE REPORTS

CASE 1.— E. B. was a 21-day-old white male born at term after an uneventful pregnancy. Delivery was by breech. He was found at birth to have multiple congenital anomalies (Fig. 1). His head size increased after birth, and he failed to gain weight adequately. He was discovered to be spastic. There was no history of a similar condition in the family.

Physical examination revealed the patient to have an oddly shaped head. The forehead was prominent. The eyes were widely spaced and the nasal bridge depressed. The sutures were widely open. Examination of the eyes revealed exotropia and a poor pupillary reflex. He had a high-arched palate. He

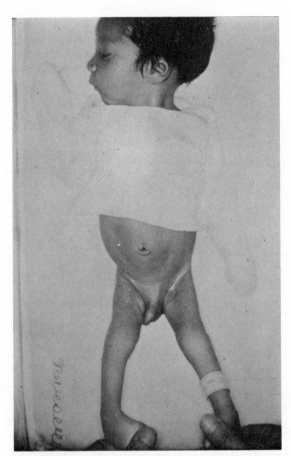

Fig. 1.— E. B. (Case Report). The lower extremities are in positions typical of the multiple dislocations of the Larsen syndrome.

had a striking genu recurvatum (Fig. 2). Club feet were present bilaterally. The hips were dislocated. Generally, the child appeared decerebrate, and there was bilateral spasticity of the extremities. Transillumination was grossly abnormal. The entire cranial vault transilluminated, indicating hydranencephaly. The EEG was flat.

Ventriculogram (Fig. 3) confirmed the presence of hydranencephaly. The only demonstrable brain tissue was at the base of the posterior fossa. The remainder of the intracranial cavity was filled with fluid. Roentgenographic examination of

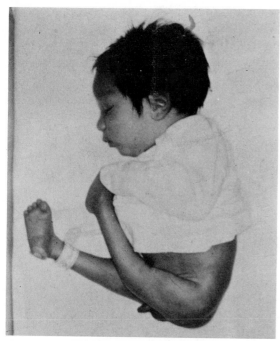

Fig. 2.–E. B. Lateral view illustrating striking genu recurvatum.

the extremities revealed bilateral dislocation of the hip and knees. There were bilateral distal femoral fractures (Fig. 4).

This particular combination of the Larsen syndrome with hydranencephaly has not been previously reported.

CASE 2.–R. H., a 23-month-old white male was referred for evaluation of multiple congenital anomalies. The child was the product of a normal pregnancy and delivery and weighed 7 lb, 5½ oz at birth. He was the first child of an 18-year-old mother and a 20-year-old father. There was no consanguinity. At birth, the infant was noted to have multiple dislocations of joints. He had severe bilateral dislocation of the knees, with lateral and anterior displacement of the tibiae on the femora. Both hips were dislocated, as were the elbows. He had bilateral club feet with valgus deformity. He also had long fingers with short palms. There was an umbilical hernia. His head was large and had frontal bossing and a broad, low nasal bridge.

Development was normal. He held his head up at 2 weeks, sat at 5½ months, and at 23 months had a vocabulary of eight or nine words. Hearing and vision were normal. He has been observed to have an exotropia. There was no family history of joint problems or a similar condition.

Physical examination revealed a boy with facial characteristics typical of the Larsen syndrome. He also had macrocephaly and a broad nasal bridge with hypertelorism. The ears were abnormal in shape. He had a high-arched palate and micrognathia. The uvula was bifid. Chest revealed pectus carinatum. He had an umbilical hernia. There was no deformity of the spine; major problems were in the extremities. He had dislocations of both elbows and shoulder joints. There was severe bilateral dislocation of the knee joint, with multiple operative scars from surgical correction. The feet also were involved. He had long fingers and the palms had extra creases. An intravenous pyelogram was normal, as were the ECG and EEG.

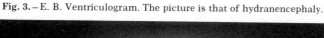

Fig. 3.–E. B. Ventriculogram. The picture is that of hydranencephaly.

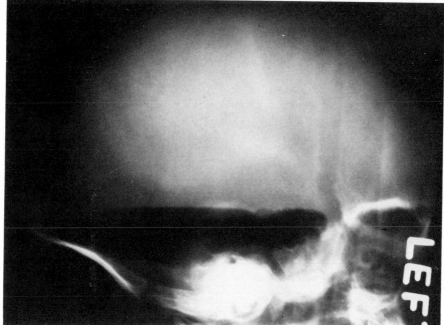

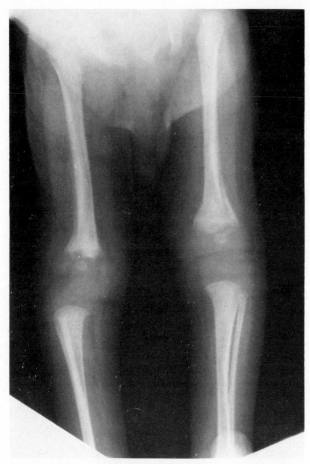

Fig. 4. – E. B. Roentgenogram of the lower extremities. The dislocations at the knee are apparent, as is some fragmentation at the ends of the femora.

GENETICS

This disorder appears to be inherited as an autosomal recessive trait. Several instances of multiple affected siblings have been observed. It also is possible that this is a dominantly inherited condition. The mother of one patient[4] had a depressed nasal bridge. Furthermore, a woman with congenital dislocation of the knees, hyperextension of the elbows and a depressed nasal bridge had three children with bilateral dislocations of the knee, each by a different father.[5]

TREATMENT

Orthopedic treatment has been directed chiefly to the knee joint. Foot deformities have been treated both surgically and by wedging plaster casts.

REFERENCES

1. McFarland, B. L.: Congenital dislocation of the knee, J. Bone Joint Surg. 11:281, 1929.
2. Gorlin, R. J., and Pindborg, J. J.: *Syndromes of the Head and Neck* (New York: McGraw-Hill Book Co., 1964).
3. Larsen, L. J., Schottstaedt, E. R., and Bost, F. C.: Multiple congenital dislocations associated with characteristic facial abnormality, J. Pediatr. 37:574, 1950.
4. Latta, R. J., Graham, C. B., Aase, J., Schan, S. M., and Smith, D. W.: Larsen's syndrome: A skeletal dysplasia with multiple joint dislocations and unusual facies, J. Pediatr. 78:291–298, 1971.
5. McFarlane, A. L.: A report on four cases of congenital genu recurvatum occurring in one family, Br. J. Surg.: 34:388, 1947.

Dysmorphic Syndromes with Prominent Facial and Ocular Abnormalities

TREACHER COLLINS SYNDROME
Mandibulofacial Dysostosis, Franceschetti Syndrome, Syndrome of Franceschetti and Klein

CARDINAL CLINICAL FEATURES

Antimongoloid slant, colobomas of lower lids, malar and mandibular hypoplasia, macrostomia, anomalous hypoplastic ears.

CLINICAL PICTURE

The first report on this syndrome was probably that of Thompson,[1] who described two patients in 1846 in whom malformations of the ears were associated with maxillary and mandibular hypoplasia. Berry[2] in 1889 described a mother and daughter with colobomas of the lower lids. In 1900 Dr. Treacher Collins,[3] a British ophthalmologist, described a patient with colobomas of the outer portion of the lower lids associated with hypoplastic malar bones. The syndrome has since been known in the English literature as the Treacher Collins syndrome. Franceschetti and Klein,[4] in an exhaustive review of the disorder in 1949, first used the term mandibulofacial dysostosis. These authors clearly described the complete syndrome as it is known today. It is often referred to in European literature as the Franceschetti or Franceschetti-Klein syndrome. The syndrome appears to represent defective development of the first visceral arch between the fifth and eighth months of fetal life.

The ocular features of this syndrome are characteristic. There is an antimongoloid slant (Fig. 1) and a notch, or coloboma, in the outer portion of each lower eyelid. The angular position of the eyes may be asymmetrical. Ectropion of the lower lid is common, and the lids may, in some cases, not be completely closed. Eyelashes are sparse or absent on the lower lid. The frontonasal angle is absent. The malar bones are hypoplastic. These patients have micrognathia and macrostomia (see Fig. 1), and there are malocclusions and abnormal positioning of the teeth. They may have a cleft or high-arched palate.

Often there are malformations of the external ear. The ear may be large, protruding or low set (Fig. 2);

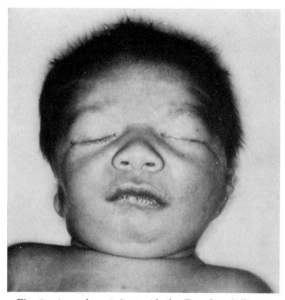

Fig. 1.—A newborn infant with the Treacher Collins syndrome. He had an antimongoloid slant, mandibular hypoplasia, macrostomia and a bulbous nose. This baby also had a cleft palate and glossoptosis. (Courtesy of Dr. David Allan of Mercy Hospital, San Diego.)

Fig. 2.—Same patient as in Figure 1. The receding chin and characteristic face are evident. His ear was low set and rotated. He had a small preauricular tag.

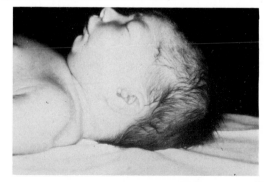

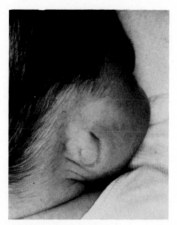

Fig. 3.—Same patient as in Figures 1 and 2. The other ear of this baby, illustrating extreme microtia, or a rudimentary ear.

more commonly there is microtia (Fig. 3). Preauricular skin tags frequently are seen. The external ear canal is narrow and may be absent. Conductive hearing loss is common. There may be malformation of the middle and inner ear, but total deafness is rare. Facial asymmetry may be present. A blind fistula may be seen between the corner of the mouth and the ear, and alternatively there may be grooves on the cheeks representing the residual facial clefts. A tongue of hair usually extends onto the cheek. Franceschetti and Klein described the profile of these patients as being fish- or bird-like.

Mentality in these patients usually is normal; retardation has been reported in about 5% of patients.

In Franceschetti and Klein's review,[4] they noted a patient reported by Nager[5] who had the characteristic facial features and, in addition, bilateral absence of the thumb and radioulnar synostosis. Mandibulofacial dysostosis with limb anomalies sometimes is referred to as Nager acrofacial dysostosis. Hypoplasia of a thumb also is seen.

Roentgenographic examination[6] reveals that the cranium in Treacher Collins syndrome is of normal size. It appears large because of the relatively small facial bones and mandible. The maxillae are hypoplastic, and the zygomatic arches are incomplete or absent. The mandible may be straightened, and its angle absent. Occasional deformities of the vertebrae are found.

DIFFERENTIAL DIAGNOSIS

The differential diagnosis of this syndrome includes the Goldenhar syndrome (see chapter, this section), in which there are colobomas of the upper lids. Other ocular anomalies include epibulbar dermoids and preauricular tags and atresia of the external ear canal. These patients regularly have vertebral abnormalities.

Hemifacial microsomia is similar, but all the abnormalities are unilateral. These include aplasia or deformity of the ear, preauricular skin tags, hypoplasia of the mandible and maxilla and ocular abnormalities. A family has been reported in which ectopia lentis was transmitted dominantly, along with mandibulofacial dysostosis and an antimongoloid obliquity but no coloboma.[7]

CASE REPORT

Six members of the S. family were seen in the Genetic Counseling Clinic of University Hospital. The problem for evaluation was a cleft palate in the mother, a daughter (M. S.) and baby son and an unusual facial appearance of the father and this daughter (M. S.).

M. S. was a 7-year-old girl who had the facial characteristics of the Treacher Collins syndrome (Fig. 4). She had developed normally intellectually. She had an antimongoloid slant and colobomas of the lower lids, on which there were very few lashes. The mandible was small and there was malar hypopla-

Fig. 4.—M. S. (Case Report). A 7-year-old girl with the Treacher Collins syndrome. She had an antimongoloid slant and colobomas of the lower lids. Eyelashes on the lower lids were very sparse. The mandible was small and the malar bones hypoplastic. The mouth was relatively large.

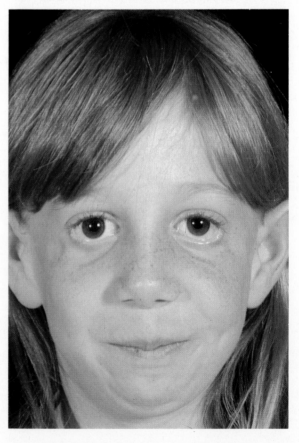

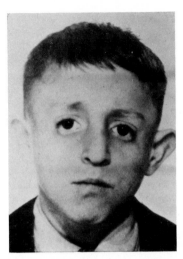

Fig. 5.—The father of M. S. as a child. He had an asymmetrical face with malar hypoplasia. Ears were large, low and prominent.

sia. In addition, she had a cleft of the soft palate. X-rays revealed a hypoplastic mandible and zygomatic arches.

The 31-year-old mother was healthy except for the presence of a cleft of the soft palate.

The father, age 30, had a facies similar to that of M. S.: an antimongoloid slant, colobomas of the outer portion of the lower lids, absent lower eyelashes, sunken cheeks, hypoplasia of the maxillae and a tongue of hair protruding onto the left cheek (Figs. 5 and 6). His face was asymmetrical, and there was malocclusion of the teeth. The ears were low set and large, and he had had conductive deafness since early childhood, for which he used a hearing aid.

The 2-year-old son had multiple congenital anomalies and

Fig. 6.—The father of M. S. as an adult. The position of the eyes was antimongoloid. There were no eyelashes on the lower lids. The colobomas were in the characteristic position in the outer part of the lower lids. He had a mild asymmetry of the face.

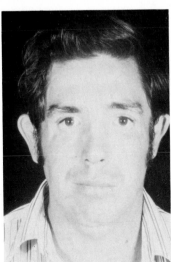

Fig. 7.—The grandfather of M. S. He had a small mandible and a small, malformed ear. The coloboma also is visible. This profile—in which the forehead and receding mandible slope from the peaked nose, the usual frontonasal depression is absent and the cheeks are flattened—has been described by Franceschetti as a fish-like or bird-like face.

delayed developmental milestones. He had blepharophimosis, and a cleft soft palate. He also had simian creases; proximally placed thumbs; long, slender fingers; and an anomaly of the right ear. Hearing tests revealed bilateral deafness.

The two other daughters were normal.

Pictures of the paternal grandparents indicated that the grandfather also had the facial characteristics of the Treacher Collins syndrome (Fig. 7). Pictures of the great grandparents appeared normal.

It was concluded that the father, the daughter (M. S.) and the deceased grandfather had the Treacher Collins syndrome. The mother, the daughter (M. S.) and the 2-year-old son had clefts of the soft palate. Thus, two dominantly inherited disor-

Fig. 8.—Pedigree of the S. family, indicating the dominant transmission of the syndrome. Two other dominant conditions were found in this kindred.

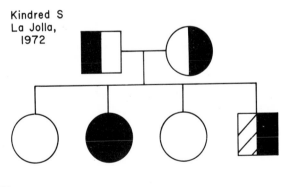

Kindred S
La Jolla,
1972

◐■ Treacher Collins Syndrome

■◑ Soft Cleft Palate

▨ Blepharophimosis

ders were found in this family (Fig. 8). The blepharophimosis in the son was thought to represent a third dominant, arising presumably by new mutation.

GENETICS

The Treacher Collins syndrome is transmitted as an autosomal dominant. The syndrome has been reported in a number of kindreds. About 60% of all reported cases represent new mutations. There appears to be complete penetrance of this gene, but there is a certain amount of variation in phenotypic expression.

TREATMENT

The ear anomalies of this syndrome should alert the physician to test as early as possible for deafness. Tomographic x-ray examination of the middle and inner ear should be done before surgery on the external ear deformity. Plastic surgery may improve the facial appearance.

REFERENCES

1. Thompson, A.: Article VI. Notice of several cases of malformation of the external ear, together with experiments on the state of hearing in such persons, Monthly J. Med. Sci. 75:420, 1846.
2. Berry, G. A.: Note on a congenital defect (?coloboma) of the lower lid, Ophthalmol. Hosp. Rep. 12:255, 1889.
3. Treacher Collins, E.: 8. Case with symmetrical congenital notches in the outer part of each lower lid and defective development of the malar bones, Trans. Ophthalmol. Soc. 20:190, 1900.
4. Franceschetti, A., and Klein, D.: The mandibulo-facial dysostosis. A new hereditary syndrome, Acta Ophthalmol. 27:143, 1949.
5. Nager, F. R., and DeReynier, J. P.: Das Gehörorgan bei den angeborenen Kopfmissbildungen, Pract. Oto-rhino-laryngol. (Suppl. 2) 10:1, 1948.
6. Stovin, J. J., Lyon, J. A., Jr., and Clemens, R. L.: Mandibulofacial dysostosis, Radiology 74:225, 1960.
7. Kirkham, T. H.: Mandibulofacial dysostosis with ectopia lentis, Am. J. Ophthalmol. 70:947, 1970.
8. Fernandez, A. O., and Ronis, M. L.: The Treacher Collins syndrome, Arch. Otolaryngol. 80:505, 1964.
9. Rovin, S., Dachi, S. F., Borenstein, D. B., and Cotter, W. B.: Mandibulofacial dysostosis, a familial study of five generations, J. Pediatr. 65:215, 1964.
10. Monnet, P., Boule, N., Neumann, E., Magnard, G. Y., and Humbert, G.: Deux cas de dysostose mandibulo-faciale ou syndrome de Franceschetti, Pediatrie 15:537, 1960.
11. Vatré, J. -L.: Etude génétique et classification clinique de 154 cas de dysostose mandibulo-faciale (syndrome de Franceschetti), avec description de leurs associations malformatives, J. Genet. Hum. 19:17, 1971.
12. Böök, J. A., and Fraccaro, M.: Genetical investigations in a North-Swedish population. Mandibulo-facial dysostosis, Acta Genet. 5:327, 1955.
13. Ombredanne, M.: Aplasies de l'oreille dans les syndromes de Franceschetti, Ann. Otolaryngol. 87:309, 1970.

WAARDENBURG SYNDROME

CARDINAL CLINICAL FEATURES

Lateral displacement of the medial canthi, broad nasal bridge, white forelock, heterochromia iridis, congenital nerve deafness.

CLINICAL PICTURE

Waardenburg[1] described the syndrome in 1948. A similar case had been reported previously in 1916 by van der Hoeve.[2] Among patients with this syndrome there is a considerable degree of variation in the expression of its features. It also is true that the various individual features may be common or rare throughout any individual kindred. The most constant feature is a lateral displacement of the medial canthi (dystopia canthorum), which always is present when one or more of the other abnormalities is present.[3] A broad nasal bridge is seen in about 80% of patients, although sometimes this is more apparent than real. The eyes may seem wide set at first but the interpupillary distance is normal, as is that between the outer canthi. The amount of sclera visible medial to the cornea is considerably reduced. The palpebral fissures are shortened. There also is a lateral displacement of the inferior lacrimal puncta. The superior puncta are in approximately normal position, but the inferior may be as far lateral as the cornea. Actually the superior puncta also are displaced along the lid margin, but the sharp downward turning of this lid carries it medially to a relatively normal distance from the midline.[3] Poor lacrimal conduction and tearing may be chronic complaints consequent to the abnormal punctal positions. These patients also may have hyperplasia of the medial portion of the eyebrows.[4]

A white forelock often is the finding that first alerts the physician to the possibility of the Waardenburg syndrome (Fig. 1). However, this partial albinism has been reported in only about 20% of some series.[4] In a number of patients it may be present only at birth, becoming pigmented early in infancy. On the other hand, these patients may develop premature graying of the hair, eyebrows or cilia. This may be seen as early as 7 years of age,[4] and gray eyebrows in the absence of a forelock may be the only equivalent seen. Partington[5] described a patient who had a white forelock as a child and whose hair turned completely gray at 23 years of age. Di George and colleagues described a similar patient.[4] Feingold and colleagues[6] described a newborn infant who had a white forelock that disappeared at 6 weeks. This infant developed iris bicolor at 3 months. His intercanthal distance was measured to increase over the first 6 months. In the adult female a white forelock is often obscured by the use of hair dye or, conversely, a normal woman may affect a white or gray forelock by the use of dyes. In adult males a white forelock may be lost to baldness.

Heterochromia iridis (Fig. 2) has been reported in 25% of cases.[4] However, it is probably true that in some families isochromic pale blue eyes are a manifestation of the syndrome, e. g., in black families, in which the eyes would not normally be blue. Some patients have hypoplasia of the iris stroma. Partial heterochromias, in one or both eyes, also are seen. Pigmentary changes in the fundus are sometimes observed.[7]

Pigmentary changes of the skin also may be seen in this syndrome[4] and may appear as vitiligo. In some patients there may be patches of what has been called albino hair. Depigmented skin may be seen anywhere on the head, trunk or extremities. There may be diffuse partial albinism of the entire body.[8]

Deafness is the major problem in this syndrome.[9, 10] In fact, a significant value of the other elements of the syndrome is to alert the physician to an early assessment of the auditory status of the patient. There may be a considerable variation in the impairment of hear-

Fig. 1.—Spectacular white forelock in a newborn infant who had a broad nasal bridge, laterally displaced medial canthi and a broad nose.

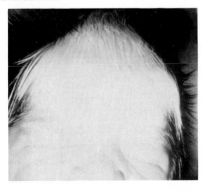

443

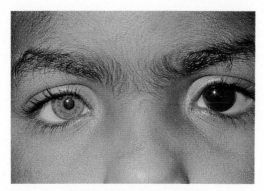

Fig. 2.—Heterochromia iridis.

Fig. 4.—C. B. (Case Report). A patient with the Waardenburg syndrome. She was retarded, whereas most patients are not. She had a small white forelock. These patients have a broad nasal bridge but, as illustrated, this patient's entire nose was broad. The inner canthi of the eyes were widely separated.

ing. It may be mild or profound. Deafness has been estimated to occur in about 20% of patients with this syndrome, and conversely this syndrome has been estimated to account for about 2% of congenital deafness.[4] Deafness usually is bilateral. It is perceptive or sensorineural and may cause mutism. Histologic examination has revealed an absence of the organ of Corti.[9] Deafness is known to occur in a variety of albino animals (Fig. 3), and is particularly common in white, blue-eyed cats. Histologic changes in the animals examined have been similar to those found in the Waardenburg syndrome.

Cleft lip and palate have been reported in this syndrome.[11] One patient was reported who had esophageal atresia.[9] The facial appearance of these patients may be characteristic.[3, 4, 9] The lips may be very full. The jaw often has a massive appearance,[9] and the face appears square. The tip of the nose may appear thin because of a decreased flaring of the alae. The tip often is rounded and upturned, making the columella visible head on.[4]

DIFFERENTIAL DIAGNOSIS

The differential diagnosis of the Waardenburg syndrome includes isolated lateral displacement of the inner canthus and true hypertelorism, in which the inner and outer canthal and interpupillary distances are increased. Dystopia canthorum also may be seen in the orofacial digital syndrome, which is thought to be a sex-linked dominant disorder lethal in the male. Patients with the orofacial digital syndrome also have hypoplasia of the alae nasi, clefts of the palate or lip, oral frenula, occasionally a lobulated hamartomatous tongue and digital asymmetry.

Fig. 3.—Heterochromia iridis is common in the Old English sheepdog. The white face is the preferred configuration for this breed; heterochromia often accompanies this white forelock equivalent.

Fig. 5.—C. B. The white forelock is visible, as well as the low-set ear.

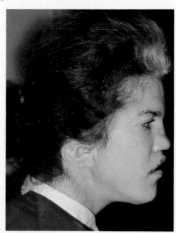

CASE REPORT*

C. B. was admitted at 19 years of age to the Fairview State Hospital in Costa Mesa for behavior problems, which were mainly temper tantrums and running away from home. She had been deaf and mute since birth and communicated only by sign language.

She was the product of a normal pregnancy and delivery, the second child of a 21-year-old mother and 22-year-old father. The mother was Rh negative and the father Rh positive, and the infant had pronounced neonatal jaundice. Development of motor skills was slightly delayed. The child was in special classes in school until a few months prior to her admission. There was no family history of a similar problem.

Physical examination revealed a well-nourished white female who was deaf and mute. She had a broad flat nose and lateral displacement of the inner canthi (Fig. 4). Her interpupillary distance was normal. She had a white forelock (Fig. 5). There was no heterochromia. The eyes were brown and she had low-set ears.

GENETICS

The Waardenburg syndrome is transmitted as an autosomal dominant.[3] In Waardenburg's study of segregation, he found 83 normal and 87 affected siblings, which is consistent with the expected 1:1 ratio. Five cases out of 16 reported by Waardenburg[3] were considered to represent fresh mutations. On the other hand, the degree of penetrance in this syndrome has been variable, making calculations of mutation rate particularly difficult. The patients first described were Dutch, but it is clear that this syndrome is widely distributed among the races of man.[4, 10, 12]

TREATMENT

The syndrome is compatible with a normal life span. Mental retardation is not characteristic, although a

---

*We are indebted to Dr. Charles Fish of Fairview State Hospital, Costa Mesa, California, for permission to photograph this patient and report on her.

patient with unrecognized severe deafness and mutism may develop abnormally. Early recognition of hearing loss and a vigorous approach to correction with hearing aids[9] can be important for learning.

REFERENCES

1. Waardenburg, P. J.: Dystopia punctorum lacrimarum, blepharophimosis en partiële irisatrophie bij een doofstomme, Ned. Tijdschr. Geneeskd. 92:3463, 1948.
2. van der Hoeve, J.: Abnorme Länge der Tränenröhrchen mit Ankyloblepharon, Klin. Monatsbl. Augenheilkd. 56: 232, 1916.
3. Waardenburg, P. J.: A new syndrome combining developmental anomalies of the eyelids, eyebrows and nose root with pigmentary defects of the iris and head hair and with congenital deafness, Am. J. Hum. Genet. 3:195, 1951.
4. Di George, A. M., Olmsted, R. W., and Harley, R. D.: Waardenburg's syndrome. A syndrome of heterochromia of the irides, lateral displacement of the medial canthi and lacrimal puncta, congenital deafness, and other characteristic associated defects, J. Pediatr. 57:649, 1960.
5. Partington, M. W.: An English family with Waardenburg's syndrome, Arch. Dis. Child. 34:154, 1959.
6. Feingold, M., Robinson, M. J., and Gellis, S. S.: Waardenburg's syndrome during the first year of life, J. Pediatr. 71: 874, 1967.
7. Goldberg, M. G.: Waardenburg syndrome with fundus and other anomalies, Arch. Ophthalmol. 76:797, 1966.
8. Klein, D.: Albinisme partiel (leucisme) accompagné de surdimutité, d'ostéomyodysplasie, de raideurs articulaires congénitales multiples et d'autres malformations congénitales, Arch. Julius Klaus Stiftung. 22:336, 1947.
9. Fisch, L.: Deafness as part of an hereditary syndrome, J. Laryngol. Otol. 73:355, 1959.
10. Houghton, N. I.: Waardenburg's syndrome with deafness as the presenting symptom. Report of two cases, N. Z. Med. J. 63:83, 1964.
11. Giacoia, J. P., and Klein, S. W.: Waardenburg's syndrome with bilateral cleft lip, Am. J. Dis. Child. 117:344, 1969.
12. Chewkheng, L., Chen, A. T., and Tan, K. H.: A Chinese family with Waardenburg's syndrome, Am. J. Ophthalmol. 65:174, 1968.

PIERRE ROBIN SYNDROME
Pierre Robin Anomalad

CARDINAL CLINICAL FEATURES

Micrognathia, glossoptosis, cleft palate, ocular abnormalities, respiratory distress.

CLINICAL PICTURE

This disorder has an unusual eponymic designation in that it bears the first and last names of Pierre Robin, who published on the subject in 1923.[1] The syndrome is characterized by three defects: micrognathia, glossoptosis and cleft palate. The infant usually breathes better in the prone position than in the supine. This is helpful in glossoptosis as the tongue tends to fall forward through gravity. Babies with hypoplasia of the mandible have serious problems with breathing. They may present as a respiratory emergency in the neonatal period. Choking spells are common, as are recurrent bouts of cyanosis.

Death in this syndrome may result from respiratory obstruction or complicating pneumonia. Hypoxia, hypercapnia, cor pulmonale and pulmonary edema have been reported.[2] Difficult feeding (Fig. 1), nutritional deficiency and failure to thrive or growth retardation may be seen, as well as death due to inanition.

Most children with hypoplasia of the mandible breathe through the mouth. This has been said to create an adenoid facies, but the face of the patient with the Pierre Robin syndrome is so much more striking than that of a child with large adenoids that the term is not helpful. To readers of the comic strips of a previous generation, the term "Andy Gump facies" is helpful, and the designation is in the literature, along with that of bird-like facies or shrew-face. The base of the nose often is flattened. The fundamental defect is the small mandible. Robin recognized that the glossoptosis was secondary to this hypotrophy.[3] The tongue is of normal size, not large; it is simply in a space too small for it. Glossoptosis may be visualized in the lateral roentgenogram, and the degree of encroachment on the pharyngeal airway may be assessed in this fashion. The palatine anomalies associated with this syndrome may vary considerably. There may be a large, classic U-shaped cleft involving both the hard and soft palate, along with a bifid uvula. At the other extreme would be a high-arched palate. Patients with cleft palate tend to have recurrent otitis media. The ears are often low set and the pinnae small and poorly shaped. A few patients have had hearing loss.

The association of ocular lesions with this syndrome was recognized by Smith and Stowe.[4] A variety of lesions have been encountered, ranging in severity from congenital glaucoma to esotropia. Congenital cataracts have been seen, as have microphthalmus and coloboma of the choroid in one patient. Also described was a patient with the Moebius syndrome. It has been recommended that every infant with the Pierre Robin syndrome have a complete ophthalmologic examination under anesthesia.

Many other anomalies have been seen in babies with this condition. These include congenital heart disease,

Fig. 1.—P. G. Pierre Robin syndrome. This infant with rather mild micrognathia had such difficulty in feeding that he presented to hospital with an admission diagnosis of marasmus.

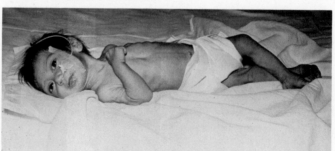

446

club feet, syndactyly and choanal atresia. Hydrocephalus and cerebral cortical atrophy have been reported. Mental retardation was present in eight of 28 patients in a series.

CASE REPORT

R. A. was a 4-day-old female infant, admitted to University Hospital because of a hypoplastic mandible and associated abnormalities. The patient had been delivered at term after an uneventful pregnancy. Birth weight was 5 lb, 4 oz. At birth the patient was flaccid, apneic and only minimally responsive to stimulation. It was noted at that time that she had a severely hypoplastic mandible. In addition, there were glossoptosis and complete absence of the soft palate. She was resuscitated using positive pressure ventilation, sodium bicarbonate and Nalline, and there was improvement in tone and responsiveness as well as initiation of spontaneous respirations. The patient had no respiratory distress thereafter except for a brief episode of grunting and retraction when placed on her back at 24 hours. She could not take feeding by nipple without reflux through the nose.

A 6-year-old sister was noted to have had a hypoplastic mandible, which had been slowly correcting itself with age, and a high-arched palate. The father stated that many members of his side of the family had had small jaws.

Physical examination revealed the characteristic appearance of the syndrome. At age 4 days the patient weighed 5 lb, 5 oz. She developed respiratory distress when placed on her back, which was relieved by placement on her side or in a prone position. The mandible was severely hypoplastic (Figs. 2 and 3). The tongue was very small and retrolocated (Fig. 4). The hard palate was completely intact, but the soft palate and uvula were absent. There was mild pectus excavatum. The liver was palpable 1 cm below the costal margin. Dermatoglyphics were normal. Physical examination was otherwise within normal limits. Blood gases were normal.

Chest film revealed a resolving right middle lobe pneumonia, which later resolved completely. Skull films were normal.

Fig. 2.—R. A. (Case Report). Pierre Robin syndrome. Profile illustrates the markedly foreshortened jaw.

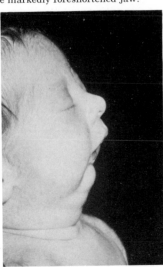

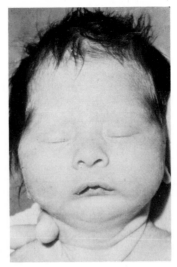

Fig. 3.—R. A. The micrognathia also is evident in the frontal position.

Films of the airway revealed impingement by the dorsally placed tongue. The proximal tibial ossification centers were missing, indicating slight prematurity.

As she was unable to feed by nipple, she was fed by gavage and was kept in the prone position. Her hospital course was otherwise uneventful, and she was discharged after 2 weeks. She was seen again at 2 months of age, at which time she appeared perfectly normal except for her small mandible.

GENETICS

Two affected brothers were reported by Smith and Stowe.[4] The mother stated that this disorder had been in her family for four generations. This would be consistent with autosomal dominant inheritance with variable penetrance.

It is generally agreed that this syndrome or anomalad is heterogeneous. It may occur in any infant with mandibular hypoplasia. Many of those with eye abnormalities have the Stickler Syndrome.[8]

Fig. 4.—R. A. Glossoptosis and cleft palate.

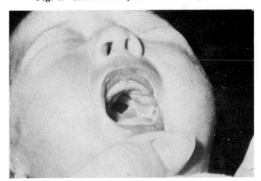

TREATMENT

If these infants can be kept alive and permitted to grow, differential growth invariably takes place. The ultimate cosmetic result usually is good. Symptoms are eliminated at least by the end of the first year. It is not known whether or not vigorous attention to ensuring normal oxygenation will prevent cerebral manifestations.

The extent of therapeutic maneuvers depends on the severity of the symptoms. In some patients tracheostomy is life saving and must be continued over a number of months. At the other extreme, skilled nursing care and management in the prone position are sufficient. Tube feeding may be very helpful. A number of operations have been devised to pull the tongue forward. None has seemed satisfactory in patients we have seen, although some excellent results have been reported. Similarly, it does not appear to add anything to cover the cleft with a prosthesis, although this too has its adherents. Once growth has taken place and acute symptomatology is in the past, the palate can readily be repaired. Ocular defects such as glaucoma or retinal detachment should be attended to promptly.

REFERENCES

1. Robin, P.: La chute de la base de la langue considérée comme une nouvelle cause de gêne dans la respiration naso-pharyngienne, Bull. Acad. Nat. Med. 89:37, 1923.
2. Jeresaty, R. M., Huszar, R. T., and Subrata, B.: Pierre Robin syndrome. Cause of respiratory obstruction, cor pulmonale, and pulmonary edema, Am. J. Dis. Child. 117:710, 1969.
3. Robin, P.: Glossoptosis due to atresia and hypotrophy of the mandible, Am. J. Dis. Child. 48:541, 1934.
4. Smith, J. L., and Stowe, F. R.: The Pierre Robin syndrome (glossoptosis, micrognathia, cleft palate). A review of 39 cases with emphasis on associated ocular lesions, Pediatrics 27:128, 1961.
5. Dennison, W. M.: The Pierre Robin syndrome, Pediatrics 36:336, 1965.
6. Routledge, R. T.: The Pierre Robin syndrome: A surgical emergency in the neonatal period, Br. J. Plast. Surg. 13:204, 1960.
7. Pruzansky, S., and Richmond, J. B.: Growth of mandible in infants with micrognathia, Am. J. Dis. Child. 88:29, 1954.
8. Stickler, G. B., Belau, P. G., Farrell, F. J., Jones, J. D., Pugh, D. G., Steinberg, A. G., and Ward, L. E.: Hereditary progressive arthro-ophthalmopathy, May. Clin. Proc. 40:433, 1965.

GOLDENHAR SYNDROME
Oculoauriculovertebral Syndrome

Cardinal Clinical Features

Epibulbar dermoid or lipodermoid, vertebral anomalies, preauricular appendages and blind fistulas, ear anomalies, facial asymmetry.

Clinical Picture

The syndrome is relatively common and has probably been recognized for some time. Its first systematic description was by Goldenhar[1] in 1952, who distinguished it from other abnormalities of structures arising from the first visceral arch.[2]

Ocular manifestations are prominent. The epibulbar dermoids, which may be bilateral in as many as 60% of cases, are usually located at the limbus, or corneal margin, in the lower outer quadrant of the eye (Fig. 1). In some patients a lipodermoid is found; these occur in the upper outer quadrant. A single patient may have both dermoid and lipodermoid, even in the same eye. A coloboma of the upper lid (Figs. 1 and 2) is common and usually unilateral. Patients with this syndrome also may have microcornea, microphthalmia or anophthalmia. Iris atrophy, polar cataracts or colobomas of the eyebrow are seen, as well as ptosis or antimongoloid obliquity. Colobomas also may involve the iris or the choroid.

The face usually is asymmetrical. There is unilateral hypoplasia of the maxilla and mandible, leading to micrognathia. Malar hypoplasia and micrognathia may produce a rather parrot-like appearance. There may be frontal bossing of the skull. Patients may have hypoplasia or atresia of the nostril. Macrostomia may be striking.

A variety of malformations occur around the ear (Fig. 3). Auricular appendices are the most constant finding. They are located in front of the tragus on a line from the tragus to the angle of the mouth. This is along the line of fusion of the maxillary and mandibular processes. Blind-ended fistulas often are seen in the same area. The ear may be small (microtia) and/or deformed or low set. Atresia or stenosis of the external auditory meatus may occur, and there may be conductive deafness.

Vertebral anomalies (Fig. 4) include hemivertebrae,

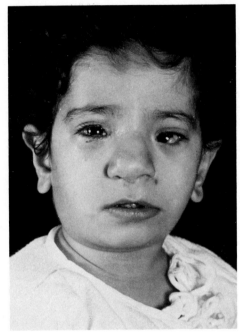

Fig. 1.—Epibulbar dermoid of left eye and colobomas of upper eyelid in right eye.

rib anomalies, occipitalization of the atlas, supernumerary vertebrae, cuneiform vertebrae, spina bifida, aplasia or lumbarization of sacral vertebrae. Roentgenographic examination may reveal hypoplasia of the bones of the thumb.

The muscles of the tongue or palate may be unilaterally hypoplastic or paralyzed. Muscles of facial expression, masseters or other muscles may be hypoplastic or aplastic on the involved side. Harelip and cleft palate are relatively frequent.

Most patients are of normal intelligence. Mental retardation has been reported in about 10% of patients. Epilepsy also has been reported.

Other abnormalities observed in patients with this syndrome are umbilical or inguinal hernia, pulmonary agenesis or hypoplasia and congenital heart disease.

449

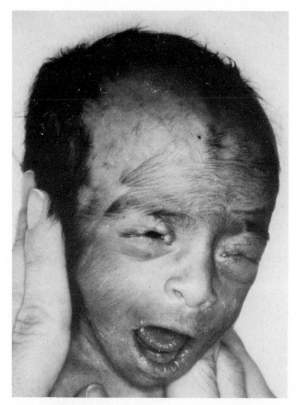

Fig. 2.—A newborn infant with the Goldenhar syndrome. The colobomas of the lids were so prominent that she appeared to have her eyes open even when the lids were closed.

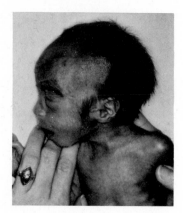

Fig. 3.—In patient shown in Figure 2, a tuft of hair anterior to the anomalous ear area illustrates the inhibition of regression of fetal facial hair that accompanies anomalous ear development.

ular hypoplasia, defects of the external ear canal and conductive deafness.

CASE REPORTS

CASE 1.—B. L. was a 16-month-old Mexican girl who was admitted to University Hospital for evaluation of a heart murmur and a history of a seizure disorder.

She was the product of a full-term, uncomplicated pregnancy and delivery. There was no history of infection or medication during gestation. Birth weight was 5 lb, 3 oz. At birth the infant was noted to have a harelip and facial asymmetry. The lip was repaired early in life. There was no family history of birth defect or consanguinity.

Her growth and development appeared normal. She smiled at 6 weeks, sat alone at 6 months, stood at 15 months and walked at 16 months. Speech amounted to a few words, and hearing was normal.

Examination revealed asymmetry of the face and skull (Fig. 5). There was frontal bossing. An epibulbar dermoid was prominent in the left infralateral area of the left eye (Fig. 6). The left ear was large and low set. There was a scar extending

Fig. 4.—Roentgenogram of a patient with the Goldenhar syndrome, illustrating vertebral anomalies.

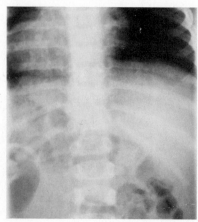

DIFFERENTIAL DIAGNOSIS

The term hemifacial microsomia has been used by Gorlin and Pindborg[4] to describe patients with unilateral microtia and macrostomia, as well as failure of formation of the mandibular ramus and condyle. They suggested that the Goldenhar syndrome may be a variant of this syndrome, and transitional forms have been seen. Therefore, there may not be two distinct syndromes. On the other hand, patients with what is called hemifacial microsomia do not have ocular dermoids or lipodermoids, colobomas of the lids or vertebral anomalies. Any of the other features of the Goldenhar syndrome occur regularly.

Mandibulofacial dysostosis is known as the Treacher Collins syndrome in the English literature, after the author who published one of the earlier descriptions of it. In Europe it is sometimes referred to as the syndrome of Franceschetti and Klein, who first referred to it as mandibulofacial dysostosis. These patients have antimongoloid slanting of the palpebral fissures, unusual malar hypoplasia and colobomas of the lower lid. This is in contrast to the upper lid colobomas of the Goldenhar syndrome. They also may have partial or total absence of the eyelashes of the lower lid, mandib-

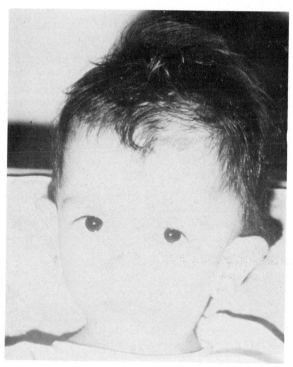

Fig. 5.—B. L. (Case 1). A 16-month-old Mexican girl with the Goldenhar syndrome. Her face was asymmetrical. She had frontal bossing, a large, low-set left ear and malar hypoplasia.

from the angle of the mouth to the midcheek and a blind-ended fistula at midcheek (Fig. 7).

The chest was clear. Examination of the heart revealed a grade III/IV systolic murmur. Physical examination was otherwise unremarkable.

Roentgenographic survey of the skeleton showed fusion anomalies of T_{11} and T_{12}. A mild degree of cranial facial disproportion could be seen on the films.

Chromosomal karyotype was that of a female 46,XX. Audiometric testing was normal, as was the EEG.

The child was diagnosed as having the Goldenhar syndrome and mild pulmonary stenosis.

CASE 2.—K. M., a 13-month-old white female, was admitted to University Hospital for evaluation of a cyanotic spell. She

Fig. 6.—B. L. Epibulbar dermoid at the lower outer quadrant of the left eye.

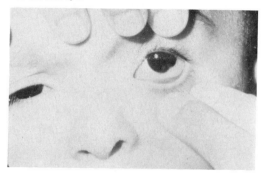

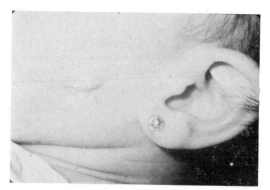

Fig. 7.—B. L. Close-up of the left ear. The skin tag was in the area of the blind fistula at midcheek.

was the product of a 37-week gestation and had a birth weight of 3 lb, 8 oz. She required resuscitation after delivery. She developed cardiac failure in the early neonatal period. Numerous congenital anomalies were diagnosed as those of the Goldenhar syndrome.

She failed to thrive. Cardiac catheterization revealed a ventricular septal defect. At 6 months of age an episode of bronchiolitis precipitated congestive heart failure. She was maintained on Digoxin, 10 μg/kg/day. She had had with increasing frequency episodes of cyanosis, characterized by anger, breath holding and opisthotonos, lasting about 1 minute.

Examination revealed a small, thin girl with mild respiratory distress. She had a peculiar facies with low-set ears, a skin tag in front of the right ear and an upper epibulbar dermoid of the left eye. She was acyanotic. Peripheral pulses were normal. There was a grade II/III holosystolic murmur along the left sternal border. It was loud and single. The liver was palpable 0.5 cm below the right costal margin. The ECG showed right axis deviation and left atrial and ventricular hypertrophy. On x-ray there was massive cardiomegaly, an increased pulmonary blood flow and a right-sided aortic arch. Films of the entire spine in posteroanterior and lateral projections revealed no osseous abnormalities.

The diagnosis was the Goldenhar syndrome with congenital heart disease, ventricular septal defect and an aortic arch on the right.

GENETICS

This syndrome is not known to be genetically transmitted. Chromosomes have been found to be normal in other cases.[4] The syndrome has been reported in a father and his two children in a family in which the mother was the first cousin of the father. Two involved sisters have been born of healthy, unrelated parents. Syndromes of abnormality around the first arch have been thought[2] to arise from an abnormal embryonic vascular supply to the first arch. Such a defect could be genetic in origin or teratogenic.

TREATMENT

Early diagnosis of the Goldenhar syndrome is of considerable interest. This is particularly important because the structural abnormalities should alert the physician to test hearing. Hearing loss, if detected, can be improved by the use of hearing aids early in life in the presence of conductive deafness. Plastic surgery can readily be performed for the auricular anomalies or those of the eyelid and usually is postponed until growth of the structure is completed. The epibulbar dermoids should be removed. They may regrow, and recurrence has been observed as long as 12 years after the original surgery.[1]

REFERENCES

1. Goldenhar, M.: Association malformatives de l'oeil et de l'oreille, en particulier le syndrome dermoid epibulbaire-appendices auriculaires-fistula auris congenita et ses relations avec la dysostose mandibulo-faciale, J. Genet. Hum. 1:243, 1952.

2. McKenzie, J.: The first arch syndrome, Arch. Dis. Child. 33: 477, 1958.

3. Gorlin, R. J., and Pindborg, J. J.: Oculo-auriculo-vertebral Dysplasia, in *Syndromes of the Head and Neck* (New York: McGraw-Hill Book Co., 1964), p. 419.

4. Sugar, H. S.: The oculoauricularvertebral dysplasia syndrome of Goldenhar, J. Ophthalmol. 62:678, 1966.

5. Gorlin, R. J., Jue, K. L., Jacobsen, U., and Goldschmidt, E.: Oculoauriculovertebral dysplasia, J. Pediatr. 63:991, 1963.

6. Berkman, M. D., and Feingold, M.: Oculoauriculovertebral dysplasia (Goldenhar's syndrome), Oral Surg. 25:408, 1968.

7. Smithells, R. W.: Oculo-auriculo-vertebral syndrome (Goldenhar's syndrome), Dev. Med. Child. Neurol. 6:406, 1964.

An Approach to the Child with Abnormal Stature

AN APPROACH TO THE CHILD WITH ABNORMAL STATURE

Malformation syndromes and metabolic disease frequently present with disturbances in linear growth. Patients most frequently referred to the dysmorphologist or the pediatric endocrinologist are children with shortness of stature. A considerable amount can be learned by a close quantitative assessment of how the patient deviates from normality.

The first decision that has to be made is whether or not growth in height or weight is abnormal. For most of us this is best done by plotting the actual measurements on a chart. Those we find most convenient for this purpose are reproduced in Figure 1.

It is also useful to have some rules of thumb. There are some that are easy to keep in your head. These formulas describe the way children normally grow. The average newborn infant is 20 inches long. In his first year he will grow half again as long as that, or 10 more inches, to make a 1-year-old height of 30 inches. In the second year he will grow half as much as in the first year, or 5 more inches, to make a total of 35 inches at 2 years. Thereafter, until age of 12 years, his height will increase at about half that rate, or 2.5 inches each year. Thus, at 4 years of age, the average height will be 40 inches, or double the height at birth. At 8 years the average height will be 50 inches, and at 12 years, 60 inches, or triple the birth height. In girls the adolescent growth spurt occurs between 10 and 12 years, and in boys between 12 and 14.

Once the child's height and weight have been recorded on the chart and it has been established that there is an abnormality, it is possible to begin to do more with these measurements. It is important not to be satisfied simply with the fact that the height and weight are below the third percentile. This much information is often evident at a glance. We have seen physicians' records and hospital charts in which this was the only information available. We have even seen workups for failure to thrive in which it was only recorded that the measurements were below the third percentile, but no actual measurements were entered on the chart. The beauty of the charts is that they permit quantitative comparisons. Extrapolate the height over to the 50th percentile and see how old a child this tall should be. This is his height age. This can be compared with bone age.

Discrepancies are very useful in diagnosis. For instance, the patient with hypopituitarism has a symmetrical disturbance in growth. The height age, weight age and bone age all match. It is only the chronologic age that does not fit. On the other hand, when the weight age is significantly lower than the height age, the etiology is almost certainly in the gastrointestinal tract. We have seen this pattern also in a patient with the diencephalic syndrome. In this disorder the emaciation and loss of subcutaneous tissue lead to a discrepancy between weight age and height age. The child with a malabsorption syndrome grows like this. So do many of the children in developing countries in which a combination of malnutrition and intestinal infection markedly influence growth. Emotional deprivation in the first year of life often manifests itself in this type of a growth curve, probably reflecting a deprivation of intake of food as well. Similar analyses can be done on the child whose height age is behind his weight age. This may be because malformations of the bones prevent their normal growth.

Assessment of the height age and weight age is simplified by the use of Table 1. There are of course marked variations in the heights and weights of normal, healthy children. These are given in Figure 1. The values given in this table, however, represent the 50th percentile. They are particularly useful in providing the height age of a child with a measured height.

Table 1 also gives the mean data for the upper and lower segments of the body and the upper to lower ratios. These values are extremely useful in the diagnosis of abnormalities of growth. Changes in the proportions of the body accompany increases in height. The normal newborn infant has an upper to lower ratio of 1.7. This ratio is obtained by measuring the total height and measuring the lower segment from the top of the symphysis pubis to the sole of the foot, and then subtracting the latter from the total height to obtain the upper segment measurement. The ratio is obtained by dividing the upper segment measurement by the lower segment measurement. As the child grows, most of the increase in height is in the legs; thus, the ratio decreases. By 10 years of age, the upper to lower ratio is about 1.0, and the average white adult's ratio is about 0.93.[1] These values are somewhat different from the data in the chart that were derived from the tables of Engelbach.[2] As indicated in Figure 2, these ratios

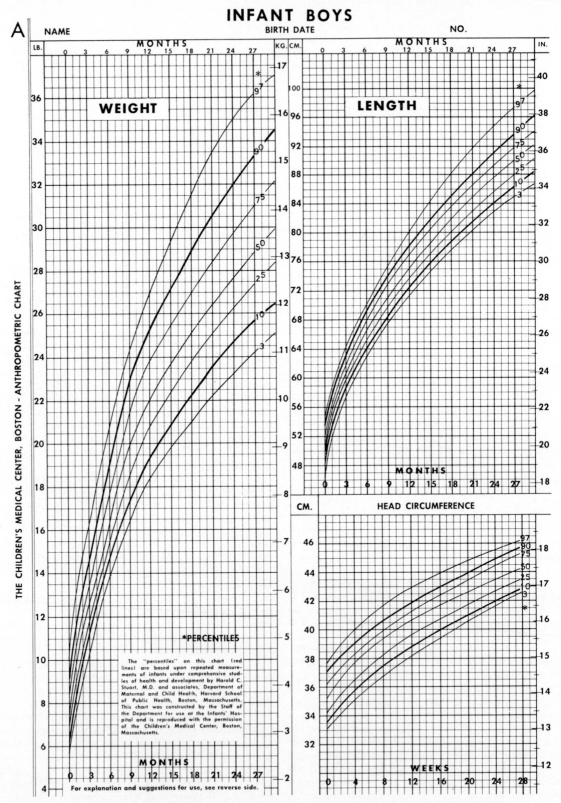

INFANT BOYS

NAME   BIRTH DATE   NO.

WEIGHT

LENGTH

HEAD CIRCUMFERENCE

*PERCENTILES

The "percentiles" on this chart (red lines) are based upon repeated measurements of infants under comprehensive studies of health and development by Harold C. Stuart, M.D. and associates, Department of Maternal and Child Health, Harvard School of Public Health, Boston, Massachusetts. This chart was constructed by the Staff of the Department for use at the Infants' Hospital and is reproduced with the permission of the Children's Medical Center, Boston, Massachusetts.

For explanation and suggestions for use, see reverse side.

THE CHILDREN'S MEDICAL CENTER, BOSTON - ANTHROPOMETRIC CHART

Fig. 1.—Anthropometric charts for assessing the height and weight of children. **A**, infant boys. (*Continued.*)

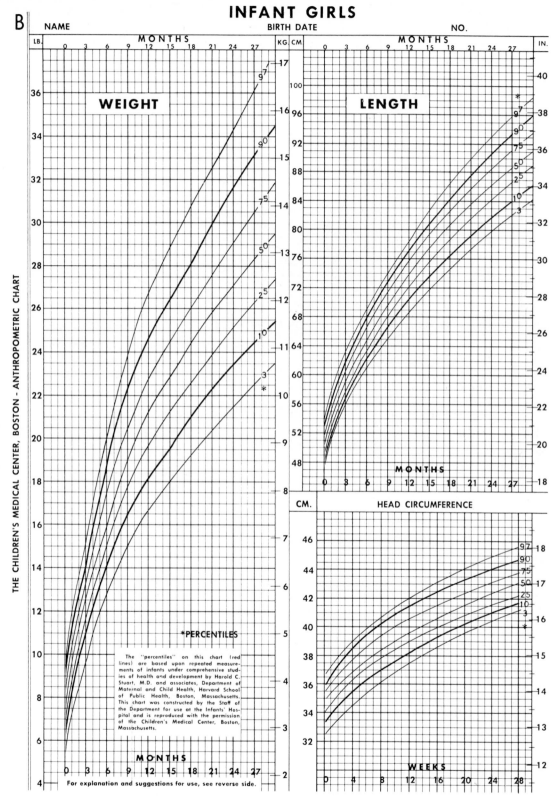

INFANT GIRLS

NAME   BIRTH DATE   NO.

THE CHILDREN'S MEDICAL CENTER, BOSTON - ANTHROPOMETRIC CHART

WEIGHT

LENGTH

HEAD CIRCUMFERENCE

*PERCENTILES

The "percentiles" on this chart (red lines) are based upon repeated measurements of infants under comprehensive studies of health and development by Harold C. Stuart, M.D. and associates, Department of Maternal and Child Health, Harvard School of Public Health, Boston, Massachusetts. This chart was constructed by the Staff of the Department for use at the Infants' Hospital and is reproduced with the permission of the Children's Medical Center, Boston, Massachusetts.

For explanation and suggestions for use, see reverse side.

Fig. 1 (cont.).—B, infant girls. (*Continued.*)

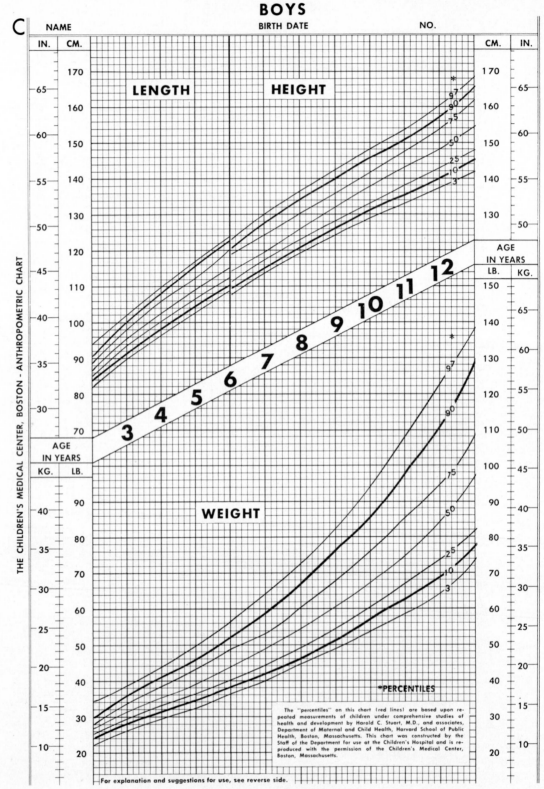

Fig. 1 (cont.).—C, boys. *(Continued.)*

458

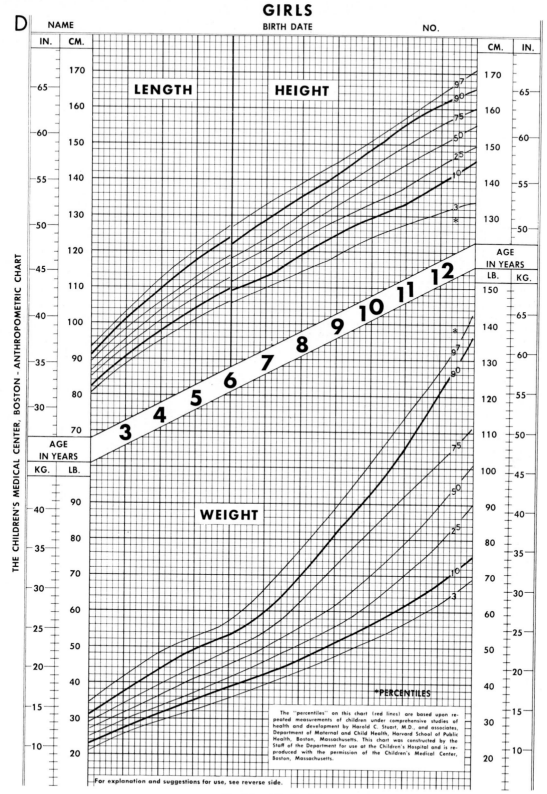

GIRLS

NAME   BIRTH DATE   NO.

LENGTH   **HEIGHT**

THE CHILDREN'S MEDICAL CENTER, BOSTON - ANTHROPOMETRIC CHART

AGE IN YEARS

WEIGHT

*PERCENTILES

The "percentiles" on this chart (red lines) are based upon repeated measurements of children under comprehensive studies of health and development by Harold C. Stuart, M.D., and associates, Department of Maternal and Child Health, Harvard School of Public Health, Boston, Massachusetts. This chart was constructed by the Staff of the Department for use at the Children's Hospital and is reproduced with the permission of the Children's Medical Center, Boston, Massachusetts.

For explanation and suggestions for use, see reverse side.

Fig. 1 (cont.).—D, girls.

TABLE 1.–STANDARDS FOR HEIGHT, UPPER AND LOWER SEGMENTS AND OTHER MEASUREMENTS AT DIFFERENT AGES*†

			MALE					FEMALE		
AGE (YR)	HEIGHT (IN.)	WEIGHT (LB)	UPPER SEGMENT (IN.)	LOWER SEGMENT (IN.)	U/L	HEIGHT (IN.)	WEIGHT (LB)	UPPER SEGMENT (IN.)	LOWER SEGMENT (IN.)	U/L
Birth	20.0	7.5	12.6	7.4	1.70	20.0	7.0	12.6	7.4	1.70
1/12	21.1	9.4	13.2	8.0	1.67	21.5	9.0	13.1	8.0	1.69
2/12	22.2	11.3	13.9	8.3	1.67	22.5	11.0	14.7	8.4	1.68
3/12	23.3	13.1	14.5	8.8	1.64	23.5	13.0	14.1	8.8	1.67
4/12	24.3	15.0	15.1	9.2	1.64	24.4	14.5	15.2	9.2	1.65
5/12	25.4	16.8	15.7	9.7	1.62	25.2	16.0	15.6	9.6	1.62
6/12	26.5	18.7	16.4	10.1	1.62	25.8	17.0	15.9	9.9	1.61
7/12	27.1	19.5	16.7	10.4	1.61	26.4	18.0	16.2	10.2	1.59
8/12	27.7	20.4	17.0	10.7	1.59	27.0	19.0	16.5	10.5	1.57
9/12	28.3	21.2	17.3	11.0	1.57	27.6	19.8	16.8	10.8	1.56
10/12	28.8	22.1	17.6	11.2	1.57	28.1	20.5	17.0	11.1	1.53
11/12	29.4	22.0	17.9	11.5	1.56	28.7	21.2	17.3	11.4	1.52
1	30.0	23.8	18.2	11.8	1.54	29.2	21.9	17.6	11.6	1.52
1 1/12	30.4	24.3	18.4	12.0	1.53	29.6	22.5	17.8	11.8	1.51
1 2/12	30.7	24.8	18.5	12.2	1.52	30.0	23.0	18.0	12.0	1.50
1 3/12	31.1	25.3	18.7	12.4	1.51	30.4	23.5	18.2	12.2	1.49
1 4/12	31.4	25.9	18.9	12.5	1.51	30.8	24.0	18.4	12.4	1.48
1 5/12	31.8	26.4	19.1	12.7	1.50	31.2	24.5	18.6	12.6	1.48
1 6/12	32.2	26.9	19.3	12.9	1.50	31.5	25.0	18.7	12.8	1.46
1 7/12	32.6	27.3	19.5	13.1	1.49	31.9	25.5	18.8	13.1	1.44
1 8/12	33.0	27.7	19.7	13.3	1.48	32.3	25.9	19.0	13.3	1.43
1 9/12	33.3	28.1	19.7	13.6	1.45	32.7	26.4	19.2	13.5	1.42
1 10/12	33.7	28.4	19.9	13.8	1.44	33.1	26.8	19.4	13.7	1.42
1 11/12	34.1	28.8	20.1	14.0	1.44	33.5	27.2	19.6	13.9	1.41
2	34.4	29.2	20.2	14.2	1.42	33.9	27.6	19.8	14.1	1.40
2 1/12	34.7	29.6	20.3	14.4	1.41	34.2	28.1	19.9	14.3	1.39
2 2/12	35.0	30.0	20.4	14.6	1.40	34.6	28.5	20.1	14.5	1.39
2 3/12	35.3	30.4	20.6	14.7	1.40	34.9	28.9	20.2	14.7	1.37
2 4/12	35.7	30.7	20.8	14.9	1.40	35.3	29.3	20.4	14.9	1.37
2 5/12	36.0	31.1	20.9	15.1	1.38	35.6	29.7	20.5	15.1	1.36
2 6/12	36.3	31.5	21.0	15.3	1.37	35.9	30.1	20.6	15.3	1.35
2 7/12	36.6	31.8	21.1	15.5	1.36	36.2	30.5	20.7	15.5	1.34
2 8/12	36.9	32.2	21.2	15.7	1.35	36.5	30.9	20.8	15.6	1.34
2 9/12	37.2	32.5	21.3	15.9	1.34	36.8	31.3	21.0	15.8	1.33
2 10/12	37.4	32.9	21.4	16.0	1.34	37.1	31.7	21.2	15.9	1.33
2 11/12	37.7	33.2	21.5	16.2	1.33	37.4	32.1	21.3	16.1	1.32
3	38.0	33.5	21.7	16.3	1.33	37.6	32.5	21.3	16.3	1.31
3 1/12	38.2	33.9	21.7	16.5	1.32	37.9	33.0	21.4	16.5	1.30
3 2/12	38.5	34.3	21.9	16.6	1.32	38.2	33.4	21.6	16.6	1.30
3 3/12	38.7	34.7	22.0	16.7	1.32	38.5	33.8	21.7	16.8	1.29
3 4/12	39.0	35.1	22.2	16.8	1.32	38.7	34.2	21.8	16.9	1.29
3 5/12	39.2	35.5	22.3	16.9	1.32	39.0	34.6	21.9	17.1	1.28
3 6/12	39.4	35.9	22.3	17.1	1.30	39.2	35.0	21.9	17.3	1.27
3 7/12	39.7	36.3	22.4	17.3	1.29	39.5	35.4	22.0	17.5	1.26
3 8/12	39.9	36.7	22.4	17.5	1.28	39.7	35.7	22.1	17.6	1.26
3 9/12	40.2	37.0	22.5	17.7	1.27	40.0	36.1	22.2	17.8	1.25
3 10/12	40.4	37.4	22.5	17.9	1.26	40.2	36.4	22.3	17.9	1.25
3 11/12	40.7	37.8	22.6	18.1	1.25	40.5	36.8	22.4	18.1	1.24

Table 1.—(*Continued*)

4 years	40.9	38.1	22.7	18.2	1.25	40.7	37.2	22.4	18.3	1.22
4 1/12	41.1	38.5	22.8	18.3	1.25	41.0	37.7	22.5	18.5	1.22
4 2/12	41.4	38.9	22.9	18.5	1.24	41.2	38.1	22.6	18.6	1.22
4 3/12	41.6	39.3	23.0	18.6	1.24	41.5	38.6	22.7	18.8	1.21
4 4/12	41.8	39.7	23.0	18.8	1.22	41.7	39.0	22.8	18.9	1.21
4 5/12	42.1	40.1	23.1	18.9	1.22	42.0	39.5	22.9	19.1	1.20
4 6/12	42.2	40.5	23.2	19.0	1.22	42.2	40.0	23.0	19.2	1.20
4 7/12	42.4	40.9	23.2	19.2	1.21	42.5	40.4	23.1	19.4	1.19
4 8/12	42.7	41.3	23.3	19.4	1.20	42.7	40.8	23.1	19.6	1.18
4 9/12	42.9	41.6	23.4	19.5	1.20	42.9	41.2	23.2	19.7	1.18
4 10/12	43.1	42.0	23.5	19.6	1.20	43.1	41.5	23.2	19.9	1.17
4 11/12	43.3	42.4	23.5	19.8	1.19	43.3	41.9	23.3	20.0	1.17
5	43.6	42.8	23.6	20.0	1.19	43.5	42.3	23.3	20.2	1.15
5 1/12	43.8	43.2	23.7	20.1	1.18	43.7	42.8	23.3	20.4	1.14
5 2/12	44.0	43.7	23.8	20.2	1.18	43.9	43.3	23.4	20.5	1.14
5 3/12	44.3	44.1	23.9	20.4	1.17	44.2	43.8	23.5	20.7	1.14
5 4/12	44.5	44.6	23.9	20.6	1.16	44.4	44.3	23.6	20.8	1.13
5 5/12	44.7	45.0	24.0	20.7	1.16	44.6	44.8	23.7	20.9	1.13
5 6/12	45.0	45.5	24.1	20.9	1.15	44.9	45.3	23.8	21.1	1.13
5 7/12	45.2	45.9	24.2	21.0	1.15	45.1	45.8	23.9	21.2	1.13
5 8/12	45.4	46.4	24.2	21.2	1.14	45.3	46.3	24.0	21.3	1.13
5 9/12	45.6	46.8	24.3	21.3	1.14	45.6	46.8	24.1	21.5	1.12
5 10/12	45.9	47.3	24.4	21.5	1.13	45.8	47.3	24.1	21.7	1.11
5 11/12	46.1	47.7	24.5	21.6	1.13	46.0	47.8	24.2	21.8	1.11
6	46.3	48.2	24.5	21.8	1.12	46.3	48.3	24.3	22.0	1.10
6 1/12	46.5	48.7	24.6	21.9	1.12	46.5	48.8	24.4	22.1	1.10
6 2/12	46.7	49.2	24.6	22.1	1.11	46.7	49.3	24.5	22.2	1.10
6 3/12	46.9	49.7	24.7	22.2	1.11	46.9	49.8	24.5	22.4	1.09
6 4/12	47.1	50.2	24.7	22.4	1.10	47.1	50.4	24.6	22.5	1.09
6 5/12	47.3	50.7	24.8	22.5	1.10	47.3	50.9	24.7	22.6	1.09
6 6/12	47.5	51.2	24.8	22.7	1.09	47.5	51.4	24.7	22.8	1.08
6 7/12	47.7	51.7	24.9	22.8	1.09	47.7	51.9	24.8	22.9	1.08
6 8/12	47.9	52.2	24.9	23.0	1.08	47.9	52.4	24.9	23.0	1.08
6 9/12	48.1	52.7	25.0	23.1	1.08	48.1	53.0	24.9	23.2	1.07
6 10/12	48.3	53.2	25.0	23.3	1.07	48.3	53.5	25.0	23.3	1.07
6 11/12	48.5	53.7	25.1	23.4	1.07	48.5	54.0	25.1	23.4	1.07
7	48.7	54.2	25.2	23.5	1.07	48.7	54.5	25.1	23.6	1.06
7 1/12	48.9	54.8	25.3	23.6	1.07	48.9	55.1	25.2	23.7	1.06
7 2/12	49.1	55.4	25.3	23.8	1.06	49.1	55.7	25.3	23.8	1.06
7 3/12	49.3	55.9	25.4	23.9	1.06	49.3	56.3	25.3	24.0	1.05
7 4/12	49.5	56.5	25.4	24.1	1.05	49.5	57.0	25.4	24.1	1.05
7 5/12	49.7	57.1	25.5	24.2	1.05	49.7	57.6	25.5	24.2	1.05
7 6/12	49.9	57.6	25.6	24.3	1.05	49.9	58.2	25.5	24.4	1.05
7 7/12	50.1	58.1	25.6	24.5	1.04	50.1	58.8	25.6	24.5	1.04
7 8/12	50.3	58.7	25.6	24.7	1.04	50.3	59.4	25.7	24.6	1.04
7 9/12	50.5	59.3	25.7	24.8	1.04	50.5	60.1	25.8	24.7	1.04
7 10/12	50.7	59.8	25.7	25.0	1.03	50.7	60.7	25.8	24.9	1.04
7 11/12	50.9	60.4	25.8	25.1	1.03	50.9	61.3	25.9	25.0	1.04
8	51.1	61.0	25.9	25.2	1.03	51.1	61.9	26.0	25.1	1.04
8 1/12	51.3	61.6	26.0	25.3	1.03	51.3	62.5	26.1	25.2	1.04
8 2/12	51.5	62.2	26.1	25.4	1.03	51.5	63.1	26.2	25.3	1.04
8 3/12	51.7	62.8	26.2	25.5	1.03	51.7	63.8	26.3	25.4	1.04
8 4/12	51.8	63.4	26.2	25.6	1.02	51.9	64.4	26.4	25.5	1.04
8 5/12	52.0	64.1	26.3	25.7	1.02	52.0	65.1	26.4	25.6	1.03
8 6/12	52.2	64.7	26.4	25.8	1.02	52.2	65.7	26.5	25.7	1.03
8 7/12	52.4	65.3	26.5	25.9	1.02	52.4	66.4	26.5	25.9	1.02
8 8/12	52.6	65.9	26.6	26.0	1.02	52.6	67.0	26.6	26.0	1.02

Table 1.—(*Continued*)

Age										
8 9/12	52.8	66.5	26.7	26.1	1.02	52.7	67.7	26.6	26.1	1.02
8 10/12	52.9	67.2	26.7	26.2	1.02	52.9	68.3	26.7	26.2	1.02
8 11/12	53.1	67.8	26.8	26.3	1.02	53.1	68.9	26.8	26.3	1.02
9	53.3	68.4	26.9	26.4	1.02	53.3	69.6	26.9	26.4	1.02
9 1/12	53.5	69.1	27.0	26.5	1.02	53.5	70.3	27.0	26.5	1.02
9 2/12	53.7	69.8	27.1	26.6	1.02	53.7	71.0	27.1	26.6	1.02
9 3/12	53.9	70.5	27.1	26.8	1.01	53.9	71.7	27.2	26.7	1.02
9 4/12	54.0	71.2	27.1	26.9	1.01	54.1	72.4	27.3	26.8	1.02
9 5/12	54.2	71.9	27.2	27.0	1.01	54.2	73.1	27.3	26.9	1.01
9 6/12	54.4	72.6	27.2	27.2	1.00	54.4	73.8	27.4	27.0	1.01
9 7/12	54.6	73.3	27.4	27.3	1.00	54.6	74.6	27.4	27.2	1.01
9 8/12	54.7	74.0	27.3	27.4	1.00	54.8	75.3	27.5	27.3	1.01
9 9/12	54.9	74.7	27.3	27.6	0.99	54.9	76.0	27.5	27.4	1.00
9 10/12	55.1	75.4	27.4	27.7	0.99	55.1	76.7	27.6	27.5	1.00
9 11/12	55.3	76.1	27.5	27.8	0.99	55.3	77.4	27.7	27.6	1.00
10	55.5	76.8	27.6	27.9	0.99	55.5	78.1	27.8	27.7	1.00
10 1/12	55.7	77.5	27.7	28.0	0.99	55.7	79.0	27.9	27.8	1.00
10 2/12	55.8	78.3	27.7	28.1	0.99	55.9	79.8	28.0	27.9	1.00
10 3/12	56.0	79.0	27.8	28.2	0.99	56.1	80.7	28.1	28.0	1.00
10 4/12	56.1	79.7	27.8	28.3	0.98	56.4	81.5	28.2	28.2	1.00
10 5/12	56.3	80.5	27.9	28.4	0.98	56.6	82.4	28.3	28.3	1.00
10 6/12	56.4	81.2	27.9	28.5	0.98	56.8	83.2	28.4	28.4	1.00
10 7/12	56.6	81.9	28.0	28.6	0.98	57.0	84.1	28.4	28.6	0.99
10 8/12	56.7	82.7	28.1	28.6	0.98	57.3	84.9	28.5	28.8	0.99
10 9/12	56.9	83.4	28.2	28.7	0.98	57.5	85.8	28.6	28.9	0.99
10 10/12	57.0	84.1	28.2	28.8	0.98	57.7	86.6	28.7	29.0	0.99
10 11/12	57.2	84.9	28.3	28.9	0.98	57.9	87.5	28.8	29.1	0.99
11	57.4	85.6	28.4	29.0	0.98	58.1	88.4	28.9	29.2	0.99
11 1/12	57.6	86.4	28.5	29.1	0.98	58.3	89.4	29.0	29.3	0.99
11 2/12	57.8	87.2	28.6	29.2	0.98	58.5	90.4	29.1	29.4	0.99
11 3/12	58.0	88.0	28.7	29.3	0.98	58.7	91.4	29.2	29.5	0.99
11 4/12	58.1	88.8	28.7	29.4	0.98	59.0	92.4	29.4	29.6	0.99
11 5/12	58.3	89.6	28.8	29.5	0.98	59.2	93.4	29.5	29.7	0.99
11 6/12	58.5	90.4	28.9	29.6	0.98	59.4	94.4	29.6	29.8	0.99
11 7/12	58.7	91.2	29.0	29.7	0.98	59.6	95.4	29.7	29.9	0.99
11 8/12	58.8	92.0	29.1	29.7	0.98	59.9	96.4	29.8	30.1	0.99
11 9/12	59.0	92.8	29.2	29.8	0.98	60.1	97.4	29.9	30.2	0.99
11 10/12	59.2	93.6	29.3	29.9	0.98	60.3	98.4	30.0	30.3	0.99
11 11/12	59.4	94.4	29.4	30.0	0.98	60.5	99.4	30.1	30.4	0.99
12	59.6	95.2	29.5	30.1	0.98	60.7	100.4	30.2	30.5	0.99
12 1/12	59.8	96.1	29.6	30.2	0.98	60.9	101.2	30.3	30.6	0.99
12 2/12	60.0	97.0	29.7	30.3	0.98	61.1	102.0	30.4	30.7	0.99
12 3/12	60.2	97.9	29.8	30.4	0.98	61.3	102.9	30.5	30.8	0.99
12 4/12	60.4	98.7	29.9	30.5	0.98	61.4	103.7	30.6	30.8	0.99
12 5/12	60.6	99.6	30.0	30.6	0.98	61.6	104.6	30.7	30.9	0.99
12 6/12	60.8	100.5	30.1	30.7	0.98	61.8	105.4	30.8	31.0	0.99
12 7/12	61.0	101.3	30.1	30.9	0.97	61.9	106.3	30.9	31.0	1.00
12 8/12	61.2	102.2	30.2	31.0	0.97	62.1	107.1	31.0	31.1	1.00
12 9/12	61.4	103.1	30.3	31.1	0.97	62.3	108.0	31.1	31.2	1.00
12 10/12	61.6	103.9	30.4	31.2	0.97	62.4	108.8	31.2	31.2	1.00
12 11/12	61.8	104.8	30.5	31.3	0.97	62.6	109.7	31.3	31.3	1.00
13	62.0	105.7	30.6	31.4	0.97	62.8	110.5	31.4	31.4	1.00
13 1/12	62.2	106.8	30.7	31.5	0.97	62.9	111.3	31.5	31.4	1.00
13 2/12	62.5	107.9	30.8	31.7	0.97	63.0	112.1	31.5	31.5	1.00
13 3/12	62.7	109.0	30.9	31.8	0.97	63.1	112.9	31.6	31.5	1.00
13 4/12	63.0	110.2	31.1	31.9	0.97	63.2	113.7	31.6	31.6	1.00
13 5/12	63.2	111.3	31.1	32.1	0.97	63.3	114.5	31.7	31.6	1.00
13 6/12	63.5	112.4	31.3	32.2	0.97	63.4	115.3	31.7	31.7	1.00

Table 1.—(Continued)

Age											
13	7/12	63.7	113.5	31.4	32.3	0.97	63.6	116.1	31.9	31.7	1.01
13	8/12	64.0	114.7	31.5	32.5	0.97	63.7	116.9	32.0	31.7	1.01
13	9/12	64.2	115.8	31.6	32.6	0.97	63.8	117.7	32.0	31.8	1.01
13	10/12	64.5	116.9	31.8	32.7	0.97	63.9	118.5	32.1	31.8	1.01
13	11/12	64.7	118.0	31.9	32.8	0.97	64.0	119.3	32.2	31.8	1.01
14		64.9	119.1	32.0	32.9	0.97	64.1	120.1	32.2	31.9	1.01
14	1/12	65.1	120.2	32.1	33.0	0.97	64.2	120.6	32.3	31.9	1.01
14	2/12	65.3	121.3	32.2	33.1	0.97	64.2	121.2	32.3	31.9	1.01
14	3/12	65.5	122.4	32.3	33.2	0.97	64.3	121.7	32.3	32.0	1.01
14	4/12	65.7	123.5	32.4	33.3	0.97	64.4	122.2	32.3	32.1	1.01
14	5/12	65.9	124.6	32.5	33.4	0.97	64.4	122.8	32.3	32.1	1.01
14	6/12	66.1	125.7	32.6	33.5	0.97	64.5	123.3	32.4	32.1	1.01
14	7/12	66.4	126.8	32.9	33.5	0.98	64.6	123.9	32.5	32.1	1.01
14	8/12	66.6	127.9	33.0	33.6	0.98	64.6	124.4	32.5	32.1	1.01
14	9/12	66.8	129.0	33.1	33.7	0.98	64.7	125.0	32.5	32.2	1.01
14	10/12	67.0	130.1	33.2	33.8	0.98	64.8	125.5	32.6	32.2	1.01
14	11/12	67.2	131.2	33.3	33.9	0.98	64.8	126.1	32.6	32.3	1.01
15		67.4	132.2	33.4	34.0	0.98	64.9	126.6	32.6	32.3	1.01
15	1/12	67.5	133.1	33.5	34.0	0.99	64.9	126.9	32.6	32.3	1.01
15	2/12	67.7	133.9	33.6	34.1	0.99	64.9	127.2	32.6	32.3	1.01
15	3/12	67.8	134.7	33.7	34.1	0.99	64.9	127.5	32.6	32.3	1.01
15	4/12	67.9	135.5	33.7	34.2	0.99	65.0	127.9	32.7	32.3	1.01
15	5/12	68.1	136.3	33.9	34.2	0.99	65.0	128.2	32.7	32.3	1.01
15	6/12	68.2	137.1	33.9	34.3	0.99	65.0	128.5	32.7	32.3	1.01
15	7/12	68.3	137.9	34.0	34.3	0.99	65.1	128.8	32.7	32.4	1.01
15	8/12	68.5	138.7	34.1	34.4	0.99	65.1	129.2	32.7	32.4	1.01
15	9/12	68.6	139.5	34.2	34.4	0.99	65.1	129.5	32.7	32.4	1.01
15	10/12	68.7	140.3	34.2	34.5	0.99	65.2	129.8	32.8	32.4	1.01
15	11/12	68.9	141.1	34.3	34.6	0.99	65.2	130.2	32.8	32.4	1.01
16		69.0	141.9	34.4	34.6	0.99	65.2	130.5	32.8	32.4	1.01
16	1/12	69.0	142.4	34.4	34.6	0.99	65.2	130.7	32.8	32.4	1.01
16	2/12	69.1	142.9	34.4	34.7	0.99	65.2	131.0	32.8	32.4	1.01
16	3/12	69.1	143.3	34.4	34.7	0.99	65.2	131.2	32.8	32.4	1.01
16	4/12	69.2	143.8	34.5	34.7	0.99	65.2	131.5	32.8	32.4	1.01
16	5/12	69.2	144.3	34.5	34.7	0.99	65.2	131.7	32.8	32.4	1.01
16	6/12	69.3	144.8	34.5	34.8	0.99	65.2	132.0	32.8	32.4	1.01
16	7/12	69.3	145.2	34.5	34.8	0.99	65.2	132.2	32.8	32.4	1.01
16	8/12	69.3	145.7	34.5	34.8	0.99	65.2	132.5	32.8	32.4	1.01
16	9/12	69.4	146.2	34.6	34.8	0.99	65.2	132.7	32.8	32.4	1.01
16	10/12	69.4	146.7	34.6	34.8	0.99	65.2	133.0	32.8	32.4	1.01
16	11/12	69.5	147.1	34.6	34.9	0.99	65.2	133.2	32.8	32.4	1.01
17		69.5	147.6	34.6	34.9	0.99	65.2	133.5	32.8	32.4	1.01

*The data at each age represent the means.
†We are indebted to Dr. Robert Blizzard for providing us with this table, which is currently used in the Pediatric Endocrine Clinic of the Johns Hopkins Hospital. It is reproduced, by permission, from Wilkins, L.: *The Diagnosis and Treatment of Endocrine Disorders in Childhood and Adolescence* (3d ed.; Springfield, Ill.: Charles C Thomas, Publishers, 1966). The data originally came from Simmons (Iowa) and Stuart and Meredith (Boston), as recorded in Watson and Lowrey, *Growth and Development of Children* (4th ed.; Chicago: Year Book Medical Publishers, Inc., 1962).

have shifted downward with increasing length of the legs.

Figure 2 is based on studies carried out in 1959 in Baltimore.[1] Negro children at all ages have a significantly lower upper to lower ratio than white children have. The secular changes in length are of course continuing, so that the ratios to be obtained in 1973 would

be even lower. Even so, we thought it of interest to redraw the classic diagram of Stratz,[3] done in 1902, on the basis of 1959 proportions (Fig. 3). Figure 3 illustrates an interesting phenomenon not evident in the original Stratz drawing. The legs increase progressively until they are the length of those of the 15 year old illustrated. Then paradoxically they decrease. They do

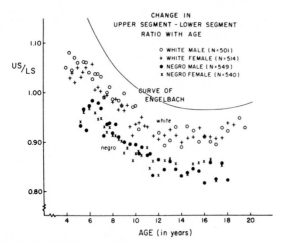

Fig. 2. – Upper segment-lower segment ratio (US/LS) in over 2,100 school children measured in Baltimore in 1959. (Reprinted with permission from McKusick.[1])

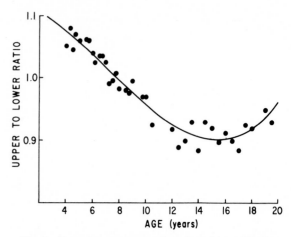

Fig. 4. – Upper-lower segment data for white males from Figure 2 were supplied to a calculator/plotter apparatus, which was asked to compute the best fit line. The cubic regression curve drawn was a slightly better fit than the quadratic. The equation that describes the line is: $Y = 1.2414 - 0.00561 X - 0.00209 X^2 + 0.00010 X^3$.

not in life, of course, but the proportion decreases, indicating that after fusion of the epiphyses in the legs the trunk continues to grow. This is clear from an appraisal of the data in Figure 2. The curve for the ratio, at least of the white population studied, definitely goes up. These data were fed into a computer asked to fit various lines to the data. Statistically, the best fit was cubic (Fig. 4). The curve the computer traced clearly rose after adolescence. Actually, careful longitudinal measurements have indicated that growth of the trunk may continue even up to 30 years of age. This may be only 0.5 cm after 20.

The upper to lower ratios are especially significant in clinical medicine in the diagnosis of short-limbed dwarfism, such as achondroplasia, in which the upper to lower ratio is markedly increased. For these purposes the data in the table are adequate. In hypothyroidism there is a delay in bone growth and maturation of the skeleton so that the ratio remains immature, or

Fig. 3. – Change in skeletal proportions with age. (The classic drawing made by Stratz in 1902[3] was redrawn by Eileen O'Farrell on the basis of data collected by McKusick[1] in 1959.)

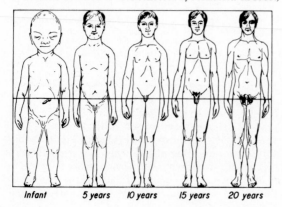

Infant   5 years   10 years   15 years   20 years

high. In short-trunked dwarfism, such as the Morquio syndrome or costovertebral dysplasia, the upper to lower ratio is decreased. This also may be seen in dysostosis multiplex. In the Marfan syndrome the patient is tall and the upper to lower ratio is decreased. An adult with this syndrome should have a ratio approximating 0.83.

Other measurements that may be useful in the Marfan syndrome are the hand to height percentage, which should be greater than 11%, and the foot to height percentage, which should be greater than 15%. The middle finger is often 1.5 times greater than the length of its metacarpal. More precise measurements of arachnodactyly have been made on roentgenograms. The metacarpal index is made on the right hand. The length in millimeters of the second, third, fourth and fifth metacarpals is measured, as is the width at the midpoint of each shaft. This value is divided into the length and the values for the four metacarpals averaged. The index is generally under 8 normally and over 9 in the Marfan syndrome.

Charts on the growth of the head are available from the University of Colorado (Fig. 5). Head circumference is measured with a flexible tape measure over the most prominent portion of the occiput and just above the supraorbital ridges. Growth is very rapid in the early months of life. There is an increase of 5 cm in the first 4 months and a second 5 cm by the end of the first year. By 18 years of age there is only another 10-cm increase. A head circumference of more than 2 s. D. below the mean usually is a sign of mental retardation. The exceptions to this rule usually are patients such as those with congenital heart disease in which there may be a symmetrical retardation of growth and a head circumference proportional to the rest of the body.

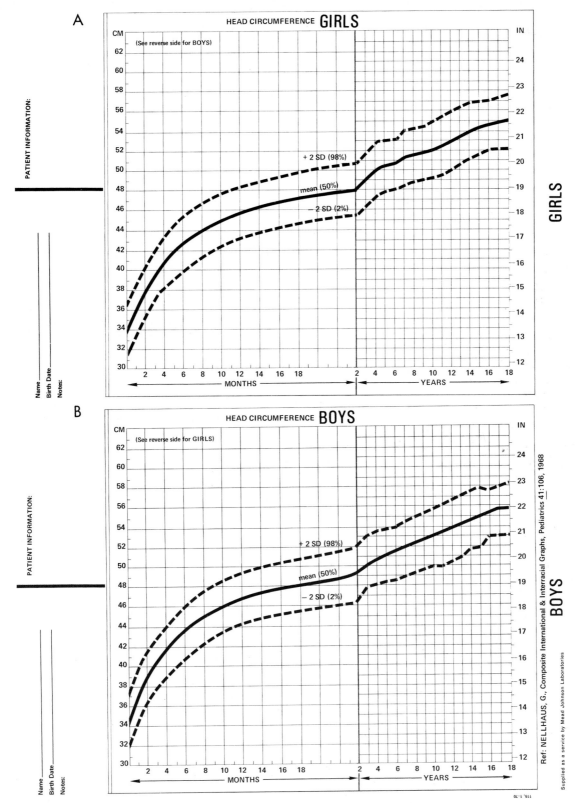

Fig. 5.—Head circumference charts for **A,** girls, and **B,** boys.

Ref: NELLHAUS, G., Composite International & Interracial Graphs, Pediatrics 41:106, 1968

Supplied as a service by Mead Johnson Laboratories

465

Intrauterine retardation of growth is seen in a number of dysmorphic syndromes. The charts of Lubchenco *et al.*[4] (Fig. 6) are useful in assessing the appropriateness of an infant's growth at the time of his birth to his gestational age. The weight to length ratio provides an index of the adequacy of intrauterine nutrition.

Intrauterine growth retardation (IUGR) must be distinguished from prematurity. This is easily done in the patient with the de Lange syndrome (see section on Distinct Syndromes with Very Short Stature) or in trisomy 18 (see section on Cytogenetic Syndromes). Retardation of intrauterine growth has many etiologies. Some children with IUGR have microcephaly. Many of these patients continue to grow poorly. The bird-headed dwarfs of Seckel[5] probably fall into this category. Of course, intrauterine rubella or cytomegalovirus infection also can lead to this type of syndrome. In addition, there is a larger group of microcephalic patients with IUGR whose case does not fit a specific syndrome. Such patients often have minor anomalies, such as a high-arched palate, clinodactyly or a simian crease. A syndrome also is seen in patients with intrauterine growth retardation in which the cranium appears relatively large and the facial bones are small. These patients do not have hydrocephalus but, rather, a relatively normal cranium on a dwarfed body. The Silver syndrome or Silver-Russell syndrome of short stature (see section on Unilateral Diseases) may represent a heterogeneous group of infants with IUGR. Warkany[6] has pointed out that hemihypertrophy does not appear to be an appropriate designation for a child with bilateral retardation of growth in whom the retardation is greater on one side than the other. In general, patients with IUGR continue to grow poorly and may end up quite short. Prognosis as to intellectual development in this group of patients also should be guarded. However, the group is sufficiently heterogeneous that predictions for growth and development should not be made. It is far better to adopt an attitude of waiting to see how the child will grow and develop.

In evaluating a child with short stature, it is important to know the heights of parents, grandparents and siblings. It also is important to determine the child's bone age. Roentgenograms of the hand, wrist and hemiskeleton are examined for maturation of the ossification centers and compared with x-ray standards such as those in the Greulich and Pyle Atlas.[7] It is common to see a child who is healthy but small for his age in whom all of the causes of short stature are easy to rule out and in whom the bone age is below 2 s. D. Such a child is most likely a late bloomer. In delayed maturation or delayed adolescence, puberty obviously will be delayed, but when it occurs it is normal. Mature height is reached in such a person at a later than average age. However, the ultimate height may be perfectly normal. It is very useful in cases of this sort to predict the ultimate height using Tables 2 and 3. In order to use these data we must know the sex, chronologic age, skeletal (bone) age and height. For example, let us consider a 12-year-old boy who has a height of 49 inches and a bone age of 7 years. In Table 2A find 49 inches in the left-hand column and follow this row

Fig. 6.—Intrauterine growth. (Reprinted with permission from Lubchenco.[4])

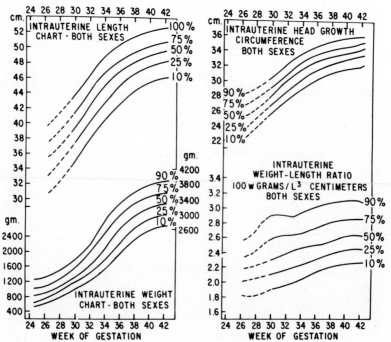

TABLE 2A.—AVERAGE BOYS. PERCENTAGES AND ESTIMATED MATURE HEIGHTS FOR BOYS WITH SKELETAL AGES WITHIN ONE YEAR OF THEIR CHRONOLOGICAL AGES: SKELETAL AGES 7 THROUGH 12 YEARS*

Skeletal Age	7-0	7-3	7-6	7-9	8-0	8-3	8-6	8-9	9-0	9-3	9-6	9-9	10-0	10-3	10-6	10-9	11-0	11-3	11-6	11-9	12-0	12-3	12-6	12-9
% of Mature Height	69.5	70.2	70.9	71.6	72.3	73.1	73.9	74.6	75.2	76.1	76.9	77.7	78.4	79.1	79.5	80.0	80.4	81.2	81.8	82.7	83.4	84.3	85.3	86.3
Ht. (inches)																								
42	60.4																							
43	61.9	61.3	60.6	60.1																				
44	63.3	62.7	62.1	61.5	60.9	60.2																		
45	64.7	64.1	63.5	62.8	62.2	61.6	60.9	60.3																
46	66.2	65.5	64.9	64.2	63.6	62.9	62.2	61.7	61.2	60.4														
47	67.6	67.0	66.3	65.6	65.0	64.3	63.6	63.0	62.5	61.8	61.1	60.5												
48	69.1	68.4	67.7	67.0	66.4	65.7	65.0	64.3	63.8	63.1	62.4	61.8	61.2	60.7	60.4	60.0								
49	70.5	69.8	69.1	68.4	67.8	67.0	66.3	65.7	65.2	64.4	63.7	63.1	62.5	61.9	61.6	61.3	60.9	60.3						
50	71.9	71.2	70.5	69.8	69.2	68.4	67.7	67.0	66.5	65.7	65.0	64.4	63.8	63.2	62.9	62.5	62.2	61.6	61.1	60.5				
51	73.4	72.6	71.9	71.2	70.5	69.8	69.0	68.4	67.8	67.0	66.3	65.6	65.1	64.5	64.2	63.8	63.4	62.8	62.3	61.7	61.1	60.5	59.8	
52	74.8	74.1	73.3	72.6	71.9	71.1	70.4	69.7	69.1	68.3	67.6	66.9	66.3	65.7	65.4	65.0	64.7	64.0	63.6	62.9	62.3	61.7	61.0	60.3
53	76.3	75.5	74.8	74.0	73.3	72.5	71.7	71.0	70.5	69.6	68.9	68.2	67.6	67.0	66.7	66.3	65.9	65.3	64.8	64.1	63.5	62.9	62.1	61.4
54	77.7	76.9	76.2	75.4	74.7	73.9	73.1	72.4	71.8	71.0	70.2	69.5	68.9	68.3	67.9	67.5	67.2	66.5	66.0	65.3	64.7	64.1	63.3	62.6
55	79.1	78.3	77.6	76.8	76.1	75.2	74.4	73.7	73.1	72.3	71.5	70.8	70.2	69.5	69.2	68.8	68.4	67.7	67.2	66.5	65.9	65.2	64.5	63.7
56	80.6	79.8	79.0	78.2	77.5	76.6	75.8	75.1	74.5	73.6	72.8	72.1	71.4	70.8	70.4	70.0	69.7	69.0	68.5	67.7	67.1	66.4	65.6	64.9
57			80.4	79.6	78.8	78.0	77.1	76.4	75.8	74.9	74.1	73.4	72.7	72.1	71.7	71.3	70.9	70.2	69.7	68.9	68.3	67.6	66.8	66.0
58					80.2	79.3	78.5	77.7	77.1	76.2	75.4	74.6	74.0	73.3	73.0	72.5	72.1	71.4	70.9	70.1	69.5	68.8	68.0	67.2
59						80.7	79.8	79.1	78.5	77.5	76.7	75.9	75.3	74.6	74.2	73.8	73.4	72.7	72.1	71.3	70.7	70.0	69.2	68.4
60								80.4	79.8	78.8	78.0	77.2	76.5	75.9	75.5	75.0	74.6	73.9	73.3	72.6	71.9	71.2	70.3	69.5
61										80.2	79.3	78.5	77.8	77.1	76.7	76.3	75.9	75.1	74.6	73.8	73.1	72.4	71.5	70.7
62											80.6	79.8	79.1	78.4	78.0	77.5	77.1	76.4	75.8	75.0	74.3	73.5	72.7	71.8
63													80.4	79.6	79.2	78.8	78.4	77.6	77.0	76.2	75.5	74.7	73.9	73.0
64														80.9	80.5	80.0	79.6	78.8	78.2	77.4	76.7	75.9	75.0	74.2
65																	80.8	80.0	79.5	78.6	77.9	77.1	76.2	75.3
66																			80.7	79.8	79.1	78.3	77.4	76.5
67																					80.3	79.5	78.5	77.6
68																						80.7	79.7	78.8
69																							80.9	80.0

*In order to use Tables 2A–C and Tables 3A–D, find the vertical column with the child's current height. Then find the horizontal column representing the child's skeletal age. The number at which these two columns intersect provides a prediction as to the height at maturity. (Reprinted with permission from Bayley and Pinneau.[8])

467

TABLE 2B.–AVERAGE BOYS. PERCENTAGES AND ESTIMATED MATURE HEIGHTS FOR BOYS WITH SKELETAL AGES WITHIN ONE YEAR OF THEIR CHRONOLOGICAL AGES: SKELETAL AGES 13 YEARS TO MATURITY*

Skeletal Age	13-0	13-3	13-6	13-9	14-0	14-3	14-6	14-9	15-0	15-3	15-6	15-9	16-0	16-3	16-6	16-9	17-0	17-3	17-6	17-9	18-0	18-3	18-6
% of Mature Height	87.6	89.0	90.2	91.4	92.7	93.8	94.8	95.8	96.8	97.3	97.6	98.0	98.2	98.5	98.7	98.9	99.1	99.3	99.4	99.5	99.6	99.8	100.0
Ht. (inches)																							
53	60.5																						
54	61.6	60.7																					
55	62.8	61.8	61.0	60.2																			
56	63.9	62.9	62.1	61.3	60.4																		
57	65.1	64.0	63.2	62.4	61.5	60.8	60.1																
58	66.2	65.2	64.3	63.5	62.6	61.8	61.2	60.5															
59	67.4	66.3	65.4	64.6	63.6	62.9	62.2	61.6	61.0	60.6	60.5	60.2	60.1										
60	68.5	67.4	66.5	65.6	64.7	64.0	63.3	62.6	62.0	61.7	61.5	61.2	61.1	60.9	60.8	60.7	60.5	60.4	60.4	60.3	60.2	60.1	60.0
61	69.6	68.5	67.6	66.7	65.8	65.0	64.3	63.7	63.0	62.7	62.5	62.2	62.1	61.9	61.8	61.7	61.6	61.4	61.4	61.3	61.2	61.1	61.0
62	70.8	69.7	68.7	67.8	66.9	66.1	65.4	64.7	64.1	63.7	63.5	63.3	63.1	62.9	62.8	62.7	62.6	62.4	62.4	62.3	62.2	62.1	62.0
63	71.9	70.8	69.8	68.9	68.0	67.2	66.5	65.8	65.1	64.7	64.5	64.3	64.2	64.0	63.8	63.7	63.6	63.4	63.4	63.3	63.3	63.1	63.0
64	73.1	71.9	71.0	70.0	69.0	68.2	67.5	66.8	66.1	65.8	65.6	65.3	65.2	65.0	64.8	64.7	64.6	64.4	64.4	64.3	64.3	64.1	64.0
65	74.2	73.0	72.1	71.1	70.1	69.3	68.6	67.8	67.2	66.8	66.6	66.3	66.2	66.0	65.9	65.7	65.6	65.5	65.4	65.3	65.3	65.1	65.0
66	75.3	74.2	73.2	72.2	71.2	70.4	69.6	68.9	68.2	67.8	67.6	67.3	67.2	67.0	66.9	66.7	66.6	66.5	66.4	66.3	66.3	66.1	66.0
67	76.5	75.3	74.3	73.3	72.3	71.4	70.7	69.9	69.2	68.9	68.6	68.4	68.2	68.0	67.9	67.7	67.6	67.5	67.4	67.3	67.3	67.1	67.0
68	77.6	76.4	75.4	74.4	73.4	72.5	71.7	71.0	70.3	69.9	69.7	69.4	69.2	69.0	68.9	68.8	68.6	68.5	68.4	68.3	68.3	68.1	68.0
69	78.8	77.5	76.5	75.5	74.4	73.6	72.8	72.0	71.3	70.9	70.7	70.4	70.3	70.0	69.9	69.8	69.6	69.5	69.4	69.3	69.3	69.1	69.0
70	79.9	78.7	77.6	76.6	75.5	74.6	73.8	73.1	72.3	71.9	71.7	71.4	71.3	71.1	70.9	70.8	70.6	70.5	70.4	70.4	70.3	70.1	70.0
71		79.8	78.7	77.7	76.6	75.7	74.9	74.1	73.4	73.0	72.7	72.4	72.3	72.1	71.9	71.8	71.6	71.5	71.4	71.4	71.3	71.1	71.0
72		80.9	79.8	78.8	77.7	76.8	75.9	75.2	74.4	74.0	73.8	73.5	73.3	73.1	73.0	72.8	72.7	72.5	72.4	72.4	72.3	72.1	72.0
73			80.9	79.9	78.7	77.8	77.0	76.2	75.4	75.0	74.8	74.5	74.3	74.1	74.0	73.8	73.7	73.5	73.4	73.4	73.3	73.1	73.0
74					79.8	78.9	78.1	77.2	76.4	76.0	75.8	75.5	75.4	75.1	75.0	74.8	74.7	74.5	74.4	74.4	74.3	74.1	74.0
75					80.9	80.0	79.1	78.3	77.5	77.1	76.8	76.5	76.4	76.1	76.0	75.8	75.7	75.5	75.5	75.4	75.3	75.2	75.0
76							80.2	79.3	78.5	78.1	77.9	77.6	77.4	77.2	77.0	76.8	76.7	76.5	76.5	76.4	76.3	76.2	76.0
77								80.4	79.5	79.1	78.9	78.6	78.4	78.2	78.0	77.9	77.7	77.5	77.5	77.4	77.3	77.2	77.0
78									80.6	80.2	79.9	79.6	79.4	79.2	79.0	78.9	78.7	78.5	78.5	78.4	78.3	78.2	78.0

*Reprinted with permission from Bayley and Pinneau.[8]

TABLE 2C.—RETARDED BOYS. PERCENTAGES AND ESTIMATED MATURE HEIGHTS FOR BOYS WITH SKELETAL AGES ONE YEAR OR MORE RETARDED FOR THEIR CHRONOLOGICAL AGES: SKELETAL AGES 6 THROUGH 13 YEARS*

Skeletal Age	6-0	6-3	6-6	6-9	7-0	7-3	7-6	7-9	8-0	8-3	8-6	8-9	9-0	9-3	9-6	9-9	10-0	10-3	10-6	10-9	11-0	11-3	11-6	11-9	12-0	12-3	12-6	12-9	13-0
% of Mature Height	68.0	69.0	70.0	70.9	71.8	72.8	73.8	74.7	75.6	76.5	77.3	77.9	78.6	79.4	80.0	80.7	81.2	81.6	81.9	82.1	82.3	82.7	83.2	83.9	84.5	85.2	86.0	86.9	88.0
Ht. (inches)																													
41	60.3																												
42	61.8	60.9	60.0																										
43	63.2	62.3	61.4	60.6																									
44	64.7	63.8	62.9	62.1	61.3	60.4																							
45	66.2	65.2	64.3	63.5	62.7	61.8	61.0	60.2																					
46	67.6	66.7	65.7	64.9	64.1	63.2	62.3	61.6	60.8	60.1																			
47	69.1	68.1	67.1	66.3	65.5	64.6	63.7	62.9	62.2	61.4	60.8	60.3																	
48	70.6	69.6	68.6	67.7	66.9	65.9	65.0	64.3	63.5	62.7	62.1	61.6	61.1	60.5	60.0														
49	72.1	71.0	70.0	69.1	68.3	67.3	66.4	65.6	64.8	64.1	63.4	62.9	62.3	61.7	61.3	60.7	60.3	60.0											
50	73.5	72.5	71.4	70.5	69.6	68.7	67.8	66.9	66.1	65.4	64.7	64.2	63.6	63.0	62.5	62.0	61.6	61.3	61.1	60.9	60.8	60.5	60.1						
51	75.0	73.9	72.9	71.9	71.0	70.1	69.1	68.3	67.5	66.7	66.0	65.5	64.9	64.2	63.8	63.2	62.8	62.5	62.3	62.1	62.0	61.7	61.3	60.8	60.4				
52	76.5	75.4	74.3	73.3	72.4	71.4	70.5	69.6	68.8	68.0	67.3	66.8	66.2	65.5	65.0	64.4	64.0	63.7	63.5	63.3	63.2	62.9	62.5	62.0	61.5	61.0	60.5		
53	77.9	76.8	75.7	74.8	73.8	72.8	71.8	71.0	70.1	69.3	68.6	68.0	67.4	66.8	66.3	65.7	65.3	65.0	64.7	64.6	64.4	64.1	63.7	63.2	62.7	62.2	61.6	61.0	60.2
54	79.4	78.3	77.1	76.2	75.2	74.2	73.2	72.3	71.4	70.6	69.9	69.3	68.7	68.0	67.5	66.9	66.5	66.2	65.9	65.8	65.6	65.3	64.9	64.4	63.9	63.4	62.8	62.1	61.4
55	80.9	79.7	78.6	77.6	76.6	75.5	74.5	73.6	72.8	71.9	71.2	70.6	70.0	69.3	68.8	68.2	67.7	67.4	67.2	67.0	66.8	66.5	66.1	65.6	65.1	64.6	64.0	63.3	62.5
56			80.0	79.0	78.0	76.9	75.9	75.0	74.1	73.2	72.4	71.9	71.2	70.5	70.0	69.4	69.0	68.6	68.4	68.2	68.0	67.7	67.3	66.7	66.3	65.7	65.1	64.4	63.6
57				80.4	79.4	78.3	77.3	76.3	75.4	74.5	73.7	73.2	72.5	71.8	71.3	70.6	70.2	69.9	69.6	69.4	69.3	68.9	68.5	67.9	67.5	66.9	66.3	65.6	64.8
58					80.8	79.7	78.6	77.6	76.7	75.8	75.0	74.5	73.8	73.0	72.5	71.9	71.4	71.1	70.8	70.6	70.5	70.1	69.7	69.1	68.6	68.1	67.4	66.7	65.9
59							79.9	79.0	78.0	77.1	76.3	75.7	75.1	74.3	73.8	73.1	72.7	72.3	72.0	71.9	71.7	71.3	70.9	70.3	69.8	69.2	68.6	67.9	67.0
60								80.3	79.4	78.4	77.6	77.0	76.3	75.6	75.0	74.4	73.9	73.5	73.3	73.1	72.9	72.6	72.1	71.5	71.0	70.4	69.8	69.0	68.2
61									80.7	79.7	78.9	78.3	77.6	76.8	76.3	75.6	75.1	74.8	74.5	74.3	74.1	73.8	73.3	72.7	72.2	71.6	70.9	70.2	69.3
62											80.2	79.6	78.9	78.1	77.5	76.8	76.4	76.0	75.7	75.5	75.3	75.0	74.5	73.9	73.4	72.8	72.1	71.3	70.5
63												80.9	80.2	79.3	78.8	78.1	77.6	77.2	76.9	76.7	76.5	76.2	75.7	75.1	74.6	73.9	73.3	72.5	71.6
64														80.6	80.0	79.3	78.8	78.4	78.1	78.0	77.8	77.4	76.9	76.3	75.7	75.1	74.4	73.6	72.7
65																80.5	80.0	79.7	79.4	79.2	79.0	78.6	78.1	77.5	76.9	76.3	75.6	74.8	73.9
66																		80.9	80.6	80.4	80.2	79.8	79.3	78.7	78.1	77.5	76.8	75.9	75.0
67																							80.5	79.9	79.3	78.6	77.9	77.1	76.1

*Reprinted with permission from Bayley and Pinneau.[8]

TABLE 3A.–AVERAGE GIRLS. PERCENTAGES AND ESTIMATED MATURE HEIGHTS FOR GIRLS WITH SKELETAL AGES WITHIN ONE YEAR OF THEIR CHRONOLOGICAL AGES: SKELETAL AGES 6 THROUGH 11 YEARS*

Skeletal Age	6-0	6-3	6-6	6-10	7-0	7-3	7-6	7-10	8-0	8-3	8-6	8-10	9-0	9-3	9-6	9-9	10-0	10-3	10-6	10-9	11-0	11-3	11-6	11-9
% of Mature Height Ht. (inches)	72.0	72.9	73.8	75.1	75.7	76.5	77.2	78.2	79.0	80.1	81.0	82.1	82.7	83.6	84.4	85.3	86.2	87.4	88.4	89.6	90.6	91.0	91.4	91.8
37	51.4																							
38	52.8	52.1	51.5																					
39	54.2	53.5	52.8	51.9	51.5	51.0																		
40	55.6	54.9	54.2	53.3	52.8	52.3	51.8	51.2																
41	56.9	56.2	55.6	54.6	54.2	53.6	53.1	52.4	51.9	51.2														
42	58.3	57.6	56.9	55.9	55.5	54.9	54.4	53.7	53.2	52.4	51.9	51.2												
43	59.7	59.0	58.3	57.3	56.8	56.2	55.7	55.0	54.4	53.7	53.1	52.4	52.0	51.4										
44	61.1	60.4	59.6	58.6	58.1	57.5	57.0	56.3	55.7	54.9	54.3	53.6	53.2	52.6	52.1	51.6	51.0							
45	62.5	61.7	61.0	59.9	59.4	58.8	58.3	57.5	57.0	56.2	55.6	54.8	54.4	53.8	53.3	52.8	52.2	51.5						
46	63.9	63.1	62.3	61.3	60.8	60.1	59.6	58.8	58.2	57.4	56.8	56.0	55.6	55.0	54.5	53.9	53.4	52.6	52.0					
47	65.3	64.5	63.7	62.6	62.1	61.4	60.9	60.1	59.5	58.7	58.0	57.2	56.8	56.2	55.7	55.1	54.5	53.8	53.2	52.5	51.9	51.6	51.4	51.2
48	66.7	65.8	65.0	63.9	63.4	62.7	62.2	61.4	60.8	59.9	59.3	58.5	58.0	57.4	56.9	56.3	55.7	54.9	54.3	53.6	53.0	52.7	52.5	52.3
49	68.1	67.2	66.4	65.2	64.7	64.1	63.5	62.7	62.0	61.2	60.5	59.7	59.3	58.6	58.1	57.4	56.8	56.1	55.4	54.7	54.1	53.8	53.6	53.4
50	69.4	68.6	67.8	66.6	66.1	65.4	64.8	63.9	63.3	62.4	61.7	60.9	60.5	59.8	59.2	58.6	58.0	57.2	56.6	55.8	55.2	54.9	54.7	54.5
51	70.8	70.0	69.1	67.9	67.4	66.7	66.1	65.2	64.6	63.7	63.0	62.1	61.7	61.0	60.4	59.8	59.2	58.4	57.7	56.9	56.3	56.0	55.8	55.6
52	72.2	71.3	70.5	69.2	68.7	68.0	67.4	66.5	65.8	64.9	64.2	63.3	62.9	62.2	61.6	61.0	60.3	59.5	58.8	58.0	57.4	57.1	56.9	56.6
53	73.6	72.7	71.8	70.6	70.0	69.3	68.7	67.8	67.1	66.2	65.4	64.6	64.1	63.4	62.8	62.1	61.5	60.6	60.0	59.2	58.5	58.2	58.0	57.7
54		74.1	73.2	71.9	71.3	70.6	69.9	69.1	68.4	67.4	66.7	65.8	65.3	64.6	64.0	63.3	62.6	61.8	61.1	60.3	59.6	59.3	59.1	58.8
55			74.5	73.2	72.7	71.9	71.2	70.3	69.6	68.7	67.9	67.0	66.5	65.8	65.2	64.5	63.8	62.9	62.2	61.4	60.7	60.4	60.2	59.9
56				74.6	74.0	73.2	72.5	71.6	70.9	69.9	69.1	68.2	67.7	67.0	66.4	65.7	65.0	64.1	63.3	62.5	61.8	61.5	61.3	61.0
57						74.5	73.8	72.9	72.2	71.2	70.4	69.4	68.9	68.2	67.5	66.8	66.1	65.2	64.5	63.6	62.9	62.6	62.4	62.1
58								74.2	73.4	72.4	71.6	70.6	70.1	69.4	68.7	68.0	67.3	66.4	65.6	64.7	64.0	63.7	63.5	63.2
59									74.7	73.7	72.8	71.9	71.3	70.6	69.9	69.2	68.4	67.5	66.7	65.8	65.1	64.8	64.6	64.3
60										74.9	74.1	73.1	72.6	71.8	71.1	70.3	69.6	68.7	67.9	67.0	66.2	65.9	65.6	65.4
61												74.3	73.8	73.0	72.3	71.5	70.8	69.8	69.0	68.1	67.3	67.0	66.7	66.4
62														74.2	73.5	72.7	71.9	71.0	70.1	69.2	68.4	68.1	67.8	67.5
63															74.6	73.9	73.1	72.1	71.3	70.3	69.5	69.2	68.9	68.6
64																	74.2	73.2	72.4	71.4	70.6	70.3	70.0	69.7
65																		74.4	73.5	72.5	71.7	71.4	71.1	70.8
66																			74.7	73.7	72.9	72.5	72.2	71.9
67																				74.8	74.0	73.6	73.3	73.0
68																						74.7	74.4	74.1

*Reprinted with permission from Bayley and Pinneau.[8]

TABLE 3B.—AVERAGE GIRLS. PERCENTAGES AND ESTIMATED MATURE HEIGHTS FOR GIRLS WITH SKELETAL AGES WITHIN ONE YEAR OF THEIR CHRONOLOGICAL AGES: SKELETAL AGES 12 THROUGH 18 YEARS*

Skeletal Age → / Ht. (inches) ↓	12-0	12-3	12-6	12-9	13-0	13-3	13-6	13-9	14-0	14-3	14-6	14-9	15-0	15-3	15-6	15-9	16-0	16-3	16-6	16-9	17-0	17-6	18-0
% of Mature Height	92.2	93.2	94.1	95.0	95.8	96.7	97.4	97.8	98.0	98.3	98.6	98.8	99.0	99.1	99.3	99.4	99.6	99.6	99.7	99.8	99.9	99.95	100.0
47	51.0																						
48	52.1	51.5	51.0																				
49	53.1	52.6	52.1	51.6	51.1																		
50	54.2	53.6	53.1	52.6	52.2	51.7	51.3	51.1	51.0														
51	55.3	54.7	54.2	53.7	53.2	52.7	52.4	52.1	52.0	51.9	51.7	51.6	51.5	51.5	51.4	51.3	51.2	51.2	51.2	51.1	51.1	51.0	51.0
52	56.4	55.8	55.3	54.7	54.3	53.8	53.4	53.2	53.1	52.9	52.7	52.6	52.5	52.5	52.4	52.3	52.2	52.2	52.2	52.1	52.1	52.0	52.0
53	57.5	56.9	56.3	55.8	55.3	54.8	54.4	54.2	54.1	53.9	53.8	53.6	53.5	53.5	53.4	53.3	53.2	53.2	53.2	53.1	53.1	53.0	53.0
54	58.6	57.9	57.4	56.8	56.4	55.8	55.4	55.2	55.1	54.9	54.8	54.7	54.5	54.5	54.4	54.3	54.2	54.2	54.2	54.1	54.1	54.0	54.0
55	59.7	59.0	58.4	57.9	57.4	56.9	56.5	56.2	56.1	56.0	55.8	55.7	55.6	55.5	55.4	55.3	55.2	55.2	55.2	55.1	55.1	55.0	55.0
56	60.7	60.1	59.5	58.9	58.5	57.9	57.5	57.3	57.1	57.0	56.8	56.7	56.6	56.5	56.4	56.3	56.2	56.2	56.2	56.1	56.1	56.0	56.0
57	61.8	61.2	60.6	60.0	59.5	58.9	58.5	58.3	58.2	58.0	57.8	57.7	57.6	57.5	57.4	57.3	57.2	57.2	57.2	57.1	57.1	57.0	57.0
58	62.9	62.2	61.6	61.1	60.5	60.0	59.5	59.3	59.2	59.0	58.8	58.7	58.6	58.5	58.4	58.4	58.2	58.2	58.2	58.1	58.1	58.0	58.0
59	64.0	63.3	62.7	62.1	61.6	61.0	60.6	60.3	60.2	60.0	59.8	59.7	59.6	59.5	59.4	59.4	59.2	59.2	59.2	59.1	59.1	59.0	59.0
60	65.1	64.4	63.8	63.2	62.6	62.0	61.6	61.3	61.2	61.0	60.9	60.7	60.6	60.5	60.4	60.4	60.2	60.2	60.2	60.1	60.1	60.0	60.0
61	66.2	65.5	64.8	64.2	63.7	63.1	62.6	62.4	62.2	62.1	61.9	61.7	61.6	61.6	61.4	61.4	61.2	61.2	61.2	61.1	61.1	61.0	61.0
62	67.2	66.5	65.9	65.3	64.7	64.1	63.7	63.4	63.3	63.1	62.9	62.8	62.6	62.6	62.4	62.4	62.2	62.2	62.2	62.1	62.1	62.0	62.0
63	68.3	67.6	67.0	66.3	65.8	65.1	64.7	64.4	64.3	64.1	63.9	63.8	63.6	63.6	63.4	63.4	63.3	63.3	63.2	63.1	63.1	63.0	63.0
64	69.4	68.7	68.0	67.4	66.8	66.2	65.7	65.4	65.3	65.1	64.9	64.8	64.6	64.6	64.5	64.4	64.3	64.3	64.2	64.1	64.1	64.0	64.0
65	70.5	69.7	69.1	68.4	67.8	67.2	66.7	66.5	66.3	66.1	65.9	65.8	65.7	65.6	65.5	65.4	65.3	65.3	65.2	65.1	65.1	65.0	65.0
66	71.6	70.8	70.1	69.5	68.9	68.3	67.8	67.5	67.3	67.1	66.9	66.8	66.7	66.6	66.5	66.4	66.3	66.3	66.2	66.1	66.1	66.0	66.0
67	72.7	71.9	71.2	70.5	69.9	69.3	68.8	68.5	68.4	68.2	68.0	67.8	67.7	67.6	67.5	67.4	67.3	67.3	67.2	67.1	67.1	67.0	67.0
68	73.8	73.0	72.3	71.6	71.0	70.3	69.8	69.5	69.4	69.2	69.0	68.8	68.7	68.6	68.5	68.4	68.3	68.3	68.2	68.1	68.1	68.0	68.0
69	74.8	74.0	73.3	72.6	72.0	71.4	70.8	70.6	70.4	70.2	70.0	69.8	69.7	69.6	69.5	69.4	69.3	69.3	69.2	69.1	69.1	69.0	69.0
70			74.4	73.7	73.1	72.4	71.9	71.6	71.4	71.2	71.0	70.9	70.7	70.6	70.5	70.4	70.3	70.3	70.2	70.1	70.1	70.0	70.0
71				74.7	74.1	73.4	72.9	72.6	72.4	72.2	72.0	71.9	71.7	71.6	71.5	71.4	71.3	71.3	71.2	71.1	71.1	71.0	71.0
72						74.5	73.9	73.6	73.5	73.2	73.0	72.9	72.7	72.7	72.5	72.4	72.3	72.3	72.2	72.1	72.1	72.0	72.0
73							74.9	74.6	74.5	74.3	74.0	73.9	73.7	73.7	73.5	73.4	73.3	73.3	73.2	73.1	73.1	73.0	73.0
74												74.9	74.7	74.7	74.5	74.4	74.3	74.3	74.2	74.1	74.1	74.0	74.0

*Reprinted with permission from Bayley and Pinneau.[8]

TABLE 3C.–RETARDED GIRLS. PERCENTAGES AND ESTIMATED MATURE HEIGHTS FOR GIRLS WITH SKELETAL AGES ONE YEAR OR MORE RETARDED FOR THEIR CHRONOLOGICAL AGES: SKELETAL AGES 6 THROUGH 11 YEARS*

Skeletal Age	6-0	6-3	6-6	6-10	7-0	7-3	7-6	7-10	8-0	8-3	8-6	8-10	9-0	9-3	9-6	9-9	10-0	10-3	10-6	10-9	11-0	11-3	11-6	11-9
% of Mature Height	73.3	74.2	75.1	76.3	77.0	77.9	78.8	79.7	80.4	81.3	82.3	83.6	84.1	85.1	85.8	86.6	87.4	88.4	89.6	90.7	91.8	92.2	92.6	92.9
Ht. (inches)																								
38	51.8	51.2																						
39	53.2	52.6	51.9	51.1																				
40	54.6	53.9	53.3	52.4	51.9	51.3																		
41	55.9	55.3	54.6	53.7	53.2	52.6	52.0	51.4	51.0															
42	57.3	56.6	55.9	55.0	54.5	53.9	53.3	52.7	52.2	51.7	51.0													
43	58.7	58.0	57.3	56.4	55.8	55.2	54.6	54.0	53.5	52.9	52.2	51.4	51.1											
44	60.0	59.3	58.6	57.7	57.1	56.5	55.8	55.2	54.7	54.1	53.5	52.6	52.3	51.7	51.3									
45	61.4	60.6	59.9	59.0	58.4	57.8	57.1	56.5	56.0	55.4	54.7	53.8	53.5	52.9	52.4	52.0	51.5							
46	62.8	62.0	61.3	60.3	59.7	59.1	58.4	57.7	57.2	56.6	55.9	55.0	54.7	54.1	53.6	53.1	52.6	52.0	51.3					
47	64.1	63.3	62.6	61.6	61.0	60.3	59.6	59.0	58.5	57.8	57.1	56.2	55.9	55.2	54.8	54.3	53.8	53.2	52.5	51.8	51.2	51.0		
48	65.5	64.7	63.9	62.9	62.3	61.6	60.9	60.2	59.7	59.0	58.3	57.4	57.1	56.4	55.9	55.4	54.9	54.3	53.6	52.9	52.3	52.1	51.8	51.7
49	66.9	66.0	65.2	64.2	63.6	62.9	62.2	61.5	60.9	60.3	59.5	58.6	58.3	57.6	57.1	56.6	56.1	55.4	54.7	54.0	53.4	53.1	52.9	52.7
50	68.2	67.4	66.6	65.5	64.9	64.2	63.5	62.7	62.2	61.5	60.8	59.8	59.5	58.8	58.3	57.7	57.2	56.6	55.8	55.1	54.5	54.2	54.0	53.8
51	69.6	68.7	67.9	66.8	66.2	65.5	64.7	64.0	63.4	62.7	62.0	61.0	60.6	59.9	59.4	58.9	58.4	57.7	56.9	56.2	55.6	55.3	55.1	54.9
52	70.9	70.1	69.2	68.2	67.5	66.8	66.0	65.2	64.7	64.0	63.2	62.2	61.8	61.1	60.6	60.0	59.5	58.8	58.0	57.3	56.6	56.4	56.2	56.0
53	72.3	71.4	70.6	69.5	68.8	68.0	67.3	66.5	65.9	65.2	64.4	63.4	63.0	62.3	61.8	61.2	60.6	60.0	59.2	58.4	57.7	57.5	57.2	57.1
54	73.7	72.8	71.9	70.8	70.1	69.3	68.5	67.8	67.2	66.4	65.6	64.6	64.2	63.5	62.9	62.4	61.8	61.1	60.3	59.5	58.8	58.6	58.3	58.1
55		74.1	73.2	72.1	71.4	70.6	69.8	69.0	68.4	67.7	66.8	65.8	65.4	64.6	64.1	63.5	62.9	62.2	61.4	60.6	59.9	59.7	59.4	59.2
56			74.6	73.4	72.7	71.9	71.1	70.3	69.7	68.9	68.0	67.0	66.6	65.8	65.3	64.7	64.1	63.3	62.5	61.7	61.0	60.7	60.5	60.3
57				74.7	74.0	73.2	72.3	71.5	70.9	70.1	69.3	68.2	67.8	67.0	66.4	65.8	65.2	64.5	63.6	62.8	62.1	61.8	61.6	61.4
58						74.5	73.6	72.8	72.1	71.3	70.5	69.4	69.0	68.2	67.6	67.0	66.4	65.6	64.7	63.9	63.2	62.9	62.6	62.4
59							74.9	74.0	73.4	72.6	71.7	70.6	70.2	69.3	68.8	68.1	67.5	66.7	65.8	65.0	64.3	64.0	63.7	63.5
60									74.6	73.8	72.9	71.8	71.3	70.5	69.9	69.3	68.7	67.9	67.0	66.2	65.4	65.1	64.8	64.6
61											74.1	73.0	72.5	71.7	71.1	70.4	69.8	69.0	68.1	67.3	66.4	66.2	65.9	65.7
62												74.2	73.7	72.9	72.3	71.6	70.9	70.1	69.2	68.4	67.5	67.2	67.0	66.7
63													74.9	74.0	73.4	72.7	72.1	71.3	70.3	69.5	68.6	68.3	68.0	67.8
64															74.6	73.9	73.2	72.4	71.4	70.6	69.7	69.4	69.1	68.9
65																	74.4	73.5	72.5	71.7	70.8	70.5	70.2	70.0
66																		74.7	73.7	72.8	71.9	71.6	71.3	71.0
67																			74.8	73.9	73.0	72.7	72.4	72.1
68																					74.1	73.8	73.4	73.2
69																						74.8	74.5	74.3

*Reprinted with permission from Bayley and Pinneau.[8]

472

TABLE 3D.—RETARDED GIRLS. PERCENTAGES AND ESTIMATED MATURE HEIGHTS FOR GIRLS WITH SKELETAL AGES ONE YEAR OR MORE RETARDED FOR THEIR CHRONOLOGICAL AGES: SKELETAL AGES 12 THROUGH 17 YEARS*

Skeletal Age	12-0	12-3	12-6	12-9	13-0	13-3	13-6	13-9	14-0	14-3	14-6	14-9	15-0	15-3	15-6	15-9	16-0	16-3	16-6	16-9	17-0
% of Mature Height	93.2	94.2	94.9	95.7	96.4	97.1	97.7	98.1	98.3	98.6	98.9	99.2	99.4	99.5	99.6	99.7	99.8	99.9	99.9	99.95	100.0
Ht. (inches)																					
48	51.5	51.0																			
49	52.6	52.0	51.6	51.2																	
50	53.6	53.1	52.7	52.2	51.9	51.5	51.2	51.0													
51	54.7	54.1	53.7	53.3	52.9	52.5	52.2	52.0	51.9	51.7	51.6	51.4	51.3	51.3	51.2	51.2	51.1	51.1	51.1	51.0	51.0
52	55.8	55.2	54.8	54.3	53.9	53.6	53.2	53.0	52.9	52.7	52.6	52.4	52.3	52.3	52.2	52.2	52.1	52.1	52.1	52.0	52.0
53	56.9	56.3	55.8	55.4	55.0	54.6	54.2	54.0	53.9	53.8	53.6	53.4	53.3	53.3	53.2	53.2	53.1	53.1	53.1	53.0	53.0
54	57.9	57.3	56.9	56.4	56.0	55.6	55.3	55.0	54.9	54.8	54.6	54.4	54.3	54.3	54.2	54.2	54.1	54.1	54.1	54.0	54.0
55	59.0	58.4	58.0	57.5	57.1	56.6	56.3	56.1	56.0	55.8	55.6	55.4	55.3	55.3	55.2	55.2	55.1	55.1	55.1	55.0	55.0
56	60.1	59.4	59.0	58.5	58.1	57.7	57.3	57.1	57.0	56.8	56.6	56.5	56.3	56.3	56.2	56.2	56.1	56.1	56.1	56.0	56.0
57	61.2	60.5	60.1	59.6	59.1	58.7	58.3	58.1	58.0	57.8	57.6	57.5	57.3	57.3	57.2	57.2	57.1	57.1	57.1	57.0	57.0
58	62.2	61.6	61.1	60.6	60.2	59.7	59.4	59.1	59.0	58.8	58.6	58.5	58.3	58.3	58.2	58.2	58.1	58.1	58.1	58.0	58.0
59	63.3	62.6	62.2	61.7	61.2	60.8	60.4	60.1	60.0	59.8	59.7	59.5	59.4	59.3	59.2	59.2	59.1	59.1	59.1	59.0	59.0
60	64.4	63.7	63.2	62.7	62.3	61.8	61.4	61.2	61.0	60.9	60.7	60.5	60.4	60.3	60.2	60.2	60.1	60.1	60.1	60.0	60.0
61	65.5	64.8	64.3	63.7	63.3	62.8	62.4	62.2	62.1	61.9	61.7	61.5	61.4	61.3	61.2	61.2	61.1	61.1	61.1	61.0	61.0
62	66.5	65.8	65.3	64.8	64.3	63.9	63.5	63.2	63.1	62.9	62.7	62.5	62.4	62.3	62.2	62.2	62.1	62.1	62.1	62.0	62.0
63	67.6	66.9	66.4	65.8	65.3	64.9	64.5	64.2	64.1	63.9	63.7	63.5	63.4	63.3	63.3	63.2	63.1	63.1	63.1	63.0	63.0
64	68.7	67.9	67.4	66.9	66.4	65.9	65.5	65.2	65.1	64.9	64.7	64.5	64.4	64.3	64.3	64.2	64.1	64.1	64.1	64.0	64.0
65	69.7	69.0	68.5	67.9	67.4	66.9	66.5	66.3	66.1	65.9	65.7	65.5	65.4	65.3	65.3	65.2	65.1	65.1	65.1	65.0	65.0
66	70.8	70.1	69.5	69.0	68.5	68.0	67.6	67.3	67.1	66.9	66.7	66.5	66.4	66.3	66.3	66.2	66.1	66.1	66.1	66.0	66.0
67	71.9	71.1	70.6	70.0	69.5	69.0	68.6	68.3	68.2	68.0	67.7	67.5	67.4	67.3	67.3	67.2	67.1	67.1	67.1	67.0	67.0
68	73.0	72.2	71.7	71.1	70.5	70.0	69.6	69.3	69.2	69.0	68.8	68.6	68.4	68.3	68.3	68.2	68.1	68.1	68.1	68.0	68.0
69	74.0	73.2	72.7	72.1	71.6	71.1	70.6	70.3	70.2	70.0	69.8	69.6	69.4	69.3	69.3	69.2	69.1	69.1	69.1	69.0	69.0
70		74.3	73.8	73.1	72.6	72.1	71.6	71.4	71.2	71.0	70.8	70.6	70.4	70.4	70.3	70.2	70.1	70.1	70.1	70.0	70.0
71			74.8	74.2	73.6	73.1	72.7	72.4	72.2	72.0	71.8	71.6	71.4	71.4	71.3	71.2	71.1	71.1	71.1	71.0	71.0
72					74.7	74.2	73.7	73.4	73.3	73.0	72.8	72.6	72.4	72.4	72.3	72.2	72.1	72.1	72.1	72.0	72.0
73							74.7	74.4	74.3	74.0	73.8	73.6	73.4	73.4	73.3	73.2	73.1	73.1	73.1	73.0	73.0
74											74.8	74.6	74.4	74.4	74.3	74.2	74.1	74.1	74.1	74.0	74.0

*Reprinted with permission from Bayley and Pinneau.[8]

across to 7 years in the skeletal age column. It can be predicted that he will reach an adult height of 70.5 inches. This is reasonably reassuring. The patient is a typical example of delayed maturation. Another 12-year-old boy had a height of 50 inches but a skeletal age of 11 years and 9 months. His adult height should be 60.5 inches. This is the pattern seen in constitutional or genetic short stature. Such an individual often has a strong family history of shortness of stature.

These charts are not used in certain syndromes in which there may be a specific reason for a marked inhibition of growth. They are most useful in the patient without any underlying illness that would explain short stature. In general, if the bone age is delayed, the predicted ultimate height may be normal. It is certain-

ly worth looking up in the tables. On the other hand, if the chronologic age is equal to the bone age in a patient whose height is short for his age, the height also will ultimately be short.

It is also possible to predict adult height, in the case of girls, from the height at menarche and the age at which menarche occurs. A nomogram has been devised for this purpose by Frisch and Nagel[9] from longitudinal data on 181 girls (Fig. 7). A girl who reaches menarche at 11 years of age and a height of 63 inches will be 66.9 inches ± 0.75 inches as an adult. Another girl who reaches 63 inches at a menarchial age of 15 years will end up only 64.5 ± 0.75 inches. This method appears to be as reliable as other methods of predicting adult height. It has the advantage that it does not

Fig. 7.—Nomogram for the prediction of adult height from the height and age at menarche. To use the nomogram, the height at menarche and the age at menarche are connected with a straight edge that extends over to the third scale where the predicted height at 18 years of age may be read. One standard error of estimate (s.e.e.) was 1.9 cm; ± 1 s.e.e. is illustrated next to 172 cm. Ninety-five per cent confidence limits are ± 3.7 cm. (Nomogram devised by Frisch and Nagel[9] and reproduced from J. Pediatr. with permission.)

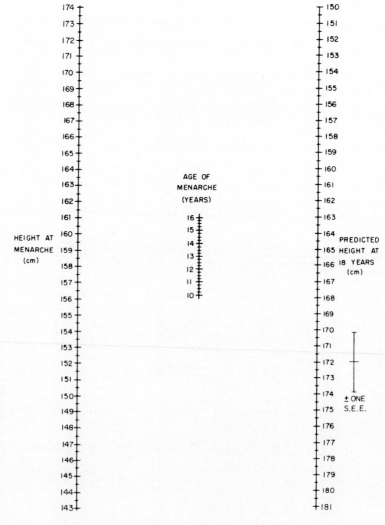

require x-ray examination. Of course, it cannot be used prior to menarche and is restricted to the female.

The stature of an individual is a sensitive bioassay of the multifactorial events that influence human growth. Stature is polygenic. It also is environmentally determined. Abnormalities in growth are challenging exercises in problem solving. An accurate assessment of the details of deviation from normality may promptly define the diagnosis.

REFERENCES

1. McKusick, V. A.: *Heritable Disorders of Connective Tissue* (4th ed.; St. Louis: C. V. Mosby Co., 1972).
2. Engelbach, W.: *Endocrine Medicine* (Springfield, Ill.: Charles C Thomas, 1932), Vol. I, p. 304.
3. Stratz, C. H.: *Der Korper des Kindes* (Stuttgart: Ferdinand Enke, 1902).
4. Lubchenco, L. O., Hansman, C., and Boyd, E.: Intrauterine growth as estimated from live births at gestational ages from 26 to 42 weeks, Pediatrics 37:403, 1966.
5. Seckel, H. P.: *Bird-headed Dwarfs* (Springfield, Ill: Charles C Thomas, 1960).
6. Warkany, J.: *Congenital Malformations* (Chicago: Year Book Medical Publishers, Inc., 1971), p. 134.
7. Greulich, W. W., and Pyle, S. I.: *Radiographic Atlas of Skeletal Development of the Hand and Wrist* (Palo Alto: Stanford Univ. Press, 1950).
8. Bayley, N., and Pinneau, S. R.: Tables for predicting adult height from skeletal age: Revised for use with the Greulich-Pyle hand standards, J. Pediatr. 40:423, 1952.
9. Frisch, R. E., and Nagel, J. S.: Prediction of adult height of girls from age of menarche and height at menarche, J. Pediatr. 85:838, 1974.

Other Useful Measurements

OTHER USEFUL MEASUREMENTS

Kenneth Lyons Jones, Jr., and William L. Nyhan

In the diagnosis of disorders of morphogenesis, we frequently must act on an impression of a degree of deviation from normality. However, when actual measurement of the deviation is possible, we are on a sounder basis.

The following measurements are often useful in the recognition of various patterns of human malformation.

Ocular hypertelorism or hypotelorism.—This is defined as an increased or decreased width between the eyes, respectively, and is a function of the interpupillary distance (C in Fig. 1.) Normative data are set forth in Figure 2. Reliable interpupillary measurements often are difficult to obtain in childhood, since accuracy requires measurement between the center of each pupil, with the eyes fixed. Therefore, Figure 3 can be used to obtain the interpupillary distance using two fixed points, the inner canthal distance (B in Fig. 1) and the outer canthal distance (A in Fig. 1).

Lateral displacement of the medial canthi.—This is defined as an increase in the distance between the inner canthi (B in Fig. 1). Normative data are set forth in Figure 4.

Certain syndromes, such as the Waardenburg syndrome, are associated with lateral displacement of the medial canthal, whereas others, such as the Aarskog

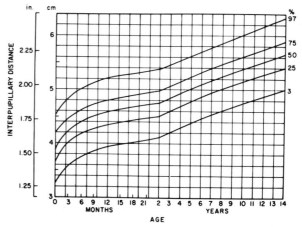

Fig. 2.—Interpupillary distance. (Reprinted with permission from Feingold and Bossert.[5])

syndrome, include ocular hypertelorism as one feature in an over-all pattern of malformation. The presence of a widow's peak often is a helpful clue to the presence of ocular hypertelorism.[1]

Palpebral fissure size (Fig. 5).—The palpebral fissures, measured from medial to lateral canthi, may be short as a consequence of a lateral displacement of the medial canthi, as in familial blepharophimosis, or as a function of decreased growth of the eye, as in the fetal alcohol syndrome.

Ear length (Fig. 6).—This is measured by the maximum distance from the superior aspect to the inferior aspect of the ear. Aase *et al.*[2] documented decreased ear length as a consistent feature in newborn infants with the Down syndrome. In their series no full-term infant with the Down syndrome had an ear length greater than 3.4 cm (mean 3 cm), and no normal full-term infant had an ear length less than 3.2 cm (mean 3.8 cm).

Anterior and Posterior Fontanels:

Anterior fontanel (Fig. 7).—Mean anterior fontanel size is an average of the length (anterior-posterior dimension) plus width (transverse dimension). An abnormally enlarged anterior fontanel can be related not only to increased intracranial pressure but also is seen

Fig. 1.—Useful distances relevant to the eyes: *A*, outer canthal distance, *B*, inner canthal distance and *C*, interpupillary distance. (Reprinted with permission from Feingold and Bossert.[5])

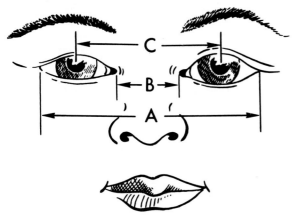

479

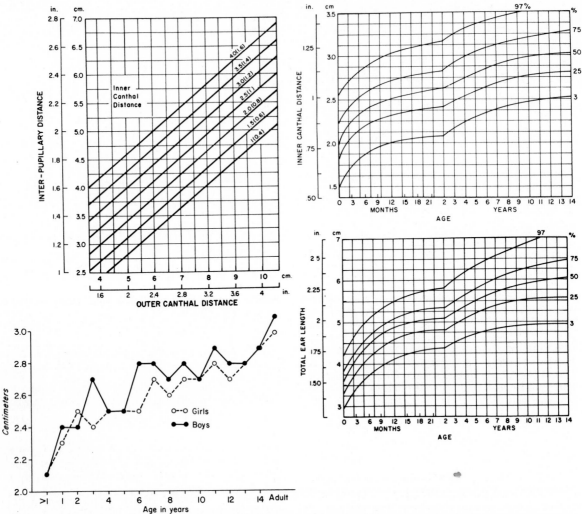

Fig. 3 (top left).—Graph for calculation of interpupillary distance. In order to use this graph, inner and outer canthal distances are measured. They are plotted on the diagonal line and the abscissa, respectively. The point at which these intersect is then drawn to the ordinate, giving the interpupillary distance. (Reprinted with permission from Feingold and Bossert.[5])

Fig. 4 (top right).—Inner canthal distance. (Reprinted with permission from Feingold and Bossert.[5])

Fig. 5 (bottom left).—Palpebral fissure size. (Reprinted with permission from Smith.[6])

Fig. 6 (bottom right).—Ear length. (Reprinted with permission from Feingold and Bossert.[5])

in other skeletal disorders in which development of the bones of the skull is delayed, such as cleidocranial dysostosis and pyknodysostosis, as well as in chromosomal abnormalities.[3] It also can be seen with immaturity. An abnormally small anterior fontanel can be secondary to decreased growth of the brain, as in primary microcephaly, or primary, as in craniosynostosis of the sagittal or coronal sutures.

Posterior fontanel.—Popich and Smith[4] found that only 3% of normal newborn infants had a mean posterior fontanel size larger than 0.5 cm. Congenital hypothyroidism should be considered in any newborn infant whose posterior fontanel is enlarged.

Total hand length.—This is measured from the distal flexion crease of the wrist to the tip of the middle finger. Total hand size is depicted in Figure 8. The length of the middle finger as a percentage of total hand size remains essentially constant from birth to 14 years of age at 42–43%.[5]

Internipple distance.—Abnormalities in the internipple distance are best indicated by the internipple distance as a percentage of the chest circumference (Fig. 9).

Recognition of a diminished upper to lower segment ratio (see section on An Approach to the Child with Abnormal Stature) is important relative to the diagno-

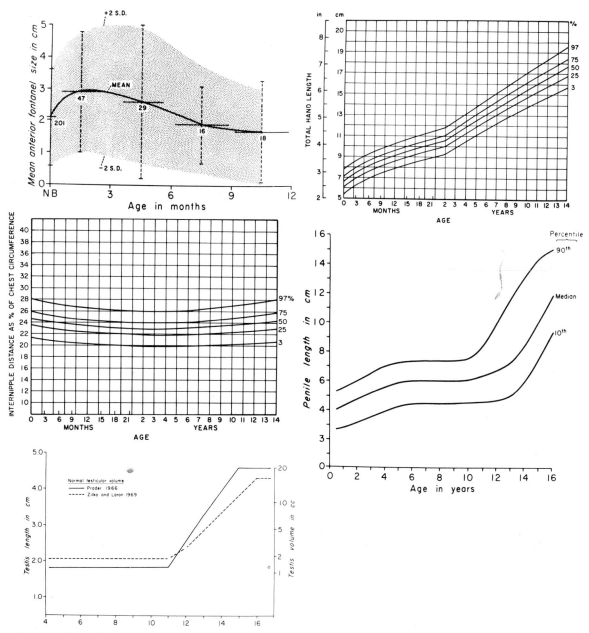

Fig. 7 (top left).—Mean anterior fontanel size. Beyond the newborn period, the data are averaged at 3-month intervals with the number of individuals in each group indicated. (Reprinted with permission from Popich and Smith.[3])

Fig. 8 (top right).—Total hand length. (Reprinted with permission from Feingold and Bossert.[5])

Fig. 9 (center left).—Internipple distance as a percentage of chest circumference. (Reprinted with permission from Feingold and Bossert.[5])

Fig. 10 (center right).—Penile length. (Reprinted with permission from Caldwell and Smith.[7])

Fig. 11 (bottom).—Testicular length. (Reprinted with permission from Caldwell and Smith.[7])

sis of the XXY Klinefelter syndrome (see section on Cytogenetic syndromes) and the Marfan syndrome (see section on Disorders of Connective Tissue).

Penile length (Fig. 10).—Measurement is made from the pubopenile skin junction to the tip of the glans, with the phallus fully stretched.

Testicular length (Fig. 11).

REFERENCES

1. Smith, D. W., and Cohen, M. M., Jr.: Widow's peak scalp anomaly: Origin and relevance to ocular hypertelorism, Lancet 2:1127, 1973.
2. Aase, J. M., Wilson, A. C., and Smith, D. W.: Small ears in Down's syndrome: A helpful diagnostic aid, J. Pediatr. 82: 845, 1973.
3. Popich, G. A., and Smith, D. W.: Fontanels: Range of normal size, J. Pediatr. 80:749, 1972.
4. Popich, G. A., and Smith, D. W.: Large fontanels in congenital hypothyroidism. A potential clue toward earlier recognition, J. Pediatr. 80:753, 1972.
5. Feingold, M., and Bossert, W. H.: Normal values for selected physical parameters: An aid to syndrome delineation, Birth Defects 10:1, 1974.
6. Smith, D. W.: Recognizable Patterns of Human Malformation (Philadelphia: W. B. Saunders Co., 1970).
7. Caldwell, P. D., and Smith, D. W.: The XXY (Klinefelter's) syndrome in childhood: Detection and treatment, J. Pediatr. 80:250, 1972.

Relationship of Advanced Parental Age to the Occurrence of Genetic Disorders in Progeny

RELATIONSHIP OF ADVANCED PARENTAL AGE TO THE OCCURRENCE OF GENETIC DISORDERS IN PROGENY

1. Increased maternal age at the time of the birth of a child increases the risk of nondisjunction. Examples are 21 trisomy (the Down syndrome), E trisomy, the Klinefelter syndrome (XXY) and the Turner syndrome (XO).

2. Advanced age of the maternal grandmother at the time of the birth of a mother also may be the reason for the production of a grandchild with the Down syndrome due to 21 trisomy. In this case the mother's age is not a factor, and she may well be a young mother of a Down patient. She may have multiple offspring with 21 trisomy. Examination of the mother's chromosomes will reveal her to be a mosaic.[1]

3. Increased paternal age at the time of conception leads to a greater risk of a new dominant mutation. Examples that have been studied are achondroplasia, the Apert syndrome[2] and the Marfan syndrome.

The production of gametes differs in females from that in males. Ova remain in the dictyotene stage of first meiotic prophase from the time of birth of a female infant for many years prior to ovulation. This has been thought on intuitive grounds to favor the development of nondisjunction. Conceptualizations of overripe or rotten eggs do easily catch the imagination. Actually it has been found that aging of eggs in Xenopus[3] causes degeneration of the meiotic spindle fibers, and this leads to meiotic nondisjunction. If aging of human oocytes interferes with the formation of the fibers of the meiotic spindle, then nondisjunction can be thought of not along the lines of sticky chromosomes but as a failure to find the right channels of even segregation into two identical gametes. Aging in the male differs from that of the female in that meiotic divisions and the formation of sperm take place continuously. DNA is actively being read out throughout a male's reproductive life. One can imagine how this might lead to an increase in copying errors as the system ages. This is consistent with the observed association of an increased incidence of new dominant mutations with increased paternal age.

4. Advanced age on the part of the maternal grandfather at the time of the birth of a mother who is a new mutation carrier of an X-linked disorder is important. A famous example is that of the Royal Family of England. Queen Victoria is often credited with having initiated the hemophilia that plagues that family because her sons were the first hemophiliacs in the kindred. When Queen Victoria was born, in 1819, her father, King Edward Augustus, was 52 years old. Thus, the cause of the new mutation was really probably his advanced age.

REFERENCES

1. Aarskog, D.: Down's syndrome transmitted through maternal mosaicism, Acta Pediatr. Scand. 58:609, 1969.
2. Blank, C. E.: Apert's syndrome (a type of acrocephalo-syndactyly). Observations on a British series of 39 cases, Ann. Hum. Genet. 24:151, 1960.
3. Mikamo, K.: Intrafollicular overripeness and teratologic development, Cytogenetics 7:212, 1968.

INDEX